Lecture Notes in Artificial Intelligence   8078

Subseries of Lecture Notes in Computer Science

LNAI Series Editors

Randy Goebel
  University of Alberta, Edmonton, Canada
Yuzuru Tanaka
  Hokkaido University, Sapporo, Japan
Wolfgang Wahlster
  DFKI and Saarland University, Saarbrücken, Germany

LNAI Founding Series Editor

Joerg Siekmann
  DFKI and Saarland University, Saarbrücken, Germany

Weiru Liu V.S. Subrahmanian Jef Wijsen (Eds.)

Scalable Uncertainty Management

7th International Conference, SUM 2013
Washington, DC, USA, September 16-18, 2013
Proceedings

 Springer

Volume Editors

Weiru Liu
Queen's University Belfast
School of Electronics, Electrical Engineering and Computer Science
Belfast BT9 5BN, UK
E-mail: w.liu@qub.ac.uk

V.S. Subrahmanian
University of Maryland
Department of Computer Science
College Park, MD 20742, USA
E-mail: vs@umiacs.umd.edu

Jef Wijsen
Université de Mons
Département d'Informatique
7000 Mons, Belgium
E-mail: jef.wijsen@umons.ac.be

ISSN 0302-9743   e-ISSN 1611-3349
ISBN 978-3-642-40380-4   e-ISBN 978-3-642-40381-1
DOI 10.1007/978-3-642-40381-1
Springer Heidelberg New York Dordrecht London

Library of Congress Control Number: 2013945725

CR Subject Classification (1998): I.2, H.4, H.3, H.5, C.2, H.2, F.4.1

LNCS Sublibrary: SL 7 – Artificial Intelligence

Typesetting: Camera-ready by author, data conversion by Scientific Publishing Services, Chennai, India

Printed on acid-free paper

Springer is part of Springer Science+Business Media (www.springer.com)

Preface

Information systems are becoming increasingly complex, involving massive amounts of data coming from different sources. Information is often inconsistent, incomplete, heterogeneous, and pervaded with uncertainty. The annual International Conference on Scalable Uncertainty Management (SUM) has grown out of this wide-ranging interest in the management of uncertainty and inconsistency in databases, the Web, the Semantic Web, and artificial intelligence applications.

The series of SUM conferences provides an international forum for the communication of research advances in the management of uncertain, incomplete, or inconsistent information. Previous SUM conferences have been held in Washington DC (2007 and 2009), Naples (2008), Toulouse (2010), Dayton (2011), and Marburg (2012).

This volume contains the papers presented at the 7th International Conference on Scalable Uncertainty Management (SUM 2013), which was held in Washington DC, USA, during September 16–18, 2013. The call for papers solicited submissions in two categories: regular research papers and short papers reporting on interesting work in progress or providing system descriptions. The call for papers resulted in 57 submissions, among which 47 regular papers and 10 short papers. Each paper was reviewed by at least three Program Committee members. Based on the review reports and discussions, 29 papers were accepted for publication and presentation at the conference, among which 26 regular papers and three short papers.

The conference program also included invited lectures by three world-leading researchers: Christos Faloutsos of Carnegie Mellon University, Steve Eubank of Virginia Tech, and Rama Chellappa of University of Maryland.

A conference such as this can only succeed as a team effort. We would like to thank several people and institutions: the authors of submitted papers, the invited speakers, and the conference participants; the members of the Program Committee and the external referees; Alfred Hofmann and Springer for providing assistance and advice in the preparation of the proceedings; the University of Maryland Institute for Advanced Computer Studies for providing local facilities; Jonathan Hourez for mastering the conference website; the creators and maintainers of the conference management system EasyChair. All of them made the success of SUM 2013 possible.

July 2013

Weiru Liu
V.S. Subrahmanian
Jef Wijsen

Organization

SUM 2013 was organized by the University of Maryland Institute for Advanced Computer Studies.

General Chair

V.S. Subrahmanian   University of Maryland, USA

Program Committee Chairs

Weiru Liu   Queen's University Belfast, UK
Jef Wijsen   University of Mons, Belgium

Program Committee

Leila Amgoud   IRIT, France
Chitta Baral   Arizona State University, USA
Nahla Ben Amor   Institut supérieur de gestion de Tunis, Tunisia
Salem Benferhat   University of Artois, France
Leopoldo Bertossi   Carleton University, Canada
Yaxin Bi   University of Ulster, UK
Loreto Bravo   University of Concepcion, Chili
Laurence Cholvy   ONERA, Toulouse, France
Jan Chomicki   SUNY Buffalo, USA
Fabio Gagliardi Cozman   University of Sao Paulo, Brazil
Alfredo Cuzzocrea   Università della Calabria, Italy
Thierry Denoeux   Université de Technologie de Compigne, France
Jürgen Dix   TU Clausthal, Germany
Didier Dubois   IRIT, France
Thomas Eiter   TU Vienna, Austria
Zied Elouedi   Institut supérieur de gestion de Tunis, Tunisia
Lluis Godo   IIIA, Spain
Nikos Gorogiannis   University College London, UK
John Grant   Towson University, USA
Sergio Greco   Università della Calabria, Italy
Anthony Hunter   University College London, UK
Gabriele Kern-Isberner   Technische Universität Dortmund, Germany
Kathryn Laskey   George Mason University, USA
Jonathan Lawry   University of Bristol, UK

Churn-Jung Liau	Academia Sinica, Taiwan
Sebastian Link	University of Auckland, New Zealand
Peter Lucas	University of Nijmegen, Netherlands
Thomas Lukasiewicz	University of Oxford, UK
Jianbing Ma	Queen's University Belfast, UK
Zongmin Ma	Northeastern University, China
Thomas Meyer	Centre for AI Research, CSIR and UKZN, South Africa
Serafin Moral	Universidad de Granada, Spain
Kedian Mu	Peking University, China
Dan Olteanu	Oxford University, UK
Jeff Z. Pan	Aberdeen University, UK
Simon Parsons	City University New York, USA
Gabriella Pasi	Università degli Studi di Milano Bicocca, Italy
Olivier Pivert	IRISA-ENSSAT, France
David Poole	University of British Columbia, Canada
Henri Prade	IRIT, France
Andrea Pugliese	Università della Calabria, Italy
Guilin Qi	Southeast University, China
Prakash Shenoy	University of Kansas, USA
Guillermo Ricardo Simari	Universidad Nacional del Sur, Bahia Blanca, Argentina
Umberto Straccia	ISTI-CNR, Italy
Nic Wilson	University College Cork, Ireland, and Queen's University Belfast, UK

External Reviewers

Teresa Alsinet
Gloria Bordogna
Julien Brunel
Nils Bulling
Minh Dao-Tran
Sergio Flesca
Filippo Furfaro
Jhonatan Garcia
Arjen Hommersom
Louise Leenen
Cristian Molinaro

Francesco Parisi
Antonino Rullo
Daria Stepanova
Ivan Varzinczak
Marina Velikova
Songxin Wang
Yining Wu
Guohui Xiao
Xiaowang Zhang
Yuting Zhao

Invited Talks

Influence Propagation in Large Graphs—Theorems, Algorithms, and Case Studies
Christos Faloutsos (Carnegie Mellon University)

Given the specifics of a virus (or product, or hashtag) how quickly will it propagate on a contact network? Will it create an epidemic, or will it quickly die out? The way a virus/product/meme propagates on a graph is important, because it can help us design immunization policies (if we want to stop it) or marketing policies (if we want it to succeed). We present some surprising results on the so-called "epidemic threshold", we discuss the effects of time-varying contact networks, and we present fast algorithms to achieve near-optimal immunization.

Using Network Reliability Polynomials to Characterize Contact Networks for Infectious Disease Epidemiology
Steve Eubank (Virginia Tech)

It is well-known that the structure of host-host contact networks can play an important role in determining the spread of infectious disease. Especially over the past decade, there have been many attempts to infer human contact networks at scales from urban regions to continents, and to simulate epidemics on the resulting networks. Moreover, since both pharmaceutical and non-pharmaceutical interventions can be represented as changes in network structure, simulated epidemics can be used to evaluate hypothetical combinations of interventions. Unfortunately, it is difficult to understand the simulated epidemics' sensitivity to details in the network structure. Results, for example those relating degree distribution to outbreak dynamics, typically make unwarranted assumptions about independence or symmetries in the network that introduce hard-to-control errors. Understanding this sensitivity to network structure is crucial for answering several related questions:

- How closely must the inferred networks match the modeled system for inferences about interventions to be useful?
- Can we take a short cut to evaluating interventions that eliminates the need for simulations by characterizing networks directly?
- Given a network, what is the optimal intervention under constrained resources? If we cannot optimize, can we at least develop useful rules of thumb?

This talk will review 50± year-old concepts of network reliability and describe how they can be extended and applied in the context of epidemiology. I will introduce a class of reliability polynomials and demonstrate several useful representations for them; discuss briefly the computational complexity of evaluating the polynomials exactly; and illustrate the use of scalable, distributed simulation for efficient approximation. I will show how to identify the contacts that are the most important targets for intervention and, more generally, how to characterize and compare networks in terms that are immediately relevant to epidemiology. Some representations of the reliability polynomial are well-suited to analytical reasoning about graph structure. I will illustrate this with a brief discussion of the phenomenon of "crossing reliability". In the context of outbreak interventions, the possibility that reliability polynomials cross implies that the relative ranking of interventions depends on the host-host transmissibility. I will discuss what kinds of structural changes induce crossing reliability, and the magnitude of the resulting difference in reliability.

The Evolution of Probabilistic Models and Uncertainty Analysis in Computer Vision Research
Rama Chellappa (University of Maryland)

During the past three decades, probabilistic methods and uncertainty analysis have been slowly but steadily integrated into computer vision research. During the early years, as more emphasis was given to geometry and probabilistic inference over geometric representations was challenging, the role of probabilistic inference was minimal. Since the introduction of Markov random fields, robust methods and error bounds, many computer vision problems are lending themselves for more rigorous analysis. In this talk, I will illustrate these ideas by highlighting the role played by MRFs in image analysis, error bounds for the structure from motion problem and some recent works on probabilistic inference on manifolds for activity recognition.

Table of Contents

Argumentation

Belief Functions, Possibility Theory and their Applications

Databases

Intelligent Data Analytics

Logics, Description Logic, and Semantic Web

Analysis of Dialogical Argumentation via Finite State Machines

Anthony Hunter

Department of Computer Science, University College London,
Gower Street, London WC1E 6BT, UK

Abstract. Dialogical argumentation is an important cognitive activity by which agents exchange arguments and counterarguments as part of some process such as discussion, debate, persuasion and negotiation. Whilst numerous formal systems have been proposed, there is a lack of frameworks for implementing and evaluating these proposals. First-order executable logic has been proposed as a general framework for specifying and analysing dialogical argumentation. In this paper, we investigate how we can implement systems for dialogical argumentation using propositional executable logic. Our approach is to present and evaluate an algorithm that generates a finite state machine that reflects a propositional executable logic specification for a dialogical argumentation together with an initial state. We also consider how the finite state machines can be analysed, with the minimax strategy being used as an illustration of the kinds of empirical analysis that can be undertaken.

1 Introduction

Dialogical argumentation involves agents exchanging arguments in activities such as discussion, debate, persuasion, and negotiation [1]. Dialogue games are now a common approach to characterizing argumentation-based agent dialogues (e.g. [2–12]). Dialogue games are normally made up of a set of communicative acts called moves, and a protocol specifying which moves can be made at each step of the dialogue. In order to compare and evaluate dialogical argumentation systems, we proposed in a previous paper that first-order executable logic could be used as common theoretical framework to specify and analyse dialogical argumentation systems [13].

In this paper, we explore the implementation of dialogical argumentation systems in executable logic. For this, we focus on propositional executable logic as a special case, and investigate how a finite state machine (FSM) can be generated as a representation of the possible dialogues that can emanate from an initial state. The FSM is a useful structure for investigating various properties of the dialogue, including conformance to protocols, and application of strategies. We provide empirical results on generating FSMs for dialogical argumentation, and how they can be analysed using the minimax strategy. We demonstrate through preliminary implementation that it is computationally viable to generate the FSMs and to analyse them. This has wider implications in

W. Liu, V.S. Subrahmanian, and J. Wijsen (Eds.): SUM 2013, LNAI 8078, pp. 1–14, 2013.

using executable logic for applying dialogical argumentation in practical uncertainty management applications, since we can now empirically investigate the performance of the systems in handling inconsistency in data and knowledge.

2   Propositional Executable Logic

In this section, we present a propositional version of the executable logic which we will show is amenable to implementation. This is a simplified version of the framework for first-order executable logic in [13].

We assume a set of atoms which we use to form propositional formulae in the usual way using disjunction, conjunction, and negation connectives. We construct modal formulae using the $\boxplus$, $\boxminus$, $\oplus$, and $\ominus$ modal operators. We only allow literals to be in the scope of a modal operator. If α is a literal, then each of $\oplus\alpha$, $\ominus\alpha$, $\boxplus\alpha$, and $\boxminus\alpha$ is an **action unit**. Informally, we describe the meaning of action units as follows: $\oplus\alpha$ means that the action by an agent is to add the literal α to its next private state; $\ominus\alpha$ means that the action by an agent is to delete the literal α from its next private state; $\boxplus\alpha$ means that the action by an agent is to add the literal α to the next public state; and $\boxminus\alpha$ means that the action by an agent is to delete the literal α from the next public state.

We use the action units to form **action formulae** as follows using the disjunction and conjunction connectives: (1) If ϕ is an action unit, then ϕ is an action formula; And (2) If α and β are action formulae, then $\alpha \vee \beta$ and $\alpha \wedge \beta$ are action formulae. Then, we define the action rules as follows: If ϕ is a classical formula and ψ is an action formula then $\phi \Rightarrow \psi$ is an **action rule**. For instance, $\mathsf{b(a)} \Rightarrow \boxplus\mathsf{c(a)}$ is an action rule (which we might use in an example where b denotes belief, and c denotes claim, and a is some information).

Implicit in the definitions for the language is the fact that we can use it as a meta-language [14]. For this, the object-language will be represented by terms in this meta-language. For instance, the object-level formula $\mathsf{p(a,b)} \rightarrow \mathsf{q(a,b)}$ can be represented by a term where the object-level literals $\mathsf{p(a,b)}$ and $\mathsf{q(a,b)}$ are represented by constant symbols, and $\rightarrow$ is represented by a function symbol. Then we can form the atom $\mathtt{belief}(\mathsf{p(a,b)} \rightarrow \mathsf{q(a,b)})$ where $\mathtt{belief}$ is a predicate symbol. Note, in general, no special meaning is ascribed the predicate symbols or terms. They are used as in classical logic. Also, the terms and predicates are all ground, and so it is essentially a propositional language.

We use a state-based model of dialogical argumentation with the following definition of an execution state. To simplify the presentation, we restrict consideration in this paper to two agents. An execution represents a finite or infinite sequence of execution states. If the sequence is finite, then t denotes the terminal state, otherwise $t = \infty$.

Definition 1. *An* **execution** *e is a tuple $e = (s_1, a_1, p, a_2, s_2, t)$, where for each $n \in \mathbb{N}$ where $0 \leq n \leq t$, $s_1(n)$ is a set of ground literals, $a_1(n)$ is a set of ground action units, $p(n)$ is a set of ground literals, $a_2(n)$ is a set of ground action units, $s_2(n)$ is a set of ground literals, and $t \in \mathbb{N} \cup \{\infty\}$. For each $n \in \mathbb{N}$, if $0 \leq n \leq t$, then an* **execution state** *is $e(n) = (s_1(n), a_1(n), p(n), a_2(n), s_2(n))$ where $e(0)$*

is the **initial state**. *We assume* $a_1(0) = a_2(0) = \emptyset$. *We call* $s_1(n)$ *the private state of agent 1 at time* n, $a_1(n)$ *the action state of agent 1 at time* n, $p(n)$ *the public state at time* n, $a_2(n)$ *the action state of agent 2 at time* n, $s_2(n)$ *the private state of agent 2 at time* n.

In general, there is no restriction on the literals that can appear in the private and public state. The choice depends on the specific dialogical argumentation we want to specify. This flexibility means we can capture diverse kinds of information in the private state about agents by assuming predicate symbols for their own beliefs, objectives, preferences, arguments, etc, and for what they know about other agents. The flexibility also means we can capture diverse information in the public state about moves made, commitments made, etc.

Example 1. The first 5 steps of an infinite execution where each row in the table is an execution state where **b** denotes belief, and **c** denotes claim.

n	$s_1(n)$	$a_1(n)$	$p(n)$	$a_2(n)$	$s_2(n)$
0	b(a)				b(¬a)
1	b(a)	⊞c(a),⊟c(¬a)			b(¬a)
2	b(a)		c(a)	⊞c(¬a),⊟c(a)	b(¬a)
3	b(a)	⊞c(a),⊟c(¬a)	c(¬a)		b(¬a)
4	b(a)		c(a)	⊞c(¬a),⊟c(a)	b(¬a)
5	...	...		...	...

We define a system in terms of the action rules for each agent, which specify what moves the agent can potentially make based on the current state of the dialogue. In this paper, we assume agents take turns, and at each time point the actions are from the head of just one rule (as defined in the rest of this section).

Definition 2. *A* **system** *is a tuple* $(Rules_x, Initials)$ *where* $Rules_x$ *is the set of action rules for agent* $x \in \{1, 2\}$, *and Initials is the set of initial states.*

Given the current state of an execution, the following definition captures which rules are fired. For agent x, these are the rules that have the condition literals satisfied by the current private state $s_x(n)$ and public state $p(n)$. We use classical entailment, denoted $\models$, for satisfaction, but other relations could be used (e.g. Belnap's four valued logic). In order to relate an action state in an execution with an action formula, we require the following definition.

Definition 3. *For an action state* $a_x(n)$, *and an action formula* ϕ, $a_x(n)$ **satisfies** ϕ, *denoted* $a_x(n) \mathbin{\vdash\!\sim} \phi$, *as follows.*

1. $a_x(n) \mathbin{\vdash\!\sim} \alpha$ *iff* $\alpha \in a_x(n)$ *when* α *is an action unit*
2. $a_x(n) \mathbin{\vdash\!\sim} \alpha \wedge \beta$ *iff* $a_x(n) \mathbin{\vdash\!\sim} \alpha$ *and* $a_x(n) \mathbin{\vdash\!\sim} \beta$
3. $a_x(n) \mathbin{\vdash\!\sim} \alpha \vee \beta$ *iff* $a_x(n) \mathbin{\vdash\!\sim} \alpha$ *or* $a_x(n) \mathbin{\vdash\!\sim} \beta$

For an action state $a_x(n)$, *and an action formula* ϕ, $a_x(n)$ **minimally satisfies** ϕ, *denoted* $a_x(n) \Vdash \phi$, *iff* $a_x(n) \mathbin{\vdash\!\sim} \phi$ *and for all* $X \subset a_x(n)$, $X \mathbin{\not\vdash\!\sim} \phi$.

Example 2. Consider the execution in Example 1. For agent 1 at n = 1, we have $a_1(1) \Vdash \boxplus c(a) \wedge \boxminus c(\neg a)$.

We give two constraints on an execution to ensure that they are well-behaved. The first (propagated) ensures that each subsequent private state (respectively each subsequent public state) is the current private state (respectively current public state) for the agent updated by the actions given in the action state. The second (engaged) ensures that an execution does not have one state with no actions followed immediately by another state with no actions (otherwise the dialogue can lapse) except at the end of the dialogue where neither agent has further actions.

Definition 4. *An execution* $(s_1, a_1, p, a_2, s_2, t)$ *is* **propagated** *iff for all* $x \in \{1, 2\}$, *for all* $n \in \{0, \ldots, t-1\}$, *where* $a(n) = a_1(n) \cup a_2(n)$

1. $s_x(n+1) = (s_x(n) \setminus \{\phi \mid \ominus\phi \in a_x(n)\}) \cup \{\phi \mid \oplus\phi \in a_x(n)\}$
2. $p(n+1) = (p(n) \setminus \{\phi \mid \boxminus\phi \in a(n)\}) \cup \{\phi \mid \boxplus\phi \in a(n)\}$

Definition 5. *Let* $e = (s_1, a_1, p, a_2, s_2, t)$ *be an execution and* $a(n) = a_1(n) \cup a_2(n)$. *e is* **finitely engaged** *iff (1)* $t \neq \infty$; *(2) for all* $n \in \{1, \ldots, t-2\}$, *if* $a(n) = \emptyset$, *then* $a(n+1) \neq \emptyset$ *(3)* $a(t-1) = \emptyset$; *and (4)* $a(t) = \emptyset$. *e is* **infinitely engaged** *iff (1)* $t = \infty$; *and (2) for all* $n \in \mathbb{N}$, *if* $a(n) = \emptyset$, *then* $a(n+1) \neq \emptyset$.

The next definition shows how a system provides the initial state of an execution and the actions that can appear in an execution. It also ensures turn taking by the two agents.

Definition 6. *Let* $S = (Rules_x, Initials)$ *be a system and* $e = (s_1, a_1, p, a_2, s_2, t)$ *be an execution. S* **generates** *e iff (1) e is propogated; (2) e is finitely engaged or infinitely engaged; (3)* $e(0) \in Initials$; *and (4) for all* $m \in \{1, \ldots, t-1\}$

1. *If m is odd, then* $a_2(m) = \emptyset$ *and either* $a_1(m) = \emptyset$ *or there is an* $\phi \Rightarrow \psi \in Rules_1$ *s.t.* $s_1(m) \cup p(m) \models \phi$ *and* $a_1(m) \Vdash \psi$
2. *If m is even, then* $a_1(m) = \emptyset$ *and either* $a_2(m) = \emptyset$ *or there is an* $\phi \Rightarrow \psi \in Rules_2$ *s.t.* $s_1(m) \cup p(m) \models \phi$ *and* $a_2(m) \Vdash \psi$

Example 3. We can obtain the execution in Example 1 with the following rules: (1) $b(a) \Rightarrow \boxplus c(a) \wedge \boxminus c(\neg a)$; And (2) $b(\neg a) \Rightarrow \boxplus c(\neg a) \wedge \boxminus c(a)$.

3   Generation of Finite State Machines

In [13], we showed that for any executable logic system with a finite set of ground action rules, and an initial state, there is an FSM that consumes exactly the finite execution sequences of the system for that initial state. That result assumes that each agent makes all its possible actions at each step of the execution. Also that result only showed that there exist these FSMs, and did not give any way of obtaining them.

In this paper, we focus on propositional executable logic where the agents take it in turn, and only one head of one action rule is used, and show how we can construct an FSM that represents the set of executions for an initial state for a system. For this, each state is a tuple $(r, s_1(n), p(n), s_2(n))$, and each letter in the alphabet is a tuple $(a_1(n), a_2(n))$, where n is an execution step and r is the agent holding the turn when $n < t$ and r is 0 when $n = t$.

Definition 7. *An FSM $M = (States, Trans, Start, Term, Alphabet)$ represents a system $S = (Rules_x, Initials)$ for an initial state $I \in Initials$ iff*

$(1) States = \{(y, s_1(n), p(n), s_2(n)) \mid$ *there is an execution* $e = (s_1, a_1, p, a_2, s_2, t)$
$\quad$ *s.t.* S *generates* e *and* $I = (s_1(0), a_1(0), p(0), a_2(0), s_2(0))$
$\quad$ *and there is an* $n \leq t$ *s.t.* $y = 0$ *when* $n = t$
$\quad\quad$ *and* $y = 1$ *when* $n < t$ *and* n *is odd*
$\quad\quad$ *and* $y = 2$ *when* $n < t$ *and* n *is even* $\}$

$(2) Term = \{(y, s_1(n), p(n), s_2(n)) \in States \mid y = 0\}$

$(3) Alphabet = \{(a_1(n), a_2(n)) \mid$ *there is an* $n \leq t$ *and there is an execution* e
$\quad$ *s.t.* S *generates* e *and* $e(0) = I$ *and* $e = (s_1, a_1, p, a_2, s_2, t).\}$

$(4) Start = (1, s_1(0), p(0), s_2(0))$ *where* $I = (s_1(0), a_1(0), p(0), a_2(0), s_2(0))$

(5) *Trans is the smallest subset of* $States \times Alphabet \times States$ *s.t. for all executions* e *and for all* $n < t$ *there is a transition* $\tau \in Trans$ *such that*

$$\tau = ((x, s_1(n), p(n), s_2(n)), (a_1(n), a_2(n)), (y, s_1(n+1), p(n+1), s_2(n+1)))$$

where x is 1 when n is odd, x is 2 when n is even, y is 1 when $n+1 < t$ and n is odd, y is 2 when $n+1 < t$ and n is even, and y is 0 when $n+1 = t$.

Example 4. Let M be the following FSM where $\sigma_1 = (1, \{b(a)\}, \{\}, \{b(\neg a)\})$; $\sigma_2 = (2, \{b(a)\}, \{c(a)\}, \{b(\neg a)\})$; $\sigma_3 = (1, \{b(a)\}, \{c(\neg a)\}, \{b(\neg a)\})$. $\tau_1 = (\{\boxplus c(a), \boxminus c(\neg a)\}, \emptyset)$; and $\tau_2 = (\emptyset, \{\boxplus c(\neg a), \boxminus c(a)\})$. M represents the system in Ex 1.

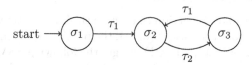

Proposition 1. *For each $S = (Rules_x, Initials)$, then there is an FSM M such that M represents S for an initial state $I \in Initials$.*

Definition 8. *A string ρ reflects an execution $e = (s_1, a_1, p, a_2, s_2, t)$ iff ρ is the string $\tau_1 \ldots \tau_{t-1}$ and for each $1 \leq n < t$, τ_n is the tuple $(a_1(n), a_2(n))$.*

Proposition 2. *Let $S = (Rules_x, Initials)$ be a system. and let M be an FSM that represents S for $I \in Initials$.*

1. *for all ρ s.t. M accepts ρ, there is an e s.t. S generates e and $e(0) = I$ and ρ reflects e,*
2. *for all finite e s.t. S generates e and $e(0) = I$, then there is a ρ such that M accepts ρ and ρ reflects e.*

So for each initial state for a system, we can obtain an FSM that is a concise representation of the executions of the system for that initial state. In Figure 1, we provide an algorithm for generating these FSMs. We show correctness for the algorithm as follows.

Proposition 3. *Let $S = (Rules_x, Initials)$ be a system and let $I \in Initials$. If M represents S w.r.t. I and $\mathsf{BuildMachine}(Rules_x, I) = M'$, then $M = M'$.*

An FSM provides a more efficient representation of all the possible executions than the set of executions for an initial state. For instance, if there is a set of states that appear in some permutation of each of the executions then this can be more compactly represented by an FSM. And if there are infinite sequences, then again this can be more compactly represented by an FSM.

Once we have an FSM of a system with an initial state, we can ask obvious simple questions such as is termination possible, is termination guaranteed, and is one system subsumed by another? So by translating a system into an FSM, we can harness substantial theory and tools for analysing FSMs.

Next we give a couple of very simple examples of FSMs obtained from executable logic. In these examples, we assume that agent 1 is trying to win an argument with agent 2. We assume that agent 1 has a goal. This is represented by the predicate $g(c)$ in the private state of agent 1 for some argument c. In its private state, each agent has zero or more arguments represented by the predicate $n(c)$, and zero or more attacks $e(d, c)$ from d to c. In the public state, each argument c is represented by the predicate $a(c)$. Each agent can add attacks $e(d, c)$ to the public state, if the attacked argument is already in the public state (i.e. $a(c)$ is in the public state), and the agent also has the attacker in its private state (i.e. $n(d)$ is in the private state). We have encoded the rules so that after an argument has been used as an attacker, it is removed from the private state of the agent so that it does not keep firing the action rule (this is one of a number of ways that we can avoid repetition of moves).

Example 5. For the following action rules, with the initial state where the private state of agent 1 is $\{g(a), n(a), n(c), e(c, b)\}$, the public state is empty, and the private state of agent 2 is $\{n(b), e(b, a)\}$), we get the following FSM, with the states below and the transitions: $\tau_1 = (\{\boxplus a(a), \ominus n(a)\}, \emptyset)$; $\tau_2 = (\emptyset, \{\boxplus a(b, a), \ominus n(b)\})$; $\tau_3 = (\{\boxplus a(c, b), \ominus n(c)\}, \emptyset)$; and $\tau_4 = (\emptyset, \emptyset)$.

$$g(a) \wedge n(a) \Rightarrow \boxplus a(a) \wedge \ominus n(a)$$
$$a(a) \wedge n(b) \wedge e(b, a) \Rightarrow \boxplus a(b, a) \wedge \ominus n(b)$$
$$a(b) \wedge n(c) \wedge e(c, b) \Rightarrow \boxplus a(c, b) \wedge \ominus n(c)$$

$$\sigma_1 = (1, \{g(a), n(a), n(c), e(c, b)\}, \{\}, \{n(b), e(b, a)\})$$
$$\sigma_2 = (2, \{g(a), n(c), e(c, b)\}, \{a(a)\}, \{n(b), e(b, a)\})$$
$$\sigma_3 = (1, \{g(a), n(c), e(c, b)\}, \{a(a), a(b, a)\}, \{e(b, a)\})$$
$$\sigma_4 = (2, \{g(a), e(c, b)\}, \{a(a), a(b), a(c), a(c, b), a(b, a)\}, \{e(b, a)\})$$
$$\sigma_5 = (1, \{g(a), e(c, b)\}, \{a(a), a(b), a(c), a(c, b), a(b, a)\}, \{e(b, a)\})$$
$$\sigma_6 = (0, \{g(a), e(c, b)\}, \{a(a), a(b), a(c), a(c, b), a(b, a)\}, \{e(b, a)\})$$

$$\boxed{c} \longrightarrow \boxed{b} \longrightarrow \boxed{a}$$

This terminal state therefore contains the above argument graph, and hence the goal argument a is in the grounded extension of the graph (as defined in [15]).

Example 6. For the following action rules, with the initial state where the private state of agent 1 is $\{g(a), n(a)\}$, the public state is empty, and the private state of agent 2 is $\{n(b), n(c), e(b, a), e(c, a)\}$), we get the following FSM, with the states below and the transitions: $\tau_1 = (\{\boxplus a(a), \ominus n(a)\}, \emptyset)$; $\tau_2 = (\emptyset, \{\boxplus a(b, a), \ominus n(b)\})$; $\tau_3 = (\emptyset, \{\boxplus a(c, a), \ominus n(c)\})$; and $\tau_4 = (\emptyset, \emptyset)$.

$$g(a) \wedge n(a) \Rightarrow \boxplus a(a) \wedge \ominus n(a)$$
$$a(a) \wedge n(b) \wedge e(b, a) \Rightarrow \boxplus a(b, a) \wedge \ominus n(b)$$
$$a(a) \wedge n(c) \wedge e(c, a) \Rightarrow \boxplus a(c, a) \wedge \ominus n(c)$$

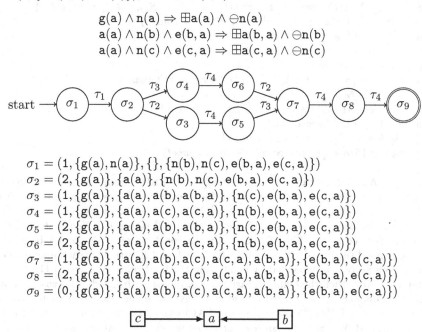

$$\sigma_1 = (1, \{g(a), n(a)\}, \{\}, \{n(b), n(c), e(b, a), e(c, a)\})$$
$$\sigma_2 = (2, \{g(a)\}, \{a(a)\}, \{n(b), n(c), e(b, a), e(c, a)\})$$
$$\sigma_3 = (1, \{g(a)\}, \{a(a), a(b), a(b, a)\}, \{n(c), e(b, a), e(c, a)\})$$
$$\sigma_4 = (1, \{g(a)\}, \{a(a), a(c), a(c, a)\}, \{n(b), e(b, a), e(c, a)\})$$
$$\sigma_5 = (2, \{g(a)\}, \{a(a), a(b), a(b, a)\}, \{n(c), e(b, a), e(c, a)\})$$
$$\sigma_6 = (2, \{g(a)\}, \{a(a), a(c), a(c, a)\}, \{n(b), e(b, a), e(c, a)\})$$
$$\sigma_7 = (1, \{g(a)\}, \{a(a), a(b), a(c), a(c, a), a(b, a)\}, \{e(b, a), e(c, a)\})$$
$$\sigma_8 = (2, \{g(a)\}, \{a(a), a(b), a(c), a(c, a), a(b, a)\}, \{e(b, a), e(c, a)\})$$
$$\sigma_9 = (0, \{g(a)\}, \{a(a), a(b), a(c), a(c, a), a(b, a)\}, \{e(b, a), e(c, a)\})$$

$$\boxed{c} \longrightarrow \boxed{a} \longleftarrow \boxed{b}$$

The terminal state therefore contains the above argument graph, and hence the goal argument a is in the grounded extension of the graph.

In the above examples, we have considered a formalisation of dialogical argumentation where agents exchange abstract arguments and attacks. It is straightforward to formalize other kinds of example to exchange a wider range of moves, richer content (e.g. logical arguments composed of premises and conclusion [10]), and richer notions (e.g. value-based argumentation [16]).

```
01 BuildMachine(Rules_x, I)
02    Start = (1, S_1, P, S_2) where I = (S_1, A_1, P, A_2, S_2)
03    States_1 = NewStates_1 = {Start}
04    States_2 = Trans_1 = Trans_2 = ∅
05    x = 1, y = 2
06    While NewStates_x ≠ ∅
07      NextStates = NextTrans = ∅
08      For (x, S_1, P, S_2) ∈ NewStates_x
09        Fired = {ψ | φ ⇒ ψ ∈ Rules_x and S_x ∪ P ⊨ φ}
10        If Fired == ∅
11        Then NextTrans = NextTrans ∪ {((x, S_1, P, S_2), (∅, ∅), (y, S_1, P, S_2))}
12        Else for A ∈ Disjuncts(Fired)
13          NewS = S_x \ {α | ⊖α ∈ A} ∪ {α | ⊕α ∈ A}
14          NewP = P \ {α | ⊟α ∈ A} ∪ {α | ⊞α ∈ A}
15          If x == 1, NextState = (2, NewS, P, S_2) and Label = (A, ∅)
16          Else NextState = (1, S_1, P, NewS) and Label = (∅, A)
17          NextStates = NextStates ∪ {NextState}
18          NextTrans = NextTrans ∪ {((x, S_1, P, S_2), Label, NextState)}
19      If x == 1, then x = 2 and y = 1, else x = 1 and y = 2
20      NewStates_x = NextStates \ States_x
21      States_x = States_x ∪ NextStates
22      Trans_x = Trans_x ∪ NextTrans
23    Close = {σ'' | (σ, τ, σ'), (σ', τ, σ'') ∈ Trans_1 ∪ Trans_2}
24    Trans = MarkTrans(Trans_1 ∪ Trans_2, Close)
25    States = MarkStates(States_1 ∪ States_2, Close)
26    Term = MarkTerm(Close)
27    Alphabet = {τ | (σ, τ, σ') ∈ States}
28    Return (States, Trans, Start, Term, Alphabet)
```

Fig. 1. An algorithm for generating an FSM from a system $S = (Rules_x, Initials)$ and an initial state I. The subsidiary function Disjuncts($Fired$) is $\{\{\psi_1^1, .., \psi_{k_1}^1\}, .., \{\psi_1^i, .., \psi_{k_i}^1\} \mid ((\psi_1^1 \wedge .. \wedge \psi_{k_1}^1) \vee .. \vee (\psi_1^i \wedge .. \wedge \psi_{k_i}^1)) \in Fired)\}$. For turn-taking, for agent x, $State_x$ is the set of expanded states and $NewStates_x$ is the set of unexpanded states. Lines 02-05 set up the construction with agent 1 being the agent to expand the initial state. At lines 06-18, when it is turn of x, each unexpanded state in $NewStates_x$ is expanded by identifying the fired rules. At lines 10-11, if there are no fired rules, then the empty transition (i.e. $(∅, ∅)$) is obtained, otherwise at lines 12-17, each disjunct for each fired rule gives a next state and transition that is added to $NextStates$ and $NextTrans$ accordingly. At lines 19-22, the turn is passed to the other agent, and $NewStates_x$, $States_x$, and $Trans_x$ updated. At line 23, the terminal states are identified from the transitions. At line 24, the MarkTrans function returns the union of the transitions for each agent but for each $\sigma = (x, S1, P, S2) \in Term$, σ is changed to $(0, S1, P, S2)$ in order to mark it as a terminal state in the FSM. At line 25, the MarkStates function returns the union of the states for each agent but for each $\sigma = (x, S1, P, S2) \in Term$, σ is changed to $(0, S1, P, S2)$, and similarly at line 26, MarkTerm function returns the set $Close$ but with each state being of the form $(0, S1, P, S2)$.

4   Minimax Analysis of Finite State Machines

Minimax analysis is applied to two-person games for deciding which moves to make. We assume two players called MIN and MAX. MAX moves first, and they take turns until the game is over. An **end function** determines when the game is over. Each state where the game has ended is an **end state**. A **utility function** (i.e. a payoff function) gives the outcome of the game (eg chess has win, draw, and loose). The **minimax strategy** is that MAX aims to get to an end state that maximizes its utility regardless of what MIN does

We can apply the minimax strategy to the FSM machines generated for dialogical argumentation as follows: (1) Undertake breadth-first search of the FSM; (2) Stop searching at a node on a branch if the node is an end state according to the end function (note, this is not necessarily a terminal state in the FSM); (3) Apply the utility function to each leaf node n (i.e. to each end state) in the search tree to give the value $value(n)$ of the node; (4) Traverse the tree in post-order, and calculate the value of each non-leaf node as follows where the non-leaf node n is at depth d and with children $\{n_1, .., n_k\}$:

- If d is odd, then $value(n)$ is the maximum of $value(n_1)$,.., $value(n_k)$.
- If d is even, then $value(n)$ is the minimum of $value(n_1)$,.., $value(n_k)$.

There are numerous types of dialogical argumentation that can be modelled using propositional executable logic and analysed using the minimax strategy. Before we discuss some of these options, we consider some simple examples where we assume that the search tree is exhaustive, (so each branch only terminates when it reaches a terminal state in the FSM), and the utility function returns 1 if the goal argument is in the grounded extension of the graph in the terminal state, and returns 0 otherwise.

Example 7. From the FSM in Example 5, we get the minimax search tree in Figure 2a, and from the FSM in Example 6, we get the minimax search tree in Figure 2b. In each case, the terminal states contains an argument graph in which the goal argument is in the grounded extension of the graph. So each leaf of the minimax tree has a utility of 1, and each non-node has the value 1. Hence, agent 1 is guaranteed to win each dialogue whatever agent 2 does.

The next example is more interesting from the point of view of using the minimax strategy since agent 1 has a choice of what moves it can make and this can affect whether or not it wins.

Example 8. In this example, we assume agent 1 has two goals **a** and **b**, but it can only present arguments for one of them. So if it makes the wrong choice it can loose the game. The executable logic rules and resulting FSM are as follows where $\tau_1 = (\{\boxplus a(b), \ominus n(b), \ominus g(a)\}, \emptyset)$, $\tau_2 = (\{\boxplus a(a), \ominus n(a), \ominus g(b)\}, \emptyset)$, $\tau_3 = (\emptyset, \{\boxplus a(c, a), \ominus n(c)\})$, and $\tau_4 = (\emptyset, \emptyset)$. For the minimax tree (given in Figure 2c) the left branch results in an argument graph in which the goal is not in the grounded extension, whereas the right branch terminates in an argument graph in which the goal is in the grounded extension. By a minimax analysis, agent 1 wins.

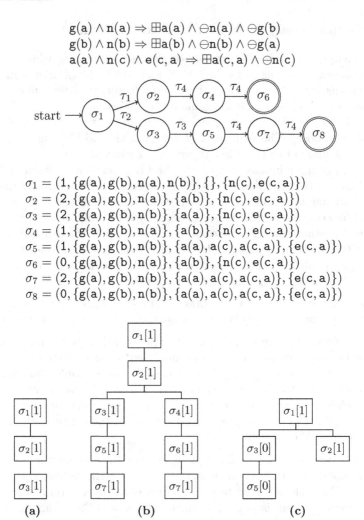

$$g(a) \wedge n(a) \Rightarrow \boxplus a(a) \wedge \ominus n(a) \wedge \ominus g(b)$$
$$g(b) \wedge n(b) \Rightarrow \boxplus a(b) \wedge \ominus n(b) \wedge \ominus g(a)$$
$$a(a) \wedge n(c) \wedge e(c,a) \Rightarrow \boxplus a(c,a) \wedge \ominus n(c)$$

$\sigma_1 = (1, \{g(a), g(b), n(a), n(b)\}, \{\}, \{n(c), e(c,a)\})$
$\sigma_2 = (2, \{g(a), g(b), n(a)\}, \{a(b)\}, \{n(c), e(c,a)\})$
$\sigma_3 = (2, \{g(a), g(b), n(b)\}, \{a(a)\}, \{n(c), e(c,a)\})$
$\sigma_4 = (1, \{g(a), g(b), n(a)\}, \{a(b)\}, \{n(c), e(c,a)\})$
$\sigma_5 = (1, \{g(a), g(b), n(b)\}, \{a(a), a(c), a(c,a)\}, \{e(c,a)\})$
$\sigma_6 = (0, \{g(a), g(b), n(a)\}, \{a(b)\}, \{n(c), e(c,a)\})$
$\sigma_7 = (2, \{g(a), g(b), n(b)\}, \{a(a), a(c), a(c,a)\}, \{e(c,a)\})$
$\sigma_8 = (0, \{g(a), g(b), n(b)\}, \{a(a), a(c), a(c,a)\}, \{e(c,a)\})$

Fig. 2. Minimax trees for Examples 7 and 8. Since each terminal state in an FSM is a copy of the previous two states, we save space by not giving these copies in the search tree. The minimax value for a node is given in the square brackets within the node. (a) is for Example 5, (b) is for Example 6 and (c) is for Example 8

We can use any criterion for identifying the end state. In the above, we have used the **exhaustive end function** giving an end state (i.e. the leaf node in the search tree) which is a terminal state in the FSM followed by two empty transitions. If the branch does not come to a terminal state in the FSM, then it is an infinite branch. We could use a **non-repetitive end function** where the search tree stops when there are no new nodes to visit. For instance, for example 4, we could use the non-repetitive end function to give a search tree that contains one branch $\sigma_1, \sigma_2, \sigma_3$ where σ_1 is the root and σ_3 is the leaf. Another simple option is a **fixed-depth end function** which has a specified maximum depth for any branch of the search tree. More advanced options for end functions include **concession end function** when an agent has a loosing position, and it knows that it cannot add anything to change the position, then it concedes.

There is also a range of options for the utility function. In the examples, we have used grounded semantics to determine whether a goal argument is in the grounded extension of the argument graph specified in the terminal public state. A refinement is the **weighted utility function** which weights the utility assigned by the grounded utility function by $1/d$ where d is the depth of the leaf. The aim of this is to favour shorter dialogues. Further definitions for utility functions arise from using other semantics such as preferred or stable semantics and richer formalisms such as valued-based argumentation [16].

5   Implementation Study

In this study, we have implemented three algorithms: The generator algorithm for taking an initial state and a set of action rules for each agent, and outputting the fabricated FSM; A breadth-first search algorithm for taking an FSM and a choice of termination function, and outputting a search tree; And a minimax assignment algorithm for taking a search tree and a choice of utility function, and outputting a minimax tree. These implemented algorithms were used together so that given an initial state and rules for each agent, the overall output was a minimax tree. This could then be used to determine whether or not agent 1 had a winning strategy (given the initial state). The implementation incorporates the exhaustive termination function, and two choices of utility function (grounded and weighted grounded).

The implementation is in Python 2.6 and was run on a Windows XP PC with Intel Core 2 Duo CPU E8500 at 3.16 GHz and 3.25 GB RAM. For the evaluation, we also implemented an algorithm for generating tests inputs. Each test input comprised an initial state, and a set of action rules for each agent. Each initial state involved 20 arguments randomly assigned to the two agents and up to 20 attacks per agent. For each attack in an agent's private state, the attacker is an argument in the agent's private state, and the attacked argument is an argument in the other agent's private state.

The results are presented in the following table. Each row is produced from 100 runs. Each run (i.e. a single initial state and action rules for each agent), was timed. If the time exceeded 100 seconds for the generator algorithm, the run was terminated

Average no. attacks	Average no. FSM nodes	Average no. FSM transitions	Average no. tree nodes	Average run time	Median run time	No. of runs timed out
9.64	6.29	9.59	31.43	0.27	0.18	0
11.47	16.01	39.48	1049.14	6.75	0.18	1
13.29	12.03	27.74	973.84	9.09	0.18	2
14.96	12.50	27.77	668.65	6.41	0.19	13
16.98	19.81	49.96	2229.64	25.09	0.20	19
18.02	19.01	47.81	2992.24	43.43	0.23	30

As can be seen from these results, up to about 15 attacks per agent, the implementation runs in negligible time. However, above 15 attacks per agent, the time did increase markedly, and a substantially minority of these timed out. To indicate the size of the larger FSMs, consider the last line of the table where the runs had an average of 18.02 attacks per agent: For this set, 8 out of 100 runs had 80+ nodes in the FSM. Of these 8 runs, the number of states was between 80 and 163, and the number of transitions was between 223 and 514.

The algorithm is somewhat naive in a number of respects. For instance, the algorithm for finding the grounded extension considers every subset of the set of arguments (i.e. 2^{20} sets). Clearly more efficient algorithms can be developed or calculation subcontracted to a system such as ASPARTIX [17]. Nonetheless, there are interesting applications where 20 arguments would be a reasonable, and so we have shown that we can analyse such situations successfully using the Minimax strategy, and with some refinement of the algorithms, it is likely that larger FSMs can be constructed and analysed.

Since the main aim was to show that FSMs can be generated and analysed, we only used a simple kind of argumentation dialogue. It is straightforward to develop alternative and more complex scenarios, using the language of propositional executable logic e.g. for capturing beliefs, goals, uncertainty etc, for specifying richer behaviour.

6   Discussion

In this paper, we have investigated a uniform way of presenting and executing dialogical argumentation systems based on a propositional executable logic. As a result different dialogical argumentation systems can be compared and implemented more easily than before. The implementation is generic in that any action rules and initial states can be used to generate the FSM and properties of them can be identified empirically.

In the examples in this paper, we have assumed that when an agent presents an argument, the only reaction the other agent can have is to present a counterargument (if it has one) from a set that is fixed in advance of the dialogue. Yet when agents argue, one agent can reveal information that can be used by the other agent to create new arguments. We illustrate this in the context of logical arguments. Here, we assume that each argument is a tuple $\langle \Phi, \psi \rangle$ where Φ is a set of formulae that entails a formula ψ. In Figure 3a, we see an argument graph instantiated with logical arguments. Suppose arguments A_1, A_3 and A_4 are presented by agent 1, and arguments A_2, A_5 and A_6 are presented by agent 2. Since agent 1 is being exhaustive in the arguments it presents, agent 2 can get a formula that it can use to create a counterargument. In Figure 3b, agent 1 is selective in the arguments it presents, and as a result, agent 2 lacks a formula in order to construct the counterarguments it needs. We can model this argumentation in propositional executable logic, generate the corresponding FSM, and provide an analysis in terms of minimax strategy that would ensure that agent 1 would provide A_4 and not A_3, thereby ensuring that it behaves more intelligently. We can capture each of these arguments as a proposition and use the minimax strategy in our implementation to obtain the tree in Figure 3b.

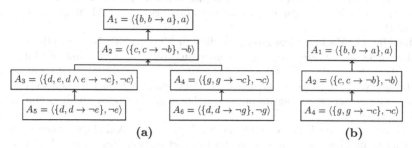

Fig. 3. Consider the following knowledgebases for each agent $\Delta_1 = \{b, d, e, g, b \to a, d \wedge e \to \neg c, g \to \neg c\}$ and $\Delta_2 = \{c, c \to \neg b, d \to \neg e, d \to \neg g\}$. (a) Agent 1 is exhaustive in the arguments posited, thereby allowing agent 2 to construct arguments that cause the root to be defeated. (b)Agent is selective in the arguments posited, thereby ensuring that the root is undefeated.

General frameworks for dialogue games have been proposed [18, 8]. They offer insights on dialogical argumentation systems, but they do not provide sufficient detail to formally analyse or implement specific systems. A more detailed framework, that is based on situation calculus, has been proposed by Brewka [19], though the emphasis is on modelling the protocols for the moves made in dialogical argumentation based on the public state rather than on strategies based on the private states of the agents.

The minimax strategy has been considered elsewhere in models of argumentation (such as for determining argument strength [20] and for marking strategies for dialectical trees [21], for deciding on utterances in a specific dialogical argumentation [22]). However, this paper appears to be the first empirical study of using the minimax strategy in dialogical argumentation.

In future work, we will extend the analytical techniques for imperfect games where only a partial search tree is constructed before the utility function is applied, and extend the representation with weights on transitions (e.g. weights based on tropical semirings to capture probabilistic transitions) to explore the choices of transition based on preference or uncertainty.

References

1. Besnard, P., Hunter, A.: Elements of Argumentation. MIT Press (2008)
2. Amgoud, L., Maudet, N., Parsons, S.: Arguments, dialogue and negotiation. In: European Conf. on Artificial Intelligence (ECAI 2000), pp. 338–342. IOS Press (2000)
3. Black, E., Hunter, A.: An inquiry dialogue system. Autonomous Agents and Multi-Agent Systems 19(2), 173–209 (2009)

4. Dignum, F., Dunin-Keplicz, B., Verbrugge, R.: Dialogue in team formation. In: Dignum, F.P.M., Greaves, M. (eds.) Agent Communication. LNCS (LNAI), vol. 1916, pp. 264–280. Springer, Heidelberg (2000)
5. Fan, X., Toni, F.: Assumption-based argumentation dialogues. In: Proceedings of International Joint Conference on Artificial Intelligence (IJCAI 2011), pp. 198–203 (2011)
6. Hamblin, C.: Mathematical models of dialogue. Theoria 37, 567–583 (1971)
7. Mackenzie, J.: Question begging in non-cumulative systems. Journal of Philosophical Logic 8, 117–133 (1979)
8. McBurney, P., Parsons, S.: Games that agents play: A formal framework for dialogues between autonomous agents. Journal of Logic, Language and Information 11, 315–334 (2002)
9. McBurney, P., van Eijk, R., Parsons, S., Amgoud, L.: A dialogue-game protocol for agent purchase negotiations. Journal of Autonomous Agents and Multi-Agent Systems 7, 235–273 (2003)
10. Parsons, S., Wooldridge, M., Amgoud, L.: Properties and complexity of some formal inter-agent dialogues. J. of Logic and Comp. 13(3), 347–376 (2003)
11. Prakken, H.: Coherence and flexibility in dialogue games for argumentation. J. of Logic and Comp. 15(6), 1009–1040 (2005)
12. Walton, D., Krabbe, E.: Commitment in Dialogue: Basic Concepts of Interpersonal Reasoning. SUNY Press (1995)
13. Black, E., Hunter, A.: Executable logic for dialogical argumentation. In: European Conf. on Artificial Intelligence (ECAI 2012), pp. 15–20. IOS Press (2012)
14. Wooldridge, M., McBurney, P., Parsons, S.: On the meta-logic of arguments. In: Parsons, S., Maudet, N., Moraitis, P., Rahwan, I. (eds.) ArgMAS 2005. LNCS (LNAI), vol. 4049, pp. 42–56. Springer, Heidelberg (2006)
15. Dung, P.: On the acceptability of arguments and its fundamental role in non-monotonic reasoning, logic programming and n-person games. Artificial Intelligence 77(2), 321–357 (1995)
16. Bench-Capon, T.: Persuasion in practical argument using value based argumentation frameworks. Journal of Logic and Computation 13(3), 429–448 (2003)
17. Egly, U., Gaggl, S., Woltran, S.: Aspartix: Implementing argumentation frameworks using answer-set programming. In: Garcia de la Banda, M., Pontelli, E. (eds.) ICLP 2008. LNCS, vol. 5366, pp. 734–738. Springer, Heidelberg (2008)
18. Maudet, N., Evrard, F.: A generic framework for dialogue game implementation. In: Proc. 2nd Workshop on Formal Semantics & Pragmatics of Dialogue, University of Twente, pp. 185–198 (1998)
19. Brewka, G.: Dynamic argument systems: A formal model of argumentation processes based on situation calculus. J. Logic & Comp. 11(2), 257–282 (2001)
20. Matt, P., Toni, F.: A game-theoretic measure of argument strength for abstract argumentation. In: Hölldobler, S., Lutz, C., Wansing, H. (eds.) JELIA 2008. LNCS (LNAI), vol. 5293, pp. 285–297. Springer, Heidelberg (2008)
21. Rotstein, N., Moguillansky, M., Simari, G.: Dialectical abstract argumentation. In: Proceedings of IJCAI, pp. 898–903 (2009)
22. Oren, N., Norman, T.: Arguing using opponent models. In: McBurney, P., Rahwan, I., Parsons, S., Maudet, N. (eds.) ArgMAS 2009. LNCS, vol. 6057, pp. 160–174. Springer, Heidelberg (2010)

What Can Argumentation Do for Inconsistent Ontology Query Answering?

Madalina Croitoru[1] and Srdjan Vesic[2,*]

[1] INRIA, LIRMM, Univ. Montpellier 2, France
[2] CRIL - CNRS, France

Abstract. The area of inconsistent ontological knowledge base query answering studies the problem of inferring from an inconsistent ontology. To deal with such a situation, different semantics have been defined in the literature (e.g. AR, IAR, ICR). Argumentation theory can also be used to draw conclusions under inconsistency. Given a set of arguments and attacks between them, one applies a particular semantics (e.g. stable, preferred, grounded) to calculate the sets of accepted arguments and conclusions. However, it is not clear what are the similarities and differences of semantics from ontological knowledge base query answering and semantics from argumentation theory. This paper provides the answer to that question. Namely, we prove that: (1) sceptical acceptance under stable and preferred semantics corresponds to ICR semantics; (2) universal acceptance under stable and preferred semantics corresponds to AR semantics; (3) acceptance under grounded semantics corresponds to IAR semantics. We also prove that the argumentation framework we define satisfies the rationality postulates (e.g. consistency, closure).

1 Introduction

Ontological knowledge base query answering problem has received renewed interest in the knowledge representation community (and especially in the Semantic Web domain where it is known as the ontology based data access problem [17]). It considers a consistent ontological knowledge base (made from facts and rules) and aims to answer if a query is entailed by the knowledge base (KB). Recently, this question was also considered in the case where the KB is *inconsistent* [16,8]. Maximal consistent subsets of the KB, called *repairs*, are then considered and different *semantics* (based on classical entailment on repairs) are proposed in order to compute the set of accepted formulae.

Argumentation theory is also a well-known method for dealing with inconsistent knowledge [5,2]. Logic-based argumentation [6] considers constructing arguments from inconsistent knowledge bases, identifying attacks between them and selecting acceptable arguments and their conclusions. In order to know which arguments to accept, one applies a particular *argumentation semantics*.

* The major part of the work on this paper was carried out while Srdjan Vesic was affiliated with the Computer Science and Communication Research Unit at the University of Luxembourg. During this period, Srdjan Vesic's project was supported by the National Research Fund, Luxembourg, and cofunded under the Marie Curie Actions of the European Commission (FP7-COFUND). At the time when the authors were finishing the work on this paper, Srdjan Vesic was a CRNS researcher affiliated with CRIL.

W. Liu, V.S. Subrahmanian, and J. Wijsen (Eds.): SUM 2013, LNAI 8078, pp. 15–29, 2013.

This paper starts from the observation that both inconsistent ontological KB query answering and instantiated argumentation theory deal with the same issue, which is reasoning under inconsistent information. Furthermore, both communities have several mechanisms to select acceptable conclusions and they both call them *semantics*. The *research questions* one could immediately ask are: Is there a link between the semantics used in inconsistent ontological KB query answering and those from argumentation theory? Is it possible to instantiate Dung's ([15]) abstract argumentation theory in a way to implement the existing semantics from ontological KB query answering? If so, which semantics from ontological KB query answering correspond to which semantics from argumentation theory? Does the proposed instantiation of Dung's abstract argumentation theory satisfy the rationality postulates [10]?

There are several benefits from answering those questions. First, it would allow to *import some results* from argumentation theory to ontological query answering and vice versa, and more generally open the way to the Argumentation Web [19]. Second, it might be possible to use these results in order to *explain* to users how repairs are constructed and why a particular conclusion holds in a given semantics by constructing and evaluating arguments in favour of different conclusions [14]. Also, on a more theoretical side, proving a link between argumentation theory and the results in the knowledge representation community would be a step forward in understanding the *expressibility* of Dung's abstract theory for logic based argumentation [21].

The paper is organised as follows. In Section 2 the ontological query answering problem is explained and the logical language used throughout the paper is introduced. The end of this section introduces the existing semantics proposed in the literature to deal with inconsistent knowledge bases. Then, in Section 3, we define the basics of argumentation theory. Section 4 proves the links between the extensions obtained under different argumentation semantics in this instantiated logical argumentation setting and the repairs of the ontological knowledge base. We show the equivalence between the semantics from inconsistent ontological KB query answering area and those defined in argumentation theory in Section 5. Furthermore, the argumentation framework thus defined respects the rationality postulates (Section 6). The paper concludes with Section 7. Some proofs are omitted due to the space restrictions, but they are all available in the technical report that can also be found online at `http://hal-lirmm.ccsd.cnrs.fr/docs/00/81/26/30/PDF/TR-Vesic-Croitoru.pdf`.

2   Ontological Conjunctive Query Answering

The main goal of this section is to introduce the syntax and semantics of the $\mathcal{SRC}$ language [3,4], which is used in this paper due to its relevance in the context of the ontological KB query answering. Note that the goal of the present paper is not to change or criticise the definitions from this area; we simply present the existing work. Our goal is to study the link between the existing work in this area and the existing work in argumentation theory. In the following, we give a general setting knowledge representation language which can then be instantiated according to properties on rules or constraints and yield equivalent languages to those used by [16] and [8].

A knowledge base is a 3-tuple $\mathcal{K} = (\mathcal{F}, \mathcal{R}, \mathcal{N})$ composed of three finite sets of formulae: a set $\mathcal{F}$ of facts, a set $\mathcal{R}$ of rules and a set $\mathcal{N}$ of constraints. Let us formally define what we accept as $\mathcal{F}$, $\mathcal{R}$ and $\mathcal{N}$.

Facts Syntax. Let $\mathbf{C}$ be a set of constants and $\mathbf{P} = P_1 \cup P_2 \ldots \cup P_n$ a set of predicates of the corresponding arity $i = 1, \ldots, n$. Let $\mathbf{V}$ be a countably infinite set of *variables*. We define the set of *terms* by $\mathbf{T} = \mathbf{V} \cup \mathbf{C}$. As usual, given $i \in \{1 \ldots n\}$, $p \in P_i$ and $t_1, \ldots, t_i \in \mathbf{T}$ we call $p(t_1, \ldots, t_i)$ an *atom*. If γ is an atom or a conjunction of atoms, we denote by $var(\gamma)$ the set of variables in γ and by $term(\gamma)$ the set of terms in γ. A *fact* is the existential closure of an atom or an existential closure of a conjunction of atoms. (Note that there is no negation or disjunction in the facts.) As an example, consider $\mathbf{C} = \{Tom\}$, $\mathbf{P} = P_1 \cup P_2$, with $P_1 = \{cat, mouse\}$, $P_2 = \{eats\}$ and $\mathbf{V} = \{x_1, x_2, x_3, \ldots\}$. Then, $cat(Tom)$, $eats(Tom, x_1)$ are examples of atoms and $\gamma = cat(Tom) \wedge mouse(x_1) \wedge eats(Tom, x_1)$ is an example of a conjunction of atoms. It holds that $var(\gamma) = \{x_1\}$ and $term(\gamma) = \{Tom, x_1\}$. As an example of a fact, consider $\exists x_1(cat(Tom) \wedge mouse(x_1) \wedge eats(Tom, x_1))$.

An *interpretation* is a pair $I = (\triangle, .^I)$ where $\triangle$ is the interpretation domain (possibly infinite) and $.^I$, the interpretation function, satisfies:

1. For all $c \in \mathbf{C}$, we have $c^I \in \triangle$,
2. For all i and for all $p \in P_i$, we have $p^I \subseteq \triangle^i$,
3. If $c, c' \in \mathbf{C}$ and $c \neq c'$ then $c^I \neq c'^I$.

Let γ be an atom or a conjunction of atoms or a fact. We say that γ is *true* under interpretation I iff there is a function ι which maps the terms (variables and constants) of γ into $\triangle$ such that for all constants c, it holds that $\iota(c) = c^I$ and for all atoms $p(t_1, \ldots t_i)$ appearing in γ, it holds that $(\iota(t_1), \ldots, \iota(t_i)) \in p^I$. For a set F containing any combination of atoms, conjunctions of atoms and facts, we say that F is *true* under interpretation I iff there is a function ι which maps the terms (variables and constants) of all formulae in F into $\triangle$ such that for all constants c, it holds that $\iota(c) = c^I$ and for all atoms $p(t_1, \ldots t_i)$ appearing in formulae of F, it holds that $(\iota(t_1), \ldots, \iota(t_i)) \in p^I$. Note that this means that for example sets $F_1 = \{\exists x(cat(x) \wedge dog(x))\}$ and $F_2 = \{\exists x(cat(x)), \exists x(dog(x))\}$ are true under exactly the same set of interpretations. Namely, in both cases, variable x is mapped to an object of $\triangle$. On the other hand, there are some interpretations under which set $F_3 = \{\exists x_1(cat(x_1)), \exists x_2(dog(x_2))\}$ is true whereas F_1 and F_2 are not.

If γ is true in I we say that I is a model of γ. Let γ' be an atom, a conjunction of atoms or a fact. We say that γ is a logical consequence of γ' (γ' entails γ, denoted $\gamma' \models \gamma$) iff all models of γ are models of γ'. If a set F is true in I we say that I is a model of F. We say that a formula γ is a logical consequence of a set F (denoted $F \models \gamma$) iff all models of F are models of γ. We say that a set G is a logical consequence of set F (denoted $F \models G$) if and only if all models of F are models of G. Two sets F and G are logically equivalent (denoted $F \equiv G$) if and only if $F \models G$ and $G \models F$.

Given a set of variables $\mathbf{X}$ and a set of terms $\mathbf{T}$, a *substitution* σ of $\mathbf{X}$ by $\mathbf{T}$ is a mapping from $\mathbf{X}$ to $\mathbf{T}$ (denoted $\sigma : \mathbf{X} \to \mathbf{T}$). Given an atom or a conjunction of atoms γ, $\sigma(\gamma)$ denotes the expression obtained from γ by replacing each occurrence of $x \in \mathbf{X} \cap var(\gamma)$ by $\sigma(x)$. If a fact F is the existential closure of a conjunction γ then we define $\sigma(F)$ as the existential closure of $\sigma(\gamma)$. Finally, let us define *homomorphism*.

Let F and F' be atoms, conjunctions of atoms or facts (it is not necessarily the case that F and F' are of the same type, e.g. F can be an atom and F' a conjunction of atoms). Let σ be a substitution such that $\sigma : var(F) \to term(F')$. We say that σ is a homomorphism from F to F' if and only if the set of atoms appearing in $\sigma(F)$ is a subset of the set of atoms appearing in $\sigma(F')$. For example, let $F = cat(x_1)$ and $F' = cat(Tom) \wedge mouse(Jerry)$. Let $\sigma : var(F) \to term(F')$ be a substitution such that $\sigma(x_1) = Tom$. Then, σ is a homomorphism from F to F' since the atoms in $\sigma(F)$ are $\{cat(Tom)\}$ and the atoms in $\sigma(F')$ are $\{cat(Tom), mouse(Jerry)\}$.

It is known that $F' \models F$ if and only if there is a homomorphism from F to F' [12].

Rules. A rule R is a formula $\forall x_1, \ldots, \forall x_n \, \forall y_1, \ldots, \forall y_m \, (H(x_1, \ldots, x_n, y_1, \ldots, y_m)$ $\to \exists z_1, \ldots \exists z_k \, C(y_1, \ldots, y_m, z_1, \ldots z_k))$ where H, the hypothesis, and C, the conclusion, are atoms or conjunctions of atoms, $n, m, k \in \{0, 1, \ldots\}$, $x_1, \ldots, x_n$ are the variables appearing in H, $y_1, \ldots, y_m$ are the variables appearing in both H and C and $z_1, \ldots, z_k$ the new variables introduced in the conclusion (for example $\forall x_1 (cat(x_1) \to miaw(x_1))$ or $\forall x_1 ((mouse(x_1) \to \exists z_1 (cat(z_1) \wedge eats(z_1, x_1)))))$.

Reasoning consists of applying rules on the set and thus inferring new knowledge. A rule $R = (H, C)$ is *applicable* to set $\mathcal{F}$ if and only if there exists $\mathcal{F}' \subseteq \mathcal{F}$ such that there is a homomorphism σ from the hypothesis of $\mathcal{R}$ to the conjunction of elements of $\mathcal{F}'$. For example, rule $\forall x_1 (cat(x_1) \to miaw(x_1))$ is applicable to set $\{cat(Tom)\}$, since there is a homomorphism from $cat(x_1)$ to $cat(Tom)$. If rule R is applicable to set F, the application of R to F according to σ produces a set $F \cup \{\sigma(C)\}$. In our example, the produced set is $\{cat(Tom), miaw(Tom)\}$. We then say that the new set (which includes the old one and adds the new information to it) is an *immediate derivation* of F by R. This new set is denoted by $R(F)$. Applying a rule on a set produces a new set.

Let F be a subset of $\mathcal{F}$ and let $\mathcal{R}$ be a set of rules. A set F_n is called an $\mathcal{R}$-*derivation* of F if there is a sequence of sets (*derivation sequence*) $(F_0, F_1, \ldots, F_n)$ such that:

- $F_0 \subseteq F$
- F_0 is $\mathcal{R}$-consistent
- for every $i \in \{1, \ldots, n-1\}$, it holds that F_i is an immediate derivation of F_{i-1}
- (no formula in $\mathcal{F}_n$ contains a conjunction and $\mathcal{F}_n$ is an immediate derivation of $\mathcal{F}_{n-1}$) or F_n is obtained from F_{n-1} by conjunction elimination.

Conjunction elimination is the following procedure: while there exists at least one conjunction in at least one formula, take an arbitrary formula φ containing a conjunction. If φ is of the form $\varphi = \psi \wedge \psi'$ then exchange it with two formulae ψ and ψ'. If φ is of the form $\exists x(\psi \wedge \psi')$ then exchange it with two formulae $\exists x(\psi)$ and $\exists x(\psi')$. The idea is just to start with an $\mathcal{R}$-consistent set and apply (some of the) rules. The only technical detail is that the conjunctions are eliminated from the final result. So if the last set in a sequence does not contain conjunctions, nothing is done. Else, we eliminate those conjunctions. This technicality is needed in order to stay as close as possible to the procedures used in the literature in the case when the knowledge base is consistent.

Given a set $\{F_0, \ldots, F_k\} \subseteq \mathcal{F}$ and a set of rules $\mathcal{R}$, the closure of $\{F_0, \ldots, F_k\}$ with respect to $\mathcal{R}$, denoted $\mathrm{Cl}_{\mathcal{R}}(\{F_0, \ldots, F_k\})$, is defined as the smallest set (with respect to $\subseteq$) which contains $\{F_0, \ldots, F_k\}$, and is closed for $\mathcal{R}$-derivation (that is, for every $\mathcal{R}$-derivation F_n of $\{F_0, \ldots, F_k\}$, we have $F_n \subseteq \mathrm{Cl}_{\mathcal{R}}(\{F_0, \ldots, F_k\})$). Finally, we say that a set $\mathcal{F}$ and a set of rules $\mathcal{R}$ *entail* a fact G (and we write $\mathcal{F}, \mathcal{R} \models G$) iff the closure of the facts by all the rules entails G (i.e. if $\text{Cl}_{\mathcal{R}}(\mathcal{F}) \models G$).

As an example, consider a set of facts $\mathcal{F} = \{cat(Tom), small(Tom)\}$ and the rule set $\mathcal{R} = \{R_1 = \forall x_1(cat(x_1) \rightarrow miaw(x_1) \wedge animal(x_1)), R_2 = \forall x_1(miaw(x_1) \wedge small(x_1) \rightarrow cute(x_1))\}$. Then, F_0, F_1, F_2 is a derivation sequence, where $F_0 = \{cat(Tom), small(Tom)\}$, $F_1 = R_1(F_0) = \{cat(Tom), small(Tom), miaw(Tom) \wedge animal(Tom)\}$, $F_2 = \{cat(Tom), small(Tom), miaw(Tom) \wedge animal(Tom), cute(Tom)\}$ and $F_3 = \{cat(Tom), small(Tom), miaw(Tom), animal(Tom), cute(Tom)\}$.

We conclude the presentation on rules in $\mathcal{SRC}$ by a remark on performing union on facts when they are viewed as sets of atoms. In order to preserve semantics the union is done by renaming variables. For example, let us consider a fact $F_1 = \{\exists x cat(x)\}$ and a fact $F_2 = \{\exists x animal(x)\}$. Then the fact $F = F_1 \cup F_2$ is the union of the two fact after variable naming has been performed: $F = \{\exists x_1 cat(x_1), \exists x_2 animal(x_2)\}$.

Constraints. A constraint is a formula $\forall x_1 \ldots \forall x_n (H(x_1, \ldots, x_n) \rightarrow \bot)$, where H is an atom or a conjunction of atoms and $n \in \{0, 1, 2, \ldots\}$. Equivalently, a constraint can be written as $\neg(\exists x_1, \ldots, \exists x_n H(x_1, \ldots x_n))$. As an example of a constraint, consider $\forall x_1(cat(x_1) \wedge dog(x_1) \rightarrow \bot)$. $H(x_1, \ldots, x_n)$ is called the hypothesis of the constraint.

Given a knowledge base $\mathcal{K} = (\mathcal{F}, \mathcal{R}, \mathcal{N})$, a set $\{F_1, \ldots, F_k\} \subseteq \mathcal{F}$ is said to be *inconsistent* if and only if there exists a constraint $N \in \mathcal{N}$ such that $\{F_1, \ldots, F_k\} \models H_N$, where H_N denotes the existential closure of the hypothesis of N. A set is consistent if and only if it is not inconsistent. A set $\{F_1, \ldots, F_k\} \subseteq \mathcal{F}$ is $\mathcal{R}$-*inconsistent* if and only if there exists a constraint $N \in \mathcal{N}$ such that $\text{Cl}_{\mathcal{R}}(\{F_1, \ldots, F_k\}) \models H_N$, where H_N denotes the existential closure of the hypothesis of N.

A set of facts is said to be $\mathcal{R}$-*consistent* if and only if it is not $\mathcal{R}$-inconsistent. A knowledge base $(\mathcal{F}, \mathcal{R}, \mathcal{N})$ is said to be *consistent* if and only if $\mathcal{F}$ is $\mathcal{R}$-consistent. A knowledge base is *inconsistent* if and only if it is not consistent.

Example 1. Let us consider the following knowledge base $\mathcal{K} = (\mathcal{F}, \mathcal{R}, \mathcal{N})$, with: $\mathcal{F} = \{cat(Tom), bark(Tom)\}$, $\mathcal{R} = \{\forall x_1(cat(x_1) \rightarrow miaw(x_1))\}$, $\mathcal{N} = \{\forall x_1(bark(x_1) \wedge miaw(x_1) \rightarrow \bot)\}$. The only rule in the knowledge base is applicable to the set $\{cat(Tom), bark(Tom)\}$ and its immediate derivation produces the set $\{cat(Tom), bark(Tom), miaw(Tom)\}$. Since $\text{Cl}_{\mathcal{R}}(\mathcal{F}) \models \exists x_1(bark(x_1) \wedge miaw(x_1))$ the KB is inconsistent.

Given a knowledge base, one can ask a conjunctive query in order to know whether something holds or not. Without loss of generality we consider in this paper boolean conjunctive queries (which are facts). As an example of a query, take $\exists x_1 cat(x_1)$. The answer to query α is positive if and only if $\mathcal{F}, \mathcal{R} \models \alpha$.

2.1 Query Answering over Inconsistent Ontological Knowledge Bases

Notice that (like in classical logic), if a knowledge base $\mathcal{K} = (\mathcal{F}, \mathcal{R}, \mathcal{N})$ is inconsistent, then everything is entailed from it. In other words, every query is true. Thus, the approach we described until now is not robust enough to deal with inconsistent information. However, there are cases when the knowledge base is inconsistent; this phenomenon has attracted particular attention during the recent years [8,16]. For example, the set $\mathcal{F}$ may be obtained by combining several sets of facts, coming from different agents. In this paper, we study a general case when $\mathcal{K}$ is inconsistent without making any hypotheses about the origin of this inconsistency. Thus, our results can be applied to an inconsistent base independently of how it is obtained.

A common solution [8,16] is to construct maximal (with respect to set inclusion) consistent subsets of $\mathcal{K}$. Such subsets are called *repairs*. Formally, given a knowledge base $\mathcal{K} = (\mathcal{F}, \mathcal{R}, \mathcal{N})$, define:

$$Repair(\mathcal{K}) = \{\mathcal{F}' \subseteq \mathcal{F} \mid \mathcal{F}' \text{ is maximal for } \subseteq \mathcal{R}\text{-consistent set}\}$$

We now mention a very important technical detail. In some papers, a *set* of formulae is identified with the *conjunction* of those formulae. This is not of particular significance when the knowledge base is *consistent*. However, in case of an *inconsistent* knowledge base, this makes a big difference. Consider for example $\mathcal{K}_1 = (\mathcal{F}_1, \mathcal{R}_1, \mathcal{N}_1)$ with $\mathcal{F}_1 = \{dog(Tom), cat(Tom)\}$, $\mathcal{R}_1 = \emptyset$ and $\mathcal{N}_1 = \{\forall x_1 (dog(x_1) \wedge cat(x_1) \rightarrow \bot)\}$, compared with $\mathcal{K}_2 = (\mathcal{F}_2, \mathcal{R}_2, \mathcal{N}_2)$ with $\mathcal{F}_2 = \{dog(Tom) \wedge cat(Tom)\}$, $\mathcal{R}_2 = \emptyset$ and $\mathcal{N}_2 = \{\forall x_1 (dog(x_1) \wedge cat(x_1) \rightarrow \bot)\}$. In this case, according to the definition of a repair, $\mathcal{K}_1$ would have two repairs and $\mathcal{K}_2$ would have no repairs at all. We could proceed like this, but we find it confusing given the existing literature in this area. This is why, in order to be completely precise, from now on we suppose that $\mathcal{F}$ does not contain conjunctions. Namely, $\mathcal{F}$ is supposed to be a set composed of of atoms and of existential closures of atoms.

Once the repairs calculated, there are different ways to calculate the set of facts that follow from an inconsistent knowledge base. For example, we may want to accept a query if it is entailed in *all repairs* (AR semantics).

Definition 1. *Let* $\mathcal{K} = (\mathcal{F}, \mathcal{R}, \mathcal{N})$ *be a knowledge base and let* α *be a query. Then* α *is* **AR-entailed** *from* $\mathcal{K}$, *written* $\mathcal{K} \models_{AR} \alpha$ *iff for every repair* $A' \in Repair(\mathcal{K})$, *it holds that* $\mathrm{Cl}_{\mathcal{R}}(A') \models \alpha$.

Another possibility is to check whether the query is entailed from the *intersection of closed repairs* (ICR semantics).

Definition 2. *Let* $\mathcal{K} = (\mathcal{F}, \mathcal{R}, \mathcal{N})$ *be a knowledge base and let* α *be a query. Then* α *is* **ICR-entailed** *from* $\mathcal{K}$, *written* $\mathcal{K} \models_{ICR} \alpha$ *iff* $\bigcap_{A' \in Repair(\mathcal{K})} \mathrm{Cl}_{\mathcal{R}}(A') \models \alpha$.

Example 2 (Example 1 Cont.). $Repair(\mathcal{K}) = \{R_1, R_2\}$ with $R_1 = \{cat(Tom)\}$ and $R_2 = \{bark(Tom)\}$. $\mathrm{Cl}_{\mathcal{R}}(R_1) = \{cat(Tom), miaw(Tom)\}$, $\mathrm{Cl}_{\mathcal{R}}(R_2) = \{bark(Tom)\}$. It is *not* the case that $\mathcal{K} \models_{ICR} cat(Tom)$.

Finally, another possibility is to consider the *intersection of all repairs* and then close this intersection under the rules (IAR semantics).

Definition 3. *Let* $\mathcal{K} = (\mathcal{F}, \mathcal{R}, \mathcal{N})$ *be a knowledge base and let* α *be a query. Then* α *is* **IAR-entailed** *from* $\mathcal{K}$, *written* $\mathcal{K} \models_{IAR} \alpha$ *iff* $\mathrm{Cl}_{\mathcal{R}}(\bigcap_{A' \in Repair(\mathcal{K})}) \models \alpha$.

The three semantics can yield different results [16,8]:

Example 3. (ICR and IAR different from AR) Consider $\mathcal{K} = (\mathcal{F}, \mathcal{R}, \mathcal{N})$, with: $\mathcal{F} = \{havecat(Tom), haveMouse(Jerry)\}$, intuitively, we have a cat (called Tom) and a mouse (called Jerry); $\mathcal{R} = \{\forall x_1 (haveCat(x_1) \rightarrow haveAnimal(x_1)),$ $\forall x_2 (haveMouse(x_2) \rightarrow haveAnimal(x_2))\}$; $\mathcal{N} = \{\forall x_1 \forall x_2 (haveCat(x_1) \wedge haveMouse(x_2) \rightarrow \bot)\}$, meaning that we cannot have both a cat and a mouse (since the cat would eat the mouse). There are two repairs: $R_1 = \{haveCat(Tom)\}$ and $R_2 = \{haveMouse(Jerry)\}$. $\text{Cl}_{\mathcal{R}}(R_1) = \{haveCat(Tom), haveAnimal(Tom)\}$ and $\text{Cl}_{\mathcal{R}}(R_2) = \{haveMouse(Jerry), haveAnimal(Jerry)\}$. Consider a query $\alpha = \exists x_1 \, haveAnimal(x_1)$ asking whether we have an animal. It holds that $\mathcal{K} \models_{AR} \alpha$ since $\text{Cl}_{\mathcal{R}}(R_1) \models \alpha$ and $\text{Cl}_{\mathcal{R}} \models \alpha$, but neither $\mathcal{K} \models_{ICR} \alpha$ (since $\text{Cl}_{\mathcal{R}}(R_1) \cap \text{Cl}_{\mathcal{R}}(R_2) = \emptyset$) nor $\mathcal{K} \models_{IAR} \alpha$ (since $R_1 \cap R_2 = \emptyset$).

Example 4. (AR and ICR different from IAR) Consider $\mathcal{K} = (\mathcal{F}, \mathcal{R}, \mathcal{N})$, with: $\mathcal{F} = \{cat(Tom), dog(Tom)\}$, $\mathcal{R} = \{\forall x_1 (cat(x_1) \rightarrow animal(x_1)),$ $\forall x_2 (dog(x_2) \rightarrow animal(x_2))\}$, $\mathcal{N} = \{\forall x (cat(x) \wedge dog(x) \rightarrow \bot)\}$.

We have $\mathcal{R}epair(\mathcal{K}) = \{R_1, R_2\}$ with $R_1 = \{cat(Tom)\}$ and $R_2 = \{dog(Tom)\}$. $\text{Cl}_{\mathcal{R}}(R_1) = \{cat(Tom), animal(Tom)\}$, $\text{Cl}_{\mathcal{R}}(R_2) = \{dog(Tom), animal(Tom)\}$.

It is *not* the case that $\mathcal{K} \models_{IAR} \exists x(animal(x))$ (since $R_1 \cap R_2 = \emptyset$). However, $\mathcal{K} \models_{AR} \exists x(animal(x))$. This is due to the fact that $\text{Cl}_{\mathcal{R}}(R_1) \models \exists x(animal(x))$ and $\text{Cl}_{\mathcal{R}}(R_2) \models \exists x(animal(x))$. Also, we have $\mathcal{K} \models_{ICR} \exists x(animal(x))$ since $\text{Cl}_{\mathcal{R}}(R_1) \cap \text{Cl}_{\mathcal{R}}(R_2) = \{animal(Tom)\}$.

3   Argumentation over Inconsistent Ontological Knowledge Bases

This section shows that it is possible to define an instantiation of Dung's abstract argumentation theory [15] used to reason with an inconsistent ontological KB.

We first define the notion of an argument. For a set of formulae $\mathcal{G} = \{G_1, \ldots, G_n\}$, notation $\bigwedge G$ is used as an abbreviation for $G_1 \wedge \ldots \wedge G_n$.

Definition 4. *Given a knowledge base* $\mathcal{K} = (\mathcal{F}, \mathcal{R}, \mathcal{N})$*, an argument* a *is a tuple* $a = (F_0, F_1, \ldots, F_n)$ *where:*

- $(F_0, \ldots, F_{n-1})$ *is a derivation sequence with respect to* $\mathcal{K}$
- F_n *is an atom, a conjunction of atoms, the existential closure of an atom or the existential closure of a conjunction of atoms such that* $F_{n-1} \models F_n$*.*

Example 5 (Example 2 Cont.). Consider $a = (\{cat(Tom)\}, \{cat(Tom), miaw(Tom)\}, miaw(Tom))$ and $b = (\{bark(Tom)\}, bark(Tom))$ as two examples of arguments.

This is a straightforward way to define an argument when dealing with $\mathcal{SRC}$language, since this way, an *argument* corresponds to a *derivation*.

To simplify the notation, from now on, we suppose that we are given a fixed knowledge base $\mathcal{K} = (\mathcal{F}, \mathcal{R}, \mathcal{N})$ and do not explicitly mention $\mathcal{F}$, $\mathcal{R}$ nor $\mathcal{N}$ if not necessary. Let $a = (F_0, ..., F_n)$ be an argument. Then, we denote $\text{Supp}(a) = F_0$ and $\text{Conc}(a) = F_n$. Let $S \subseteq \mathcal{F}$ a set of facts, $Arg(S)$ is defined as the set of all arguments a such that $\text{Supp}(a) \subseteq S$. Note that the set $Arg(S)$ is also dependent on the set of rules and the set of constraints, but for simplicity reasons, we do not write $\text{Arg}(S, \mathcal{R}, \mathcal{N})$ when it is clear to which $\mathcal{K} = (\mathcal{F}, \mathcal{R}, \mathcal{N})$ we refer to. Finally, let $\mathcal{E}$ be a set of arguments. The base of $\mathcal{E}$ is defined as the union of the argument supports: $\text{Base}(\mathcal{E}) = \bigcup_{a \in \mathcal{E}} \text{Supp}(a)$.

Arguments may attack each other, which is captured by a binary attack relation $\text{Att} \subseteq \text{Arg}(\mathcal{F}) \times \text{Arg}(\mathcal{F})$. Recall that the repairs are the subsets of $\mathcal{F}$ while the set $\mathcal{R}$ is always taken as a whole. This means that the authors of the semantics used to deal with an inconsistent ontological KB envisage the set of facts as inconsistent and the set of rules as consistent. When it comes to the attack relation, this means that we only need the so called "assumption attack" since, roughly speaking, all the inconsistency "comes from the facts".

Definition 5. *Let $\mathcal{K} = (\mathcal{F}, \mathcal{R}, \mathcal{N})$ be a knowledge base and let a and b be two arguments. The argument a attacks argument b, denoted $(a, b) \in$ Att, if and only if there exists $\varphi \in \text{Supp}(b)$ such that the set $\{Conc(a), \varphi\}$ is $\mathcal{R}$-inconsistent.*

This attack relation is not symmetric. To see why, consider the following example. Let $\mathcal{F} = \{p(m), q(m), r(m)\}$, $\mathcal{R} = \emptyset$, $\mathcal{N} = \{\forall x_1(p(x_1) \wedge q(x_1) \wedge r(x_1) \rightarrow \perp)\}$. Let $a = (\{p(m), q(m)\}, p(m) \wedge q(m))$, $b = (\{r(m)\}, r(m))$. We have $(a, b) \in$ Att and $(b, a) \notin$ Att. Note that using attack relations which are not symmetric is very common in argumentation literature. Moreover, symmetric attack relation have been criticised for violating some desirable properties [1].

Definition 6. *Given a knowledge base $\mathcal{K} = (\mathcal{F}, \mathcal{R}, \mathcal{N})$, the corresponding argumentation framework $\mathcal{AF}_K$ is a pair $(\mathcal{A} = \text{Arg}(\mathcal{F}), \text{Att})$ where $\mathcal{A}$ is the set of arguments that can be constructed from $\mathcal{F}$ and Att is the corresponding attack relation as specified in Definition 5. Let $\mathcal{E} \subseteq \mathcal{A}$ and $a \in \mathcal{A}$. We say that $\mathcal{E}$ is conflict free iff there exists no arguments $a, b \in \mathcal{E}$ such that $(a, b) \in$ Att. $\mathcal{E}$ defends a iff for every argument $b \in \mathcal{A}$, if we have $(b, a) \in$ Att then there exists $c \in \mathcal{E}$ such that $(c, b) \in$ Att. $\mathcal{E}$ is admissible iff it is conflict free and defends all its arguments. $\mathcal{E}$ is a complete extension iff $\mathcal{E}$ is an admissible set which contains all the arguments it defends. $\mathcal{E}$ is a preferred extension iff it is maximal (with respect to set inclusion) admissible set. $\mathcal{E}$ is a stable extension iff it is conflict-free and for all $a \in \mathcal{A} \setminus \mathcal{E}$, there exists an argument $b \in \mathcal{E}$ such that $(b, a) \in$ Att. $\mathcal{E}$ is a grounded extension iff $\mathcal{E}$ is a minimal (for set inclusion) complete extension. If a semantics returns exactly one extension for every argumentation framework, then it is called a single-extension semantics. For an argumentation framework $AS = (\mathcal{A}, \text{Att})$ we denote by $Ext_x(AS)$ (or by $Ext_x(\mathcal{A}, \text{Att})$) the set of its extensions with respect to semantics x. We use the abbreviations c, p, s, and g for respectively complete, preferred, stable and grounded semantics. An argument is sceptically accepted if it is in all extensions, credulously accepted if it is in at least one extension and rejected if it is not in any extension.*

Finally, we introduce two definitions allowing us to reason over such an argumentation framework. The output of an argumentation framework is usually defined [10, Definition 12] as the set of conclusions that appear in all the extensions (under a given semantics).

Definition 7 (Output of an argumentation framework). *Let* $\mathcal{K} = (\mathcal{F}, \mathcal{R}, \mathcal{N})$ *be a knowledge base and* $\mathcal{AF}_K$ *the corresponding argumentation framework. The output of* $\mathcal{AF}_K$ *under semantics* x *is defined as:*

$$\mathtt{Output}_x(\mathcal{AF}_K) = \bigcap_{\mathcal{E} \in \mathtt{Ext}_x(\mathcal{AF}_K)} \mathtt{Concs}(\mathcal{E}).$$

When $\mathtt{Ext}_x(\mathcal{AF}_K) = \emptyset$, *we define* $\mathtt{Output}(\mathcal{AF}_K) = \emptyset$ *by convention.*

Note that the previous definition asks for existence of a conclusion in every extension. This kind of acceptance is usually referred to as *sceptical* acceptance. We say that a query α is sceptically accepted if it is a logical consequence of the output of $\mathcal{AF}_K$:

Definition 8 (Sceptical acceptance of a query). *Let* $\mathcal{K} = (\mathcal{F}, \mathcal{R}, \mathcal{N})$ *be a knowledge base and* $\mathcal{AF}_K$ *the corresponding argumentation framework. A query* α *is sceptically accepted under semantics* x *if and only if* $\mathtt{Output}_x(\mathcal{AF}_K) \models \alpha$.

It is possible to make an alternative definition, which uses the notion of universal acceptance instead of sceptical one. According to universal criteria, a query α is accepted if it is a logical consequence of conclusions of every extension:

Definition 9 (Universal acceptance of a query). *Let* $\mathcal{K} = (\mathcal{F}, \mathcal{R}, \mathcal{N})$ *be a knowledge base and* $\mathcal{AF}_K$ *the corresponding argumentation framework. A query* α *is universally accepted under semantics* x *if and only if for every extension* $\mathcal{E}_i \in \mathtt{Ext}_x(\mathcal{AF}_K)$, *it holds that* $\mathtt{Concs}(\mathcal{E}_i) \models \alpha$.

In general, universal and sceptical acceptance of a query do not coincide. Take for instance the KB from Example 3, construct the corresponding argumentation framework, and compare the sets of universally and sceptically accepted queries under preferred semantics. Note that for single-extension semantics (e.g. grounded), the notions of sceptical and universal acceptance coincide. So we simply use word "accepted" in this context.

Definition 10 (Acceptance of a query). *Let* $\mathcal{K} = (\mathcal{F}, \mathcal{R}, \mathcal{N})$ *be a knowledge base,* $\mathcal{AF}_K$ *the corresponding argumentation framework,* x *a single-extension semantics and let* $\mathcal{E}$ *be the unique extension of* $\mathcal{AF}_K$. *A query* α *is accepted under semantics* x *if and only if* $\mathtt{Concs}(\mathcal{E}) \models \alpha$.

4   Equivalence between Repairs and Extensions

In this section, we prove two links between the repairs of an ontological KB and the corresponding argumentation framework: Theorem 1 shows that the repairs of the KB correspond exactly to the stable (and preferred, since in this instantiation the stable and the preferred semantics coincide) extensions of the argumentation framework; Theorem 2 proves that the intersection of all the repairs of the KB corresponds to the grounded extension of the argumentation framework.

Theorem 1. *Let $\mathcal{K} = (\mathcal{F}, \mathcal{R}, \mathcal{N})$ be a knowledge base, $\mathcal{AF}_K$ the corresponding argumentation framework and $x \in \{s, p\}$[1]. Then:*

$$\texttt{Ext}_x(\mathcal{AF}_K) = \{\texttt{Arg}(A') \mid A' \in \mathcal{R}epair(\mathcal{K})\}$$

Proof. The plan of the proof is as follows:

1. We prove that $\{\texttt{Arg}(A') \mid A' \in \mathcal{R}epair(\mathcal{K})\} \subseteq \texttt{Ext}_s(\mathcal{AF}_K)$.
2. We prove that $\texttt{Ext}_p(\mathcal{AF}_K) \subseteq \{\texttt{Arg}(A') \mid A' \in \mathcal{R}epair(\mathcal{K})\}$.
3. Since every stable extension is a preferred one [15], we can proceed as follows. From the first item, we have that $\{\texttt{Arg}(A') \mid A' \in \mathcal{R}epair(\mathcal{K})\} \subseteq \texttt{Ext}_p(\mathcal{AF}_K)$, thus the theorem holds for preferred semantics. From the second item we have that $\texttt{Ext}_s(\mathcal{AF}_K) \subseteq \{\texttt{Arg}(A') \mid A' \in \mathcal{R}epair(\mathcal{K})\}$, thus the theorem holds for stable semantics.

1. We first show $\{\texttt{Arg}(A') \mid A' \in \mathcal{R}epair(\mathcal{K})\} \subseteq \texttt{Ext}_s(\mathcal{AF}_K)$. Let $A' \in \mathcal{R}epair(\mathcal{K})$ and let $\mathcal{E} = \texttt{Arg}(A')$. Let us prove that $\mathcal{E}$ is a stable extension of $(\texttt{Arg}(\mathcal{F}), \texttt{Att})$.
   We first prove that $\mathcal{E}$ is conflict-free. By means of contradiction we suppose the contrary, i.e. let $a, b \in \mathcal{E}$ such that $(a, b) \in \texttt{Att}$. From the definition of attack, there exists $\varphi \in \texttt{Supp}(b)$ such that $\{\texttt{Conc}(a), \varphi\}$ is $\mathcal{R}$-inconsistent. Thus $\texttt{Supp}(a) \cup \{\varphi\}$ is $\mathcal{R}$-inconsistent; consequently A' is $\mathcal{R}$-inconsistent, contradiction. Therefore $\mathcal{E}$ is conflict-free.
   Let us now prove that $\mathcal{E}$ attacks all arguments outside the set. Let $b \in \texttt{Arg}(\mathcal{F}) \setminus \texttt{Arg}(A')$ and let $\varphi \in \texttt{Supp}(b)$, such that $\varphi \notin A'$. Let A'_c be the set obtained from A' by conjunction elimination and let $a = (A', A'_c, \bigwedge A'_c)$. We have $\varphi \notin A'$, so, due to the set inclusion maximality for the repairs, $\{\bigwedge A'_c, \varphi\}$ is $\mathcal{R}$-inconsistent. Therefore, $(a, b) \in \texttt{Att}$. Consequently, $\mathcal{E}$ is a stable extension.
2. We now need to prove that $\texttt{Ext}_p(\mathcal{AF}_K) \subseteq \{\texttt{Arg}(A') \mid A' \in \mathcal{R}epair(\mathcal{K})\}$. Let $\mathcal{E} \in \texttt{Ext}_p(\mathcal{AF}_K)$ and let us prove that there exists a repair A' such that $\mathcal{E} = \texttt{Arg}(A')$. Let $S = \texttt{Base}(\mathcal{E})$. Let us prove that S is $\mathcal{R}$-consistent. Aiming to a contradiction, suppose that S is $\mathcal{R}$-inconsistent. Let $S' \subseteq S$ be such that (1) S' is $\mathcal{R}$-inconsistent and (2) every proper set of S' is $\mathcal{R}$-consistent. Let us denote $S' = \{\varphi_1, \varphi_2, ..., \varphi_n\}$. Let $a \in \mathcal{E}$ be an argument such that $\varphi_n \in \texttt{Supp}(a)$. Let S'_c be the set obtained from $S' \setminus \{\varphi\}$ by conjunction elimination and let $a' = (S' \setminus \{\varphi_n\}, S'_c, \bigwedge S'_c)$. We have that $(a', a) \in \texttt{Att}$. Since $\mathcal{E}$ is conflict free, then $a' \notin \mathcal{E}$. Since $\mathcal{E}$ is an admissible set, there exists $b \in \mathcal{E}$ such that $(b, a') \in \texttt{Att}$. Since b attacks a' then there exists $i \in \{1, 2, ..., n-1\}$ such that $\{\texttt{Conc}(b), \varphi_i\}$ is $\mathcal{R}$-inconsistent. Since $\varphi_i \in \texttt{Base}(\mathcal{E})$, then there exists $c \in \mathcal{E}$ such that $\varphi_i \in \texttt{Supp}(c)$. Thus $(b, c) \in \texttt{Att}$, contradiction. So it must be that S is $\mathcal{R}$-consistent.
   Let us now prove that there exists no $S' \subseteq \mathcal{F}$ such that $S \subsetneq S'$ and S' is $\mathcal{R}$-consistent. We use the proof by contradiction. Thus, suppose that S is not a maximal $\mathcal{R}$-consistent subset of $\mathcal{F}$. Then, there exists $S' \in \mathcal{R}epair(\mathcal{K})$, such that $S \subsetneq S'$. We have that $\mathcal{E} \subseteq \texttt{Arg}(S)$, since $S = \texttt{Base}(\mathcal{E})$. Denote $\mathcal{E}' = \texttt{Arg}(S')$. Since $S \subsetneq S'$ then $\texttt{Arg}(S) \subsetneq \mathcal{E}'$. Thus, $\mathcal{E} \subsetneq \mathcal{E}'$. From the first part of the proof, $\mathcal{E}' \in \texttt{Ext}_s(\mathcal{AF}_K)$.

---

[1] Recall that *s* stands for *stable* and *p* for *preferred* semantics.

Consequently, $\mathcal{E}' \in \text{Ext}_p(\mathcal{AF}_K)$. We also know that $\mathcal{E} \in \text{Ext}_p(\mathcal{AF}_K)$. Contradiction, since no preferred set can be a proper subset of another preferred set. Thus, we conclude that $\text{Base}(\mathcal{E}) \in \mathcal{R}epair(\mathcal{K})$.

Let us show that $\mathcal{E} = \text{Arg}(\text{Base}(\mathcal{E}))$. It must be that $\mathcal{E} \subseteq \text{Arg}(S)$. Also, we know (from the first part) that $\text{Arg}(S)$ is a stable and a preferred extension, thus the case $\mathcal{E} \subsetneq \text{Arg}(s)$ is not possible.

3. Now we know that $\{\text{Arg}(A') \mid A' \in \mathcal{R}epair(\mathcal{K})\} \subseteq \text{Ext}_s(\mathcal{AF}_K)$ and $\text{Ext}_p(\mathcal{AF}_K)$ $\subseteq \{\text{Arg}(A') \mid A' \in \mathcal{R}epair(\mathcal{K})\}$. The theorem follows from those two facts, as explained at the beginning of the proof.

To prove Theorem 2, we use the following lemma which says that if there are no rejected arguments under preferred semantics, then the grounded extension is equal to the intersection of all preferred extensions. Note that this result holds for every argumentation framework (not only for the one studied in this paper, where arguments are constructed from an ontological knowledge base). Thus, we only suppose that we are given a set and a binary relation on it (called attack relation).

Lemma 1. *Let $AS = (\mathcal{A}, Att)$ be an argumentation framework and* GE *its grounded extension.*

$$\text{If } \mathcal{A} \subseteq \bigcup_{\mathcal{E}_i \in \text{Ext}_p(AS)} \mathcal{E}_i \quad \text{then} \quad \text{GE} = \bigcap_{\mathcal{E}_i \in \text{Ext}_p(AS)} \mathcal{E}_i.$$

We can now, using the previous result, show the link between the intersection of repairs and the grounded extension.

Theorem 2. *Let $\mathcal{K} = (\mathcal{F}, \mathcal{R}, \mathcal{N})$ be a knowledge base and $\mathcal{AF}_K$ the corresponding argumentation framework. Denote the grounded extension of $\mathcal{AF}_K$ by* GE. *Then:*

$$\text{GE} = \text{Arg}(\bigcap_{A' \in \mathcal{R}epair(\mathcal{K})} A').$$

5   Semantics Equivalence

This section presents the main result of the paper. It shows the links between semantics from argumentation theory (stable, preferred, grounded) and semantics from inconsistent ontology KB query answering (ICR, AR, IAR). More precisely, we show that: (1) sceptical acceptance under stable and preferred semantics corresponds to ICR semantics; (2) universal acceptance under stable and preferred semantics corresponds to AR semantics; (3) acceptance under grounded semantics corresponds to IAR semantics. The proof of Theorem 3 is based on Theorem 1 and the proof of Theorem 4 is derived from Theorem 2.

Theorem 3. *Let $\mathcal{K} = (\mathcal{F}, \mathcal{R}, \mathcal{N})$ be a knowledge base, let $\mathcal{AF}_K$ be the corresponding argumentation framework and let α be a query. Let $x \in \{s, p\}$ be stable or preferred semantics. Then:*

– $\mathcal{K} \models_{ICR} \alpha$ *iff α is sceptically accepted under semantics x.*
– $\mathcal{K} \models_{AR} \alpha$ *iff α is universally accepted under semantics x.*

Theorem 4. *Let* $\mathcal{K} = (\mathcal{F}, \mathcal{R}, \mathcal{N})$ *be a knowledge base, let* $\mathcal{AF}_K$ *be the corresponding argumentation framework and let* α *be a query. Then:*

$$\mathcal{K} \models_{IAR} \alpha \text{ iff } \alpha \text{ is accepted under grounded semantics.}$$

Proof. Let us denote the grounded extension of $\mathcal{AF}_K$ by GE and the intersection of all repairs by $\texttt{Ioar} = \bigcap_{A' \in \mathcal{R}epair(\mathcal{K})} A'$. From Definition 10, we have:

$$\alpha \text{ is accepted under grounded semantics iff } \texttt{Concs(GE)} \models \alpha. \tag{1}$$

From Theorem 2, we have:
$$\texttt{GE} = \texttt{Arg(Ioar)}. \tag{2}$$

Note also that for every set of facts $\{F_1, \ldots, F_n\}$ and for every query α, we have that $\texttt{Cl}_{\mathcal{R}}(\{F_1, \ldots, F_n\}) \models \alpha$ if and only if $\texttt{Concs(Arg}(\{F_1, \ldots, F_n\})) \models \alpha$. Thus,

$$\texttt{Cl}_{\mathcal{R}}(\texttt{Ioar}) \models \alpha \text{ if and only if } \texttt{Concs(Arg(Ioar))} \models \alpha. \tag{3}$$

From (2) and (3) we have that:

$$\texttt{Cl}_{\mathcal{R}}(\texttt{Ioar}) \models \alpha \text{ if and only if } \texttt{Concs(GE)} \models \alpha. \tag{4}$$

From Definition 3, one obtains:

$$\texttt{Cl}_{\mathcal{R}}(\texttt{Ioar}) \models \alpha \text{ if and only if } \mathcal{K} \models_{IAR} \alpha. \tag{5}$$

The theorem now follows from (1), (4) and (5).

6   Postulates

In this section, we prove that the framework we propose in this paper satisfies the rationality postulates for instantiated argumentation frameworks [10]. We first prove the indirect consistency postulate.

Proposition 1 (Indirect consistency). *Let* $\mathcal{K} = (\mathcal{F}, \mathcal{R}, \mathcal{N})$ *be a knowledge base,* $\mathcal{AF}_K$ *the corresponding argumentation framework and* $x \in \{s, p, g\}$. *Then:*

– *for every* $\mathcal{E}_i \in \texttt{Ext}_x(\mathcal{AF}_K)$, $\texttt{Cl}_{\mathcal{R}}(\texttt{Concs}(\mathcal{E}_i))$ *is a consistent set*
– $\texttt{Cl}_{\mathcal{R}}(\texttt{Output}_x(\mathcal{AF}_K))$ *is a consistent set.*

Proof.
– Let $\mathcal{E}_i$ be a stable or a preferred extension of $\mathcal{AF}_K$. From Theorem 1, there exists a repair $A' \in \mathcal{R}epair(\mathcal{K})$ such that $\mathcal{E}_i = \texttt{Arg}(A')$. Note that $\texttt{Concs}(\mathcal{E}_i) = \texttt{Cl}_{\mathcal{R}}(A') \cup \{\alpha \mid \texttt{Cl}_{\mathcal{R}}(A) \models \alpha\}$ (this follows directly from Definition 4). Consequently, the set of $\mathcal{R}$-derivations of $\texttt{Concs}(\mathcal{E}_i)$ and the set of $\mathcal{R}$-derivations of $\texttt{Cl}_{\mathcal{R}}(A')$ coincide. Formally, $\texttt{Cl}_{\mathcal{R}}(\texttt{Cl}_{\mathcal{R}}(A')) = \texttt{Cl}_{\mathcal{R}}(\texttt{Concs}(\mathcal{E}_i))$. Since $\texttt{Cl}_{\mathcal{R}}$ is idempotent, this means that $\texttt{Cl}_{\mathcal{R}}(A') = \texttt{Cl}_{\mathcal{R}}(\texttt{Concs}(\mathcal{E}_i))$. Since $\texttt{Cl}_{\mathcal{R}}(A')$ is consistent, then $\texttt{Cl}_{\mathcal{R}}(\texttt{Concs}(\mathcal{E}_i))$ is consistent.

Let us now consider the case of grounded semantics. Denote GE the grounded extension of $\mathcal{AF}_K$. We have just seen that for every $\mathcal{E}_i \in \text{Ext}_p(\mathcal{AF}_K)$, it holds that $\text{Cl}_\mathcal{R}(\text{Concs}(\mathcal{E}_i))$ is a consistent set. Since the grounded extension is a subset of the intersection of all the preferred extensions [15], and since there is at least one preferred extension, say $\mathcal{E}_1$, then $\text{GE} \subseteq \mathcal{E}_1$. Since $\text{Cl}_\mathcal{R}(\text{Concs}(\mathcal{E}_i))$ is consistent then $\text{Cl}_\mathcal{R}(\text{Concs}(\text{GE}))$ is also consistent.

- Consider the case of stable or preferred semantics. Let us prove $\text{Cl}_\mathcal{R}(\text{Output}_x(\mathcal{AF}_K))$ is a consistent set. Recall that $\text{Output}_x(\mathcal{AF}_K) = \bigcap_{\mathcal{E}_i \in \text{Ext}_x(\mathcal{AF}_K)} \text{Concs}(\mathcal{E}_i)$. Since every knowledge base has at least one repair then, according to Theorem 1, there is at least one stable or preferred extension $\mathcal{E}_i$. From Definition 7, we have that $\text{Output}_x(\mathcal{AF}_K) \subseteq \text{Concs}(\mathcal{E}_i)$. $\text{Concs}(\mathcal{E}_i)$ is $\mathcal{R}$-consistent thus $\text{Output}_x(\mathcal{AF}_K)$ is $\mathcal{R}$-consistent. In other words, $\text{Cl}_\mathcal{R}(\text{Output}_x(\mathcal{AF}_K))$ is consistent.

Note that in the case of grounded semantics the second part of the proposition follows directly from the first one, since $\text{Cl}_\mathcal{R}(\text{Output}_g(\mathcal{AF}_K)) = \text{Cl}_\mathcal{R}(\text{Concs}(\text{GE}))$.

Since our instantiation satisfies indirect consistency then it also satisfies direct consistency. This comes from $\mathcal{R}$-consistency definition; namely, if a set is $\mathcal{R}$-consistent, then it is necessarily consistent. Thus, we obtain the following corollary.

Corollary 1 (Direct consistency). *Let* $\mathcal{K} = (\mathcal{F}, \mathcal{R}, \mathcal{N})$ *be a knowledge base,* $\mathcal{AF}_K$ *the corresponding argumentation framework and* $x \in \{s, p, g\}$. *Then:*

- *for every* $\mathcal{E}_i \in \text{Ext}_x(\mathcal{AF}_K)$, $\text{Concs}(\mathcal{E}_i)$ *is a consistent set*
- $\text{Output}_x(\mathcal{AF}_K)$ *is a consistent set.*

We can now also show that the present argumentation formalism also satisfies the closure postulate.

Proposition 2 (Closure). *Let* $\mathcal{K} = (\mathcal{F}, \mathcal{R}, \mathcal{N})$ *be a knowledge base,* $\mathcal{AF}_K$ *the corresponding argumentation framework and* $x \in \{s, p, g\}$. *Then:*

- *for every* $\mathcal{E}_i \in \text{Ext}_x(\mathcal{AF}_K)$, $\text{Concs}(\mathcal{E}_i) = \text{Cl}_\mathcal{R}(\text{Concs}(\mathcal{E}_i))$.
- $\text{Output}_x(\mathcal{AF}_K) = \text{Cl}_\mathcal{R}(\text{Output}_x(\mathcal{AF}_K))$.

7 Summary and Conclusion

This paper investigates the links between the semantics used in argumentation theory and those from the inconsistent ontological KB query answering.

Contribution of the Paper. First, we show that it is possible to instantiate Dung's abstract argumentation theory in a way to deal with inconsistency in an ontological KB. Second, we formally prove the links between the semantics from ontological KB query answering and those from argumentation theory: ICR semantics corresponds to sceptical acceptance under stable or preferred argumentation semantics, AR semantics corresponds to universal acceptance under stable / preferred argumentation semantics and IAR semantics corresponds to acceptance under grounded argumentation semantics. Third, we show that the instantiation we define satisfies the rationality postulates.

The fourth contribution of the paper is to make a bridge between the argumentation community and the knowledge representation community.

Applications of Our Work. The first possible application of our work is to import some results about semantics and acceptance from argumentation to ontological KB query answering and vice versa. Second, arguments can be used for explanatory purposes. In other words, we can use arguments and counter arguments to graphically represent and explain why different points of view are conflicting or not and why certain argument is (not) in all extensions. However, we suppose that the user understands the notion of logical consequence under first order logic when it comes to *consistent* data. For example, we suppose that the user is able to understand that if $cat(Tom) \wedge miaw(Tom)$ is present in the set, then queries $cat(Tom)$ and $\exists x cat(x)$ are both true. To sum up, we suppose that the other methods are used to explain reasoning under *consistent* knowledge and we use argumentation to explain reasoning under *inconsistent* knowledge.

Related Work. Note that this is the first work studying the link between semantics used in argumentation (stable, preferred, grounded) and semantics used in inconsistent ontological knowledge base query answering (AR, IAR, ICR). There is not much related work. However, we review some papers that study similar issues.

For instance, the link between maximal consistent subsets of a knowledge base and stable extensions of the corresponding argumentation system was shown by Cayrol [11]. That was the first work showing this type of connection between argument-based and non argument-based reasoning. This result was generalised [20] by studying the whole class of argumentation systems corresponding to maximal consistent subsets of the propositional knowledge base. The link between the ASPIC system [18] and the Argument Interchange Format (AIF) ontology [13] has recently been studied [7]. Another related paper comprises constructing an argumentation framework with ontological knowledge allowing two agents to discuss the answer to queries concerning their knowledge (even if it is inconsistent) without one agent having to copy all of their ontology to the other [9]. While those papers are in the area of our paper, none of them is related to the study of the links between different semantics for inconsistent ontological KB query answering and different argumentation semantics.s

Future Work. We plan to answer different questions, like: Can other semantics from argumentation theory yield different results? Are those results useful for inconsistent ontological KB query answering? What happens in the case when preferences are present?

References

1. Amgoud, L., Besnard, P.: Bridging the gap between abstract argumentation systems and logic. In: Godo, L., Pugliese, A. (eds.) SUM 2009. LNCS, vol. 5785, pp. 12–27. Springer, Heidelberg (2009)
2. Amgoud, L., Cayrol, C.: Inferring from inconsistency in preference-based argumentation frameworks. Journal of Automated Reasoning 29 (2), 125–169 (2002)
3. Baget, J.-F., Mugnier, M.-L.: The Complexity of Rules and Constraints. JAIR 16, 425–465 (2002)
4. Baget, J.-F., Mugnier, M.-L., Rudolph, S., Thomazo, M.: Walking the complexity lines for generalized guarded existential rules. In: Proceedings of the 22nd International Joint Conference on Artificial Intelligence (IJCAI 2011), pp. 712–717 (2011)

5. Benferhat, S., Dubois, D., Prade, H.: Argumentative inference in uncertain and inconsistent knowledge bases. In: Proceedings of the 9th Conference on Uncertainty in Artificial intelligence (UAI 1993), pp. 411–419 (1993)
6. Besnard, P., Hunter, A.: Elements of Argumentation. MIT Press (2008)
7. Bex, F.J., Modgil, S.J., Prakken, H., Reed, C.: On logical specifications of the argument interchange format. Journal of Logic and Computation (2013)
8. Bienvenu, M.: On the complexity of consistent query answering in the presence of simple ontologies. In: Proc. of AAAI (2012)
9. Black, E., Hunter, A., Pan, J.Z.: An argument-based approach to using multiple ontologies. In: Godo, L., Pugliese, A. (eds.) SUM 2009. LNCS, vol. 5785, pp. 68–79. Springer, Heidelberg (2009)
10. Caminada, M., Amgoud, L.: On the evaluation of argumentation formalisms. Artificial Intelligence Journal 171 (5-6), 286–310 (2007)
11. Cayrol, C.: On the relation between argumentation and non-monotonic coherence-based entailment. In: Proceedings of the 14th International Joint Conference on Artificial Intelligence (IJCAI 1995), pp. 1443–1448 (1995)
12. Chein, M., Mugnier, M.-L.: Graph-based Knowledge Representation and Reasoning— Computational Foundations of Conceptual Graphs. Advanced Information and Knowledge Processing. Springer (2009)
13. Chesnevar, C., McGinnis, J., Modgil, S., Rahwan, I., Reed, C., Simari, G., South, M., Vreeswijk, G., Willmott, S.: Towards an argument interchange format. Knowledge Engineering Review 21(4), 293–316 (2006)
14. Dix, J., Parsons, S., Prakken, H., Simari, G.R.: Research challenges for argumentation. Computer Science - R&D 23(1), 27–34 (2009)
15. Dung, P.M.: On the acceptability of arguments and its fundamental role in nonmonotonic reasoning, logic programming and n-person games. Artificial Intelligence Journal 77, 321–357 (1995)
16. Lembo, D., Lenzerini, M., Rosati, R., Ruzzi, M., Savo, D.F.: Inconsistency-tolerant semantics for description logics. In: Proc. of RR, pp. 103–117 (2010)
17. Lenzerini, M.: Data integration: A theoretical perspective. In: Proc. of PODS 2002 (2002)
18. Modgil, S.J., Prakken, H.: A general account of argumentation with preferences. Artificial Intelligence Journal (2013)
19. Rahwan, I., Zablith, F., Reed, C.: Laying the foundations for a world wide argument web. Artificial Intelligence 171(10-15), 897–921 (2007)
20. Vesic, S.: Maxi-consistent operators in argumentation. In: 20th European Conference on Artificial Intelligence (ECAI 2012), pp. 810–815 (2012)
21. Vesic, S., van der Torre, L.: Beyond maxi-consistent argumentation operators. In: del Cerro, L.F., Herzig, A., Mengin, J. (eds.) JELIA 2012. LNCS, vol. 7519, pp. 424–436. Springer, Heidelberg (2012)

Enforcement in Argumentation Is a Kind of Update

Pierre Bisquert, Claudette Cayrol,
Florence Dupin de Saint-Cyr, and Marie-Christine Lagasquie-Schiex

IRIT – UPS, Toulouse, France
{bisquert,ccayrol,dupin,lagasq}@irit.fr

Abstract. In the literature, enforcement consists in changing an argumentation system in order to force it to accept a given set of arguments. In this paper, we extend this notion by allowing incomplete information about the initial argumentation system. Generalized enforcement is an operation that maps a propositional formula describing a system and a propositional formula that describes a goal, to a new formula describing the possible resulting systems. This is done under some constraints about the allowed changes. We give a set of postulates restraining the class of enforcement operators and provide a representation theorem linking them to a family of proximity relations on argumentation systems.

Keywords: dynamics in argumentation, belief change.

1 Introduction

During a trial, a lawyer makes her final address to the judge; the lawyer of the opposite party, say O, is able to build the argumentation system (a graph containing arguments and attacks relation between them) corresponding to this pleading. O is also able to compute all the arguments that are accepted according to the pleading, *i.e.*, the set of consensual arguments. Suppose now that O wants to force the audience to accept another set of arguments. She has to make a change to the argumentation system, either by adding an argument or by making an objection about an argument (to remove it) in order to achieve this goal. In the literature, the operation to perform on an argumentation system in order to ensure that a given set of arguments is accepted given a set of authorized changes is called "enforcement" [3].

This enforcement may be done more or less easily, since it may involve more or less changes (costs to add/remove arguments may be introduced). The aim of the speaker will be to find the least expensive changes to make to the argumentation system.

The previous example is a particular case of a more general enforcement operator. Since we could consider cases where Agent O does not know exactly the argumentation system on which she must make a change but knows only some information about it (*e.g.* some arguments that are accepted or that are present in the system). In this more general case, the idea is to ensure that the argumentation system after change satisfies a given goal whatever the initial system is. The result of enforcement will give a characterization of the set of argumentation systems that could be obtained (taking into account a set of authorized changes).

The key idea developed in this paper is the parallel between belief update theory [19,16] and enforcement in argumentation. Enforcement consists in searching for the

W. Liu, V.S. Subrahmanian, and J. Wijsen (Eds.): SUM 2013, LNAI 8078, pp. 30–43, 2013.

argumentation systems that are closest to a given starting argumentation system, in a set of argumentation systems in which some target arguments are accepted. This gives us the parallel with preorders on worlds in belief update. Hence worlds correspond to argumentation systems while formulas should represent knowledge about these argumentation systems. In classical enforcement this knowledge is expressed in terms of a description of an initial argumentation system and a set of arguments that one wants to see accepted. This is why we propose to introduce a propositional language in which this kind of information may be expressed. This language enables us to generalize enforcement with a broader expressiveness.

Our paper is situated in the growing domain of dynamics of argumentation systems [8,7,9,3,18,17] which covers both addition and removal of arguments or interactions. It is organized as follows. We first restate abstract argumentation theory. Then we present a framework that illustrates a particular case of change in argumentation, it concerns an agent that wants to act on a given target system, this agent has a given goal and her possible actions are limited. We then recall classical enforcement. In the third section we propose a generalization of classical enforcement. Finally, we do a parallel with belief update. As classical update postulates do not allow to deal with restrictions about the authorized changes, we had to introduce a new set of postulates that characterizes generalized enforcement. All the proofs can be found in [4].

2 Framework

2.1 Abstract Argumentation

Let us consider a set Arg of symbols (denoted by lower case letters) representing a set of arguments and a relation Rel on Arg × Arg. The pair ⟨Arg, Rel⟩, called *universe*, allows us to represent the set of possible arguments together with their interactions. More precisely, Arg represents a maybe infinite set of arguments usable in a given domain (*e.g.* if the domain is a knowledge base then Arg and Rel are the set of all arguments and interactions that may be built from the formulas of the base). We can also, as in the following example borrowed from [5], assume that Arg and Rel are explicitly provided.

Example 1. *During a trial[1] concerning a defendant (Mr. X), several arguments can be involved to determine his guilt. This set of arguments i.e., the set Arg and the relation Rel are given below.*

x_0 *Mr. X is not guilty of premeditated murder of Mrs. X, his wife.*
x_1 *Mr. X is guilty of premeditated murder of Mrs. X.*
x_2 *The defendant has an alibi, his business associate has solemnly sworn that he met him at the time of the murder.*

---

[1] In real life, lawyers may be confronted to tougher problems than the one presented here. Namely objection should often be done before an argument is fully laid out in order to stop the jury forming an impression. Unfortunately, this side of real life argumentation is not yet handled in our proposal.

x_3	*The close working business relationships between Mr. X and his associate induce suspicions about his testimony.*
x_4	*Mr. X loves his wife so deeply that he asked her to marry him twice. A man who loves his wife cannot be her killer.*
x_5	*Mr. X has a reputation for being promiscuous.*
x_6	*The defendant had no interest to kill his wife, since he was not the beneficiary of the huge life insurance she contracted.*
x_7	*The defendant is a man known to be venal and his "love" for a very rich woman could be only lure of profit.*

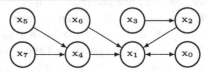

A new definition of argumentation system derives directly from a universe $\langle$Arg, Rel$\rangle$. It differs slightly from the definition of [13] by the fact that arguments and interactions are taken in the universe. In the following, we will use indifferently "argumentation system" or "argumentation graph".

Definition 1. *An argumentation graph G is a pair (A, R) where $A \subseteq$ Arg is the finite set of vertices of G called "arguments" and $R \subseteq R_A =$ Rel $\cap$ $(A \times A)$ (R_A is the restriction of Rel on A) is its set of edges, called "attacks". The set of argumentation graphs that may be built on the universe $\langle$Arg, Rel$\rangle$ is denoted by Γ. In the following, $x \in G$ when x is an argument, is a shortcut for $x \in A$.*

Example 2. *In Example 1, we consider that all the arguments of the universe are known by Agent O, but she is not sure about the content of the jury's argumentation system. She hesitates between two graphs:*

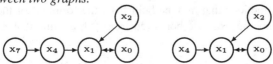

In argumentation theory, see [13], given such graphs, there are several ways to compute a set of "accepted" arguments. This computation depends on the way to select admissible groups of arguments, called "extensions"; several definitions can be considered for the "extensions", they are called "semantics". It depends also on the attribution of a status to arguments, for instance an argument can be "accepted skeptically" or respectively "credulously", if it belongs to all, respectively, to at least one extension. For sake of generality, we are not interested in a particular semantics nor on the mechanism used to instate the status of the arguments. We only consider a function $acc : \Gamma \to 2^{\text{Arg}}$ which associates with any argumentation graph G the set of arguments that have an accepted status in G according to a given semantics and a given status computation[2].

---

[2] This function could be parameterized by the precise semantics used.

We will define a propositional language $\mathscr{L}$ in order to be able to describe an argumentation system and its set of accepted arguments. Its semantics will be defined with respect to Γ. $\forall \varphi \in \mathscr{L}$, we denote by $[\varphi]$ the set of argumentation graphs such that φ is true in these graphs, namely $[\varphi] = \{G \in \Gamma \text{ s.t. } \varphi \text{ is true in } G\}$. As usual, we denote $G \models \varphi$ iff $G \in [\varphi]$ and $\varphi \models \psi$ iff $[\varphi] \subseteq [\psi]$.

For sake of simplicity in all the examples, we are going to use a restricted propositional language $\mathscr{L}_{\text{Arg}}$, only able to express conditions about the presence or the accepted status of an argument in a graph. With this language, we can only handle examples about argument addition or removal. Hence, changes about interactions won't be considered, which allows us to assume that R is always equal to R_A in all our examples.

Definition 2. *Let Γ_{Arg} be the set of argumentation graphs (A, R_A) that may be built on* Arg. *Let $\mathscr{L}_{\text{Arg}}$ be the propositional language associated with the vocabulary $\{a(x), on(x) \mid x \in \text{Arg}\}^3$, with the usual connectives $\neg, \wedge, \vee, \rightarrow, \leftrightarrow$ and constants $\perp$ and $\top$. Its semantics is defined with respect to Γ_{Arg} as follows: let $G \in \Gamma_{\text{Arg}}$*
- *the formula $\perp$ is always false in G*
- *the formula $\top$ is always true in G*
- *if $x \in$ Arg then*
  - *the formula $a(x)$ is true in G iff $x \in acc(G)$,*
  - *the formula $on(x)$ is true in G iff $x \in G$*
- *the non atomic formulas are interpreted as usual, $\neg \varphi$ is true in G if φ is not true in G, $\varphi_1 \vee \varphi_2$ is true in G if φ_1 or φ_2 is true in G, etc.*

Note that every accepted argument in a graph should belong to the graph, hence in $\mathscr{L}_{\text{Arg}}$, $\forall G \in \Gamma_{\text{Arg}}, \forall x \in \text{Arg}, G \models a(x) \rightarrow on(x)$.

Definition 3. *The* characteristic function f_{Arg} *associated with $\mathscr{L}_{\text{Arg}}$, $f_{\text{Arg}} : \Gamma_{\text{Arg}} \rightarrow \mathscr{L}_{\text{Arg}}$, is defined by:*
$$\forall G \in \Gamma_{\text{Arg}}, f_{\text{Arg}}(G) = \bigwedge_{x \in G} on(x) \wedge \bigwedge_{x \in \text{Arg} \setminus G} \neg on(x).$$

Note that, in Definition 2, the attack relation being fixed, if the set of arguments belonging to G is known then G is perfectly known. More formally, $f_{\text{Arg}}(G)$ characterizes G in a unique way:

Property 1. $\forall G \in \Gamma_{\text{Arg}}, [f_{\text{Arg}}(G)] = \{G\}$

Example 3. *The jury's system is not completely known by Agent O. It is represented in $\mathscr{L}_{\text{Arg}}$ by the formula $\varphi_{Jury} = on(x_0) \wedge on(x_1) \wedge on(x_2) \wedge on(x_4) \wedge \neg on(x_3) \wedge \neg on(x_5) \wedge \neg on(x_6) \wedge (on(x_7) \vee \neg on(x_7))$ which covers the two graphs drawn in Example 2; the disjunction between $on(x_7)$ and $\neg on(x_7)$ expresses the fact that Agent O hesitates. Moreover, x_0, x_2 and $(x_4$ or $x_7)$ are the only members of the "grounded extension" [13]. Hence, $\varphi_{Jury} \models a(x_0) \wedge a(x_2) \wedge (a(x_4) \vee a(x_7))$.*

Note that the idea to write propositional formulas for expressing acceptability of arguments was first proposed in [11]. This was done with a completely different aim, namely to generalize Dung's argumentation framework by taking into account additional constraints (expressed in logic) on the admissible sets of arguments.

---

[3] "a" stands for "accepted in G" while "on" stands for "belongs to G".

2.2 Change in Argumentation

In this section we propose a definition of change in argumentation based on the work of [9,6] and adapted to the encoding of generalized enforcement operators. [9] have distinguished four change operations. An elementary change is either adding/removing an argument with a set of attacks involving it, or adding/removing an attack. According to the restriction explained in Section 2.1, we only present in Definition 4 the operations of addition and removal on arguments. Moreover operations are only defined for specific argumentation systems of the form (A, R_A) where $R_A = \text{Rel} \cap (A \times A)$ i.e. R_A contains all the attacks concerning arguments of A that are present in the universe (Arg, Rel). Note that this definition gives only a particular example of change operations when the attack relation is fixed.

The purpose of the following definitions is the introduction of a particular framework, that will be used to illustrate enforcement. In this framework, we consider an agent that may act on a target argumentation system. This agent has a goal and should follow some constraints about the actions she has the right to do. For instance, an agent can only advance arguments that she knows. Hence some restrictions are added on the possible changes that may take place on the system. These constraints are represented by the notion of *executable operation*.

We first refine the notion of *elementary operation* within the meaning of [9] in four points: first a precise syntax is given; then we define an *allowed operation* w.r.t. a given agent's knowledge; we restrict this notion w.r.t. its feasibility on the target system (it is not possible to add an already present argument or to remove an argument which was not in the graph), it leads to the notion of *executable operation*; and finally, we study the impact of an operation on an argumentation system. Note that considering only elementary operations does not result in a loss of generality since any change can be translated into a sequence of elementary operations, called program in Definition 5.

Definition 4. *Let k be an agent and $G_k = \langle A_k, R_{A_k} \rangle$ be her argumentation system and let $G = \langle A, R_A \rangle$ be an argumentation system.*
- *An* elementary operation *is a pair $o = \langle op, x \rangle$ where $op \in \{\oplus, \ominus\}$ and $x \in \text{Arg}$.*
- *An elementary operation $\langle op, x \rangle$ is allowed for k iff $x \in A_k$.[4]*
- *An operation executable by k on G is an operation $\langle op, x \rangle$ allowed for k such that:*
  - *if $op = \oplus$ then $x \notin A$*
  - *if $op = \ominus$ then $x \in A$.*
- *An operation $o = \langle op, x \rangle$ executable by k on G provides a new argumentation system $G' = o(G) = \langle A', R_{A'} \rangle$ such that:*
  - *if $op = \oplus$ then $G' = \langle A \cup \{x\}, R_{A \cup \{x\}} \rangle$*
  - *if $op = \ominus$ then $G' = \langle A \setminus \{x\}, R_{A \setminus \{x\}} \rangle$*

Example 4. *From* Arg *and* Rel *given in Example 1, several elementary operations are syntactically correct, e.g., $\langle \oplus, \{x_2\} \rangle$ and $\langle \ominus, \{x_4\} \rangle$. Among the elementary operations, Agent O is only allowed to use those concerning arguments she knows. Since O*

---

[4] Note that in the case of an argument addition, if the attack relation had not been imposed then it would have been possible to add an argument with only a part of the known attacks and therefore to "lie by omission" or to add attacks unknown to the agent and therefore lie in an "active" way. This will be the subject of future work.

learnt all about this trial, all the elementary operations are allowed for O. $\langle \oplus, \{x_5\} \rangle$, $\langle \ominus, \{x_4\} \rangle$, $\langle \ominus, \{x_2\} \rangle$ *are some executable operations for O on the systems described by* φ_{Jury}.

Finally, we consider sequences of operations executed by an agent on an argumentation system, called *programs*, which are providing the possibility for an agent to perform several elementary operations one after the other.

Definition 5. *Let* $G = \langle A, R_A \rangle$ *be an argumentation system. A* program *p executable by an agent k on G is a finite ordered sequence of n operations* $(o_1, \cdots, o_n)$ *s.t.:*
- $n = 1 : o_1$ *is executable by k on G. Hence* $p(G) = o_1(G)$.
- $n > 1 : (o_1, \cdots, o_{n-1})$ *is a program* p' *executable by k on G such that* $p'(G) = G'$ *and* o_n *is executable by k on* G'. *Hence* $p(G) = o_n(G')$.
- *By extension, an empty sequence is also a program. Hence, for* $p = ()$, $p(G) = G$.

2.3 Enforcement

The main references about enforcement are [3,2] that address the following question : is it possible to change a given argumentation system, by applying change operations, so that a desired set of arguments becomes accepted? Baumann has specified necessary and sufficient conditions under which enforcements are possible, in the case where change operations are restricted to the addition of new arguments and new attacks. More precisely, [2] introduces three types of changes called expansions: the *normal expansion* adds new arguments and new attacks concerning at least one of the new arguments, the *weak expansion* refines the normal expansion by the addition of new arguments not attacking any old argument and the *strong expansion* refines the normal expansion by the addition of new arguments not being attacked by any old argument.

It is not the case in general that any desired set of arguments is enforceable using a particular expansion. Moreover, in some cases, several enforcements are possible, some of them requiring more effort than others. In order to capture this idea, [2] introduces the notion of characteristic which depends on a semantics and on a set of possible expansions. The characteristic of a set of arguments is defined as the minimal number of modifications (defined by the differences between the attacks on the two graphs) that are needed in order to enforce this set of arguments. This number equals 0 when each argument of the desired set is already accepted. It equals infinity if no enforcement is possible. [2] provides means to compute the characteristic w.r.t. a given type of expansion and a given semantics.

3 Towards Generalized Enforcement

Let us formalize enforcement using the definitions presented in Section 2.1. Let $G \in \Gamma$ and $X \subseteq \texttt{Arg}$. An enforcement of X on G is a graph $G' \in \Gamma$ obtained from G by applying change operations and such that $X \subseteq acc(G')$. Different enforcements of X on G can be compared using a preorder $\preceq_G$. For instance, it seems natural to look for enforcements performing a minimal change on G. Minimality can be based on a distance for instance. In that case, given two enforcements G' and G'' of X on G, $G' \preceq_G G''$ may be defined as $distance(G, G') \leq distance(G, G'')$.

This preorder $\preceq_G$ suggests to draw a parallel between the enforcement problem and an update problem. Indeed, as we will see in Section 4.1, update is also related to the same kind of preorder on worlds w.r.t. a given world. More precisely an update operator maps a knowledge base and a piece of information to a new knowledge base, where knowledge bases are expressed in terms of propositional formulas. The semantic counterpart of this mapping is defined by operations on models of formulas, *i.e.*, worlds. This gives birth to the idea that graphs are to worlds what formulas characterizing sets of graphs are to formulas characterizing sets of worlds.

Definition 2 enables us to continue the parallel. Let $S \subseteq \text{Arg}$ and $\alpha = \bigwedge_{x \in S} a(x)$. $[\alpha]$ can be considered as the set of graphs in which the elements of S are accepted. In other words, $[\alpha]$ plays the role of the set of graphs that accept S.

This leads to formalize an enforcement problem as an operator applying to propositional formulas, with a semantic counterpart working with argumentation graphs. So enforcing a propositional formula α on a propositional formula φ means enforcing α on the graphs that satisfy φ.

This setting allows us to have two generalizations of enforcement: first it is now possible to use enforcement not only to impose that a set of arguments is accepted, but also to make enforcement with any goal that can be expressed in a propositional language describing graphs. Second, the initial graph does not necessarily have to be completely known since a description in a propositional language allows for a richer expressivity. Hence, a set of graphs will be considered as representing the initial state of the argumentation system.

Let us explain more precisely the notion of *goal*: it reflects conditions that an agent would like to see satisfied in a particular argumentation system. We may consider two types of goals, namely "absolute" and "relative". An absolute goal only takes into account the resulting system after modifying the target system; it formally focuses on G'. A relative goal takes into account the target system and its resulting system; it formally focuses on (G, G'). An example of relative goal could be that the number of accepted arguments increases after enforcement. In the following, we only consider absolute goals, since relative goals are difficult to express in an update manner.

These goals could involve the arguments, the extensions, the set of extensions as well as its cardinality, the set of extensions containing a particular argument as well as its cardinality. Hence goals are represented by expressions involving these notions and that may contain classic comparison operators ($=$, $<$, $>$, etc.), quantifiers $\forall$ and $\exists$, membership ($\in$) and inclusion ($\subseteq$), union ($\cup$) and intersection ($\cap$) of sets, classical logic operators ($\wedge$, $\vee$, $\rightarrow$, $\leftrightarrow$, $\neg$). If we associate a propositional formula with an absolute goal then a goal is satisfied in a graph if the associated formula holds in this graph.

Example 5. *We know that O wants to enforce the set $\{x_1\}$. This goal can be expressed in $\mathcal{L}_{\text{Arg}}$ by the formula $a(x_1)$. To enforce Argument x_1 on the Jury's graph, O can use the program $(\langle \oplus, x_5 \rangle, \langle \oplus, x_3 \rangle)$ which has the impact shown in the following graphs. A more complex goal could be e.g., $\neg a(x_4) \vee a(x_0)$.*

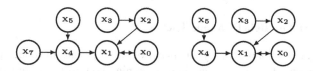

We are now able to define formally generalized enforcement.

Requirement: *Generalized enforcement is based on a propositional language $\mathscr{L}$ able to describe any argumentation system and its set of accepted arguments, and a characteristic function f associated with $\mathscr{L}$, such that $\forall G \in \Gamma$, $[f(G)] = \{G\}$.*

For instance, $\mathscr{L}_{\mathrm{Arg}}$ of Definition 2 could be used as $\mathscr{L}$. However, $\mathscr{L}_{\mathrm{Arg}}$ does not allow to express conditions about the cardinality of each extension after enforcement. $\mathscr{L}_{\mathrm{Arg}}$ is only an example that has been introduced for illustrative purpose. The following results hold for any propositional language $\mathscr{L}$.

In order to capture classical enforcement we also need to be able to restrict the ways that graphs are allowed to change. This is done by introducing a set $T \subseteq \Gamma \times \Gamma$ of allowed transitions between graphs.

Here are three examples of sets of allowed transitions that could be used:

- If the allowed changes are executable elementary operations for an agent k then $T_e^k = \{(G, G') \in \Gamma \times \Gamma, \exists o$ s.t. o is an elementary executable operation by k on G s.t. $o(G) = G'\}$.
- If the allowed changes are executable programs by an agent k then $T_p^k = \{(G, G') \in \Gamma \times \Gamma, \exists p$ s.t. p is an executable program by k on G s.t. $p(G) = G'\}$
- Baumann's normal expansion can be translated in terms of allowed transitions as follows: $T_B = \{(G, G') \in \Gamma \times \Gamma$, with $G = (A, R_A)$ and $G' = (A', R_{A'})$ s.t. $A \subsetneq A'\}$. It means that the transitions admitted by Baumann's normal expansion are restricted to the addition of a new set of arguments.

Now, we are in position to define formally a generalized enforcement operator:

Definition 6. *A generalized enforcement operator is a mapping relative to a set of authorized transitions $T \subseteq \Gamma \times \Gamma$ from $\mathscr{L} \times \mathscr{L} \to \mathscr{L}$ which associates with any formula φ giving information about a target argumentation system, and any formula α encoding a goal, a formula, denoted $\varphi \Diamond_T \alpha$, characterizing the argumentation systems in which α holds, that can be obtained by a change belonging to T.*

Example 6. *In Example 5, if Agent O wants to enforce acceptation of x_1 when x_2 and x_4 are present (w.r.t. the grounded semantics) with an executable program then she can use the following result: $\varphi_{Jury} \Diamond_{T_p^O} (a(x_1) \wedge on(x_2) \wedge on(x_4)) \models on(x_3) \wedge on(x_5)$*

Notation: *$\forall \varphi, \psi \in \mathscr{L}$, a transition in T is possible between a set of graphs satisfying φ to a set of graphs satisfying ψ, denoted $(\varphi, \psi) \models T$, iff ($[\varphi] \neq \varnothing$ and $\forall G \in [\varphi]$, $\exists G' \in [\psi]$, $(G, G') \in T$).* In other words, a transition from a given set of graphs towards another set is possible, iff there is a possible transition from *each graph* of the first set (which should not be empty) towards at least one graph of the second set.

4   Generalized Enforcement Postulates

4.1   Background on Belief Change Theory

In the field of belief change theory, the paper of AGM [1] has introduced the concept of "belief revision". Belief revision aims at defining how to integrate a new piece of information into a set of initial beliefs. Beliefs are represented by sentences of a formal language. Revision consists in adding information while preserving consistency.

A very important distinction between belief revision and belief update was first established in [19]. The difference is in the nature of the new piece of information: either it is completing the knowledge of the world or it informs that there is a change in the world. More precisely, update is a process which takes into account a physical evolution of the system while revision is a process taking into account an epistemic evolution, it is the knowledge about the world that is evolving. In this paper, we rather face an update problem, since in enforcement, the agent wants to change a graph in order to ensure that some arguments are now accepted (graphs play the role of worlds, as explained in Section 3)[5].

We need to recall some background on belief update. An update operator [19,16] is a function mapping a knowledge base φ, expressed in a propositional logic $\mathscr{L}$, representing knowledge about a system in an initial state and a new piece of information $\alpha \in \mathscr{L}$, to a new knowledge base $\varphi \diamond \alpha \in \mathscr{L}$ representing the system after this evolution. In belief update, the input α should be interpreted as the projection of the expected effects of some "explicit change", or more precisely, the expected effect of the action "make α true". The key property of belief update is Katsuno and Mendelzon's Postulate U8 which tells that models of φ are updated independently (contrarily to belief revision). We recall here the postulates of Katsuno and Mendelzon, where $\mathscr{L}$ denotes any propositional language and $[\varphi]$ denotes the set of models of the formula φ:[6] $\forall \varphi, \psi, \alpha, \beta \in \mathscr{L}$,

U1: $\varphi \diamond \alpha \models \alpha$
U2: $\varphi \models \alpha \implies [\varphi \diamond \alpha] = [\varphi]$
U3: $[\varphi] \neq \varnothing$ and $[\alpha] \neq \varnothing \implies [\varphi \diamond \alpha] \neq \varnothing$
U4: $[\varphi] = [\psi]$ and $[\alpha] = [\beta] \implies [\varphi \diamond \alpha] = [\psi \diamond \beta]$
U5: $(\varphi \diamond \alpha) \wedge \beta \models \varphi \diamond (\alpha \wedge \beta)$
U8: $[(\varphi \vee \psi) \diamond \alpha] = [(\varphi \diamond \alpha) \vee (\psi \diamond \alpha)]$
U9: if $card([\varphi]) = 1$ then $[(\varphi \diamond \alpha) \wedge \beta] \neq \varnothing \implies \varphi \diamond (\alpha \wedge \beta) \models (\varphi \diamond \alpha) \wedge \beta$ (where $card(E)$ denotes the cardinality of the set E)

These postulates allow Katsuno and Mendelzon to write the following representation theorem concerning update, namely, an operator satisfying these postulates can be defined by means of a ternary preference relation on worlds (the set of all worlds is denoted by Ω).

---

[5] A revision approach would apply to situations in which the agent learns some information about the initial argumentation system and wants to correct her knowledge about it. This would mean that the argumentation system has not changed but the awareness of the agent has evolved.

[6] Postulates U6 and U7 are not considered here since the set U1-U8 is only related to a family of partial preorders while replacing U6-U7 by U9 ensures a family of complete preorders.

Theorem 1 ([16]). *There is an operator $\diamond : \mathscr{L} \times \mathscr{L} \to \mathscr{L}$ satisfying U1, U2, U3, U4, U5, U8, U9 iff there is a* faithful *assignment that associates with each $\omega \in \Omega$ a complete preorder, denoted $\preceq_\omega$ s.t. $\forall \varphi, \alpha \in \mathscr{L}$, $[\varphi \diamond \alpha] = \bigcup_{\omega \in [\varphi]} \{\omega' \in [\alpha] \text{ s.t. } \forall \omega'' \in [\alpha], \omega' \preceq_\omega \omega''\}$*

where an assignment of a preorder[7] $\preceq_\omega$ to each $\omega \in \Omega$ is faithful iff $\forall \omega, \omega' \in \Omega$, $\omega \prec_\omega \omega'$.

This set of postulates has already been broadly discussed in the literature (see *e.g.*, [15,14,12]). U2 for instance imposes inertia which is not always suitable. Herzig [14] proposes to restrict possible updates by taking into account integrity constraints, *i.e.*, formulas that should hold before and after update. Dubois et al. [12] proposes to not impose inertia and to allow for update failure even if the formulas are consistent. This is done by introducing an unreachable world called z in order to dispose of an upper bound of the proximity from a current world to an unreachable world. In the following, as seen in Section 3, we want to restrain the possible changes. Hence we have to allow for enforcement failure. As we have seen, we choose to introduce a set of allowed transitions T which restricts possible enforcements. The idea to define an update operator based on a set of authorized transitions was first introduced by Cordier and Siegel [10]. Their proposal goes beyond our idea since they allow for a greater expressivity by using prioritized transition constraints. However, this proposal is only defined at a semantical level (in terms of preorders between worlds), hence they do not provide postulates nor representation theorem associated with their update operator. Moreover our idea to define postulates related to a set T of authorized transitions generalizes [14] since integrity constraints can be encoded with T (the converse is not possible). Consequently, we have now to adapt update postulates in order to restrict possible transitions.

4.2 Postulates Characterizing Enforcement on Graphs with Transition Constraints

We are going to define a set of rational postulates for $\diamondsuit_T$. These postulates are constraints that aim at translating the idea of enforcement. Some postulates coming from update are suitable, namely U1, since it ensures that after enforcement the constraints imposed by α are true. U2 postulate is optional, it imposes that if α already holds in a graph then enforcing α means no change. This postulate imposes inertia as a preferred change, this may not be desirable in all situations. U3 transposed in terms of graphs imposes that if a formula holds for some graphs and if the update piece of information also holds for some graphs then the result of enforcement should give a non empty set of graphs. Here, we do not want to impose that any enforcement is always possible since some graphs may be unreachable from others. So we propose to replace U3 by a postulate called E3 based on the set of authorized transitions T: $\forall \varphi, \psi, \alpha, \beta \in \mathscr{L}$

E3: $[\varphi \diamondsuit_T \alpha] \neq \varnothing$ iff $(\varphi, \alpha) \models T$

Due to the definition of $(\varphi, \alpha) \models T$, E3 handles two cases of enforcement impossibility: no possible transition and no world (*i.e.* no graph satisfying φ or α, as it will be shown in Proposition 3).

---

[7] In the following, $\prec_\omega$ is defined from $\preceq_\omega$ as usual by: $a \prec_\omega b$ iff $a \preceq_\omega b$ and not $b \preceq_\omega a$.

U4 is suitable in our setting since enforcement operators are defined semantically. U5 is also suitable for enforcement since it says that graphs enforced by α in which β already holds are graphs in which the constraints α and β are enforced. Due to the fact that we want to allow for enforcement failure, this postulate had been restricted to "complete" formulas[8].

E5: if $card([\varphi]) = 1$ then $(\varphi \Diamond_T \alpha) \wedge \beta \models \varphi \Diamond_T (\alpha \wedge \beta)$ if $card([\varphi]) = 1$ then $(\varphi \Diamond_T \alpha) \wedge \beta \models \varphi \Diamond_T (\alpha \wedge \beta)$

U8 captures the decomposability of enforcement with respect to a set of possible input attack graphs. We slightly change this postulate in order to take into account the possibility of failure, namely if enforcing something is impossible then enforcing it on a larger set of graphs is also impossible, else the enforcement can be decomposable:

E8 if $([\varphi] \neq \varnothing$ and $[\varphi \Diamond_T \alpha] = \varnothing)$ or $([\psi] \neq \varnothing$ and $[\psi \Diamond_T \alpha] = \varnothing)$
then $[(\varphi \vee \psi) \Diamond_T \alpha] = \varnothing$
else $[(\varphi \vee \psi) \Diamond_T \alpha] = [(\varphi \Diamond_T \alpha) \vee (\psi \Diamond_T \alpha)]$

Postulate U9 is a kind of converse of U5 but restricted to a "complete" formula φ *i.e.* such that, $card([\varphi]) = 1$, this restriction is required in the proof of KM theorem as well as in Theorem 2.

Note that the presence of U1 in the set of postulates characterizing an enforcement operator is not necessary since U1 can be derived from E3, E5 and E8.

Proposition 1. *U1 is implied by E3, E5 and E8.*

These postulates allow us to write the following representation theorem concerning enforcement, namely, an enforcement operator satisfying these postulates can be defined by means of the definition of a family of preorders on graphs.

Definition 7. *Given a set $T \subseteq \Gamma \times \Gamma$ of accepted transitions, an assignment respecting T is a function that associates with each $G \in \Gamma$ a complete preorder $\preceq_G$ such that $\forall G_1, G_2 \in \Gamma$, if $(G, G_1) \in T$ and $(G, G_2) \notin T$ then $G_2 \npreceq_G G_1$.*

Theorem 2. *Given a set $T \subseteq \Gamma \times \Gamma$ of accepted transitions, there is an operator $\Diamond_T : \mathcal{L} \times \mathcal{L} \to \mathcal{L}$ satisfying E3, U4, E5, E8, U9 iff there is an assignment respecting T s.t. $\forall G \in \Gamma$, $\forall \varphi, \alpha \in \mathcal{L}$,*

- $[f(G) \Diamond_T \alpha] = \{G_1 \in [\alpha]$ s.t. $(G, G_1) \in T$ and $\forall G_2 \in [\alpha]$ s.t. $(G, G_2) \in T$,
$$G_1 \preceq_G G_2\}$$

- $[\varphi \Diamond_T \alpha] = \begin{vmatrix} \varnothing & \text{if } \exists G \in [\varphi] \text{ s.t. } [f(G) \Diamond_T \alpha] = \varnothing \\ \bigcup_{G \in [\varphi]} [f(G) \Diamond_T \alpha] & \text{otherwise} \end{vmatrix}$

This result is a significant headway, but as usual for a representation theorem, it gives only a link between the existence of an assignment of preorders and the fact that an enforcement operator satisfies the postulates. It does not give any clue about how to assign these preorders i.e., how to design precisely an enforcement operator.

The following proposition establishes the fact that 5 postulates are necessary and sufficient to define an enforcement operator, namely E3, U4, E5, E8 and U9. Indeed, U1 can be derived from them (as seen in Proposition 1).

---

[8] Note that $card[\varphi] = 1$ iff $\exists G \in \Gamma$ s.t. $[\varphi] = [f(G)]$.

Proposition 2. *E3, U4, E5, E8, U9 constitute a minimal set: no postulate can be derived from the others.*

From Theorem 2 we can deduce two simple cases of impossibility: if the initial situation or the goal is impossible then enforcement is impossible (this result is a kind of converse of U3).

Proposition 3. *If $\Diamond_T$ satisfies E3, U4, E5, E8 and U9 then $([\varphi] = \varnothing$ or $[\alpha] = \varnothing \implies [\varphi\Diamond_T\alpha] = \varnothing)$.*

The following property ensures that if an enforcement is possible then a more general enforcement is also possible.

Proposition 4. *If $\Diamond_T$ satisfies E3 then $([\varphi] \neq \varnothing$ and $[\varphi\Diamond_T\alpha] \neq \varnothing \implies [\varphi\Diamond_T(\alpha \vee \beta)] \neq \varnothing)$.*

Note that there are some cases where U2 does not hold together with E3, U4, E5, E8 and U9. If U2 is imposed then the enforcement operator is associated with a preorder in which a given graph is always closer to itself than to any other graph. This is why it imposes to have a faithful assignment. In that case, the relation represented by T should be reflexive.

Definition 8. *A faithful assignment is a function that associates with each $G \in \Gamma$ a complete preorder[9] $\preceq_G$ such that $\forall G_1 \in \Gamma, G \prec_G G_1$.*

Proposition 5. *Given a reflexive relation $T \subseteq \Gamma \times \Gamma$ of accepted transitions, there is an operator $\Diamond_T : \mathcal{L} \times \mathcal{L} \to \mathcal{L}$ satisfying E3, U4, E5, E8, U9 that satisfies U2 iff there is a faithful assignment respecting T defined as in Theorem 2.*

If we remove the constraint about authorized transitions then we recover Katsuno and Mendelzon theorem, namely:

Proposition 6. *If $T = \Gamma \times \Gamma$ then $\Diamond_T$ satisfies U2, E3, U4, E5, E8, U9 iff $\Diamond$ satisfies U1, U2, U3, U4, U5, U8 and U9.*

Among the different kinds of changes proposed by Baumann, the normal expansion, *i.e.*, adding an argument with the attacks that concern it, can be encoded in our framework as follows.

Remark 1. Baumann's enforcement by normal expansion is a particular enforcement operator $\Diamond_T : \mathcal{L} \times \mathcal{L} \to \mathcal{L}$ such that $T = T_B$. Moreover, the language used is restricted as follows: the formulas that describe the initial system are restricted to $\{\varphi \in \mathcal{L}_{on}, card([\varphi]) = 1\}$ and formulas that describe the facts that should be enforced are only conjunctions of positive literals of $\mathcal{L}_a$, where $\mathcal{L}_a$ and $\mathcal{L}_{on}$ are respectively the propositional languages based only on $a(x)$ and on $on(x)$ variables.

In Baumann's framework, the formula concerning the initial graph should be complete, *i.e.*, should correspond to only one graph. The formula concerning the goal of

---

[9] In the following, $\prec_G$ is defined from $\preceq_G$ as usual by: $a \prec_G b$ iff $a \preceq_G b$ and not $b \preceq_G a$.

enforcement should describe a set of arguments that should be accepted (under a given semantics) after the change. Due to Theorem 2, there exists a family of preorders that could be defined. Baumann proposes to use the following: $G' \preceq_G G''$ iff $dist(G, G') \leq dist(G, G'')$ where $dist(G, G')$ is the number of attacks that differs in G and G'.[10]

5   Conclusion

The work of [2] gives the basics about enforcement, our approach investigates several new issues:

– we propose to take into account the ability to remove an argument, which could help to enforce a set of arguments with less effort. We also generalize what can be enforced, not only sets of arguments can be enforced but any goal that can be expressed in propositional logic is allowed.
– we enable the possibility to restrict the authorized changes. In generalized enforcement, authorized changes may be described by a set of possible transitions. Hence, the structure of the changes can be restricted (for instance to additions only or to elementary operations) as well as the arguments that are allowed to be added/removed.

Finally, our main contribution is to state that enforcement is a kind of update, which allows for an axiomatic approach. This kind of update is more general than classical update since it allows to take into account transition constraints.

In this paper, for sake of shortness, we use a simplified logical language for describing argumentation systems in our examples, this makes us focus only on changes about arguments hence allow us to consider a fixed attack relation. However our results hold on any given propositional logic, hence choosing a logic in which attacks are encoded would enable us to deal with changes on attacks. This deserves more investigation.

Another issue is to find postulates that are more specific for argumentation dynamics. Indeed, we have defined a set of postulates that may characterize changes in any kind of graphs that can be defined in propositional logic, provided that a transition function is given. Further research should take into account the particularities of the graphs representing argumentation systems (semantics notions should be introduced in the postulates). Moreover, in this paper we have focused on a representation theorem based on complete preorders between pairs of argumentation graphs ; another study would be required for partial preorders. Finally, it would be worthwhile to study what could be the counterpart of enforcement for revision instead of update.

References

1. Alchourrón, C., Gärdenfors, P., Makinson, D.: On the logic of theory change: partial meet contraction and revision functions. Journal of Symbolic Logic 50, 510–530 (1985)
2. Baumann, R.: What does it take to enforce an argument? minimal change in abstract argumentation. In: ECAI, pp. 127–132 (2012)

---

[10] Note that since Baumann's enforcement is defined on one graph and not on a set of graphs, then it is also a kind of belief revision since revision and update collapse when the initial world is completely known (this kind of belief revision won't be a pure AGM revision but rather a revision under transition constraints).

3. Baumann, R., Brewka, G.: Expanding argumentation frameworks: Enforcing and monotonicity results. In: Proc. of COMMA, pp. 75–86. IOS Press (2010)

4. Bisquert, P., Cayrol, C., de Saint Cyr, F.D., Lagasquie-Schiex, M.C.: Axiomatic approach of enforcement in argumentation. Tech. rep., IRIT, Toulouse, France (2013), ftp://ftp.irit.fr/pub/IRIT/ADRIA/rap-IRIT-2013-24.pdf

5. Bisquert, P., Cayrol, C., de Saint-Cyr, F.D., Lagasquie-Schiex, M.-C.: Change in argumentation systems: Exploring the interest of removing an argument. In: Benferhat, S., Grant, J. (eds.) SUM 2011. LNCS, vol. 6929, pp. 275–288. Springer, Heidelberg (2011)

6. Bisquert, P., Cayrol, C., de Saint-Cyr, F.D., Lagasquie-Schiex, M.-C.: Duality between Addition and Removal. In: Greco, S., Bouchon-Meunier, B., Coletti, G., Fedrizzi, M., Matarazzo, B., Yager, R.R. (eds.) IPMU 2012, Part I. CCIS, vol. 297, pp. 219–229. Springer, Heidelberg (2012)

7. Boella, G., Kaci, S., van der Torre, L.: Dynamics in argumentation with single extensions: Abstraction principles and the grounded extension. In: Sossai, C., Chemello, G. (eds.) EC-SQARU 2009. LNCS (LNAI), vol. 5590, pp. 107–118. Springer, Heidelberg (2009)

8. Boella, G., Kaci, S., van der Torre, L.: Dynamics in argumentation with single extensions: Attack refinement and the grounded extension. In: Proc. of AAMAS, pp. 1213–1214 (2009)

9. Cayrol, C., de Saint-Cyr, F.D., Lagasquie-Schiex, M.-C.: Change in abstract argumentation frameworks: Adding an argument. Journal of Artificial Intelligence Research 38, 49–84 (2010)

10. Cordier, M.-O., Siegel, P.: Prioritized transitions for updates. In: Froidevaux, C., Kohlas, J. (eds.) ECSQARU 1995. LNCS, vol. 946, pp. 142–150. Springer, Heidelberg (1995)

11. Coste-Marquis, S., Devred, C., Marquis, P.: Constrained argumentation frameworks. In: Proc. of KR, pp. 112–122 (2006)

12. Dubois, D., de Saint-Cyr, F.D., Prade, H.: Update postulates without inertia. In: Froidevaux, C., Kohlas, J. (eds.) ECSQARU 1995. LNCS (LNAI), vol. 946, pp. 162–170. Springer, Heidelberg (1995)

13. Dung, P.: On the acceptability of arguments and its fundamental role in nonmonotonic reasoning, logic programming and n-person games. Artificial Intelligence 77(2), 321–358 (1995)

14. Herzig, A.: On updates with integrity constraints. In: Belief Change in Rational Agents (2005)

15. Herzig, A., Rifi, O.: Propositional belief base update and minimal change. Artificial Intelligence 115, 107–138 (1999)

16. Katsuno, H., Mendelzon, A.: On the difference between updating a knowledge base and revising it. In: Proc. of KR, pp. 387–394 (1991)

17. Liao, B., Jin, L., Koons, R.: Dynamics of argumentation systems: A division-based method. Artificial Intelligence 175(11), 1790 (2011)

18. Moguillansky, M.O., Rotstein, N.D., Falappa, M.A., García, A.J., Simari, G.R.: Argument theory change through defeater activation. In: Proc. of COMMA, pp. 359–366. IOS Press (2010)

19. Winslett, M.: Reasoning about action using a possible models approach. In: Proc. of AAAI, pp. 89–93 (1988)

A Conditional Logic-Based
Argumentation Framework

Philippe Besnard[1], Éric Grégoire[2], and Badran Raddaoui[2]

[1] IRIT
UMR 5505 CNRS, 118 route de Narbonne
F-31065 Toulouse Cedex, France
besnard@irit.fr
[2] CRIL
Université d'Artois & CNRS, rue Jean Souvraz SP18
F-62307 Lens Cedex, France
{gregoire,raddaoui}@cril.fr

Abstract. The goal of this paper is twofold. First, a logic-based argumentation framework is introduced in the context of conditional logic, as conditional logic is often regarded as an appealing setting for knowledge representation and reasoning. Second, a concept of conditional contrariety is defined that covers usual inconsistency-based conflicts and puts in light a specific form of conflicts that often occurs in real-life: when an agent asserts an *If then* rule, it can be argued that additional conditions are actually needed to derive the conclusion.

Keywords: Conditional Logic, Logical Argumentation Theory, Conditional Contrariety.

1 Introduction

Argumentation has long been a major topic in A.I. (see e.g., [1, 2] and for more recent accounts e.g., [3]) that has concerned a large variety of application domains for more than a decade, like e.g., law [4, 5], medicine [6], negotiation [7], decision making [8] and multiagent systems [9, 10]. Two main families of computational models for argumentation have been proposed in the literature: namely, the abstract and the logic-based argumentation frameworks. Following the seminal work of [11], the first family is based on graph-oriented representations and focuses mainly on the interaction between arguments without taking the possible internal structure of the involved arguments into account. On the contrary, the logic-based approaches (e.g., [12–19]) exploit the logical internal structure of arguments and adopt inconsistency as a pivotal paradigm: any pair of conflicting arguments must be contradictory. Consequently, no conflicting arguments can be found together inside a same consistent set of formulas.

However, many natural real-life arguments and counter-arguments do not necessarily appear mutually inconsistent in usual knowledge representation modes. For example, consider the assertion *If there is a match tonight then John will*

W. Liu, V.S. Subrahmanian, and J. Wijsen (Eds.): SUM 2013, LNAI 8078, pp. 44–56, 2013.

go to the stadium encoded through a (material) implicative formula in standard logic as *MatchTonight → JohnGoesToStadium*, so that in case *MatchTonight* is true, *JohnGoesToStadium* can be deduced. Now, the following objection can be raised against that argument through the sentence *If there is a match tonight and if John has got enough money then John will go to the stadium*, which requires an additional condition for John to go the stadium if there is a match tonight. The latter sentence is a deductive consequence of the first one and taking both of them does not yield an inconsistent set. This paper aims to extend the logic-based approaches by encompassing this specific form of contrariety.

It is possible to represent the above example as a case of inconsistency-based conflict by using e.g. a modal logic of necessity and possibility, and more information: when the first sentence is augmented so that it excludes the possibility that John does not go to the stadium tonight if there is a match, whereas the other sentence allows this possibility to happen. Such an alternative representation requires all this additional or implicit information to be asserted and represented in some way. On the contrary, we provide a representation framework where the motivating example can be modeled in a way close to natural language implications and without resorting to logical inconsistency; moreover, the framework allows a form of contrariety to be recognized between the implicative formulas.

To this end, we resort to conditional logic, which allows an additional specific implicative connective to be used, in addition to standard-logic material implication. Conditional logic is actually rooted in the formalization of hypothetical or counterfactual reasoning of the form *If α were true then β would be true* and attempts to avoid some pitfalls of material implication to represent patterns of conditional or hypothetical reasoning. Actually, the conditional implication connective is often regarded as a very suitable connective to encode many implicative reasoning patterns from real-life; accordingly, conditional logic has long been investigated in many A.I. areas [20] like belief revision [21], data base and knowledge update [22] natural language semantics for handling hypothetical and counterfactual sentences [23], non-monotonic and prototypical reasoning [24, 25], causal inference [26] and logic programming [27], just to mention some seminal works.

The goal of this paper is thus twofold. First, we revisit frameworks *à la* [15, 28] to lay down the main foundations of a logic-based argumentation framework based on conditional logic. Second, we introduce a concept of conditional contrariety that encompasses both the conflicts through inconsistency and a generalization of the conflict illustrated in the motivating example. One intended benefit is that when an agent represents *If then* rules using the conditional connective, the framework allows one to argue against this rule by stating that additional conditions are required in order for the conclusion of the rule to hold. Accordingly, in this framework, the conditional implication connective is intended to be used to represent hypothetical reasoning and other implications that can be questioned within an argumentation process.

The paper is organized as follows. In the next section, the user is provided with basic elements of conditional logic MP. In section 3, conditional contrariety is motivated and introduced. Section 4 revisits the main foundational concepts of Besnard and Hunter's framework so that conditional contrariety is covered. The last section discusses some promising paths for further research. For the clarity of presentation, all proofs are given in an appendix.

2   Conditional Logics

Conditional logics are rooted in the formalization of counterfactual or hypothetical reasoning of the form 'If α were true then β would be true", for which an additional connective, called *conditional connective* and denoted $\Rightarrow$, is generally introduced. Roughly, a conditional formula $\alpha \Rightarrow \beta$ is valid when β is true in the possible worlds where α is true. Clearly, this diverges from the (material) implicative standard-logic formula $\alpha \to \beta$, since this latter one is equivalent to $\neg\alpha \lor \beta$ which is also satisfied when α is false. In the paper, we consider the well-known conditional logic MP, which can be extended to yield many of the other popular conditional logics (see e.g., [29]).

MP is an extension of the language and inference system of classical Boolean logic. It is a language of formulas, denoted $\mathcal{L}_c$. We use α, β, γ, δ, ... to denote formulas of $\mathcal{L}_c$ and Δ, Φ, Ψ, Θ, ... to denote sets of formulas of $\mathcal{L}_c$. Formulas are built in the usual way from the standard connectives $\neg$, $\land$, $\lor$, $\to$ and $\leftrightarrow$: they accommodate the conditional connective $\Rightarrow$ through the additional formation rule: if α and β are formulas, so is $\alpha \Rightarrow \beta$. $\top$ and $\bot$ represent truth and falsity, respectively. A concept of *extended literal* proves useful: an extended literal is of the form α or $\neg\alpha$ such that α is either an atom or a formula with the conditional connective as the main connective.

The inferential apparatus of MP consists of the following axioms schemas and inference rules [29], enriching standard Boolean logic to yield an inference relation denoted $\vdash_c$.

RCEA.
$$\frac{\vdash_c \alpha \leftrightarrow \beta}{\vdash_c (\alpha \Rightarrow \gamma) \leftrightarrow (\beta \Rightarrow \gamma)}$$

RCEC.
$$\frac{\vdash_c \alpha \leftrightarrow \beta}{\vdash_c (\gamma \Rightarrow \alpha) \leftrightarrow (\gamma \Rightarrow \beta)}$$

CC. $\vdash_c ((\alpha \Rightarrow \beta) \land (\alpha \Rightarrow \gamma)) \to (\alpha \Rightarrow (\beta \land \gamma))$

CM. $\vdash_c (\alpha \Rightarrow (\beta \land \gamma)) \to ((\alpha \Rightarrow \beta) \land (\alpha \Rightarrow \gamma))$

CN. $\vdash_c (\alpha \Rightarrow \top)$

MP. $\vdash_c (\alpha \Rightarrow \beta) \to (\alpha \to \beta)$

In the paper, an expression of the form $\alpha \equiv \beta$ will be a shortcut for $\alpha \vdash_c \beta$ and $\beta \vdash_c \alpha$. We make use of the *disjunctive form* of conditional formulas, defined as follows.

Definition 1. *The disjunctive form of a formula α of $\mathcal{L}_c$, denoted $DF(\alpha)$, is the first (according to the lexicographic order) formula of the form $\alpha_1 \lor \ldots \lor \alpha_n$*

that is logically equivalent with α under $\vdash_c$ and such that each α_i is a conjunction of extended literals.

3   Conditional Contrariety

Conditional contrariety (in short, contrariety) is the cornerstone concept in this paper. It is intended to encompass both logical inconsistency in MP and a form of contrariety involving a pair of conditional implicative formulas where the first one would entail the other one in standard logic if the material implication were used. Let us introduce the concept progressively and refer to items of the next formal definition through their numbering like e.g., (I), (II.a) or (II.2.).

Let α and β be two formulas of $\mathcal{L}_c$ s.t. $DF(\alpha) = \alpha_1 \vee \ldots \vee \alpha_n$ and $DF(\beta) = \beta_1 \vee \ldots \vee \beta_m$.

α *is in contrariety to* β, denoted $\alpha \bowtie \beta$, in any of the following situations.

(I.) First, α and β are mutually inconsistent in MP. Note that this also covers the standard-logic occurrences of inconsistency. Formally, whenever $\{\alpha, \beta\} \vdash_c \bot$ we have $\alpha \bowtie \beta$. Taking into account the DF of α and β, this amounts to $\forall \alpha_i, \forall \beta_j$ $\{\alpha_i, \beta_j\} \vdash_c \bot$. Let us note that if β is itself inconsistent then any formula α is in contrariety to β, in particular β is in contrariety to β.

(II.) Second, we need to address the case that requires α of the form $\phi \wedge \epsilon \Rightarrow \psi$ to be in contrariety to β of the form $\phi \Rightarrow \psi$, just as in the motivating example from the introduction. Actually, we can be more general and consider the cases where a similar situation occurs with respect to a more general class of pairs of "formulas in contrariety" that are of the form $\gamma = \gamma_1 \Rightarrow \gamma_2$ and $\delta = \delta_1 \Rightarrow \delta_2$, provided that δ and γ would be "derivable" in some sense from α and β. The class of pairs of formulas is defined through specific inferential links between their elements γ_1, γ_2, δ_1 and δ_2. First, γ and δ must be two conditionals about the same conclusion: hence, $\gamma_2 \equiv \delta_2$ (II.d). Generalizing the motivating example, we require the antecedent of the first conditional to entail the antecedent of the second one (but not conversely), formally $\gamma_1 \vdash_c \delta_1$ (II.a) and $\delta_1 \nvdash_c \gamma_1$ (II.b). Another condition is required to prevent valid formulas from being in a contrariety position: $\gamma_1 \nvdash_c \gamma_2$ (II.c).

Then, we need to make clear the inferential links between the pair of formulas α and β for which we explore a contrariety situation, and the above γ and δ formulas that are themselves in contrariety (II.1 and II.2). First, a contrariety situation occurs when β conditionally entails the last two formulas, i.e. $\beta \vdash_c \gamma \wedge \delta$ (or equivalently, taking the DF of β, this occurs when taking β_j as premises, $\forall \beta_j$) (II.2). The motivation is as follows. Remember that whenever β was inconsistent, any α contraried β. Likewise, if β allows by itself the derivation of both γ and δ that are in a contrariety position, then β is in some way self-contraried, and any α is in contrariety to β.

Finally, $\alpha \vdash_c \gamma$ while $\{\alpha, \beta\} \vdash_c \gamma \wedge \delta$ naturally covers the last $\alpha \bowtie \beta$ case. In the definition, this condition is expressed taking the DF of both α and β into account (II.1).

It is important to stress that the contrariety concept that is defined is deductively-based in all of the following senses. First, inconsistency is reached through deduction. Second, the inferential relations between elements of the pair (α, β) with elements of (γ, δ) are also of a deductive nature. Finally, the condition $\gamma_1 \vdash_c \delta_1$ is also deductive. Accordingly, we will see that contrariety is not symmetric in the general case.

Definition 2. *Let α and β be two formulas of $\mathcal{L}_c$ s.t. $DF(\alpha) = \alpha_1 \vee \ldots \vee \alpha_n$ and $DF(\beta) = \beta_1 \vee \ldots \vee \beta_m$.*

α is in contrariety to β, denoted $\alpha \bowtie \beta$,
iff $\forall \alpha_i, \forall \beta_j$

> *(I.) $\{\alpha_i, \beta_j\} \vdash_c \bot$, or*
> *(II.) There exist $\gamma_1, \gamma_2, \delta_1$ and δ_2 in $\mathcal{L}_c$ s.t.*
>> *(II.a) $\gamma_1 \vdash_c \delta_1$ and*
>> *(II.b) $\delta_1 \nvdash_c \gamma_1$ and*
>> *(II.c) $\gamma_1 \nvdash_c \gamma_2$ and*
>> *(II.d) $\gamma_2 \equiv \delta_2$*
> *where*
>> *(II.1) $\{\alpha_i, \beta_j\} \vdash_c (\gamma_1 \Rightarrow \gamma_2) \wedge (\delta_1 \Rightarrow \delta_2)$*
>> *s.t. $\alpha_i \vdash_c \gamma_1 \Rightarrow \gamma_2$, or*
>> *(II.2) $\beta_j \vdash_c (\gamma_1 \Rightarrow \gamma_2) \wedge (\delta_1 \Rightarrow \delta_2)$.*

Example 1. $a \wedge b \Rightarrow c$ is in contrariety to $a \Rightarrow c$. $a \wedge (a \wedge c \Rightarrow f)$ is in contrariety to both formulas $a \rightarrow (a \Rightarrow f)$ and $\neg a \vee c \Rightarrow f$. Also, $\neg a \wedge b$ and $a \wedge (a \wedge c \Rightarrow b \vee \neg d)$ are in contrariety to each other.

The concept of *being in contrariety to a formula* is naturally extended into a concept of *being in contrariety to a set of formulas*.

Definition 3. *Let Φ and α be a subset and a formula of $\mathcal{L}_c$, respectively. α is in contrariety to Φ, denoted $\alpha \bowtie \Phi$, iff there exists β in $\mathcal{L}_c$ s.t. $\Phi \vdash_c \beta$ and $\alpha \bowtie \beta$.*

Example 2. Let $\Phi = \{a \Rightarrow b, a \vee d \Rightarrow b \wedge c, a \Rightarrow c\}$. Let $\alpha = a \Rightarrow b \wedge c$. Note that $\Phi \vdash_c \alpha$. However, $\alpha \bowtie \Phi$ because $\alpha \bowtie a \vee d \Rightarrow b \wedge c$.

Obviously, $\bowtie$ is neither symmetric, nor antisymmetric, nor antireflexive. However, it is *monotonic* and *syntax-independent*.

Proposition 1. *Let Φ, Ψ and α be two subsets and a formula of $\mathcal{L}_c$, respectively. If $\alpha \bowtie \Phi$ then $\alpha \bowtie \Phi \cup \Psi$.*

Proposition 2. *Let Φ, α and β be a subset and two formulas of $\mathcal{L}_c$, respectively. If $\alpha \equiv \beta$ then $\alpha \bowtie \Phi$ iff $\beta \bowtie \Phi$.*

4   A Conditional-Logic Argumentation Framework

Having defined the pivotal concept of contrariety, we are now ready to revisit [15, 28]'s framework and lay down the foundations of a conditional-logic argumentation framework based on contrariety. Accordingly, we revisit and extend the following concepts, successively: arguments, conflicts, rebuttals, defeaters and argumentation trees.

In the following, we assume a subset Δ of $\mathcal{L}_c$ that can be inconsistent. All concepts will be implicitly defined relatively to Δ. Membership of formulas and inclusion of sets of formulas to $\mathcal{L}_c$ will also be implicit from now on.

4.1   Arguments

After Besnard-Hunter, an argument is made of a set formulas together with a conclusion that can be derived from the set. The usual non-contradiction condition expressed by $\Phi \nvdash \perp$ is naturally extended and replaced by a non-contrariety requirement (second item).

Definition 4. *An argument A is a pair $\langle \Phi, \alpha \rangle$ s.t.:*

1. $\Phi \subseteq \Delta$
2. $\forall \beta \; s.t. \; \Phi \vdash_c \beta, \; \beta \not\bowtie \Phi$
3. $\Phi \vdash_c \alpha$
4. $\forall \Phi' \subset \Phi, \; \Phi' \nvdash_c \alpha$

A is said to be an *argument* for α. The set Φ and the formula α are the *support* and the *conclusion* of A, respectively.

Example 3. Let $\Delta = \{(a \Rightarrow \neg d) \wedge \neg b, a \Rightarrow c, \neg a\}$. In view of Δ, some arguments are:

$$\langle \{\neg a\}, \neg(a \wedge b) \rangle,$$
$$\langle \{a \Rightarrow c\}, a \Rightarrow c \rangle,$$
$$\langle \{(a \Rightarrow \neg d) \wedge \neg b, a \Rightarrow c\}, a \Rightarrow \neg d \wedge c \rangle.$$

Note that CC is used to obtain the conclusion of the last argument.

The following result shows that the revisited concept of argument still preserves coherence, such a coherence concept being in some sense extended to $\bowtie$.

Proposition 3. *If $\langle \Phi, \alpha \rangle$ is an argument then $\alpha \not\bowtie \Phi$ and $\neg \alpha \bowtie \Phi$.*

A notion of *quasi-identical* arguments is now introduced as follows. It is intended to capture situations where two arguments can be said to make the same point on the same grounds.

Definition 5. *Two arguments $\langle \Phi, \alpha \rangle$ and $\langle \Psi, \beta \rangle$ are quasi-identical iff $\Phi = \Psi$ and $\alpha \equiv \beta$.*

Not surprisingly, provided one argument, its quasi-identical ones form an infinite set.

Proposition 4. *Let $\langle \Phi, \alpha \rangle$ be an argument. There is an infinite set of arguments of the form $\langle \Psi, \beta \rangle$ s.t. $\langle \Phi, \alpha \rangle$ and $\langle \Psi, \beta \rangle$ are quasi-identical.*

Arguments are not necessarily independent. The definition of *more conservative arguments* captures a notion of subsumption between arguments, translating situations where an argument is in some sense contained within another one.

Definition 6. *An argument $\langle \Phi, \alpha \rangle$ is more conservative than an argument $\langle \Psi, \beta \rangle$ iff $\Phi \subseteq \Psi$ and $\beta \vdash_c \alpha$.*

Example 4. The argument $\langle \{a\}, a \vee b \rangle$ is more conservative than $\langle \{\neg a \vee b, a\}, b \rangle$. Also, $\langle \{(a \Rightarrow b) \wedge c, c \to d\}, (a \Rightarrow b) \wedge d \rangle$ is more conservative than $\langle \{(a \Rightarrow b) \wedge c, c \to d\}, (a \Rightarrow b) \wedge c \wedge d \rangle$.

Proposition 5. *If $\langle \Phi, \alpha \rangle$ is more conservative than $\langle \Psi, \beta \rangle$, then $\beta \not\bowtie \Phi$, $\alpha \not\bowtie \Psi$ and $\neg \alpha \bowtie \Psi$.*

In this last result, it is worth noting that $\neg \beta \bowtie \Phi$ does not hold in full generality. A counter-example consists of the two arguments $\langle \{a\}, a \vee b \rangle$ and $\langle \{a, b\}, a \wedge b \rangle$; $\langle \{a\}, a \vee b \rangle$ is more conservative than $\langle \{a, b\}, a \wedge b \rangle$ but $\neg(a \wedge b) \not\bowtie a$.

Actually, the concept of being more conservative induces the concept of quasi-identical arguments, and conversely.

Proposition 6. *Two arguments $\langle \Phi, \alpha \rangle$ and $\langle \Psi, \beta \rangle$ are quasi-identical iff each one is more conservative than the other.*

Example 5. $\langle \{a \Rightarrow b, a \Rightarrow c\}, (a \Rightarrow b) \wedge (a \Rightarrow c) \rangle$ and $\langle \{a \Rightarrow b, a \Rightarrow c\}, a \Rightarrow b \wedge c \rangle$ are quasi-identical as each one is more conservative than the other. In fact, the proof of equivalence between $(a \Rightarrow b) \wedge (a \Rightarrow c)$ and $a \Rightarrow b \wedge c$ is obtained by applying the *CC* and *CM* axioms schemas.

The notions of quasi-identicality and of being more conservative will be used in the next Subsection to avoid some redundancy when counter-arguments need to be listed.

4.2 Conflicts between Arguments

We now revisit conflicts-related concepts in light of conditional contrariety. Let us start with *rebuttals*.

Definition 7. *A rebuttal for an argument $\langle \Phi, \alpha \rangle$ is an argument $\langle \Psi, \beta \rangle$ s.t. $\beta \bowtie \alpha$.*

Example 6. A rebuttal for $\langle \{\neg a \vee b, \neg b\}, \neg a \wedge \neg b \rangle$ is $\langle \{a\}, \neg \neg a \rangle$. Also, $\langle \{a \wedge e \Rightarrow b, a \wedge e \Rightarrow \neg c\}, \neg f \vee (a \wedge e \Rightarrow b \wedge \neg c) \rangle$ is a rebuttal for $\langle \{a \vee d \Rightarrow b \wedge \neg c, f\}, (a \vee d \Rightarrow b \wedge \neg c) \wedge f \rangle$.

Note that *CC* is used to obtain the conclusion $\neg f \vee (a \wedge e \Rightarrow b \wedge \neg c)$.

In classical-logic-based argumentation [15], if $\langle \Phi, \alpha \rangle$ is a rebuttal for $\langle \Psi, \beta \rangle$ then $\langle \Psi, \beta \rangle$ is also a rebuttal for $\langle \Phi, \alpha \rangle$. Consequently, the notion of rebuttal is symmetric. However, this property does not hold with respect to the contrariety paradigm. Let us return to Example 6, the argument $\langle \{a \lor d \Rightarrow b \land \neg c, f\}, (a \lor d \Rightarrow b \land \neg c) \land f \rangle$ is not a rebuttal for $\langle \{a \land e \Rightarrow b, a \land e \Rightarrow \neg c\}, \neg f \lor (a \land e \Rightarrow b \land \neg c) \rangle$. The notion of rebuttal defined here is thus asymmetric.

Another concept is captured by *defeaters*, which are arguments whose conclusion is in contrariety to the support of their targeted argument.

Definition 8. *A defeater for an argument $\langle \Phi, \alpha \rangle$ is an argument $\langle \Psi, \beta \rangle$ s.t. $\beta \bowtie \Phi$.*

Example 7. Some defeaters for $\langle \{a \lor \neg d \Rightarrow b \land c, f \lor \neg b, b\}, f \land (a \lor \neg d \Rightarrow b \land c) \rangle$ are listed below:
$\langle \{\neg b\}, \neg b \rangle$,
$\langle \{\neg b\}, \neg(\neg b \to b) \rangle$,
$\langle \{\neg b, \neg a \to b\}, \neg b \land a \rangle$,
$\langle \{e \land \neg d \Rightarrow b \land c\}, e \land \neg d \Rightarrow b \land c \rangle$,
$\langle \{a \Rightarrow b, a \Rightarrow c\}, a \Rightarrow b \land c \rangle$,
$\langle \{a \Rightarrow b, a \Rightarrow c\}, \neg\neg(a \Rightarrow b \land c) \rangle$,
$\langle \{a \Rightarrow b, a \Rightarrow c\}, (a \Rightarrow b) \land (a \Rightarrow c) \rangle$.

Proposition 7. *If $\langle \Psi, \beta \rangle$ is a rebuttal for $\langle \Phi, \alpha \rangle$ then $\langle \Psi, \beta \rangle$ is a defeater for $\langle \Phi, \alpha \rangle$.*

Example 8. $\langle \{(a \land e \Rightarrow b \land c) \land \neg d\}, (a \land e \Rightarrow b \land c) \land \neg d \rangle$ is a rebuttal for $\langle \{a \Rightarrow b, a \Rightarrow c\}, (a \Rightarrow b \land c) \lor d \rangle$.
Here $\{a \Rightarrow b, a \Rightarrow c\} \vdash_c (a \Rightarrow b \land c) \lor d$, then $((a \land e \Rightarrow b \land c) \land \neg d) \bowtie \{a \Rightarrow b, a \Rightarrow c\}$. Consequently, $\langle \{(a \land e \Rightarrow b \land c) \land \neg d\}, (a \land e \Rightarrow b \land c) \land \neg d \rangle$ is a defeater for $\langle \{a \Rightarrow b, a \Rightarrow c\}, (a \Rightarrow b \land c) \lor d \rangle$.

An interesting special kind of defeaters are *challenges*. As the next proposition shows, challenges capture some situations where the defeat relation is asymmetric.

Definition 9. *Let $\langle \Phi, \alpha \rangle$ and $\langle \Psi, \beta \rangle$ be two arguments. $\langle \Phi, \alpha \rangle$ is a challenge to $\langle \Psi, \beta \rangle$ iff $\alpha \bowtie \Psi$ and $\forall \gamma$ s.t. $\Psi \vdash_c \gamma$, $\gamma \not\bowtie \Phi$.*

Example 9. The argument $\langle \{a \land e \Rightarrow b, a \land e \Rightarrow c\}, a \land e \Rightarrow b \land c \rangle$ is a challenge to the argument $\langle \{a \lor d \Rightarrow b \land c\}, a \lor d \Rightarrow b \land c \rangle$.

Proposition 8. *If $\langle \Phi, \alpha \rangle$ is a challenge to $\langle \Psi, \beta \rangle$ then $\langle \Phi, \alpha \rangle$ is a defeater for $\langle \Psi, \beta \rangle$ and $\langle \Psi, \beta \rangle$ is not a defeater for $\langle \Phi, \alpha \rangle$.*

As intended, defeaters can exist even though there is no inconsistency involved. The next result shows that the support of a challenge is consistent with the support of the argument that it attacks.

Proposition 9. *If $\langle \Phi, \alpha \rangle$ is a challenge to $\langle \Psi, \beta \rangle$ then $\Phi \cup \Psi \not\vdash_c \bot$.*

Definition 10. *An argument $\langle \Psi, \beta \rangle$ is a maximally conservative defeater for $\langle \Phi, \alpha \rangle$ iff $\langle \Psi, \beta \rangle$ is a defeater for $\langle \Phi, \alpha \rangle$ such that no defeaters for $\langle \Phi, \alpha \rangle$ are strictly more conservative than $\langle \Psi, \beta \rangle$.*

We assume that there exists an enumeration which we call *canonical enumeration* of all maximally conservative defeaters for $\langle \Phi, \alpha \rangle$.

Example 10. Let us return to Example 7. Both of the following $\langle \{a \Rightarrow b, a \Rightarrow c\}, a \Rightarrow b \wedge c \rangle$, $\langle \{e \wedge \neg d \Rightarrow b \wedge c\}, e \wedge \neg d \Rightarrow b \wedge c \rangle$, $\langle \{\neg b\}, \neg b \rangle$, $\langle \{a \Rightarrow b, a \Rightarrow c\}, \neg\neg(a \Rightarrow b \wedge c) \rangle$, $\langle \{\neg b\}, \neg(\neg b \to b) \rangle$, and $\langle \{a \Rightarrow b, a \Rightarrow c\}, (a \Rightarrow b) \wedge (a \Rightarrow c) \rangle$ are maximally conservative defeaters for the argument $\langle \{a \vee \neg d \Rightarrow b \wedge c, f \vee \neg b, b\}, f \wedge (a \vee \neg d \Rightarrow b \wedge c) \rangle$.

Note that, like arguments, maximally conservative defeaters are in an infinite number, as shown by the following results.

Proposition 10. *Let $\langle \Psi, \beta \rangle$ be a maximally conservative defeater for $\langle \Phi, \alpha \rangle$. $\langle \Psi, \gamma \rangle$ is a maximally conservative defeater for $\langle \Phi, \alpha \rangle$ iff $\langle \Psi, \beta \rangle$ and $\langle \Psi, \gamma \rangle$ are quasi-identical.*

Corollary 1. *Let $\langle \Psi, \beta \rangle$ be a maximally conservative defeater for $\langle \Phi, \alpha \rangle$. There is an infinite set of maximally conservative defeaters for $\langle \Phi, \alpha \rangle$ of the form $\langle \Theta, \gamma \rangle$ such that $\langle \Psi, \beta \rangle$ and $\langle \Theta, \gamma \rangle$ are quasi-identical.*

Now, it is possible to avoid some amount of redundancy among counterarguments by ignoring the unnecessary variants of maximally conservative defeaters. To this end, we define a concept of *pertinent defeaters* as follows.

Definition 11. *Let $\langle \Psi_1, \beta_1 \rangle, \ldots, \langle \Psi_n, \beta_n \rangle, \ldots$ be the canonical enumeration of all maximally conservative defeaters for $\langle \Phi, \alpha \rangle$. $\langle \Psi_i, \beta_i \rangle$ is a pertinent defeater for $\langle \Phi, \alpha \rangle$ iff for every $j < i$, $\langle \Psi_i, \beta_i \rangle$ and $\langle \Psi_j, \beta_j \rangle$ are not quasi-identical.*

Thus, a pertinent defeater can be interpreted as the representative of a set of counter-arguments.

Example 11. Let us return to Example 10. Suppose that $\langle \{a \Rightarrow b, a \Rightarrow c\}, a \Rightarrow b \wedge c \rangle$, $\langle \{e \wedge \neg d \Rightarrow b \wedge c\}, e \wedge \neg d \Rightarrow b \wedge c \rangle$, $\langle \{\neg b\}, \neg b \rangle$, $\langle \{a \Rightarrow b, a \Rightarrow c\}, \neg\neg(a \Rightarrow b \wedge c) \rangle$, $\langle \{\neg b\}, \neg(\neg b \to b) \rangle$, $\langle \{a \Rightarrow b, a \Rightarrow c\}, (a \Rightarrow b) \wedge (a \Rightarrow c) \rangle, \ldots$ is the canonical enumeration of the maximally conservative defeaters. Both of the following $\langle \{a \Rightarrow b, a \Rightarrow c\}, a \Rightarrow b \wedge c \rangle$, $\langle \{e \wedge \neg d \Rightarrow b \wedge c\}, e \wedge \neg d \Rightarrow b \wedge c \rangle$, $\langle \{\neg b\}, \neg b \rangle$ are pertinent defeaters for the argument $\langle \{a \vee \neg d \Rightarrow b \wedge c, f \vee \neg b, b\}, f \wedge (a \vee \neg d \Rightarrow b \wedge c) \rangle$.

Clearly, an argument may have more than one pertinent defeater. The next result shows how the pertinent defeaters for the same argument differ from one another.

Proposition 11. *Any two different pertinent defeaters for the same argument have distinct supports.*

4.3   Argumentation Trees

A last basic brick of logic-based argumentation theories that we revisit is the notion of *argumentation tree* and its related topics. From a set Δ of formulas, several possibly interconnected arguments can co-exist that should be assembled to get a full understanding about the pros and cons conducting a conclusion to be accepted or rejected. Argumentation trees are intended to collect and organize those arguments.

Definition 12. *An* argumentation tree *for α is a tree T whose nodes are arguments s.t.:*

1. *The root of T is an argument for α,*
2. *For every node $\langle \Psi, \beta \rangle$ whose ancestor nodes are $\langle \Psi_1, \beta_1 \rangle, \ldots, \langle \Psi_n, \beta_n \rangle$, there exists $\gamma \in \Psi$ s.t. $\gamma \notin \Psi_i$ for $i = 1..n$,*
3. *Each child node is a pertinent defeater of its parent node.*

An argumentation tree aims to exhaustively (but implicitly) capture the way counter-arguments can take place as a dispute develops. Condition 2 requires that each counter-argument involves extra information thereby precluding cycles.

Example 12. Let us return to Example 7. Let $\alpha = f \wedge (a \vee \neg d \Rightarrow b \wedge c)$.

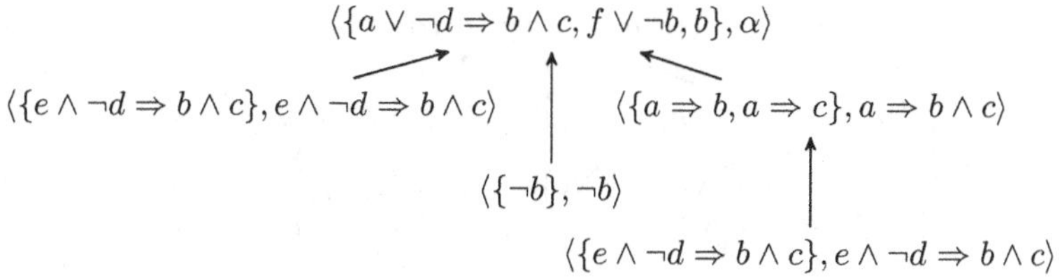

Fig. 1. Argumentation tree for $f \wedge (a \vee \neg d \Rightarrow b \wedge c)$

Proposition 12. *An argumentation tree for a formula α is finite.*

Proposition 13. *For any α s.t. $\Delta \vdash_c \alpha$, there is only a finite number of argumentation trees for α.*

Clearly, the last two properties are important in practice. They show that an argumentation tree can indeed be an effective way of representing an argumentation process.

In standard-logic argumentation [15], if Δ is consistent then all argumentation trees have exactly one node. This is not the case in contrariety-based argumentation: from a consistent knowledge base, argumentation trees that do not collapse into a single node exist.

Example 13 illustrates that attacks between arguments need not be inconsistency-based but can indeed be rooted in contrariety in conditional logic.

Example 13. Let $\Delta = \{a \Rightarrow b, \neg d, (a \wedge (d \vee \neg f)) \Rightarrow b \wedge c, a \Rightarrow c\}$.
Note that Δ is consistent. Let $\alpha = (a \Rightarrow b \wedge c) \vee \neg d$.

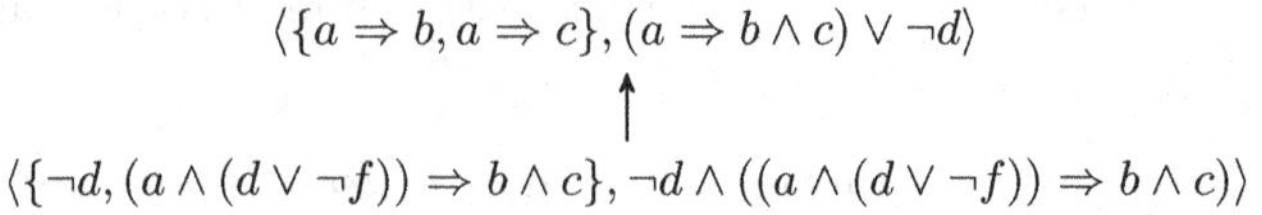

$$\langle \{a \Rightarrow b, a \Rightarrow c\}, (a \Rightarrow b \wedge c) \vee \neg d \rangle$$
$$\uparrow$$
$$\langle \{\neg d, (a \wedge (d \vee \neg f)) \Rightarrow b \wedge c\}, \neg d \wedge ((a \wedge (d \vee \neg f)) \Rightarrow b \wedge c) \rangle$$

Fig. 2. Argumentation tree for $(a \Rightarrow b \wedge c) \vee \neg d$

As several different argumentation trees for a given formula α can co-exist, the following *full argumentation tree* concept aims to represent them in a global manner by considering all pertinent defeaters and all possible attacks.

Definition 13. *Let T be an argumentation tree for α. T is a full argumentation tree for α if the children of any node A consists of all pertinent defeaters of A.*

Example 14. Let $\Delta = \{a \Rightarrow b, a, a \wedge d \Rightarrow b \wedge c, c \wedge \neg a, a \Rightarrow c, \neg c \wedge \neg a\}$. Let $\alpha = a \wedge (a \Rightarrow b \wedge c)$.

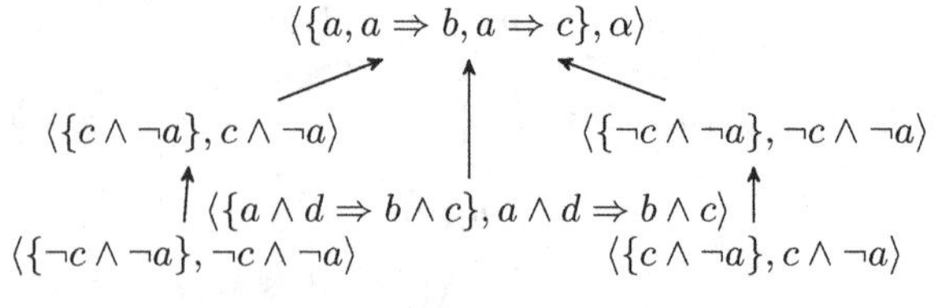

Fig. 3. Full argumentation tree for $a \wedge (a \Rightarrow b \wedge c)$

5   Perspectives and Conclusion

Conditional logic is a widespread tool in A.I. This paper is an attempt to lay down the basic bricks of logic-based argumentation in conditional logic. Interestingly, it has allowed us to put in light and encompass a specific form of conflict that often occurs in real-life argumentation: i.e., claims that additional conditions are required for the conclusion of a rule to hold. In this respect and to some extent, this paper targets some patterns of reasoning similar to those in [30, 31], where preemption operators are investigated in the framework of standard logic: preemption operators allow a logically weaker piece of information to replace a stronger one. However, the problem that we have addressed in this paper is different: the focus has been on confronting arguments, with a specific attention to comparing conditional formulas making use of the conditional connective, allowing in some sense weaker formulas of that kind to be compared to stronger ones.

In the future, we plan to investigate how to compare and rationalize argumentation trees in conditional logic, consider audience and impact-related issues on arguments and build various algorithmic tools for handling arguments and reasoning about them.

References

1. Bench-Capon, T.J.M., Dunne, P.E.: Argumentation in artificial intelligence. Artificial Intelligence (Special Issue on Argumentation) 171(10-15) (2007)
2. Rahwan, I., Simari, G.R.: Argumentation in artificial intelligence. Springer (2009)
3. Verheij, B., Szeider, S., Woltran, S. (eds.): Computational Models of Argument - Proceedings of COMMA 2012, Vienna, Austria, September 10-12. Frontiers in Artificial Intelligence and Applications, vol. 245. IOS Press (2012)
4. Prakken, H.: An argumentation framework in default logic. Annals of Mathematics and Artificial Intelligence 9(1-2), 93–132 (1993)
5. Prakken, H., Sartor, G.: A dialectical model of assessing conflicting arguments in legal reasoning. Artificial Intelligence and Law 4(3-4), 331–368 (1996)
6. Das, S.K., Fox, J., Krause, P.: A unified framework for hypothetical and practical reasoning (1): Theoretical foundations. In: Gabbay, D.M., Ohlbach, H.J. (eds.) FAPR 1996. LNCS, vol. 1085, pp. 58–72. Springer, Heidelberg (1996)
7. Parsons, S., Sierra, C., Jennings, N.R.: Agents that reason and negotiate by arguing. Journal of Logic and Computation 8(3), 261–292 (1998)
8. Ferguson, G., Allen, J.F., Miller, B.W.: Trains-95: Towards a mixed-initiative planning assistant. In: Proceedings of the Third International Conference on Artificial Intelligence Planning Systems (AIPS 1996), pp. 70–77 (1996)
9. Parsons, S., Wooldridge, M., Amgoud, L.: Properties and complexity of some formal inter-agent dialogues. Journal of Logic and Computation 13(3), 347–376 (2003)
10. McBurney, P., Parsons, S., Rahwan, I. (eds.): ArgMAS 2011. LNCS, vol. 7543. Springer, Heidelberg (2012)
11. Dung, P.M.: On the acceptability of arguments and its fundamental role in nonmonotonic reasoning, logic programming and n-person games. Artificial Intelligence 77(2), 321–358 (1995)
12. Pollock, J.L.: How to reason defeasibly. Artificial Intelligence 57, 1–42 (1992)
13. Krause, P., Ambler, S., Elvang-Gøransson, M., Fox, J.: A logic of argumentation for reasoning under uncertainty. Computational Intelligence 11, 113–131 (1995)
14. Chesnevar, C.I., Maguitman, A.G., Loui, R.P.: Logical models of argument. ACM Computing Surveys 32(4), 337–383 (2000)
15. Besnard, P., Hunter, A.: A logic-based theory of deductive arguments. Artificial Intelligence 128(1-2), 203–235 (2001)
16. Garcia, A.J., Simari, G.R.: Defeasible logic programming: An argumentative approach. Theory and Practice of Logic Programming 4(1-2), 95–138 (2004)
17. Santos, E., Martins, J.P.: A default logic based framework for argumentation. In: Proceedings of ECAI 2008, pp. 859–860 (2008)
18. Besnard, P., Grégoire, É., Piette, C., Raddaoui, B.: MUS-based generation of arguments and counter-arguments. In: Proceedings of the 11th IEEE Int. Conf. on Information Reuse and Integration (IRI 2010), pp. 239–244 (2010)

19. Gorogiannis, N., Hunter, A.: Instantiating abstract argumentation with classical logic arguments: Postulates and properties. Artif. Intell. 175(9-10), 1479–1497 (2011)
20. Crocco, G., Fariñas del Cerro, L., Herzig, A.: Conditionals: From philosophy to computer science. Studies in Logic and Computation. Oxford University Press (1995)
21. Giordano, L., Gliozzi, V., Olivetti, N.: Iterated belief revision and conditional logic. Studia Logica 70(1), 23–47 (2002)
22. Grahne, G.: Updates and counterfactuals. Journal of Logic and Computation (1), 87–117 (2002)
23. Nute, D.: Topics in conditional logic. Reidel. Dordrecht (1980)
24. Delgrande, J.P.: An approach to default reasoning based on a first-order conditional logic: Revised report. Artificial Intelligence 36(1), 63–90 (1988)
25. Kraus, S., Lehmann, D.J., Magidor, M.: Nonmonotonic reasoning, preferential models and cumulative logics. Artificial Intelligence 44(1-2), 167–207 (1990)
26. Giordano, L., Schwind, C.: Conditional logic of actions and causation. Artificial Intelligence 157(1-2), 239–279 (2004)
27. Gabbay, D.M., Giordano, L., Martelli, A., Olivetti, N., Sapino, M.L.: Conditional reasoning in logic programming. Journal of Logic Programming 44(1-3), 37–74 (2000)
28. Besnard, P., Hunter, A.: Elements of Argumentation. MIT Press (2008)
29. Chellas, B.F.: Basic conditional logic. Journal of Philosophical Logic 4(2), 133–153 (1975)
30. Besnard, P., Grégoire, É., Ramon, S.: Preemption operators. In: Raedt, L.D., Bessière, C., Dubois, D., Doherty, P., Frasconi, P., Heintz, F., Lucas, P.J.F. (eds.) ECAI. Frontiers in Artificial Intelligence and Applications, vol. 242, pp. 893–894. IOS Press (2012)
31. Besnard, P., Grégoire, É., Ramon, S.: Enforcing logically weaker knowledge in classical logic. In: Xiong, H., Lee, W.B. (eds.) KSEM 2011. LNCS, vol. 7091, pp. 44–55. Springer, Heidelberg (2011)

Modelling Uncertainty in Persuasion

Anthony Hunter

Department of Computer Science, University College London,
Gower Street, London WC1E 6BT, UK

Abstract. Participants in argumentation often have some doubts in their arguments and/or the arguments of the other participants. In this paper, we model uncertainty in beliefs using a probability distribution over models of the language, and use this to identify which are good arguments (i.e. those with support with a probability on or above a threshold). We then investigate three strategies for participants in dialogical argumentation that use this uncertainty information. The first is an exhaustive strategy for presenting a participant's good arguments, the second is a refinement of the first that selects the good arguments that are also good arguments for the opponent, and the third selects any argument as long as it is a good argument for the opponent. We show that the advantage of the second strategy is that on average it results in shorter dialogues than the first strategy, and the advantage of the third strategy is that under some general circumstances the participant can always win the dialogue.

1 Introduction

Persuasion is a complex multifaceted concept. In this paper, we consider uncertainty in persuasion which is a topic that is underdeveloped in formal models of argument. We represent the uncertainty that an agent has over its own beliefs by a probability distribution over the models of the language. The agent uses this to judge which arguments are "good arguments" (arguments with premises with a probability on or above a "good argument" threshold), and which are "good targets" (arguments with premises with a probability on or below a "good target" threshold) and as such should be attacked if an attacker exists. The idea is that if an argument is a good argument but not a good target, then the agent considers the argument but ignores any attack on it. To illustrate, consider the following arguments[1]. Suppose each is a good argument.

- A_1 "The metro is the best way to the airport."
- A_2 "There is a strike today by metro workers."

Now consider the threshold for attack. It would be reasonable in this context to take a skeptical view (because we worry about missing the flight) and set the threshold for being a good target to be above the threshold for being a good argument. So even if the threshold for being a good argument might be set to a high level, we might want the threshold for being a good target to be even higher. Therefore, for this example, we would get an argument graph with both arguments where A_2 attacks A_1.

---

[1] Note, we not proposing a formal model of argument-based decision making (c.f. [1]), but rather investigating criteria for selecting arguments and attacks to present in argumentation.

W. Liu, V.S. Subrahmanian, and J. Wijsen (Eds.): SUM 2013, LNAI 8078, pp. 57–70, 2013.
© Springer-Verlag Berlin Heidelberg 2013

As an alternative example, consider A_3 and the potential counterargument A_2 where both are good arguments. Here we might take a credulous view (because we might not worry too much about the risk or consequences of delay on the metro when going home). So we set the threshold for being a good argument as above the threshold for being a good target. In this context, we may say that even though there exists the counterargument A_2, A_3 is not a good target because the threshold for being a good target is lower in this case. In other words, there is insufficient doubt in A_3 for A_2 to attack it. In this example, this is reasonable since often some trains still run when there are problems with the service.

- A_3 "The metro is the best way to go home"

As well as considering how an agent might judge its own arguments and counterarguments using its probability distribution, we also want to consider how it can be used in dialogue strategies. For this, we let an agent have an estimate of its opponent's probability distribution. This can be used to make the argumentation more efficient and/or more efficacious. There is no point in presenting arguments that are not going to persuade an opponent, particularly when there may be alternative arguments that could bring about the required outcome. Consider the following dialogue where participant 1 (husband) wants to persuade participant 2 (wife) to buy a particular car. Argument A_5 indicates that participant 2 does not believe argument A_4, and so participant 1 has not used a good argument to persuade participant 2.

- A_4 "The car is a nice red colour, and that is the only criterion to consider, therefore we should buy it."
- A_5 "It is a nice red colour, but I don't agree that that is the only criterion to consider."

Now consider argument A_6 which participant 2 sees as a good argument but not a good target. So if participant 1 has a good estimate of the probability distribution of participant 2, then it could see A_6 as better to posit than A_4, and that this could result in a more persuasive dialogue.

- A_6 "The car is the most economical and easy car to drive out of the options available to us, and those are the criteria we want to satisfy, so we should buy the car."

In this paper, we formalise good argument and good attack, and investigate their use in persuasion dialogues.

2 Preliminaries

We review abstract argumentation [2], probabilistic logic [3], and the use of probabilistic logic in argumentation [4].

2.1 Abstract Argumentation

An **abstract argument graph** is a pair $(\mathcal{A}, \mathcal{R})$ where $\mathcal{A}$ is a set and $\mathcal{R} \subseteq \mathcal{A} \times \mathcal{A}$. Each element $A \in \mathcal{A}$ is called an **argument** and $(A, B) \in \mathcal{R}$ means that A **attacks** B (accordingly, A is said to be an **attacker** of B) and so A is a **counterargument** for B. A set of arguments $S \subseteq \mathcal{A}$ **attacks** $A_j \in \mathcal{A}$ iff there is an argument $A_i \in S$ such that A_i attacks A_j. Also, S **defends** $A_i \in S$ iff for each argument $A_j \in \mathcal{A}$, if A_j attacks A_i then S attacks A_j. A set $S \subseteq \mathcal{A}$ of arguments is **conflict-free** iff there are no arguments A_i and A_j in S such that A_i attacks A_j. Let Γ be a conflict-free set of arguments, and let Defended : $\wp(\mathcal{A}) \mapsto \wp(\mathcal{A})$ be a function such that Defended$(\Gamma) = \{A \mid \Gamma \text{ defends } A\}$. We consider the following extensions: (1) Γ is a **complete extension** iff $\Gamma = $ Defended(Γ); and (2) Γ is a **grounded extension** iff it is the minimal (w.r.t. set inclusion) complete extension. For $G = (\mathcal{A}, \mathcal{R})$, let Nodes$(G) = \mathcal{A}$ and let Grounded(G) be the grounded extension of G.

2.2 Probabilistic Logic

We use an established proposal for capturing probabilistic belief in classical propositional formulae [3]. For this, we assume that the propositional language $\mathcal{L}$ is finite. The set of models (i.e. interpretations) of $\mathcal{L}$ is denoted $\mathcal{M}^{\mathcal{L}}$. Each **model** in $\mathcal{M}^{\mathcal{L}}$ is an assignment of *true* or *false* to the formulae of the language defined in the usual way for classical logic. So for each model m, and $\psi \in \mathcal{L}$, $m(\psi) = true$ or $m(\psi) = false$. For $\phi \in \mathcal{L}$, Models(ϕ) denotes the set of models of ϕ (i.e. Models$(\phi) = \{m \in \mathcal{M}^{\mathcal{L}} \mid m(\phi) = true\}$), and for $\Delta \subseteq \mathcal{L}$, Models$(\Delta)$ denotes the set of models of Δ (i.e. Models$(\Delta) = \cap_{\phi \in \Delta}$Models$(\phi)$). Let $\Delta \models \psi$ denote Models$(\Delta) \subseteq$ Models(ψ).

Let $\mathcal{L}$ be a propositional language and let $\mathcal{M}^{\mathcal{L}}$ be the models of the language. A function $P : \mathcal{M}^{\mathcal{L}} \to [0, 1]$ is a **probability distribution** iff $\sum_{m \in \mathcal{M}^{\mathcal{L}}} P(m) = 1$. From a probability distribution, we get the **probability of a formula** $\phi \in \mathcal{L}$ as follows: $P(\phi) = \sum_{m \in \text{Models}(\phi)} P(m)$.

Example 1. Let the atoms of $\mathcal{L}$ be $\{a, b\}$, and so $\mathcal{L}$ is the set of propositional formulae formed from them. Let m_1 and m_2 be models s.t. $m_1(a) = true$, $m_1(b) = true$, $m_2(a) = true$, and $m_2(b) = false$. Now suppose $P(m_1) = 0.8$ and $P(m_2) = 0.2$. Hence, $P(a) = 1$, $P(a \wedge b) = 0.8$, $P(b \vee \neg b) = 1$, $P(\neg a \vee \neg b) = 0.2$, etc.

For any probability distribution P, if $\models \alpha$, then $P(\alpha) = 1$, and if $\models \neg(\alpha \wedge \beta)$, then $P(\alpha \vee \beta) = P(\alpha) + P(\beta)$.

2.3 Logical Arguments

We use deductive arguments based on classical logic to instantiate abstract argument graphs. Let $\Delta \subseteq \mathcal{L}$ be a set of propositional formulae and let $\vdash$ be the classical consequence relation. $\langle \Phi, \alpha \rangle$ is a **deductive argument** (or simply **argument**) iff $\Phi \subseteq \Delta$ and $\Phi \vdash \alpha$ and $\Phi \not\vdash \bot$ and there is no $\Psi \subset \Phi$ s.t. $\Psi \vdash \alpha$. For an argument $A = \langle \Phi, \alpha \rangle$, let Support$(A) = \Phi$ and Claim$(A) = \alpha$. Let Arg(Δ) be the set of deductive arguments obtained from Δ. For counterarguments, we use direct undercuts

[5, 6]. Argument A is a **direct undercut** of argument B when $\text{Claim}(A)$ is $\neg\psi$ for some $\psi \in \text{Support}(B)$. The set of direct undercuts is $\text{Ucuts}(\Delta) = \{(A, B) \mid A, B \in \text{Arg}(\Delta)$ and A is a direct undercut of $B\}$. The probability of an argument is the probability of its support.

Definition 1. *Let P be a probability distribution on $\mathcal{M}^{\mathcal{L}}$. The **probability of an argument** $\langle \Phi, \alpha \rangle \in \text{Arg}(\mathcal{L})$, denoted $P(\langle \Phi, \alpha \rangle)$, is $P(\phi_1 \wedge \ldots \wedge \phi_n)$, where $\Phi = \{\phi, \ldots, \phi_n\}$.*

Example 2. Consider the following probability distribution over models (with atoms a and b) for each participant.

Model	a	b	Participant 1	Participant 2
m_1	true	true	0.5	0.0
m_2	true	false	0.5	0.0
m_3	false	true	0.0	0.6
m_4	false	false	0.0	0.4

Let $\Delta_1 = \{a, \neg b\}$ (resp. $\Delta_2 = \{b, \neg b, b \rightarrow \neg a\}$) be the knowlegebase for participant 1 (resp. 2). Below is the probability of each argument according to each participant.

Argument	Participant 1	Participant 2
$A_1 = \langle \{a\}, a \rangle$	1.0	0.0
$A_2 = \langle \{b, b \rightarrow \neg a\}, \neg a \rangle$	0.0	0.6
$A_3 = \langle \{\neg b\}, \neg b \rangle$	0.5	0.4

It is possible for the knowledgebase to be inconsistent and yet for the participant to have a probability distribution over the models, as illustrated by Example 2.

3   Good Arguments and Good Attacks

Each agent has a knowledgebase Δ, and a probability distribution P, and these are used to identify good arguments.

Definition 2. *For a knowledgebase Δ, a probability distribution P, and a threshold $T \in [0, 1]$, the set of **good arguments** is $\text{GoodArg}(\Delta, P, T) = \{A \in \text{Arg}(\Delta) \mid P(A) \geq T\}$.*

Hence, if $T = 0$, then all arguments from the knowledgebase are good arguments (i.e. $\text{GoodArg}(\Delta, P, T) = \text{Arg}(\Delta)$). Whereas if $T = 1$, then only arguments that have premises that are certain are good arguments. Furthermore, if $T = 1$, then the premises of the good arguments are consistent together (i.e. $(\cup_{A \in \text{GoodArg}(\Delta, P, T)} \text{Support}(A)) \not\vdash \perp$), and so there are no $A, B \in \text{GoodArg}(\Delta, P, T)$ such that A is a direct undercut of B when $T = 1$.

Good targets (defined next) are arguments for which there is sufficient doubt in their premises for us to want to attack them even if an attacker exists. If an argument is not a good target, then we will ignore attacks on it. This is a form of inconsistency/conflict tolerance allowing us to focus on the more significant inconsistencies/conflicts.

Definition 3. *For a probability distribution P, a threshold $S \in [0,1]$, and a knowledge-base Δ, the set of* **good targets** *is* $\mathsf{GoodTarget}(\Delta, P, S) = \{B \in \mathsf{Arg}(\Delta) \mid P(B) \leq S\}$.

If $S = 1$, then $\mathsf{GoodTarget}(\Delta, P, S) = \mathsf{Arg}(\Delta)$, whereas if $S = 0$, then only arguments with support with zero probability are good targets. Next, we use S to select the attacks.

Definition 4. *For a knowledgebase Δ, a probability function over arguments P, and a threshold $S \in [0,1]$, the set of* **good attacks** *is* $\mathsf{GoodAttack}(\Delta, P, S) = \{(A, B) \mid (A, B) \in \mathsf{Ucuts}(\Delta) \text{ and } P(B) \leq S\}$.

Given a knowledgebase, and a probability distribution, a good graph is the set of good arguments and good attacks that can be formed.

Definition 5. *For a knowledgebase Δ, thresholds $T, S \in [0,1]$, and a probability distribution P, the* **good graph** *is an argument graph,* $\mathsf{GoodGraph}(\Delta, P, T, S) = (\mathcal{A}, \mathcal{R})$, *where $\mathcal{A} = \mathsf{GoodArg}(\Delta, P, T)$ and $\mathcal{R} = \mathsf{GoodAttack}(\Delta, P, S)$.*

For considering whether or not a specific argument is in the grounded extension of a (good) graph, we only need to consider the component containing it, as illustrated next.

Example 3. Suppose $\Delta = \{a, \neg a\}$. Let $A_1 = \langle \{a\}, a \rangle$ and $A_2 = \langle \{\neg a\}, \neg a \rangle$. Suppose we want to determine whether A_1 is in the grounded extension of the good graph. Depending on the choice of P, S, and T, the component to consider is one of G_1 to G_6 where G_1 is $(\emptyset, \emptyset)$ when $P(A_1) < T$, and the constraints for G_2 to G_6 are tabulated below.

Graph	Structure	$P(A_1)?T$	$P(A_2)?T$	$P(A_1)?S$	$P(A_2)?S$
G_2	A_1	$P(A_1) \geq T$	$P(A_2) < T$	n/a	n/a
G_3	$A_1 \quad A_2$	$P(A_1) \geq T$	$P(A_2) \geq T$	$P(A_1) > S$	$P(A_2) > S$
G_4	$A_1 \leftarrow A_2$	$P(A_1) \geq T$	$P(A_2) \geq T$	$P(A_1) \leq S$	$P(A_2) > S$
G_5	$A_1 \rightarrow A_2$	$P(A_1) \geq T$	$P(A_2) \geq T$	$P(A_1) > S$	$P(A_2) \leq S$
G_6	$A_1 \leftrightarrow A_2$	$P(A_1) \geq T$	$P(A_2) \geq T$	$P(A_1) \leq S$	$P(A_2) \leq S$

Proposition 1. *If $T > S$, then $\forall \Delta$, P, $\mathsf{GoodArg}(\Delta, P, T) \cap \mathsf{GoodTarget}(\Delta, P, S) = \emptyset$. and if $T \leq S$, then $\exists \Delta$, P s.t. $\mathsf{GoodArg}(\Delta, P, T) \cap \mathsf{GoodTarget}(\Delta, P, S) \neq \emptyset$. Also if $T = 0$ and $S = 1$, then $\forall \Delta$, P, $\mathsf{GoodArg}(\Delta, P, T) = \mathsf{GoodTarget}(\Delta, P, S)$.*

Example 4. Consider the arguments $A_1 = \langle \{a\}, a \rangle$ and $A_2 = \langle \{\neg a\}, \neg a \rangle$ generated from Δ where $T = 0$ and $S = 1$. Whatever choice is made for P, either $P(A_1) < 1$ or $P(A_2) < 1$ or both $P(A_1) < 1$ and $P(A_2) < 1$. So if $T = 0$ then both arguments are good arguments, and if $S = 1$ then each argument attacks the other.

In the following, we consider how components in good graphs are constructed via dialogical argumentation. For this, we will assume $S < T$, and so T affects the choice of arguments to present, and S affects the choice of counterarguments to present.

4 Participants

We will assume two participants called 1 and 2 where 1 wants to persuade 2 about a claim ϕ which we refer to as the **persuasion claim**. Informally, for participant 1 to persuade participant 2 to accept the persuasion claim, it needs to give an argument with claim ϕ that is in the grounded extension of the argument graph produced during the dialogue. We formalize this in the next section.

For the good argument threshold T, and the good target threshold S, we assume $S < T$ so that each participant cannot attack its own good arguments. Each participant has a **position**: Participant 1 has position $\Pi_1 = (\Delta_1, P_1, P', T, S, \phi)$ containing its knowledgebase Δ_1, its probability distribution P_1, the probability distribution P' which is an estimate of the probability distribution P_2 of the other agent, the thresholds T and S, and the persuasion claim ϕ, and participant 2 has a position $\Pi_2 = (\Delta_2, P_2, T, S)$ containing its knowledgebase Δ_2, its probability distribution P_2, and the thresholds T and S. Note, position 1 has more parameters because participant 1 has the lead role in the dialogue. Also, note each participant does not know the position of the other participant (apart from S and T).

Participant 1 can build P' as an estimate of P_2 over time, such as by learning from previous dialogues. However, participant 1 does not know whether P' is a good estimate of P_2. But, we as external observers do know Π_1 and Π_2, and so we can measure how well P' models P_2. For this, we use a rank correlation coefficient which assigns a value in $[-1, 1]$ such that when P' and P_2 completely agree on the ranking of the arguments, the coefficient is 1, and when they completely disagree on the ranking of the arguments (i.e. one is the reverse order of the other), the coefficient is -1 (as defined next).

Consider the set of arguments $\mathsf{Arg}(\Delta)$ for some Δ and the threshold S. We compare P' and P_2 in terms of how they rank each argument in $A \in \mathsf{Arg}(\Delta)$ with respect to S. Let n_a be the number of arguments that P' and P_2 agree on (i.e. $n_a = |\{A \in \mathsf{Arg}(\Delta) \mid (P'(A) > S \text{ and } P_2(A) > S) \text{ or } (P'(A) \leq S \text{ and } P_2(A) \leq S)\}|$, and let n_d be the number of arguments that P' and P_2 disagree on (i.e. $n_d = |\{A \in \mathsf{Arg}(\Delta) \mid (P'(A) > S \text{ and } P_2(A) \leq S) \text{ or } (P'(A) \leq S \text{ and } P_2(A) > S)\}|$. From this, the **rank correlation coefficient** is

$$\mathsf{Correlation}(P', P_2) = \frac{n_a - n_d}{n_a + n_d}$$

Example 5. For $\mathsf{Arg}(\Delta) = \{A_1, A_2, A_3, A_4\}$, and $S = 0.5$, let $P'(A_1) = 1$, $P'(A_2) = 0.3$, $P'(A_3) = 0.7$, $P'(A_4) = 0.4$, $P_2(A_1) = 1$, $P_2(A_2) = 0.8$, $P_2(A_3) = 0.6$, and $P_2(A_4) = 0.2$. So, the coefficient is $(3 - 1)/4 = 1/2$.

Note, the coefficient is the same as the Kendall rank correlation coefficient [7], but the way we calculate n_a and n_d is quite different.

5 Dialogical Argumentation

Participants take turns to contribute arguments and/or attacks, thereby constructing an argument graph. For this, we just record the additions to the graph as defined next.

Definition 6. *A **dialogue state** is a pair* (X, Y) *where* X *is a set of arguments, and* Y *is a set of attacks. Note,* Y *is not necessarily a subset of* $X \times X$. *A **dialogue**, denoted* D, *is a sequence of dialogue states* $[(X_1, Y_1), ..., (X_n, Y_n)]$.

We use the function D to denote a dialogue, where for an index $i \in \{1, ..., n\}$, $D(i) = (X_i, Y_i)$ is a dialogue state. For a dialogue $D = [(X_1, Y_1), ..., (X_n, Y_n)]$, $\mathsf{Len}(D) = n$ is the index of the last step, and $\mathsf{Sub}(D, i) = [(X_1, Y_1), ..., (X_i, Y_i)]$ is the first i steps. For each step of the dialogue, there is an argument graph. We define this graph recursively with the base case being the empty graph.

Definition 7. *For dialogue* D, *s.t.* $1 \leq i \leq \mathsf{Len}(D)$, *and* $D(i) = (X_i, Y_i)$, $\mathsf{Graph}(D, i)$ $= (\mathcal{A}_{i-1} \cup X_i, \mathcal{R}_{i-1} \cup Y_i)$ *is the **dialogue graph** where if* $i = 1$, *then* $(\mathcal{A}_{i-1}, \mathcal{R}_{i-1})$ *is* $(\emptyset, \emptyset)$, *and if* $i > 1$, *then* $(\mathcal{A}_{i-1}, \mathcal{R}_{i-1})$ *is* $\mathsf{Graph}(D, i - 1)$.

So for each step of the dialogue, we can construct the current state of the argument graph. So the sequence of states of the dialogue are all used to construct the current state of the graph. Clearly, this is monotonic: Arguments and attacks are added to the graph, and none are subtracted.

Example 6. Consider the following probability distribution over models for each participant, where $T = 0.5$ and $S = 0.3$.

Model	a	b	Participant 1	Participant 2
m_1	true	true	0.8	0.0
m_2	true	false	0.1	0.5
m_3	false	true	0.1	0.0
m_4	false	false	0.0	0.5

Hence, we get the following probabilities for A_1 to A_3.

Argument	Participant 1	Participant 2
$A_1 = \langle \{b, b \to a\}, a \rangle$	0.8	0.0
$A_2 = \langle \{\neg b\}, \neg b \rangle$	0.1	1.0
$A_3 = \langle \{b\}, b \rangle$	0.9	0.0

Now consider dialogue $D = [\ (\{A_1\}, \{\}), (\{A_2\}, \{(A_2, A_1)\}), (\{A_3\}, \{(A_3, A_2)\}), (\{\}, \{(A_2, A_3)\}), (\{\}, \{\})\ ]$, where $\mathsf{Len}(D) = 5$, giving the dialogue graph below.

A dialogue can be infinite since for example the contribution $(\emptyset, \emptyset)$ can be repeatedly added. So to draw the dialogue to a close, we restrict consideration to complete dialogues.

Definition 8. *A dialogue* D *is **complete**, where* $\mathsf{Len}(D) = n$, *and the persuasion claim is* ϕ, *iff*

1. $\forall i, j \in \{1, \ldots, n\}$, *if* $i \neq j$ & $D(i) = D(j)$, *then* $D(i) = (\emptyset, \emptyset)$ & $D(j) = (\emptyset, \emptyset)$
2. $\forall i \in \{1, \ldots, n - 2\}$, *if* $D(i) = (\emptyset, \emptyset)$, *then* $D(i + 1) \neq (\emptyset, \emptyset)$
3. *if* n *is even, then* $\exists A \in \mathsf{Grounded}(\mathsf{Graph}(D, n))$ *s.t.* $\mathsf{Claim}(A) = \phi$.
4. *if* n *is odd, then* $\nexists A \in \mathsf{Grounded}(\mathsf{Graph}(D, n))$ *s.t.* $\mathsf{Claim}(A) = \phi$.
5. $D(n) = (\emptyset, \emptyset)$

We explain the above conditions as follows: Condition 1 ensures that the only state that can be repeated is the empty state; Condition 2 ensures that only the last two step of the dialogue can have the empty state followed immediately by the empty state; Conditions 3 and 4 ensure that if the last step is an even step then there is an argument with the claim in the grounded extension, whereas if the last step is an odd step then there is not an argument with the claim in the grounded extension; And condition 5 ensures that the last step is the empty state.

In the rest of the paper, for each step i, if i is odd (respectively even) participant 1 (respectively participant 2) will add $D(i)$. So intuitively, at the last step n, if n is odd (respectively even), participant 1 (respectively participant 2) has conceded the dialogue (perhaps because it has nothing more to add).

Proposition 2. *Let D be a complete dialogue and let $\mathcal{A}$ be a finite set of arguments. If for each i, $X_i \subseteq \mathcal{A}$, and $Y_i \subseteq \mathcal{A} \times \mathcal{A}$, and $D(i) = (X_i, Y_i)$, then D is finite (i.e. $\mathsf{Len}(D) \in \mathbb{N}$).*

Taking a simple view of persuasion, a participant is persuaded of a claim if the dialogue graph constructed is such that there is an argument for the claim in the grounded extension of the graph. We justify this in the next section.

Definition 9. *For a complete dialogue D, where $\mathsf{Len}(D) = n$, the **outcome** of the dialogue is specified as follows: If n is even, then participant 1 **wins**, whereas if n is odd, then participant 1 **looses**.*

So if n is even, then participant 1 is successful in persuading participant 2, otherwise participant 1 is unsuccessful. The dialogue D in Example 6 is a complete dialogue, and hence participant 1 looses. In the following sections, we present and justify three strategies for constructing complete dialogues.

6   Simple Dialogues

In a simple dialogue, participant 1 can add arguments for the persuasion claim that are not in the current dialogue graph.

Definition 10. *For position Π_1, and dialogue D, a **posit contribution** by participant 1 is $(\{A\}, \{\})$ where $A \in \mathsf{GoodArg}(\Delta_1, P_1, T)$ and $A \notin \mathsf{Nodes}(\mathsf{Graph}(D, i))$ and $\mathsf{Claim}(A) = \phi$. The set of posit contributions by participant 1 for D at step i is $\mathsf{Posit}(\Pi_1, D, i)$.*

Both participants can add counterarguments. For this, the NewAttackers function identifies the good arguments that participant x has that are not in the current dialogue graph $G_i = \mathsf{Graph}(D, i)$ but that attack an argument in G_i.

$$\mathsf{NewAttackers}(\Pi_x, D, i) = \{A \in \mathsf{GoodArg}(\Delta_x, P_x, T) \mid A \notin \mathsf{Nodes}(G_i) \text{ and } \\ \exists B \in \mathsf{Grounded}(G_i) \text{ s.t. } P_x(B) \leq S \text{ and } A \text{ is a direct undercut of } B\}$$

Definition 11. *For position* Π_x, *and a dialogue* D, *a* **counter contribution** *by participant* x *is* (X_{i+1}, Y_{i+1}) *s.t.*

> *if there is an* $A \in$ NewAttackers(Π_x, D, i),
> *then* $X_{i+1} = \{A\}$ *and* $Y_{i+1} =$ NewArcs$(X_{i+1} \cup$ Nodes$(G_i), S)$
> *else* $X_{i+1} = \{\}$ *and* $Y_{i+1} =$ NewArcs$($Nodes$(G_i), S)$

where $G_i =$ Graph(D, i) *and* NewArcs$(Z, S) = \{ (A, B) \mid A, B \in Z$ *and* $P_x(A) \leq S$ *and* A *is a direct undercut of* $B\}$. *The set of counter contributions by participant* x *for* D *at step* i *is* Counter(Π_x, D, i).

So a counter contribution is zero or one argument and zero or more arcs, as illustrated in Example 6. Let Part1$(i) =$ Simple$(\Pi_1, D, i) \cup$ Counter(Π_1, D, i) (respectively Part2$(i) =$ Counter(Π_2, D, i)) be the contributions for participant 1 (respectively participant 2) at step i. The next definition ensures that the participants take turns in the contributions.

Definition 12. *For positions* Π_1 *and* Π_2, *a dialogue* D *is* **turn taking** *iff for each* $i \in \{1, \ldots,$ Len$(D)\}$, *if* i *is odd, then* $D(i) \in$ Part1(i) *and if* i *is even, then* $D(i) \in$ Part2(i).

The next definition ensures that each agent gives a contribution other than $(\emptyset, \emptyset)$ if possible (i.e. there is a non-empty contribution) and needed (i.e. for participant 1, there is not an argument for the persuasion claim in the grounded extension of the current dialogue graph, and for participant 2, there is a argument for the persuasion claim in the grounded extension of the current dialogue graph). Note, $(\emptyset, \emptyset)$ is always available as a counter contribution.

Definition 13. *For positions* Π_1 *and* Π_2, *a complete dialogue* D *is* **exhaustive** *iff for each* $i \in \{1, \ldots,$ Len$(D)\}$, *where* $G_i =$ Graph(D, i), *the following conditions hold.*

1. *If* i *is odd, and* $\exists A \in$ Grounded(G_i) *s.t.* Claim$(A) = \phi$, *then* $D(i) = (\emptyset, \emptyset)$.
2. *If* i *is odd, and* $\nexists A \in$ Grounded(G_i) *s.t.* Claim$(A) = \phi$, *and* $|$Part1$(i)| > 1$, *then* $D(i) \neq (\emptyset, \emptyset)$.
3. *If* i *is even, and* $\nexists A \in$ Grounded(G_i) *s.t.* Claim$(A) = \phi$, *then* $D(i) = (\emptyset, \emptyset)$.
4. *If* i *is even, and* $\exists A \in$ Grounded(G_i) *s.t.* Claim$(A) = \phi$, *and* $|$Part2$(i)| > 1$, *then* $D(i) \neq (\emptyset, \emptyset)$.

A **simple dialogue** is a dialogue that is turning taking and exhaustive. These definitions specify how the dialogue is constructed, and if the dialogue is complete it will terminate. The definitions ensure both agents only add good arguments and good attacks. Let SD(Π_1, Π_2) be the set of simple dialogues.

Example 7. For A_1, A_3 and A_5 from participant 1 and A_2 and A_4 from participant 2, D_1 is a simple dialogue for which Participant 1 wins.

$$
\begin{aligned}
A_1 &= \langle\{b, b \to a\}, a\rangle & D_1(1) &= (\{A_1\}, \{\}) \\
A_2 &= \langle\{c, c \to \neg b\}, \neg b\rangle & D_1(2) &= (\{A_2\}, \{(A_2, A_1)\}) \\
A_3 &= \langle\{d, d \to \neg c\}, \neg c\rangle & D_1(3) &= (\{A_3\}, \{(A_3, A_2)\}) \\
A_4 &= \langle\{\neg d\}, \neg d\rangle & D_1(4) &= (\{A_4\}, \{(A_4, A_3)\}) \\
A_5 &= \langle\{e, e \to \neg c\}, \neg c\rangle & D_1(5) &= (\{A_5\}, \{(A_5, A_2)\}) \\
& & D_1(6) &= (\{\}, \{\})
\end{aligned}
$$

Example 8. Participant 1 has A_1 and A_3 and participant 2 has A_2. $D = [(\{A_1\}, \{\}),$ $(\{A_2\}, \{(A_2, A_1)\}), (\{A_3\}, \{(A_3, A_2)\}), (\{\}, \{(A_1, A_3)\}), (\{\}, \{\})]$ is a simple dialogue that participant 1 looses.

Example 9. Participant 1 has A_1, A_3, A_5 and A_6, and participant 2 has A_2 and A_4. $D =$ $[(\{A_1\}, \{\}), (\{A_2\}, \{(A_2, A_1)\}), (\{A_3\}, \{(A_3, A_2)\}), (\{A_4\}, \{(A_4, A_3), (A_4, A_1)\}),$ $(\{A_5\}, \{(A_5, A_2)\}) (\{\}, \{\}) (\{A_6\}, \{(A_6, A_4)\}) (\{\}, \{\})]$ is a simple dialogue that participant 1 wins.

We use the joint graph (defined next) to show a type of correctness of the simple dialogues in the following result.

Definition 14. *For positions Π_1 and Π_2, the **joint graph**, is an argument graph $(\mathcal{A}, \mathcal{R})$, denoted* JointGraph$(\Pi_1, \Pi_2)$, *where* $\mathcal{A} =$ GoodArg$(\Delta_1, P_1, T) \cup$ GoodArg(Δ_2, P_2, T) *and* $\mathcal{R} = \{(A, B) \mid A, B \in \mathcal{A}$ *and* $(P_1(B) \le S$ *or* $P_2(B) \le S)$ *and* A *is a direct undercut of* B $\}$.

Proposition 3. *For positions Π_1 and Π_2, let G^* be* JointGraph(Π_1, Π_2). *For each $D \in$* SD(Π_1, Π_2), *participant 1 wins D iff there is an $A \in$* Grounded(G^*) *such that* Claim$(A) = \phi$.

So a simple dialogue just involves each participant making contributions until one or other participant concedes. Both agents are selective in the sense that they only present good arguments and good attacks. But for participant 1, there is no consideration of what might be more likely to be persuasive (such as presenting arguments that are less likely to be attacked by participant 2). We address this next.

7   Bestfirst Dialogues

The bestfirst dialogue involves participant 1 selecting its best arguments for positing first in the dialogue. Its best arguments, the bestfirst contributions, are its good arguments that it believes are not good targets for participant 2.

Definition 15. *For position Π_1, and dialogue D, the set of **bestfirst contributions** is* Bestfirst$(\Pi_1, D, i) = \{(\{A\}, Y) \in$ Simple$(\Pi_1, D, i) \cup$ Counter$(\Pi_1, D, i) \mid P'(A) > S\}$.

Definition 16. *For Π_1, and Π_2, a simple dialogue D is **bestfirst** iff for each $i \in \{1, ..., $*Len$(n)\}$, *if i is odd, and* Bestfirst$(\Pi_1, D, i) \ne \emptyset$, *then $D(i) \in$* Bestfirst(Π_1, D, i). *Let* BD(Π_1, Π_2) *be the set of bestfirst dialogues.*

Example 10. Let $D_1 = [(\{A_1\}, \{\}), (\{\}, \{\})]$ and $D_2 = [(\{A_2\}, \{\}), (\{A_3\}, \{(A_3, A_2)\}), (\{A_1\}, \{\}), (\{\}, \{\})]$. Also let Correlation$(P', P_2) = 1$. If $P'(A_1) > S$ and $P'(A_2) \le S$, then D_1 is bestfirst, and if $P'(A_1) \le S$ and $P'(A_2) > S$, then D_2 is bestfirst. In both cases, participant 1 wins.

If the correlation is positive for P' and P_2, then the next result shows that on average the bestfirst dialogues are shorter than the simple dialogues.

Proposition 4. *For the majority of positions Π_1 and Π_2, s.t. Correlation$(P', P_2) > 0$, then*

$$\left(\frac{\sum_{D \in BD(\Pi_1, \Pi_2)} \mathsf{Len}(D)}{\mid BD(\Pi_1, \Pi_2) \mid} \right) \leq \left(\frac{\sum_{D \in SD(\Pi_1, \Pi_2)} \mathsf{Len}(D)}{\mid SD(\Pi_1, \Pi_2) \mid} \right)$$

So the bestfirst dialogue captures a more efficient form of persuasion than the simple dialogue. Participant 1 presents its better arguments first, and if it does not succeed, then it will use its remaining arguments.

8   Insincere Dialogues

The insincere dialogue is characterised by the proponent selecting its arguments based on what it believes the other participant believes (and therefore selecting the arguments that are less likely to be attacked by the other participant). Note, we do not assume that participant 1 actually believes these arguments. It is being manipulative by presenting arguments that it believes that the other participant will accept.

Definition 17. *For position Π_1, and a dialogue D, the set of insincere contributions by participant 1 for D is the following where $\Pi^{insincere} = (\Delta_1, P', P', S, T, \phi)$.*

$$\mathsf{Insincere}(\Pi_1, D, i) = \mathsf{Posit}(\Pi^{insincere}, D, i) \cup \mathsf{Counter}(\Pi^{insincere}, D, i)\}$$

Definition 18. *For positions Π_1 and Π_2, a simple dialogue D is* **insincere** *iff for each $i \in \{1, \ldots, \mathsf{Len}(n)\}$, if i is odd, then $D(i) \in \mathsf{Insincere}(\Pi_1, D, i)$, and if i is even, then $D(i) \in \mathsf{Counter}(\Pi_2, D, i)$. Let $\mathsf{ID}(\Pi_1, \Pi_2)$ be the set of insincere dialogues.*

So $D \in \mathsf{ID}(\Pi_1, \Pi_2)$ iff $D \in \mathsf{SD}(\Pi^{insincere}, \Pi_2)$. The advantage for participant 1 is that it is not restricted by its own probability distribution in making its contributions. Rather, the aim for participant 1 is to present any arguments it can with the sole aim of winning the dialogue. Though one would assume that participant 1 would have a high belief in the persuasion claim ϕ (i.e. $P_1(\phi)$ is high) for it to want to resort to an insincere dialogue.

Example 11. Let $m_1(a) = true$, $m_1(b) = true$, $m_1(c) = false$, $m_2(a) = true$, $m_2(b) = false$, and $m_2(c) = true$. For positions Π_1 and Π_2, where $\Delta_1 = \{b, b \rightarrow a, c, c \rightarrow a\}$, $P_1(m_1) = 1$, $\Delta_2 = \{\neg b\}$, and $P_2(m_2) = 1$, let ϕ be a. Also, suppose $P' = P_2$. So $A_1 = \langle \{b, b \rightarrow a\}, a \rangle$ is a good argument for participant 1, but a good target for participant 2. In a simple dialogue, participant 1 only has one argument for a, and it would loose the dialogue (because participant 2 would attack with $A_2 = \langle \{\neg b\}, \neg b \rangle$). In contrast, $A_3 = \langle \{c, c \rightarrow a\}, a \rangle$ is not a good argument for participant 1, but for participant 2, it is a good argument and not a good target. So, A_3 is an insincere contribution for participant 1, and it would win the insincere dialogue using it.

In the next example, we let $\Delta_1 = \mathcal{L}$. Since participant 1 is prepared to say anything that participant 2 believes, this just means that it is prepared to present any argument A available in the language $\mathcal{L}$ as long as the recipient believes it.

Example 12. Let $\Delta_1 = \mathcal{L}$ where $A_1, A_3 \in \mathsf{Arg}(\Delta_1)$ and $A_2 \in \mathsf{Arg}(\Delta_2)$, and assume the following regarding the probability distributions.

$$P_1(A_1) > T; P'(A_1) > T; P'(A_1) > S; P_2(A_1) < S$$
$$P_1(A_2) < T; P'(A_2) < T; P'(A_2) < S; P_2(A_2) < S$$
$$P_1(A_3) < T; P'(A_3) > T; P'(A_3) > S; P_2(A_3) > S$$

So P' only differs from P_2 on A_1. Hence, $D = [\, (\{A_1\}, \{\}), (\{A_2\}, \{(A_2, A_1)\}),$ $(\{A_3\}, \{(A_3, A_2)\}), (\{\}, \{\}) \,]$ is an insincere dialogue that participant 1 wins.

The following definition of openness of a position just means that there is at least one atom in the language for which there are no strong arguments for or against it. In effect, it means that participant 2 has not got a position so constrained that it is impossible to persuade it.

Definition 19. *A position Π_2 is* **open** *iff there is an atom $\psi \in \mathcal{L}$, s.t. for all $A \in$* $\mathsf{GoodArgs}(\Delta_2, P_2, T)$, $\mathsf{Claim}(A) \neq \psi$ *and* $\mathsf{Claim}(A) \neq \neg\psi$.

Proposition 5. *Let Π_1 and Π_2 be positions s.t. $\Delta_1 = \mathcal{L}$, the persuasion claim is ϕ, and $\mathsf{Correlation}(P', P_2) = 1$. For any $D \in \mathsf{ID}(\Pi_1, \Pi_2)$ if either ($P_2(\neg\phi) \leq S$ and $S < 0.5$) or Π_2 is open, then participant 1 wins D.*

The above result is a situation where the participant 1 has a very good model of participant 2. We can generalise the result to imperfect models of the opponent so that with high probably that participant 1 wins.

The idea of an insincere strategy is important; If a protocol for a argumentative dialogue allows for this strategy, then the above result shows that a participant can dominate in a quite negative way. It can manipulate the opponent, and the opponent may be oblivious to this manipulation. We are not proposing that we want to build agents who use the insincere strategy. But, we may want to build agents who are aware that there are other agents who do use the insincere strategy and protect against it. So we need to formalise and investigate the insincere strategy and developments of it.

9 Discussion

In this paper, we have introduced good arguments, good targets, and good attacks. We therefore provide a new approach to constructing argument graphs, drawing on probability theory, that allows us to drop arguments if there is too much doubt in them, and to drop attacks if there is insufficient doubt in them. There are other proposals that drop attacks (e.g. preference-based argumentation frameworks [8], value-based argumentation frameworks [9], and weighted argumentation frameworks [10]), but they do not drop arguments other than by attacking them, and they are not based on a quantitative theory of uncertainty. There are proposals for using probability theory in argumentation (e.g. [4, 11–15]) but they do not drop arguments or attacks, and there is a possibility theory approach [16]) but it is not based on argument graphs.

Our approach has been influenced by Amgoud et al [17] (a detailed protocol for exchanging logical arguments using preference-based argumentation). We go beyond that by providing a way to select arguments and counterarguments to be used, and for strategies that use selectivity. We can allow for instance for an agent to present the arguments it has greatest belief in and it thinks the other agent has high belief in. We also allow for tolerance of arguments by an opponent. For instance an opponent may choose to not attack an argument if it thinks the argument is not too bad.

There are a number of papers that formalize aspects of persuasion. Most approaches are aimed at providing protocols for dialogues (for a review see [18]). Forms of correctness (e.g. the dialogue system has the same extensions under particular dialectical semantics as using the agent's knowledgebases) have been shown for a variety of systems (e.g. [19–22]). However, strategies for persuasion, in particular taking into account beliefs of the opponent are under-developed. Using selection of arguments, based on probability distributions for the agents, and for modelling one agent by another, we can formalise interesting strategies. To illustrate the potential, we consider the bestfirst strategy with a clear proven advantage, and the more complex insincere strategy.

Strategies in argumentation have been analysed using game theory [23, 24]. This mechanism design approach assumes that all the agents reveal their arguments at the same time, and the resulting argument graph is evaluated using grounded semantics. This is a one step process that does not involve logical arguments, dialogues or opponent modelling. Mechanism design has also been used for comparing strategies for logic-based dialogical argumentation that may involve lying [25]. This complements our work since they do not consider the uncertainty of beliefs or modelling the opponent.

Finally, audience modelling has been considered in value-based argumentation frameworks [9, 26] and in deductive argumentation [27, 28]. However, they have not been harnessed in strategies in dialogical argumentation, and only [26] considers uncertainty in the form of a probability assignment that an argument will promote a particular "value" with an agent, which is a different kind of uncertainty to that considered here.

In conclusion, we provide a novel framework for modelling uncertainty in argumentation, and use this to give three examples of strategy for dialogical argumentation. In future work, we will develop further strategies, and investigate learning the probability distributions from previous interactions.

References

1. Amgoud, L., Prade, H.: Using arguments for making and explaining decisions. Artificial Intelligence 173(3-4), 413–436 (2009)
2. Dung, P.: On the acceptability of arguments and its fundamental role in nonmonotonic reasoning, logic programming, and n-person games. Artificial Intelligence 77, 321–357 (1995)
3. Paris, J.: The Uncertain Reasoner's Companion: A Methematical Perspective. Cambridge University Press (1994)
4. Hunter, A.: A probabilistic approach to modelling uncertain logical arguments. International Journal of Approximate Reasoning 54(1), 47–81 (2013)
5. Elvang-Gransson, M., Krause, P., Fox, J.: Acceptability of arguments as logical uncertainty. In: Moral, S., Kruse, R., Clarke, E. (eds.) ECSQARU 1993. LNCS, vol. 747, pp. 85–90. Springer, Heidelberg (1993)
6. Cayrol, C.: On the relation between argumentation and non-monotonic coherence-based entailment. In: Proceedings of the Fourteenth International Joint Conference on Artificial Intelligence (IJCAI 1995), pp. 1443–1448 (1995)

7. Kendall, M.: A new measure of rank correlation. Biometrika 30(1-2), 81–93 (1938)
8. Amgoud, L., Cayrol, C.: A reasoning model based on the production of acceptable arguments. Annals of Mathematics and Artificial Intelligence 34, 197–216 (2002)
9. Bench-Capon, T.: Persuasion in practical argument using value based argumentationframeworks. Journal of Logic and Computation 13(3), 429–448 (2003)
10. Dunne, P.E., Hunter, A., McBurney, P., Parsons, S., Wooldridge, M.: Weighted argument systems: Basic definitions, algorithms, and complexity results. Artificial Intelligence 175(2), 457–486 (2011)
11. Haenni, R.: Cost-bounded argumentation. International Journal of Approximate Reasoning 26(2), 101–127 (2001)
12. Dung, P., Thang, P.: Towards (probabilistic) argumentation for jury-based dispute resolution. In: Computational Models of Argument (COMMA 2010), pp. 171–182. IOS Press (2010)
13. Li, H., Oren, N., Norman, T.: Probabilistic argumentation frameworks. In: Modgil, S., Oren, N., Toni, F. (eds.) TAFA 2011. LNCS, vol. 7132, pp. 1–16. Springer, Heidelberg (2012)
14. Thimm, M.: A probabilistic semantics for abstract argumentation. In: Proceedings of the European Conference on Artificial Intelligence (ECAI 2012), pp. 750–755 (2012)
15. Hunter, A.: Some foundations for probabilistic argumentation. In: Proceedings of the International Comference on Computational Models of Argument (COMMA 2012), pp. 117–128 (2012)
16. Alsinet, T., Chesñevar, C., Godo, L., Simari, G.: A logic programming framework for possibilistic argumentation: Formalization and logical properties. Fuzzy Sets and Systems 159(10), 1208–1228 (2008)
17. Amgoud, L., Maudet, N., Parsons, S.: Arguments, dialogue and negotiation. In: Fourteenth European Conference on Artifcial Intelligence (ECAI 2000), pp. 338–342. IOS Press (2000)
18. Prakken, H.: Formal sytems for persuasion dialogue. Knowledge Engineering Review 21(2), 163–188 (2006)
19. Prakken, H.: Coherence and flexibility in dialogue games for argumentation. Journal of Logic and Computation 15(6), 1009–1040 (2005)
20. Black, E., Hunter, A.: An inquiry dialogue system. Autonomous Agents and Multi-Agent Systems 19(2), 173–209 (2009)
21. Fan, X., Toni, F.: Assumption-based argumentation dialogues. In: Proceedings of International Joint Conference on Artificial Intelligence (IJCAI 2011), pp. 198–203 (2011)
22. Caminada, M., Podlaszewski, M.: Grounded semantics as persuasion dialogue. In: Computational Models of Argument (COMMA 2012), pp. 478–485 (2012)
23. Rahwan, I., Larson, K.: Mechanism design for abstract argumentation. In: Proceedings of the 7th International Joint Conference on Autonomous Agents and Multiagent Systems (AAMAS 2008, IFAAMAS), pp. 1031–1038 (2008)
24. Rahwan, I., Larson, K., Tohmé, F.: A characterisation of strategy-proofness for grounded argumentation semantics. In: Proceedings of the 21st International Joint Conference on Artificial Intelligence (IJCAI 2009), pp. 251–256 (2009)
25. Fan, X., Toni, F.: Mechanism design for argumentation-based persuasion. In: Computational Models of Argument (COMMA 2012), pp. 322–333 (2012)
26. Oren, N., Atkinson, K., Li, H.: Group persuasion through uncertain audience modelling. In: Proceedings of the International Comference on Computational Models of Argument (COMMA 2012), pp. 350–357 (2012)
27. Hunter, A.: Towards higher impact argumentation. In: Proceedings of the 19th National Conference on Artificial Intelligence (AAAI 2004), pp. 275–280. MIT Press (2004)
28. Hunter, A.: Making argumentation more believable. In: Proceedings of the 19th National Conference on Artificial Intelligence (AAAI 2004), pp. 269–274. MIT Press (2004)

On the Implementation of a Multiple Output Algorithm for Defeasible Argumentation

Teresa Alsinet[1], Ramón Béjar[1], Lluis Godo[2], and Francesc Guitart[1]

[1] Department of Computer Science – University of Lleida
Jaume II, 69 – 25001 Lleida, Spain
{tracy,ramon,fguitart}@diei.udl.cat
[2] Artificial Intelligence Research Institute (IIIA-CSIC)
Campus UAB - 08193 Bellaterra, Barcelona, Spain
godo@iiia.csic.es

Abstract. In a previous work we defined a recursive warrant semantics for Defeasible Logic Programming based on a general notion of collective conflict among arguments. The main feature of this recursive semantics is that an output of a program is a pair consisting of a set of warranted and a set of blocked formulas. A program may have multiple outputs in case of circular definitions of conflicts among arguments. In this paper we design an algorithm for computing each output and we provide an experimental evaluation of the algorithm based on two SAT encodings defined for the two main combinatorial subproblems that arise when computing warranted and blocked conclusions for each output.

1   Introduction and Motivation

Defeasible Logic Programming (DeLP) [8] is a formalism that combines techniques of both logic programming and defeasible argumentation. As in logic programming, knowledge is represented in DeLP using facts and rules; however, DeLP also provides the possibility of representing defeasible knowledge under the form of weak (defeasible) rules, expressing reasons to believe in a given conclusion. In DeLP, a conclusion succeeds in a program if it is warranted, i.e., if there exists an argument (a consistent set of defeasible rules) that, together with non-defeasible rules and facts, entails the conclusion, and moreover, this argument is found to be undefeated by a dialectical analysis procedure. This builds a dialectical tree containing all arguments that challenge this argument, and all counterarguments that challenge those arguments, and so on, recursively.

In [1] we defined a new recursive semantics for DeLP based on a general notion of collective (non-binary) conflict among arguments. In this framework, called *Recursive DeLP* (R-DeLP for short), an output (or extension) of a program is a pair consisting of a set of warranted and a set of blocked formulas. Arguments for both warranted and blocked formulas are recursively based on warranted formulas but, while warranted formulas do not generate any collective conflict, blocked conclusions do. Formulas that are neither warranted nor blocked correspond to rejected formulas. The key feature that our warrant recursive semantics addresses is the *closure under subarguments postulate* recently proposed by Amgoud[4], claiming that if an argument is excluded from an output, then all the arguments built on top of it should also be excluded from that output.

W. Liu, V.S. Subrahmanian, and J. Wijsen (Eds.): SUM 2013, LNAI 8078, pp. 71–77, 2013.
© Springer-Verlag Berlin Heidelberg 2013

Then, in case of circular definitions of conflict among arguments, the recursive semantics for warranted conclusions may result in multiple outputs for R-DeLP programs.

In this paper, after overviewing in Section 2 the main elements of the warrant recursive semantics for R-DeLP, in Section 3 we design an algorithm for computing every output for R-DeLP programs with multiple outputs, and in Section 4 we present empirical results. These are obtained with an implementation of the algorithm based on two SAT encodings defined in [2] for the two main combinatorial subproblems that arise when computing warranted and blocked conclusions for each output for an R-DeLP program, so that we can take profit of existing state-of-the-art SAT solvers for solving instances of big size.

2 Preliminaries on R-DeLP

The *language* of R-DeLP [1], denoted $\mathcal{L}$, is inherited from the language of logic programming, including the notions of atom, literal, rule and fact. Formulas are built over a finite set of propositional variables $\{p, q, \ldots\}$ which is extended with a new (negated) atom "$\sim p$" for each original atom p. Atoms of the form p or $\sim p$ will be referred as literals.[1] *Formulas* of $\mathcal{L}$ consist of rules of the form $Q \leftarrow P_1 \wedge \ldots \wedge P_k$, where $Q, P_1, \ldots, P_k$ are literals. A fact will be a rule with no premises. We will also use the name *clause* to denote a rule or a fact. The R-DeLP framework is based on the propositional logic $(\mathcal{L}, \vdash)$ where the inference operator $\vdash$ is defined by instances of the modus ponens rule of the form: $\{Q \leftarrow P_1 \wedge \ldots \wedge P_k, P_1, \ldots, P_k\} \vdash Q$. A set of clauses Γ will be deemed as *contradictory*, denoted $\Gamma \vdash \perp$, if , for some atom q, $\Gamma \vdash q$ and $\Gamma \vdash \sim q$.

An R-DeLP *program* $\mathcal{P}$ is a tuple $\mathcal{P} = (\Pi, \Delta)$ over the logic $(\mathcal{L}, \vdash)$, where $\Pi, \Delta \subseteq \mathcal{L}$, and $\Pi \nvdash \perp$. Π is a finite set of clauses representing strict knowledge (information we take for granted they hold true), Δ is another finite set of clauses representing the defeasible knowledge (formulas for which we have reasons to believe they are true).

The notion of *argument* is the usual one. Given an R-DeLP program $\mathcal{P}$, an argument for a literal (conclusion) Q of $\mathcal{L}$ is a pair $\mathcal{A} = \langle A, Q \rangle$, with $A \subseteq \Delta$ such that $\Pi \cup A \nvdash \perp$, and A is minimal (with respect to set inclusion) such that $\Pi \cup A \vdash Q$. If $A = \emptyset$, then we will call $\mathcal{A}$ a s-argument (s for strict), otherwise it will be a d-argument (d for defeasible). The notion of *subargument* is referred to d-arguments and expresses an incremental proof relationship between arguments which is defined as follows. Let $\langle B, Q \rangle$ and $\langle A, P \rangle$ be two d-arguments such that the minimal sets (with respect to set inclusion) $\Pi_Q \subseteq \Pi$ and $\Pi_P \subseteq \Pi$ such that $\Pi_Q \cup B \vdash Q$ and $\Pi_P \cup A \vdash P$ verify that $\Pi_Q \subseteq \Pi_P$. Then, $\langle B, Q \rangle$ is a *subargument* of $\langle A, P \rangle$, written $\langle B, Q \rangle \sqsubset \langle A, P \rangle$, when either $B \subset A$ (strict inclusion for defeasible knowledge), or $B = A$ and $\Pi_Q \subset \Pi_P$ (strict inclusion for strict knowledge). More generally, we say that $\langle B, Q \rangle$ is a *subargument of a set of arguments* G, written $\langle B, Q \rangle \sqsubset G$, if $\langle B, Q \rangle \sqsubset \langle A, P \rangle$ for some $\langle A, P \rangle \in G$. A literal Q of $\mathcal{L}$ is called *justifiable conclusion* with respect to $\mathcal{P}$ if there exists an argument for Q, i.e. there exists $A \subseteq \Delta$ such that $\langle A, Q \rangle$ is an argument.

The warrant recursive semantics for R-DeLP is based on the following notion of collective conflict. Let $\mathcal{P} = (\Pi, \Delta)$ be an R-DeLP program and let $W \subseteq \mathcal{L}$ be a set

---

[1] For a given literal Q, we write $\sim Q$ to denote "$\sim q$" if $Q = q$ and "q" if $Q = \sim q$.

of conclusions. We say that a set of arguments $\{\langle A_1, Q_1 \rangle, \dots, \langle A_k, Q_k \rangle\}$ *minimally conflicts* with respect to W iff the two following conditions hold: (i) the set of argument conclusions $\{Q_1, \dots, Q_k\}$ is contradictory with respect to W, i.e. it holds that $\Pi \cup W \cup \{Q_1, \dots, Q_k\} \vdash \bot$; and (ii) the set $\{\langle A_1, Q_1 \rangle, \dots, \langle A_k, Q_k \rangle\}$ is minimal with respect to set inclusion satisfying (i), i.e. if $S \subsetneq \{Q_1, \dots, Q_k\}$, then $\Pi \cup W \cup S \not\vdash \bot$.

An *output for an R-DeLP program* $\mathcal{P} = (\Pi, \Delta)$ is any pair (*Warr, Block*), where *Warr* $\cap$ *Block* $= \emptyset$ and $\{Q \mid \Pi \vdash Q\} \subseteq$ *Warr*, satisfying the following recursive constraints:

1. $P \in$ *Warr* $\cup$ *Block* iff there exists an argument $\langle A, P \rangle$ such that for every $\langle B, Q \rangle \sqsubset \langle A, P \rangle$, $Q \in$ *Warr*. In this case we say that the argument $\langle A, Q \rangle$ is *valid* with respect to *Warr*.
2. For each valid argument $\langle A, Q \rangle$:
   - $Q \in$ *Block* whenever there exists a set of valid arguments G such that (i) $\langle A, Q \rangle \not\sqsubseteq G$, and (ii) $\{\langle A, Q \rangle\} \cup G$ minimally conflicts with respect to the set $W = \{P \mid \langle B, P \rangle \sqsubset G \cup \{\langle A, Q \rangle\}\}$.
   - otherwise, $Q \in$ *Warr*.

In [1] we showed that, in case of some circular definitions of conflict among arguments, the output of an R-DeLP program may be not unique, that is, there may exist several pairs (*Warr, Block*) satisfying the above conditions for a given R-DeLP program. Following the approach of Pollock [9], circular definitions of conflict were formalized by means of what we called *warrant dependency graphs*. A warrant dependency graph represents (i) support relations of almost valid arguments with respect to valid arguments and (ii) conflict relations of valid arguments with respect to almost valid arguments. An almost valid argument is an argument based on a set of valid arguments and whose status is warranted or blocked (but not rejected), whenever every valid argument in the set is warranted, and rejected, otherwise. Then, a cycle in a warrant dependency graph represents a circular definition of conflict among a set of arguments.

3 Computing the Set of Outputs for an R-DeLP Program

From a computational point of view, an output for an R-DeLP program can be computed by means of a recursive procedure, starting with the computation of warranted conclusions from strict clauses and recursively going from warranted conclusions to defeasible arguments based on them. Next we design an algorithm implementing this procedure for computing warranted and blocked conclusions by checking the existence of conflicts between valid arguments and cycles at some warrant dependency graph.

The algorithm R-DeLP outputs first computes the set of warranted conclusions form the set of strict clauses Π. Then, computes the set *VA* of valid arguments with respect to the strict part, i.e. arguments with an empty set of subarguments. The recursive procedure extension receives as input the current partially computed output (W, B) and the set of valid arguments *VA* and dynamically updates the set *VA* depending on new warranted and blocked conclusions and the appearance of cycles in some warrant dependence graph. When a cycle is found in a warrant dependence graph, each valid argument of the cycle can lead to a different output. Then, the procedure extension selects one valid argument of the cycle and recursively computes the resulting output by warranting the selected argument. The procedure `extension` finishes when the status for every valid argument of the current output is computed. When an R-DeLP program has multiple outputs, each output is stored in the set of outputs O.

Algorithm `R-DeLP outputs`

Input $\mathcal{P} = (\Pi, \Delta)$: An R-DeLP program

Output O: Set of outputs for $\mathcal{P}$

Variables
    (W, B): Current output for $\mathcal{P}$
    VA: Set of valid arguments w.r.t. the current set of warranted conclusions W

Method
    $O := \emptyset$;
    $W := \{Q \mid \Pi \vdash Q\}$;
    $B := \emptyset$;
    $VA := \{\langle A, Q\rangle \mid \langle A, Q\rangle$ is valid w.r.t. $W\}$;
    `extension`$((W, B), VA, O)$

end algorithm `R-DeLP outputs`

Procedure `extension` (**input** (W, B); **input** VA; **input_output** O)

Variables
    W_{ext}: Extended set of warranted conclusions
    VA_{ext}: Extended set of valid arguments
    is_leaf: Boolean

Method
    $is\_leaf :=$ `true`;
    while $(VA \neq \emptyset$ **and** $is\_leaf =$ `true`) **do**
        while $(\exists \langle A, Q\rangle \in VA \mid \neg$ `cycle`$(\langle A, Q\rangle, VA, W,$ `almost_valid`$(VA, (W, B)))$ **and**
            $\neg$ `conflict`$(\langle A, Q\rangle, VA, W,$ `not_dependent`$(\langle A, Q\rangle,$ `almost_valid`$(VA, (W, B))))$ **do**
            $W := W \cup \{Q\}$;
            $VA := VA \backslash \{\langle A, Q\rangle\} \cup \{\langle C, P\rangle \mid \langle C, P\rangle$ is valid w.r.t. $W\}$
        end while
        $I := \{\langle A, Q\rangle \in VA \mid$ `conflict`$(\langle A, Q\rangle, VA, W, \emptyset)\}$;
        $B := B \cup \{Q \mid \langle A, Q\rangle \in I\}$;
        $VA := VA \backslash I$;
        $J := \{\langle A, Q\rangle \in VA \mid$ `cycle`$(\langle A, Q\rangle, VA, W,$ `almost_valid`$(VA, (W, B)))\}$
        for each argument $(\langle A, Q\rangle \in J)$ **do**
            $W_{ext} := W \cup \{Q\}$;
            $VA_{ext} := VA \backslash \{\langle A, Q\rangle\} \cup \{\langle C, P\rangle \mid \langle C, P\rangle$ is valid w.r.t. $W_{ext}\}$;
            `extension`$((W_{ext}, B), VA_{ext}, O)$
        end for
        if $(J \neq \emptyset)$ **then** $is\_leaf :=$ `false`
    end while
    if $((W, B) \notin O$ **and** $is\_leaf =$ `true`) **then** $O := O \cup \{(W, B)\}$
end procedure `extension`

The function `almost_valid` computes the set of almost valid arguments based on some valid arguments in VA. The function `not_dependent` computes the set of almost valid arguments which do not depend on $\langle A, Q\rangle$. The function `conflict` has two different functionalities. On the one hand, the function `conflict` checks conflicts among the argument $\langle A, Q\rangle$ and the set VA of valid arguments, and thus, every valid argument involved in a conflict is blocked. On the other had, the function `conflict` checks possible conflicts among the argument $\langle A, Q\rangle$ and the set VA of valid arguments extended with the set of almost valid arguments whose supports depend on some argument in $VA \backslash \{\langle A, Q\rangle\}$, and thus, every valid argument with options to be involved in a conflict remains as valid. Finally, the function `cycle` checks the existence of a cycle in the warrant dependency graph for the set of valid arguments VA and the set of almost valid arguments based on some valid arguments in VA.

One of the main advantages of the warrant recursive semantics is from the implementation point of view. In order to determine the warrant status of an argument of a given program, warrant semantics based on dialectical trees, like DeLP [5], might explore the entire set of arguments of a program in order to present an exhaustive synthesis of the relevant chains of pros and cons for a given conclusion. In the worst case, this could be an exponential number of arguments with respect to the number of program rules. To avoid the systematic exploration, in [6] an improved algorithm for computing dialectical trees in a depth-first fashion was defined, where an evaluation criteria of arguments (based on dialectical constraints) is used as an heuristic to prune the search space. Our approach is a bit different, it does not compute the entire set of arguments for a given literal but instead the search is driven towards the computation of at most a valid argument per each literal. In fact, for every output, our algorithm can be implemented to work in polynomial space since for each literal we need to keep in memory at most one valid argument. Analogously, function not_dependent can be implemented to generate at most one almost valid argument not based on $\langle A, Q \rangle$ for a given literal. The only function that in the worst case can need to explore an exponential number of arguments is cycle, but we showed [2] that whenever cycle returns true for $\langle A, Q \rangle$, then a conflict will be detected with the set of almost valid arguments which do not depend on $\langle A, Q \rangle$. Moreover, the set of valid arguments J computed by function cycle can also be computed by checking the stability of the set of valid arguments after two consecutive iterations, so it is not necessary to explicitly compute dependency graphs.

4   Empirical Results

In order to compute the sets of warranted and blocked conclusions for every output (extension) the procedure extension computes two main queries during its execution: i) whether an argument is almost valid and ii) whether there is a conflict for a valid argument. SAT encodings were proposed in [2] to resolve them with a SAT solver. [2]

In this paper we study the average number of outputs for R-DeLP instances and the median computational cost of solving them with the R-DeLP outputs algorithm, as the instances size increase with different instances characteristics. The main algorithm has been implemented with python, and for solving the SAT encodings, we have used the solver MiniSAT [7]. An on-line web based implementation of the R-DeLP argumentation framework is available at the URL: http://arinf.udl.cat/rp-delp.

To generate R-DeLP problem instances with different sizes and characteristics, we have used the generator algorithm described in [3]. We generate test-sets of instances with different number of variables (V): $\{15, 20, 25, 30\}$[3] and with clauses with one or two literals. [4] For each number of variables, we generate three sets of instances, each

---

[2] The set *VA* of valid arguments can be easily updated whenever a new conclusion is warranted.

[3] Notice that the total number of literals is twice the number of variables.

[4] In [3] we considered test-sets of instances with a maximum clause length parameter (ML): $\{2, 4, 6\}$. Since the experimental results showed that increasing the number of literals per clause also increases the number of blocked conclusions, in this paper we have only considered the case of $ML = 2$ which should in principle favor the appearance of cycles.

one with a different ratio of clauses to variables (C/V): $\{2, 4, 6\}$. From all the clauses of an instance, a 10% of them are considered in the strict part of the program (Π) and a 90% of them are considered in the defeasible part (Δ). [5]

Table 1 shows the experimental results obtained for our test-sets. So far, we have computed the average number of outputs per instance ($\#\,O$), the average number of warrants per output, the average number of warrants in the intersection of the set of outputs and the median time for solving the instances. The results show that even for a small number of variables V and a small ratio C/V we can have instances with multiple outputs. Observe that although the average number of outputs is not too different between all the test-sets, the complexity of solving the instances seems to increase exponentially as either V or C/V increases. We believe this is mainly due to an increase in the complexity of deciding the final status (warranted or blocked) of each literal for each output.

Table 1. Experimental results for the R-DeLP outputs algorithm

V	C/V	$\#\,O$	# Warrants per output	# Warrants in intersection	Time (s.)
	2	1.04	7.14	7.1	0.906
15	4	1.31	7.28	6.93	7.96
	6	5.40	6.43	5.90	19.11
	2	1.06	9.89	9.82	1.93
20	4	1.65	9.63	9.31	28.89
	6	1.44	9.09	8.81	38.87
	2	1.10	11.47	11.38	5.09
25	4	2.44	11.72	10.74	76.31
	6	1.90	8.04	7.50	151.31
	2	1.08	12.56	12.5	9.20
30	4	1.81	12.16	11.56	142.49
	6	1.89	11.02	9.92	227.97

As future work we propose to study the average number of blocked conclusions per output and to parameterize the maximum number of literals per clause which would give us an idea if the computation time is higher because there is a relationship between the number of blocked and warranted conclusions. We also propose to extend the R-DeLP outputs algorithm to the case of multiple levels for defeasible facts and rules.

Acknowledgments. The authors acknowledge the Spanish projects ARINF (TIN2009-14704-C03-01), TASSAT (TIN2010-20967-C04-03) and EdeTRI (TIN2012-39348-C02-01).

---

[5] These ratios were selected given the experimental results in [3] with single output programs, they gave non-trivial instances in the sense of being computationally hard to solve.

References

1. Alsinet, T., Béjar, R., Godo, L.: A characterization of collective conflict for defeasible argumentation. In: COMMA 2010, pp. 27–38 (2010)
2. Alsinet, T., Béjar, R., Godo, L., Guitart, F.: Maximal ideal recursive semantics for defeasible argumentation. In: Benferhat, S., Grant, J. (eds.) SUM 2011. LNCS, vol. 6929, pp. 96–109. Springer, Heidelberg (2011)
3. Alsinet, T., Béjar, R., Godo, L., Guitart, F.: Using answer set programming for an scalable implementation of defeasible argumentation. In: ICTAI 2012, pp. 1016–1021 (2012)
4. Amgoud, L.: Postulates for logic-based argumentation systems. In: ECAI 2012 Workshop WL4AI, pp. 59–67 (2012)
5. Cecchi, L., Fillottrani, P., Simari, G.: On the complexity of DeLP through game semantics. In: NMR 2006, pp. 386–394 (2006)
6. Chesñevar, C.I., Simari, G.R., Godo, L.: Computing dialectical trees efficiently in possibilistic defeasible logic programming. In: Baral, C., Greco, G., Leone, N., Terracina, G. (eds.) LPNMR 2005. LNCS (LNAI), vol. 3662, pp. 158–171. Springer, Heidelberg (2005)
7. Eén, N., Sörensson, N.: An extensible SAT-solver. In: Giunchiglia, E., Tacchella, A. (eds.) SAT 2003. LNCS, vol. 2919, pp. 502–518. Springer, Heidelberg (2004)
8. García, A., Simari, G.R.: Defeasible Logic Programming: An Argumentative Approach. Theory and Practice of Logic Programming 4(1), 95–138 (2004)
9. Pollock, J.L.: A recursive semantics for defeasible reasoning. In: Rahwan, Simari (eds.) Argumentation in Artificial Intelligence, pp. 173–198. Springer (2009)

A Formal Characterization of the Outcomes
of Rule-Based Argumentation Systems

Leila Amgoud and Philippe Besnard

IRIT – CNRS
118, route de Narbonne
31062, Toulouse Cedex 09

Abstract. Rule-based argumentation systems are developed for reasoning about
defeasible information. As a major feature, their logical language distinguishes
between strict rules and defeasible ones. This paper presents the first study on
the outcomes of such systems under various semantics such as naive, stable, preferred, ideal and grounded. For each of these semantics, it characterizes both the
extensions and the set of plausible inferences drawn by these systems under a few
intuitive postulates.

1 Introduction

There are two major categories of instantiations of Dung's abstract argumentation
framework [4]. A category uses *deductive logics* (such as propositional logic [2,6] or
Tarskian logics [1]). The second category uses *rule-based languages* [3,5,7] which distinguish between *facts*, *strict rules* (they encode strict information), and *defeasible rules*
(they describe general behavior with exceptional cases). Despite the popularity of rule-
based argumentation systems, the results they return have not been characterized yet.
The following questions are still open:

- what are the underpinnings of the extensions under various semantics?
- do Dung's semantics return different results as at the abstract level?
- what is the number of extensions a system may have under a given semantics?
- what are the plausible conclusions with such systems?

In this paper, we answer all the above questions. We start with a knowledge base
called a *theory* (a set of facts, a set of strict rules and a set of defeasible rules), we
define a notion of a *derivation schema* which we use to generate arguments from the
theory. For the sake of generality, the attack relation is left unspecified. We extend the
list of postulates proposed in [3] with three new postulates. We investigate outputs of
rule-based argumentation systems that satisfy all the postulates. We show that naive extensions return maximal *options* of the theory (an option being a sub-theory that gathers
all the facts and strict rules, and a maximal -up to consistency- set of defeasible rules
that do not conflict with the strict part). Every maximal option gives birth to a naive
extension. Furthermore, the set of plausible conclusions under the naive semantics contains all the conclusions that are drawn from all the maximal options. Stable extensions
return maximal options but not necessarily all of them, it depends on the attack relation

W. Liu, V.S. Subrahmanian, and J. Wijsen (Eds.): SUM 2013, LNAI 8078, pp. 78–91, 2013.

at work. Should not all maximal options be picked as stable extensions, defining an attack relation that discard exactly the spurious ones turns out be tricky. The same results hold for preferred semantics. We characterize both ideal and grounded extensions.

2   Rule-Based Argumentation Systems

In what follows, we consider the language used in [3]. Let $\mathcal{L}$ is a set of *literals*, i.e., atoms or negation of atoms. The negation of an atom x from $\mathcal{L}$ is denoted $\neg x$. Three kinds of information ($x, x_1...x_n$ denoting literals in $\mathcal{L}$) are distinguished:

- *Facts*, which are elements of $\mathcal{L}$
- *Strict rules*, which are of the form $x_1, \ldots, x_n \rightarrow x$
- *Defeasible rules*, which are of the form $x_1, \ldots, x_n \Rightarrow x$

Throughout the text, rules are named $r_1, r_2, \ldots$ For each rule $r = x_1, \ldots, x_n \rightarrow x$ (as well as $r = x_1, \ldots, x_n \Rightarrow x$), the *head* of the rule is $\text{Head}(r) = x$ and the *body* of the rule is $\text{Body}(r) = \{x_1, \ldots, x_n\}$. A strict rule expresses general information that has no exception, e.g. "penguins cannot fly" whereas a defeasible rule expresses general information that may have exceptions, e.g. "birds can fly".

Definition 1 (Theory). *A theory is a triple* $\mathcal{T} = (\mathcal{F}, \mathcal{S}, \mathcal{D})$ *where* $\mathcal{F}$ *is a set of facts and* $\mathcal{S}$ *(resp.* $\mathcal{D}$*) is a set of strict (resp. defeasible) rules.*

Notation. Let $\mathcal{T} = (\mathcal{F}, \mathcal{S}, \mathcal{D})$ and let $\mathcal{T}' = (\mathcal{F}', \mathcal{S}', \mathcal{D}')$ be two theories. We say that $\mathcal{T}$ is a *sub-theory* of $\mathcal{T}'$, written $\mathcal{T} \sqsubseteq \mathcal{T}'$, iff $\mathcal{F} \subseteq \mathcal{F}'$ and $\mathcal{S} \subseteq \mathcal{S}'$ and $\mathcal{D} \subseteq \mathcal{D}'$. The relation $\sqsubset$ is the strict version of $\sqsubseteq$ (i.e., it is the case that at least one of the three inclusions is strict).

The notion of consistency is defined as follows.

Definition 2 (Consistency). *A set* $X \subseteq \mathcal{L}$ *is consistent iff* $\nexists x, y \in X$ *s.t.* $x = \neg y$. *It is* inconsistent *otherwise.*

Assumption 1. *The body of every (strict/defeasible) rule is finite and* not empty. *Moreover, for each rule* r, $\text{Body}(r) \cup \{\text{Head}(r)\}$ *is consistent. We say that* r *is consistent.*

The notion of a *derivation schema* generalizes derivations as defined in [5,8] and others. It shows how literals can follow from a theory.

Definition 3 (Derivation schema). *Let* $\mathcal{T} = (\mathcal{F}, \mathcal{S}, \mathcal{D})$ *be a theory and* $x \in \mathcal{L}$. *A* derivation schema *for* x *from* $\mathcal{T}$ *is a finite sequence* $d = \langle(x_1, r_1), \ldots, (x_n, r_n)\rangle$ *s.t.*

- $x_n = x$
- *for* $i = 1 \ldots n$,
  - $x_i \in \mathcal{F}$ *and* $r_i = \emptyset$, *or*
  - $r_i \in \mathcal{S} \cup \mathcal{D}$ *and* $\text{Head}(r_i) = x_i$ *and* $\text{Body}(r_i) \subseteq \{x_1, .., x_{i-1}\}$

$\text{Seq}(d) = \{x_1, \ldots, x_n\}.$
$\text{Facts}(d) = \{x_i \mid i \in \{1, \ldots, n\}, r_i = \emptyset\}.$
$\text{Strict}(d) = \{r_i \mid i \in \{1, \ldots, n\}, r_i \in \mathcal{S}\}.$
$\text{Def}(d) = \{r_i \mid i \in \{1, \ldots, n\}, r_i \in \mathcal{D}\}.$

Notation. In order to improve readability, we somehow abuse the notation in derivation schemata: We use the name of the rules instead of the rules themselves.

A derivation schema is not necessarily consistent (such as (7) below), as it may contain opposite literals in the form $x_i = \neg x_j$ for some i and j (this is in accordance with Definition 2).

Example 1. *Consider $\mathcal{T}_1$ such that $\mathcal{F}_1$, $\mathcal{S}_1$, $\mathcal{D}_1$ are as follows.*

$$\mathcal{F}_1 \begin{cases} p \\ q \end{cases} \qquad \mathcal{S}_1 \begin{cases} p \to s & (r_1) \\ q \to \neg s & (r_2) \\ p, s \to u & (r_3) \end{cases} \qquad \mathcal{D}_1 \begin{cases} \neg s \Rightarrow t & (r_4) \\ t, u \Rightarrow \neg v & (r_5) \\ p \Rightarrow q & (r_6) \end{cases}$$

Each of (1)–(7) below is a derivation schema from $\mathcal{T}_1$

$$\langle (p, \emptyset) \rangle \tag{1}$$
$$\langle (q, \emptyset), (\neg s, r_2) \rangle \tag{2}$$
$$\langle (p, \emptyset), (s, r_1), (u, r_3) \rangle \tag{3}$$
$$\langle (p, \emptyset), (s, r_1), (p, \emptyset), (u, r_3) \rangle \tag{4}$$
$$\langle (p, \emptyset), (q, \emptyset), (s, r_1), (u, r_3) \rangle \tag{5}$$
$$\langle (p, \emptyset), (q, r_6), (\neg s, r_2) \rangle \qquad \cdot \tag{6}$$
$$\langle (p, \emptyset), (q, \emptyset), (\neg s, r_2), (s, r_1), (u, r_3), (t, r_4), (\neg v, r_5) \rangle \tag{7}$$

A derivation schema may not be ($\subseteq$-)minimal. There are two reasons for that:

- repeating pairs (x_i, r_i) as in derivation (4) $((p, \emptyset)$ is repeated twice),
- involving literals that do not serve towards inferring the conclusion x, as is illustrated by (5) (q is of no use there). The derivation schema fails thus to be *focussed*.

Definition 4 (Minimal/focussed derivation schema). *A derivation schema for x from $\mathcal{T}$ is minimal iff none of its proper subsequences is a derivation schema for x from $\mathcal{T}$. It is focussed iff it can be reduced to a minimal one by just deleting repeated pairs (x_i, r_i).*

Property 1. *Let $\mathcal{T} = (\mathcal{F}, \mathcal{S}, \mathcal{D})$ be a theory. A derivation schema $d = \langle (x_1, r_1), \ldots, (x_n, r_n) \rangle$ from $\mathcal{T}$ is minimal iff d is focussed and the literals $x_1, \ldots, x_n$ are pairwise distinct.*

Notation. $\text{CN}(\mathcal{T})$ denotes the set of all literals that have a derivation schema from $\mathcal{T}$. We call $\text{CN}(\mathcal{T})$ the potential consequences drawn from $\mathcal{T}$ (for short, consequences) but they need not be definitive as they may be dismissed by opposite conclusions.

Property 2. *Let* $\mathcal{T} = (\mathcal{F}, \mathcal{S}, \mathcal{D})$ *be a theory.*

- $\mathcal{F} \subseteq \mathtt{CN}(\mathcal{T}) \subseteq \mathcal{F} \cup \{\mathtt{Head}(r) \mid r \in \mathcal{S} \cup \mathcal{D}\} \subseteq \mathcal{L}$
- *If* $\mathcal{T}$ *is finite, then* $\mathtt{CN}(\mathcal{T})$ *is finite*
- $\mathcal{F} = \emptyset$ *iff* $\mathtt{CN}(\mathcal{T}) = \emptyset$
- *If* d *is a derivation schema from* $\mathcal{T}$, $\mathtt{Seq}(d) \subseteq \mathtt{CN}(\mathcal{T})$

Some rules may not be *activated* (i.e., their body has no derivation schema). Let us consider the following example.

Example 2. *Let* $\mathcal{T}_2 = (\mathcal{F}_2, \mathcal{S}_2, \mathcal{D}_2)$ *such that*

$$\mathcal{F}_2 \begin{cases} p \\ q \end{cases} \qquad \mathcal{S}_2 \begin{cases} p \to t & (r_1) \\ q \to t & (r_2) \\ s \to u & (r_3) \end{cases} \qquad \mathcal{D}_2 \begin{cases} p \Rightarrow q & (r_4) \\ u \Rightarrow v & (r_5) \end{cases}$$

There are rules here whose head is not a consequence of $\mathcal{T}_2$. $\mathtt{CN}(\mathcal{T}_2) = \{p, q, t\} \subset \{p, q, t, u, v\} = \mathcal{F}_2 \cup \mathtt{Head}(\mathcal{S}_2 \cup \mathcal{D}_2)$.

It is also easy to show that $\mathtt{CN}$ is monotonic.

Property 3. *If* $\mathcal{T} \sqsubseteq \mathcal{T}'$ *then* $\mathtt{CN}(\mathcal{T}) \subseteq \mathtt{CN}(\mathcal{T}')$.

The backbone of an argumentation system is naturally the notion of *arguments*. They are built from a theory using the notion of derivation schema as follows.

Definition 5 (Argument). *Let* $\mathcal{T} = (\mathcal{F}, \mathcal{S}, \mathcal{D})$ *be a theory. An* argument *defined from* $\mathcal{T}$ *is a pair* (d, x) *s.t.*

- $x \in \mathcal{L}$
- d *is a derivation schema for* x *from* $\mathcal{T}$
- $\mathtt{Seq}(d)$ *is consistent*
- $\nexists \mathcal{T}' \sqsubset (\mathtt{Facts}(d), \mathtt{Strict}(d), \mathtt{Def}(d))$ *s.t.* $x \in \mathtt{CN}(\mathcal{T}')$

An argument (d, x) *is* strict *iff* $\mathtt{Def}(d) = \emptyset$.

Notation. If $a = (d, x)$ is an argument then $\mathtt{Conc}(a) = x$. For a set $\mathcal{E}$ of arguments, $\mathtt{Concs}(\mathcal{E}) = \{x \mid (d, x) \in \mathcal{E}\}$. $\mathtt{Arg}(\mathcal{T})$ is the set of all the arguments defined from $\mathcal{T}$. For a set $\mathcal{E}$ of arguments,

$$\mathtt{Th}(\mathcal{E}) = (\bigcup_{(d,x) \in \mathcal{E}} \mathtt{Facts}(d), \bigcup_{(d,x) \in \mathcal{E}} \mathtt{Strict}(d), \bigcup_{(d,x) \in \mathcal{E}} \mathtt{Def}(d)).$$

Theorem 1. *Let* $\mathcal{T}$ *be a theory. For all consistent sequence* $d = \langle (x_1, r_1), \ldots, (x_n, r_n) \rangle$ *from* $\mathcal{T}$, *the following two statements are equivalent:*

- (d, x) *is an argument (from* $\mathcal{T}$*)*
- d *is a focussed derivation schema from* $\mathcal{T}$ *s.t.* $x = x_n$

Definition 6 (Sub-argument). *An argument* (d, x) *is a* sub-argument *of* (d', x') *iff* $(\mathtt{Facts}(d), \mathtt{Strict}(d), \mathtt{Def}(d)) \sqsubseteq (\mathtt{Facts}(d'), \mathtt{Strict}(d'), \mathtt{Def}(d'))$.

Notation Sub(a) denotes the set of all sub-arguments of a.

Example 1 (Cont). The argument $(\langle (q, \emptyset), (\neg s, r_2) \rangle, \neg s)$ has two sub-arguments: $(\langle (q, \emptyset) \rangle, q)$ and itself. By contrast, $(\langle (q, \emptyset) \rangle, q)$ is not a sub-argument of $(\langle (p, \emptyset), (q, r_6) \rangle, q)$.

Clearly, if (d, x) is a sub-argument of (d', x') then $\mathrm{Seq}(d) \subseteq \mathrm{Seq}(d')$, but the converse is not true as shown next.

Example 2 (Cont). Arguments $a = (\langle (p, \emptyset), (t, r_1) \rangle, t)$ and $b = (\langle (p, \emptyset), (q, r_4), (t, r_2) \rangle, t)$ are s.t. $\mathrm{Seq}(a) = \{p, t\} \subseteq \{p, q, t\} = \mathrm{Seq}(b)$ but a is not a sub-argument of b.

From the monotonicity of CN, it follows that the construction of arguments is a monotonic process.

Proposition 1. *If $\mathcal{T} \sqsubseteq \mathcal{T}'$ then $\mathrm{Arg}(\mathcal{T}) \subseteq \mathrm{Arg}(\mathcal{T}')$.*

An argumentation system is defined as follows:

Definition 7 (Argumentation system). *An argumentation system (AS for short) defined over a theory $\mathcal{T} = (\mathcal{F}, \mathcal{S}, \mathcal{D})$ is a pair $\mathcal{H} = (\mathrm{Arg}(\mathcal{T}), \mathcal{R})$ where $\mathcal{R} \subseteq \mathrm{Arg}(\mathcal{T}) \times \mathrm{Arg}(\mathcal{T})$ is called an attack relation.*

In what follows, arguments are evaluated using semantics proposed in [4]. Before recalling them, let us first introduce the two requirements on which they are based.

Definition 8 (Conflict-freeness – Defence). *Let $\mathcal{H} = (\mathcal{A}, \mathcal{R})$ be an AS, $\mathcal{E} \subseteq \mathcal{A}$ and $a \in \mathcal{A}$.*

- $\mathcal{E}$ *is* conflict-free *iff $\nexists a, b \in \mathcal{E}$ s.t. $a \mathcal{R} b$.*
- $\mathcal{E}$ *defends a iff $\forall b \in \mathcal{A}$, if $b \mathcal{R} a$ then $\exists c \in \mathcal{E}$ s.t. $c \mathcal{R} b$.*

Definition 9 recalls the semantics of interest in the sequel.

Definition 9 (Acceptability semantics). *Let $\mathcal{H} = (\mathcal{A}, \mathcal{R})$ be an AS and $\mathcal{E} \subseteq \mathcal{A}$.*

- $\mathcal{E}$ *is a* naive *extensions iff it is a maximal (w.r.t. set $\subseteq$) conflict-free set.*
- $\mathcal{E}$ *is an* admissible *set iff it is conflict-free and defends all its elements.*
- $\mathcal{E}$ *is a* preferred *extension iff it is a maximal (w.r.t. set $\subseteq$) admissible set.*
- $\mathcal{E}$ *is a* stable *extension iff it is conflict-free and $\forall a \in \mathcal{A} \setminus \mathcal{E}$, $\exists b \in \mathcal{E}$ s.t. $b \mathcal{R} a$.*
- $\mathcal{E}$ *is a* grounded extension *iff it is a minimal (w.r.t. set $\subseteq$) set that is admissible and contains any argument it defends.*
- $\mathcal{E}$ *is an* ideal extension *iff it is the maximal (w.r.t. set $\subseteq$) admissible set which is part of any preferred extension.*

Notation. $\mathrm{Ext}_x(\mathcal{H})$ denotes the set of all the extensions of a system $\mathcal{H}$ under semantics x where $x \in \{n, p, s\}$ and n (resp. p, s) stands for naive (resp. preferred, stable).

Plausible conclusions are those common to all extensions.

Definition 10 (Plausible conclusions). *If* $\mathcal{H} = (\text{Arg}(\mathcal{T}), \mathcal{R})$ *is an AS built over a theory* $\mathcal{T}$, *the set of* plausible conclusions *of* $\mathcal{H}$ *is*

$$\text{Output}(\mathcal{H}) = \bigcap_{\mathcal{E}_i \in \text{Ext}_x(\mathcal{H})} \text{Concs}(\mathcal{E}_i).$$

From the above definitions, namely that of an argument, it follows that the plausible conclusions of an argumentation system are a subset of the consequences that follow wrt CN from the theory over which the system is built.

Property 4. *Let* $\mathcal{H} = (\text{Arg}(\mathcal{T}), \mathcal{R})$ *be an AS built over* $\mathcal{T}$. $\text{Output}(\mathcal{H}) \subseteq \text{CN}(\mathcal{T})$.

3   Postulates for Argumentation Systems

We present rationality postulates that any rule-based argumentation system should satisfy. The first two were already proposed in [3] and the others are new. The first postulate ensures that the set of conclusions of arguments of each extension is consistent. This is compatible with the fact that each extension represents a coherent position.

Postulate 1 (Consistency). *Let* $\mathcal{H} = (\text{Arg}(\mathcal{T}), \mathcal{R})$ *be an AS built over a theory* $\mathcal{T}$. *For all* $\mathcal{E} \in \text{Ext}_x(\mathcal{H})$, $\text{Concs}(\mathcal{E})$ *is consistent. We say that* $\mathcal{H}$ *satisfies consistency.*

It was shown in [3] that if an argumentation system $\mathcal{H}$ satisfies consistency, then its set $\text{Output}(\mathcal{H})$ of plausible conclusions is consistent as well.

Property 5 ([3]). *If an AS* $\mathcal{H}$ *satisfies consistency, then* $\text{Output}(\mathcal{H})$ *is consistent.*

The second postulate ensures that the extensions of an argumentation system are closed under strict rules. The idea is that if there is an argument with conclusion x in an extension and there exists a strict rule $x \to y$, then y should also be supported by an argument in the same extension.

Postulate 2 (Closure under strict rules). *Let* $\mathcal{H} = (\text{Arg}(\mathcal{T}), \mathcal{R})$ *be an AS built over a theory* $\mathcal{T}$. *For all* $\mathcal{E} \in \text{Ext}_x(\mathcal{H})$, $\text{Concs}(\mathcal{E}) = \text{CN}((\text{Concs}(\mathcal{E}), \mathcal{S}, \emptyset))$. *We say that* $\mathcal{H}$ *is* closed under strict rules.

It is known that if an argumentation system $\mathcal{H}$ is closed under strict rules, then its set $\text{Output}(\mathcal{H})$ is necessarily closed under strict rules.

Property 6 ([3]). *Let* $\mathcal{H}$ *be an AS built over a theory* $\mathcal{T} = (\mathcal{F}, \mathcal{S}, \mathcal{D})$. *If* $\mathcal{H}$ *is closed under strict rules, then* $\text{Output}(\mathcal{H}) = \text{CN}((\text{Output}(\mathcal{H}), \mathcal{S}, \emptyset))$.

It was also shown in [3] that a system that satisfies consistency and closure under strict rules satisfies *indirect* consistency.

Property 7 ([3]). *Let* $\mathcal{H}$ *be an AS built over a theory* $\mathcal{T} = (\mathcal{F}, \mathcal{S}, \mathcal{D})$. *If* $\mathcal{H}$ *satisfies consistency and is closed under strict rules, then for all* $\mathcal{E} \in \text{Ext}_x(\mathcal{H})$, $\text{CN}((\text{Concs}(\mathcal{E}), \mathcal{S}, \emptyset))$ *is consistent.*

We propose three new postulates. The first says that if an argument belongs to an extension, then all its sub-arguments should be in the extension. It means that an argument cannot be accepted in an extension if one of its sub-parts is rejected.

Postulate 3 (Closure under sub-arguments). *Let* $\mathcal{H} = (\text{Arg}(\mathcal{T}), \mathcal{R})$ *be an AS built over a theory* $\mathcal{T}$. *For all* $\mathcal{E} \in \text{Ext}_x(\mathcal{H})$, *if* $a \in \mathcal{E}$ *then* $\text{Sub}(a) \subseteq \mathcal{E}$. *We say that* $\mathcal{H}$ *is closed under sub-arguments.*

The following result characterizes the extensions of an argumentation system which is closed under sub-arguments.

Proposition 2. *If an AS* $\mathcal{H}$ *is closed under sub-arguments, then* $\forall \mathcal{E} \in \text{Ext}_x(\mathcal{H})$,

- $\text{Concs}(\mathcal{E}) = \text{CN}(\text{Th}(\mathcal{E}))$
- $\forall (d, x) \in \text{Arg}(\text{Th}(\mathcal{E}))$, $\text{Seq}(d) \subseteq \text{Concs}(\mathcal{E})$

Importantly, even when a system is closed under sub-arguments, the equality $\mathcal{E} = \text{Arg}(\text{Th}(\mathcal{E}))$ is not always true. This depends on the semantics as we will see later.

Proposition 3. *If an argumentation system* $\mathcal{H}$ *satisfies consistency and closure under sub-arguments, then* $\forall \mathcal{E} \in \text{Ext}_x(\mathcal{H})$, $\text{CN}(\text{Th}(\mathcal{E}))$ *is consistent.*

Since facts and strict rules are the "hard" part in a theory, it is natural that any strict argument should be in all extensions. This principle is applied in default logic [9].

Postulate 4 (Strict precedence). *Let* $\mathcal{H}$ *be an AS built over a theory* $\mathcal{T} = (\mathcal{F}, \mathcal{S}, \mathcal{D})$. *For all* $\mathcal{E} \in \text{Ext}_x(\mathcal{H})$, $\text{Arg}((\mathcal{F}, \mathcal{S}, \emptyset)) \subseteq \mathcal{E}$. *We say that* $\mathcal{H}$ *satisfies strict precedence.*

We show next that every argumentation system satisfying Postulate 4 infers all the conclusions that follow from the set of facts and the strict rules of a theory.

Proposition 4. *Let* $\mathcal{H}$ *be an AS built over a theory* $\mathcal{T} = (\mathcal{F}, \mathcal{S}, \mathcal{D})$. *If* $\mathcal{H}$ *satisfies strict precedence, then* $\mathcal{F} \subseteq \text{CN}((\mathcal{F}, \mathcal{S}, \emptyset)) \subseteq \text{Output}(\mathcal{H})$.

Next is an important result for the rest of our study: it says that if an argumentation system over a theory $\mathcal{T}$ satisfies Postulates 2, 3, 4, then the set of literals deduced from $\text{Th}(\mathcal{E})$, the theory of an extension $\mathcal{E}$, is exactly the one obtained from $\text{Th}(\mathcal{E})$ extended by all facts and strict rules of $\mathcal{T}$ which are not in $\text{Th}(\mathcal{E})$.

Proposition 5. *Let* $\mathcal{H}$ *be an argumentation system built over a theory* $\mathcal{T} = (\mathcal{F}, \mathcal{S}, \mathcal{D})$. *If* $\mathcal{H}$ *satisfies postulates 2, 3, 4, then for all* $\mathcal{E} \in \text{Ext}_x(\mathcal{H})$,

$$\text{CN}(\text{Th}(\mathcal{E})) = \text{CN}((\mathcal{F}, \mathcal{S}, \bigcup_{(d,x) \in \mathcal{E}} \text{Def}(d))).$$

The last postulate ensures a form of completeness of the extensions. It says that if the sequence of an argument is part of the conclusions of a given extension, then the argument (Definition 5 ensures consistency) should belong to the extension. Informally: If each step in the argument is good enough to be in the extension, then so is the argument itself.

Postulate 5 (Exhaustiveness). *Let* $\mathcal{H} = (\text{Arg}(\mathcal{T}), \mathcal{R})$ *be an AS built over a theory* $\mathcal{T} = (\mathcal{F}, \mathcal{S}, \mathcal{D})$. *For all* $\mathcal{E} \in \text{Ext}_x(\mathcal{H})$, *for all* $(d, x) \in \text{Arg}(\mathcal{T})$, *if* $\text{Seq}(d) \subseteq \text{Concs}(\mathcal{E})$, *then* $(d, x) \in \mathcal{E}$.

The extensions (under any semantics) of any argumentation system that satisfies exhaustiveness and closure under sub-arguments are closed in terms of arguments.

Proposition 6. *If an AS* $\mathcal{H}$ *is closed under sub-arguments and satisfies the exhaustiveness postulate, then* $\forall \mathcal{E} \in \text{Ext}_x(\mathcal{H})$, $\mathcal{E} = \text{Arg}(\text{Th}(\mathcal{E}))$.

Under some semantics like naive and stable, Postulate 5 follows from consistency and closure under sub-arguments. This is mostly the case when the attack relation is *conflict-dependent*, that is, it captures the inconsistency of the theory over which the argumentation system is built.

Definition 11 (Conflict-dependency). *Let* $\mathcal{H} = (\text{Arg}(\mathcal{T}), \mathcal{R})$ *be an argumentation system. The attack relation* $\mathcal{R}$ *is* conflict-dependent *iff for all* $(d, x), (d', x') \in \text{Arg}(\mathcal{T})$, *if* $(d, x) \, \mathcal{R} \, (d', x')$ *then* $\text{Seq}(d) \cup \text{Seq}(d')$ *is inconsistent.*

Proposition 7. *Let* $\mathcal{H} = (\text{Arg}(\mathcal{T}), \mathcal{R})$ *be an argumentation system built over a theory* $\mathcal{T}$ *s.t.* $\mathcal{R}$ *is conflict-dependent. If* $\mathcal{H}$ *satisfies consistency and closure under sub-arguments, then* $\mathcal{H}$ *satisfies exhaustiveness under naive and stable semantics.*

Finally, it is worth noticing that conflict-dependent relations do not admit self-attacking arguments.

Proposition 8. *Let* $\mathcal{H} = (\text{Arg}(\mathcal{T}), \mathcal{R})$ *be an argumentation system. If* $\mathcal{R}$ *is conflict-dependent,* $\forall a \in \text{Arg}(\mathcal{T})$ $(a, a) \notin \mathcal{R}$.

4 Outcomes of Argumentation Systems

This section analyzes the outputs of rule-based argumentation systems under the semantics recalled in Def. 9. In the sequel, we consider *only* systems that satisfy the postulates introduced in Section 3. As in [3,5,9], we assume that the "hard" part of a theory is consistent. Formally:

Assumption 2. *For all theory* $\mathcal{T} = (\mathcal{F}, \mathcal{S}, \mathcal{D})$, $\text{CN}((\mathcal{F}, \mathcal{S}, \emptyset))$ *is consistent.*

Let us first introduce a key concept: that of an *option*.

Definition 12 (Option). *Let* $\mathcal{T} = (\mathcal{F}, \mathcal{S}, \mathcal{D})$ *be a theory. An* option *of* $\mathcal{T}$ *is a sub-theory* $\mathcal{T}' = (\mathcal{F}', \mathcal{S}', \mathcal{D}')$ *of* $\mathcal{T}$ *such that:*

- $\mathcal{F}' = \mathcal{F}$ *and* $\mathcal{S}' = \mathcal{S}$ *(hence* $\mathcal{D}' \subseteq \mathcal{D}$*)*
- $\text{CN}(\mathcal{T}')$ *is consistent*
- $\forall r \in \mathcal{D} \setminus \mathcal{D}'$, $\text{CN}((\mathcal{F}, \mathcal{S}, \mathcal{D}' \cup \{r\}))$ *is inconsistent.*

Let $\text{Opt}(\mathcal{T})$ *denote the set of all options of* $\mathcal{T}$.

Example 3. *Consider* T_3 *such that* F_3, S_3, D_3 *are as follows.*

$$F_3 \begin{cases} p \\ q \\ \neg s \end{cases} \qquad\qquad S_3 \left\{ t, u, v \to s \quad (r_1) \right. \qquad\qquad D_3 \begin{cases} p \Rightarrow t & (r_2) \\ q \Rightarrow u & (r_3) \\ u \Rightarrow v & (r_4) \end{cases}$$

The theory T_3 *has three options:*

- $\mathcal{O}_1 = (F_3, S_3, \{p \Rightarrow t, q \Rightarrow u\})$
- $\mathcal{O}_2 = (F_3, S_3, \{p \Rightarrow t, u \Rightarrow v\})$
- $\mathcal{O}_3 = (F_3, S_3, \{q \Rightarrow u, u \Rightarrow v\})$

When a theory is consistent, it has a unique option: itself. This is the case in Example 2: $\text{Opt}(T_2) = \{T_2\}$.

Property 8. *Let* $T = (F, S, D)$ *be a theory.*

- $\text{Opt}(T) = \{T\}$ *iff* $\text{CN}(T)$ *is consistent.*
- *If* $\text{CN}((F, S, \emptyset))$ *is inconsistent, then* $\text{Opt}(T) = \emptyset$.
- *For all* $r \in D$, *if* $\text{CN}((F, S, \{r\}))$ *is consistent, then there exists an option* $\mathcal{O}$ *s.t.* $(F, S, \{r\}) \sqsubseteq \mathcal{O}$.

Note that the set of consequences of an option is not necessarily maximal for set inclusion as shown by Example 3.

Example 3 (Cont). We have $\text{CN}(\mathcal{O}_1) = \{p, q, \neg s, t, u\}$, $\text{CN}(\mathcal{O}_2) = \{p, q, \neg s, t\}$, and $\text{CN}(\mathcal{O}_3) = \{p, q, \neg s, u, v\}$. Thus, $\text{CN}(\mathcal{O}_2) \subseteq \text{CN}(\mathcal{O}_1)$.

Notation For a set $\mathcal{B}$ of theories, we denote its maximum as $\text{Max}(\mathcal{B}) = \{T \in \mathcal{B} \mid \nexists T' \in \mathcal{B} \text{ s.t. } \text{CN}(T) \subset \text{CN}(T')\}$. In Example 3, $\text{Max}(\text{Opt}(T_3)) = \{\mathcal{O}_1, \mathcal{O}_3\}$.

The defeasible rules of a theory do not necessarily belong to an option of the theory as shown by the following example.

Example 4. *The theory* T_4 *s.t.* $F_4 = \{p, q\}$, $S_4 = \{p \to s\}$ *and* $D_4 = \{q \Rightarrow \neg s\}$ *has a single option:* $\mathcal{O} = (F_4, S_4, \emptyset)$.

4.1   Naive Semantics

We start by characterizing the naive extensions of any argumentation system satisfying the above rationality postulates. We show that each naive extension returns a maximal option of the theory over which the system is built.

Theorem 2. *Let* $\mathcal{H} = (\text{Arg}(T), \mathcal{R})$ *be an AS built over a theory* T *s.t.* $\mathcal{R}$ *is conflict-dependent and* $\mathcal{H}$ *satisfies the postulates 1, 2, 3, and 4. For all* $\mathcal{E} \in \text{Ext}_n(\mathcal{H})$, *there exists a unique* $\mathcal{O} \in \text{Max}(\text{Opt}(T))$ *such that* $\text{Th}(\mathcal{E}) \sqsubseteq \mathcal{O}$ *and* $\text{Concs}(\mathcal{E}) = \text{CN}(\mathcal{O})$.

Note that the theory of a naive extension may be a proper subset of the corresponding maximal option. This is mainly due to the fact that an option may contain non-activated rules while arguments are minimal and thus focussed.

Example 2 (Cont). Since theory $\mathcal{T}_2$ is consistent, then it has a single (maximal) option which is the theory itself. Any AS built over $\mathcal{T}_2$ and which obeys the postulates and whose attack relation is conflict-dependent will have a single naive extension $\mathcal{E}$ with $\mathrm{Th}(\mathcal{E}) = (\mathcal{F}, \{r_1, r_2\}, \{r_5\}) \sqsubseteq \mathcal{T}_2$. Rules r_3 and r_5 are not used in arguments.

Notation For $\mathcal{E}$ naive extension of $\mathcal{H}$ s.t. $\mathcal{O}$ in $\mathrm{Max}(\mathrm{Opt}(\mathcal{T}))$ satisfies $\mathrm{Th}(\mathcal{E}) \sqsubseteq \mathcal{O}$ and

$$\mathrm{Concs}(\mathcal{E}) = \mathrm{CN}(\mathcal{O}), \text{ let } \mathrm{Option}(\mathcal{E}) \overset{\mathrm{def}}{=} \mathcal{O}.$$

We prove that no two naive extensions return the same option. Moreover, naive extensions are closed in terms of arguments.

Theorem 3. *Let* $\mathcal{H} = (\mathrm{Arg}(\mathcal{T}), \mathcal{R})$ *be an AS built over a theory* $\mathcal{T}$ *s.t.* $\mathcal{R}$ *is conflict-dependent and* $\mathcal{H}$ *satisfies the postulates 1, 2, 3, and 4.*

- *For all* $\mathcal{E}, \mathcal{E}' \in \mathrm{Ext}_n(\mathcal{H})$, *if* $\mathrm{Option}(\mathcal{E}) = \mathrm{Option}(\mathcal{E}')$, *then* $\mathcal{E} = \mathcal{E}'$
- *For all* $\mathcal{E} \in \mathrm{Ext}_n(\mathcal{H})$, $\mathcal{E} = \mathrm{Arg}(\mathrm{Option}(\mathcal{E}))$

We have shown that each naive extension captures exactly one maximal option and it supports all, and only, the consequences of that option. Theorem 4 states that every option has a corresponding naive extension. So, there is a bijection from the set of naive extensions to the set of maximal options.

Theorem 4. *Let* $\mathcal{H} = (\mathrm{Arg}(\mathcal{T}), \mathcal{R})$ *be an AS built over a theory* $\mathcal{T}$ *s.t.* $\mathcal{R}$ *is conflict-dependent and* $\mathcal{H}$ *satisfies the postulates 1, 2, 3, and 4.*

- *For all* $\mathcal{O} \in \mathrm{Max}(\mathrm{Opt}(\mathcal{T}))$, $\mathrm{Arg}(\mathcal{O}) \in \mathrm{Ext}_n(\mathcal{H})$.
- *For all* $\mathcal{O} \in \mathrm{Max}(\mathrm{Opt}(\mathcal{T}))$, $\mathcal{O} = \mathrm{Option}(\mathrm{Arg}(\mathcal{O}))$
- *For all* $\mathcal{O}, \mathcal{O}' \in \mathrm{Max}(\mathrm{Opt}(\mathcal{T}))$, *if* $\mathrm{Arg}(\mathcal{O}) = \mathrm{Arg}(\mathcal{O}')$ *then* $\mathcal{O} = \mathcal{O}'$.

Example 3 (Cont). The theory $\mathcal{T}_3$ has three options, of which only two are maximal: $\mathrm{Max}(\mathrm{Opt}(\mathcal{T})) = \{\mathcal{O}_1, \mathcal{O}_3\}$. For all argumentation system $\mathcal{H}$ built over $\mathcal{T}_3$, if the attack relation of $\mathcal{H}$ is to be conflict-dependent and the postulates satisfied, then $\mathrm{Ext}_n(\mathcal{H}) = \{\mathrm{Arg}(\mathcal{O}_1), \mathrm{Arg}(\mathcal{O}_3)\}$.

It is thus possible to delimit the number of naive extensions of any argumentation system that satisfies the four postulates.

Corollary 1. *Let* $\mathcal{H} = (\mathrm{Arg}(\mathcal{T}), \mathcal{R})$ *be an AS built over a theory* $\mathcal{T}$ *s.t.* $\mathcal{R}$ *is conflict-dependent and* $\mathcal{H}$ *satisfies the postulates 1, 2, 3, and 4. The equality* $|\mathrm{Ext}_n(\mathcal{H})| = |\mathrm{Max}(\mathrm{Opt}(\mathcal{T}))|$ *holds.*

What about the plausible conclusions that are drawn from a theory using an argumentation system that satisfies the postulates? From the previous results, it is easy to show that they are the literals that follow from all the maximal options.

Theorem 5. *Let* $\mathcal{H} = (\mathrm{Arg}(\mathcal{T}), \mathcal{R})$ *be an AS built over a theory* $\mathcal{T}$ *s.t.* $\mathcal{R}$ *is conflict-dependent and* $\mathcal{H}$ *satisfies the postulates 1, 2, 3, and 4.*

$$\mathrm{Output}(\mathcal{H}) = \bigcap_{\mathcal{O}_i \in \mathrm{Max}(\mathrm{Opt}(\mathcal{T}))} \mathrm{CN}(\mathcal{O}_i)$$

Example 3 (Cont). Any argumentation system $\mathcal{H}$ that can be built over the theory $\mathcal{T}_3$ and has a conflict-dependent attack relation and satisfies the postulates 1, 2, 3, 4 will have as output the set $\mathrm{Output}(\mathcal{H}) = \mathrm{CN}(\mathcal{O}_1) \cap \mathrm{CN}(\mathcal{O}_2) = \{p, q, \neg s, u\}$.

4.2 Stable Semantics

We now analyze the outcomes of rule-based argumentation systems under stable semantics, again considering only systems that satisfy the rationality postulates. We show
that such systems have stable extensions if the set of facts is not empty.

Theorem 6. *Let* $T = (\mathcal{F}, \mathcal{S}, \mathcal{D})$ *be a theory. Whenever* $\mathcal{H} = (\mathtt{Arg}(T), \mathcal{R})$ *is an AS
satisfying postulate 4,* $|\mathtt{Ext}_s(\mathcal{H})| = 0$ *iff* $\mathcal{F} = \emptyset$.

As for naive extensions, stable extensions of any argumentation system that satisfies
the postulates return maximal options of the theory at hand.

Theorem 7. *Let* $\mathcal{H} = (\mathtt{Arg}(T), \mathcal{R})$ *be an AS defined over a theory* T *s.t.* $\mathcal{R}$ *is conflict-
dependent and* $\mathcal{H}$ *satisfies the postulates 1, 2, 3, 4. For all* $\mathcal{E} \in \mathtt{Ext}_s(\mathcal{H})$, $\exists! \mathcal{O} \in$
$\mathtt{Max}(\mathtt{Opt}(T))$ *s.t.*

- $\mathtt{Th}(\mathcal{E}) \sqsubseteq \mathcal{O}$ *and* $\mathtt{Concs}(\mathcal{E}) = \mathtt{CN}(\mathcal{O})$.
- $\mathcal{E} = \mathtt{Arg}(\mathcal{O})$.

Two stable extensions capture distinct options.

Theorem 8. *Let* $\mathcal{H} = (\mathtt{Arg}(T), \mathcal{R})$ *be an AS defined over a theory* T *s.t.* $\mathcal{R}$ *is conflict-
dependent and* $\mathcal{H}$ *satisfies the postulates 1, 2, 3, 4.
For all* $\mathcal{E}, \mathcal{E}' \in \mathtt{Ext}_s(\mathcal{H})$, *if* $\mathtt{Option}(\mathcal{E}) = \mathtt{Option}(\mathcal{E}')$ *then* $\mathcal{E} = \mathcal{E}'$.

Corollary 2. *Let* $\mathcal{H} = (\mathtt{Arg}(T), \mathcal{R})$ *be an AS defined over a theory* $T = (\mathcal{F}, \mathcal{S}, \mathcal{D})$
s.t. $\mathcal{F} \neq \emptyset$ *and* $\mathcal{R}$ *is conflict-dependent and* $\mathcal{H}$ *satisfies Postulates 1,2,3,4. It holds that*
$1 \leq |\mathtt{Ext}_s(\mathcal{H})| \leq |\mathtt{Max}(\mathtt{Opt}(T))|$.

Theorem 7 does not guarantee that each maximal option of a theory T has a corre-
sponding stable extension. The equality $|\mathtt{Ext}_s(\mathcal{H})| = |\mathtt{Max}(\mathtt{Opt}(T))|$ depends on the
attack relation. Let $\Re_s$ be the set of *all* attack relations that are conflict-dependent and
that ensure Postulates 1, 2, 3, 4 under stable semantics. This set contains two *disjoints*
subsets of attack relations, i.e. $\Re_s = \Re_{s_1} \cup \Re_{s_2}$:

- $\Re_{s_1}$: the relations s.t. $|\mathtt{Ext}_s(\mathcal{H})| < |\mathtt{Max}(\mathtt{Opt}(T))|$
- $\Re_{s_2}$: the relations s.t. $|\mathtt{Ext}_s(\mathcal{H})| = |\mathtt{Max}(\mathtt{Opt}(T))|$

Systems that use relations in $\Re_{s_1}$ choose a proper subset of the maximal options of T
and make inferences from them. Their output sets are as follows:

Theorem 9. *Let* $\mathcal{H} = (\mathtt{Arg}(T), \mathcal{R})$ *be an argumentation system built over a theory*
T *s.t.* $\mathcal{R} \in \Re_{s_1}$. $\mathtt{Output}(\mathcal{H}) = \bigcap_{\mathcal{O}_i \in \mathcal{S}} \mathtt{CN}(\mathcal{O}_i)$ *with* $\mathcal{S} = \{\mathcal{O}_i \in \mathtt{Max}(\mathtt{Opt}(T)) \mid$
$\mathtt{Arg}(\mathcal{O}_i) \in \mathtt{Ext}_s(\mathcal{H})\}$.

These attack relations introduce a "critical discrimination" between the maximal options of a theory. Hence, great care must be exercised when designing rule-based argumentation systems based on stable semantics: The principles governing the interaction
between $\Rightarrow$ and $\mathcal{R}$ must be both rigorously and meticulously specified so as to avoid
trouble of which the following example is an easy case.

Example 5. *Let* T_5 *be s.t.* $\mathcal{F}_5 = \{p, q\}$ *and* $\mathcal{S}_5 = \emptyset$ *and* $\mathcal{D}_5 = \{p \Rightarrow s, q \Rightarrow \neg s\}$. T_5 *has two maximal options:* $\mathcal{O}_1 = (\mathcal{F}_5, \mathcal{S}_5, \{p \Rightarrow s\})$ *and* $\mathcal{O}_2 = (\mathcal{F}_5, \mathcal{S}_5, \{q \Rightarrow \neg s\})$. *For any system* $\mathcal{H} = (\text{Arg}(T_5), \mathcal{R})$ *s.t.* $\mathcal{R} \in \Re_{s_1}$, *either i)* $\text{Ext}_s(\mathcal{H}) = \{\text{Arg}(\mathcal{O}_1)\}$ *or ii)* $\text{Ext}_s(\mathcal{H}) = \{\text{Arg}(\mathcal{O}_2)\}$. *In case (i),* $s \in \text{Output}(\mathcal{H})$ *and* $\neg s \notin \text{Output}(\mathcal{H})$. *In case (ii),* $\neg s$ *is the plausible conclusion. Either choice would be arbitrary.*

Attack relations of category $\Re_{s2}$ induce a bijection between the stable extensions of an argumentation system and the maximal options of the theory over which it is built.

Theorem 10. *Let* $T = (\text{Arg}(T), \mathcal{R})$ *be an argumentation system over a theory* T *s.t.* $\mathcal{R} \in \Re_{s2}$. *For all* $\mathcal{O} \in \text{Max}(\text{Opt}(T))$, $\text{Arg}(\mathcal{O}) \in \text{Ext}_s(\mathcal{H})$.

Argumentation systems with an attack relation from $\Re_{s2}$ are *coherent*, meaning that the preferred extensions exhaust all and only the stable ones.

Theorem 11. *Let* $T = (\text{Arg}(T), \mathcal{R})$ *be an argumentation system over a theory* T *s.t.* $\mathcal{R} \in \Re_{s2}$. $\text{Ext}_s(\mathcal{H}) = \text{Ext}_p(\mathcal{H}) = \text{Ext}_n(\mathcal{H})$.

Attack relations in category $\Re_{s2}$ conform exactly to the result obtained under naive semantics: Plausible conclusions for them are already characterized in Theorem 5.

To sum up, attack relations satisfying the postulates can be split into two categories: $\Re_{s_1}$ and $\Re_{s_2}$. Relations from $\Re_{s_2}$ do not offer added value as they make the stable semantics case to collapse to the naive semantics case. For stable semantics to substantiate (as compared with naive semantics) a rule-based argumentation system, attack relations from category $\Re_{s_1}$ must be favored. However, pitfalls threaten as options are discarded, and a lot of care must be exercised when designing such a system.

4.3 Preferred Semantics

Preferred semantics was initially proposed to overcome the limitation of stable semantics which does not guarantee the existence of extensions. Indeed, any argumentation system has at least one preferred extension which may be empty. We show that in case of rule-based systems the empty set cannot be an extension.

Proposition 9. *Let* $\mathcal{H}$ *be an AS built over a theory* $T = (\mathcal{F}, \mathcal{S}, \mathcal{D})$ *s.t.* $\mathcal{H}$ *satisfies strict precedence.* $\text{Ext}_p(\mathcal{H}) = \{\emptyset\}$ *iff* $\mathcal{F} = \emptyset$.

Unlike the cases of naive and stable extensions, a preferred extension may capture only a sub-part of the consequences drawn from a maximal option.

Theorem 12. *Let* $\mathcal{H} = (\text{Arg}(T), \mathcal{R})$ *be an AS built over a theory* T *s.t.* $\mathcal{R}$ *is conflict-dependent and* $\mathcal{H}$ *satisfies the postulates 1 and 3. For all* $\mathcal{E} \in \text{Ext}_p(\mathcal{H})$, $\exists \mathcal{O} \in \text{Max}(\text{Opt}(T))$ *s.t.* $\text{Th}(\mathcal{E}) \sqsubseteq \mathcal{O}$ *and* $\text{Concs}(\mathcal{E}) \subseteq \text{CN}(\mathcal{O})$.

Two preferred extensions refer to different options.

Theorem 13. *Let* $\mathcal{H} = (\text{Arg}(T), \mathcal{R})$ *be an argumentation system s.t.* $\mathcal{R}$ *is conflict-dependent and* $\mathcal{H}$ *satisfies the postulates 1, 2, 3, and 4. Let* $\mathcal{E}, \mathcal{E}' \in \text{Ext}_p(\mathcal{H})$ *and* $\mathcal{O} \in \text{Max}(\text{Opt}(T))$. *If* $\text{Th}(\mathcal{E}) \sqsubseteq \mathcal{O}$ *and* $\text{Th}(\mathcal{E}') \sqsubseteq \mathcal{O}$, *then* $\mathcal{E} = \mathcal{E}'$.

From the previous result, it follows that the number of preferred extensions does not exceed the number of maximal options of the theory over which the system is built.

Theorem 14. *Let $\mathcal{H} = (\text{Arg}(\mathcal{T}), \mathcal{R})$ be a system built over a theory $\mathcal{T}$ s.t. $\mathcal{R}$ is conflict-dependent and $\mathcal{H}$ satisfies Postulates 1, 2, 3, and 4. $|\text{Ext}_p(\mathcal{H})| \leq |\text{Max}(\text{Opt}(\mathcal{T}))|$.*

Regarding the outputs of a rule-based argumentation system under preferred semantics, there are two cases: i) Attack relations of category $\mathcal{R}_{s2}$ lead to coherent systems whose plausible conclusions are characterized by Theorem 5. Thus, naive, stable and preferred semantics coincide. ii) Attack relations of category $\mathcal{R}_{s1}$ lead to pick up some maximal options and to reason about them. The plausible conclusions are given by Theorem 9. Thus the situation about preferred semantics is similar with that for stable semantics: For preferred semantics to offer added value over naive semantics, the attack relation chosen must discard some maximal options but it takes a lot of care to specify such an attack relation in full generality.

4.4   Grounded Semantics – Ideal Semantics

This section analyses the outcomes of rule-based systems under grounded and ideal semantics. We show that the ideal extension is exactly the set of arguments built from the *free* part of a theory. The free part of a theory $\mathcal{T} = (\mathcal{F}, \mathcal{S}, \mathcal{D})$, denoted by $\text{Free}(\mathcal{T})$, is a sub-theory $(\mathcal{F}, \mathcal{S}, \mathcal{D}')$ where $\mathcal{D}' = \cap \mathcal{D}_i$ where $(\mathcal{F}, \mathcal{S}, \mathcal{D}_i) \in \text{Opt}(\mathcal{T})$. In other words, $\mathcal{D}'$ contains all the defeasible rules that are not involved in any conflict.

Proposition 10. *Let $\mathcal{T}$ be a theory. $\text{CN}(\text{Free}(\mathcal{T}))$ is consistent.*

We show that when the attack relation satisfies a very natural requirement, then $\text{Arg}(\text{Free}(\mathcal{T}))$ is admissible (i.e., it is conflict-free and defends all its elements).

Definition 13. *Let $\mathcal{H} = (\text{Arg}(\mathcal{T}), \mathcal{R})$ be an AS over a theory $\mathcal{T}$. An attack relation $\mathcal{R}$ privileges strict arguments iff for all $a = (d, x), b = (d', x') \in \text{Arg}(\mathcal{T})$, if a is strict and $\text{Seq}(d) \cup \text{Seq}(d')$ is inconsistent, then $a\mathcal{R}b$.*

As far as we know, all the attack relations in existing rule-based argumentation systems privilege strict arguments.

Theorem 15. *Let $\mathcal{H} = (\text{Arg}(\mathcal{T}), \mathcal{R})$ be a system built over a theory $\mathcal{T}$ s.t. $\mathcal{R}$ is conflict-dependent and privileges strict arguments. $\text{Arg}(\text{Free}(\mathcal{T}))$ is admissible.*

The set $\text{Arg}(\text{Free}(\mathcal{T}))$ is part of every preferred extension.

Theorem 16. *Let $\mathcal{H} = (\text{Arg}(\mathcal{T}), \mathcal{R})$ be an AS over a theory $\mathcal{T}$ s.t. $\mathcal{R}$ is conflict-dependent and privileges strict arguments, and $\mathcal{H}$ satisfies Postulates 1, 3, 4. $\text{Arg}(\text{Free}(\mathcal{T})) \subseteq \bigcap_{\mathcal{E}_i \in \text{Ext}_p(\mathcal{H})} \mathcal{E}_i$.*

We show next that in case of attack relations of category $\Re_{s2}$, $\text{Arg}(\text{Free}(\mathcal{T}))$ is equal to the intersection of all preferred extensions. Recall that in this case, preferred extensions coincide with stable extensions and with naive ones.

Theorem 17. *Let* $\mathcal{H} = (\text{Arg}(\mathcal{T}), \mathcal{R})$ *be an AS over a theory* $\mathcal{T}$. *If* $\mathcal{R} \in \Re_{s2}$ *then* $\text{Arg}(\text{Free}(\mathcal{T})) = \bigcap_{\mathcal{E}_i \in \text{Ext}_x(\mathcal{H})} \mathcal{E}_i$.

From the previous result, it follows that when the attack relation is of category $\Re_{s2}$ and privileges strict arguments, then $\text{Arg}(\text{Free}(\mathcal{T}))$ is the ideal extension.

Theorem 18. *Let* $\mathcal{H} = (\text{Arg}(\mathcal{T}), \mathcal{R})$ *be an AS over a theory* $\mathcal{T}$. *If* $\mathcal{R} \in \Re_{s2}$ *and privileges strict arguments, then*

- $\text{Arg}(\text{Free}(\mathcal{T}))$ *is the ideal extension of* $\mathcal{H}$.
- *The grounded extension of* $\mathcal{H}$ *is a subset of* $\text{Arg}(\text{Free}(\mathcal{T}))$.

The above result shows that ideal and grounded semantics allow the inference of literals only from the free part of a theory. Note also that grounded extension is more cautious than ideal one and may miss intuitive (free) conclusions.

5   Conclusion

The paper provides the first investigation on the outputs of rule-based argumentation systems. The study is general in the sense that it keeps the attack relation unspecified. Thus, the system can be instantiated with any of the attack relations that are used in existing systems. The results show that under naive semantics, the systems return the literals that follow from all the options of the theory at hand. Stable and preferred semantics either do not provide an added value wrt naive semantics or the attack relation of a system should be formalized in a very rigorous way in order to avoid arbitrary results. Ideal semantics returns the free part of a theory whereas the grounded semantics returns a sub-part of the free part meaning that it may miss interesting conclusions.

References

1. Amgoud, L., Besnard, P.: Logical limits of abstract argumentation frameworks. Journal of Applied Non-Classical Logics (2013)
2. Amgoud, L., Cayrol, C.: Inferring from inconsistency in preference-based argumentation frameworks. Inter. J. of Automated Reasoning 29(2), 125–169 (2002)
3. Caminada, M., Amgoud, L.: On the evaluation of argumentation formalisms. Artificial Intelligence J. 171(5-6), 286–310 (2007)
4. Dung, P.: On the acceptability of arguments and its fundamental role in nonmonotonic reasoning, logic programming and n-person games. AI. J. 77(2), 321–357 (1995)
5. García, A., Simari, G.: Defeasible logic programming: an argumentative approach. Theory and Practice of Logic Programming 4(1-2), 95–138 (2004)
6. Gorogiannis, N., Hunter, A.: Instantiating abstract argumentation with classical logic arguments: Postulates and properties. Artificial Intelligence J. 175(9-10), 1479–1497 (2011)
7. Governatori, G., Maher, M., Antoniou, G., Billington, D.: Argumentation semantics for defeasible logic. J. of Logic and Computation 14(5), 675–702 (2004)
8. Marek, V., Nerode, A., Remmel, J.: A theory of nonmonotonic rule systems I. Annals of Mathematics and Artificial Intelligence 1, 241–273 (1990)
9. Reiter, R.: A logic for default reasoning. Artificial Intelligence J. 13(1-2), 81–132 (1980)

Meta-level Argumentation
with Argument Schemes

Jann Müller[1], Anthony Hunter[2], and Philip Taylor[1]

[1] SAP Next, Belfast BT3 9DT, United Kingdom
jann.mueller@sap.com
[2] University College London, London WC1E 6BT
a.hunter@cs.ucl.ac.uk

Abstract. Arguments in real-world decision making, for example in medical or engineering domains, are often based on patterns of informal argumentation, called argument schemes. In order to improve automated tool support of decision making in such domains, a formal model of argument schemes appears necessary. To address this need, we represent each argument scheme as a defeasible rule in the meta-language, so each application of an argument scheme results in a meta-level argument, and we deal with critical questions via meta-level counter-arguments. In order to understand the interactions between the object-level and meta-level arguments, we introduce bimodal graphs. The utility of the framework is demonstrated by a use case characteristic of the requirements of our partner in the aviation industry.

1 Introduction

Complex decision making processes such as engineering design are driven by argumentation between their participants. Human argumentation often involves common patterns of informal reasoning, called argument schemes. An argument scheme is, for example, the Appeal to Expert Opinion, in which one refers to a statement by a technical expert about a particular problem. Decision making processes, whether or not they involve argumentation, are supported by automated tools. The utility of such tools grows with the accuracy of their internal representation of the world. Since argument schemes are a fundamental part of complex decision making processes, they need to be formalised in software tools. Such a formalisation will greatly improve the support that automated tools may provide to engineering and other processes. In this paper we propose a formalisation of argument schemes via meta-level argumentation.

Our framework comprises three ingredients: Structured argumentation, meta-level argumentation, and bimodal graphs. Structured argumentation allows one to create arguments from defeasible rules. Meta-level argumentation has been proposed to reason about arguments [11,4,18], describing the properties of arguments and attacks. We present a meta-language for structured argumentation in which argument schemes are expressed. Arguments both on the meta- and on the object-level are given a graph-based interpretation using bimodal graphs.

W. Liu, V.S. Subrahmanian, and J. Wijsen (Eds.): SUM 2013, LNAI 8078, pp. 92–105, 2013.

The rest of this paper is organised as follows. After a brief summary of related work (Section 2), a brief overview of Dung's framework for abstract argumentation is given in Section 3. In Section 4 we present the framework for structured argumentation upon which the subsequent work is based. Abstract (graph-based) argumentation is extended in Section 5, where Bimodal Graphs are introduced as an interpretation of meta-level argumentation. Section 6 describes the meta-language used in our framework for structured argumentation. In Section 7, we use meta-ASPIC to extend structured argumentation with argument schemes. The paper concludes with some considerations on the usefulness of our model in practice, in particular for our use case partner (Section 8).

2   Related Work

This paper is based upon three different approaches: Abstract argumentation, meta-level argumentation and structured argumentation.

Abstract argumentation [8] provides a graph-based interpretation of argument graphs. Bipolar argumentation [6,12,7] is an extension of Dung's abstract argumentation framework, adding a "supports"-relation as a second relation over arguments. Dung's original framework considered this relationship only implicitly, using the concept of defence for the defeaters of an argument's defeaters. Supporting arguments allow additional extension semantics. For example, sets of arguments are considered safe if none of their members depend on (are supported by) an argument outside the extension, which results in a stronger notion of internal coherence than just being conflict-free. Whilst bipolar argumentation is appealing as it offers a range of possibilities for defining the "supports"-relation, there is no formalisation of meta-level arguments, and supports for attacks (i.e. each attack by an argument A on argument B is justified by an argument C) cannot be defined.

Meta-argumentation is concerned with using arguments to reason about arguments, rather than using arguments to reason about a domain. Earlier work on meta-level argumentation [11] has shown how several extensions to abstract argumentation can be modeled using meta-level constructs in a "pure" abstract argumentation system as defined by Dung [8]. This is achieved by translating each of these additions, such as attacks on attacks, or preferences, into a constellation of several arguments that are only connected by the "attacks" relation. The extensions of the extended abstract argumentation systems are shown to conincide with those of the resulting argument graph. However, this approach to meta-level argumentation does not provide a systematic way of instantiating abstract arguments. The examples in [11] suggest that there is a need for a systematic approach which uses structured arguments to unify the various proposals for abstract argumentation.

Argument schemes are patterns of informal reasoning often employed in discussions between humans [17]. Argument schemes simplify the argumentation process considerably, since they remove the need for making explicit every detail of an argument, as initial results from our use cases in the aerospace industry have shown. Previous research has been concerned with the representation of argument schemes in a formal setting [3,2,1], in particular for the legal domain [13,9,19]. However, these proposals not do provide meta-level argumentation as a means of reasoning about arguments. In this paper we argue that meta-level argumentation offers some benefits to formalise a range of argument schemes that involve arguments for decision making.

However, to represent argument schemes in structured argumentation, there is a need to develop the meta-level aspects of structural argumentation. We draw upon previous work on a hierarchy of meta-argumentats [18], on a logical formisation of argument schemes [10] and recently argument schemes as a component of social interaction [15]. Furthermore, to understand the interactions of arguments and meta-level arguments, we propose bimodal argument graphs.

3   Dung's Framework

Interactions between arguments can be characterised by argument graphs. This approach was first explored by Dung [8] whose definitions we will briefly recall.

Definition 1 (Argument Graph and Extensions). *A tuple $(\mathcal{A}, \mathcal{R})$ is an argument graph iff $\mathcal{A}$ is a set and $\mathcal{R} \subseteq \mathcal{A} \times \mathcal{A}$ is relation over $\mathcal{A}$, the "attacks"-relation. Let $\mathcal{G} = (\mathcal{A}, \mathcal{R})$ be an argument graph and let $S \subseteq \mathcal{A}$.*

1. *S is conflict-free iff there exist no $A_1, A_2 \in S$ such that $(A_1, A_2) \in \mathcal{R}$.*
2. *Let $A \in S$. S defends A iff for every $(B, A) \in \mathcal{R}$, there exists a $C \in S$ such that $(C, B) \in \mathcal{R}$.*
3. *S is an admissible set iff S is conflict free and defends all of its elements $A \in S$.*
4. *S is a preferred extension iff S admissible and S is maximal with respect to $\subseteq$.*
5. *Let $\mathcal{F}$ be a function of subsets of $\mathcal{A}$ such that $\mathcal{F}(S) = \{A \mid S \text{ defends } A\}$. Let E be the least fixed point of $\mathcal{F}$. E is the grounded extension of $\mathcal{G}$.*
6. *Let $\mathcal{G} = (\mathcal{A}, \mathcal{R})$ be an argument graph and let $x \in \{\text{grounded, preferred}\}$. Then, $\Sigma_x(\mathcal{G}) = \{\mathcal{E} \subseteq \mathcal{A} \mid \mathcal{E} \text{ is a } x\text{-extension of } \mathcal{G}\}$.*

See Fig. 1 for an example of an argument graph.

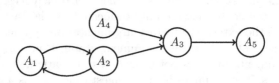

Fig. 1. An argument graph. There are two preferred extensions $\{A_1, A_4, A_5\}$ and $\{A_2, A_4, A_5\}$ and a grounded extension $\{A_4, A_5\}$. Some conflict-free sets are $\{A_1, A_3\}$, $\emptyset$, and $\{A_2, A_4\}$. The set $\{A_1, A_4, A_5\}$ defends A_1, A_4 and A_5. $\{A_1, A_2\}$ defends A_1 and A_2. $\{A_1, A_4, A_5\}$ is an admissible set and a preferred extension. The least fixed point of $\mathcal{F}$ is $\{A_4, A_5\}$.

4 Structured Argumentation

In this section we present a framework for structured argumentation that instantiates abstract argument graphs. It is a subset of ASPIC+ [14].

We only model a subset of the original ASPIC+ definitions, in order to increase the clarity of our presentation. For example, our framework does not consider an ordering of the logical language, nor does it divide the knowledge base into premises, axioms, assumptions and issues. However, the missing aspects of ASPIC+ may be added easily using the same method. Our framework uses only defeasible rules, thus avoiding some of the potential issues with strict rules in ASPIC+ [5].

Definition 2 (Logical Language). *Let $\mathcal{L}$ be a set of positive and negative literals such that, if x is a positive literal, then $\neg x$ is a negative literal.*

Definition 3 (Defeasible Rule). *Let $\mathcal{L}$ be a logical language and let $\varphi_1, \ldots, \varphi_n$, $\varphi \in \mathcal{L}$ with $n \geq 1$. Then, $\varphi_1, \ldots, \varphi_n \Rightarrow \varphi$ is a defeasible rule over $\mathcal{L}$.*

Example 1. A defeasible rule over $\mathcal{L}_A$ is $a, c, \neg k \Rightarrow \neg d$.

The letter $\mathcal{D}$ is used to denote sets of defeasible rules. Rules can be assigned a name in order to refer to them in arguments, using a naming function n. We now have the ingredients of an argumentation system.

Definition 4 (Argumentation System). *An argumentation system $AS = (\mathcal{L}, \mathcal{D}, n)$ is an argumentation system iff*

1. *$\mathcal{L}$ is a logical language*
2. *$\mathcal{D}$ is a set of defeasible rules over $\mathcal{L}$*
3. *$n : \mathcal{D} \to \mathcal{L}$ assigns names to defeasible rules.*

A knowledge base contains some elements of the logical language. These are the premises of arguments.

Definition 5 (Knowledge Base). *Let $\mathcal{L}$ be a logical language. A set $\mathcal{K}$ is a knowledge base iff $\mathcal{K} \subseteq \mathcal{L}$.*

Arguments are built by applying the defeasible rules in an argumentation system to a knowledge base. We write arguments as a sequence in square brackets: $[s]$. An argument is either a fact from the knowledge base or it is composed by applying a defeasible rule to several arguments. We will write $[q; s; r]$ to indicate that q and s are arguments and r is a defeasible rule.

Definition 6 (Argument). *Let $AS = (\mathcal{L}, \mathcal{D}, n)$ be an argumentation system, let $\mathcal{K} \subseteq \mathcal{L}$.*

1. *Every $\varphi \in \mathcal{K}$ is an argument $[\varphi]$ with $Conc([\varphi]) = \varphi$*
2. *If $A_1, \ldots, A_n$ are arguments and there exists a rule $r \in \mathcal{D}$ such that $r = Conc(A_1), \ldots, Conc(A_n) \Rightarrow \varphi$, then $A = [A_1; \ldots; A_n; r]$ is an argument with $Conc(A) = \varphi$*

Example 2. In the examples, we will omit the square brackets around arguments if there is no ambiguity. Let $\mathcal{L}_A$ be defined as above and let $\mathcal{D} = \{a, \neg k \Rightarrow \neg d\}$. Let $\mathcal{K} = \{a, \neg k, d\}$. Possible arguments are $[a]$, $[\neg k]$ and $[a; \neg k; a, \neg k \Rightarrow \neg d]$ and $[d]$

Definition 7 (Sub-argument). *Let A, B be two arguments. A is a subargument of B, short $A \sqsubseteq B$, iff $B = A$ or $B = [A_1; \ldots; A_n; r]$ and $\exists i \leq n$ such that $A \sqsubseteq A_i$.*

The auxiliary function *Rules* is defined on arguments and gives information about the rules used in an argument.

Definition 8 (Rules). *Let A be an argument. The function Rules returns the rules used in A and is defined as $Rules([\varphi]) = \emptyset$ and $Rules([A_1; \ldots; A_n; r]) = \{r\} \cup Rules(A_1) \cup \ldots \cup Rules(A_n)$.*

If the premises or conclusions of two arguments are contrary, then they attack each other. There are two kinds of attack: Undercuts resulting from attacks on defeasible rules, and rebuttals from attacks on conclusions.[1]

Definition 9 (Attack). *Let $AS = (\mathcal{L}, \mathcal{D}, n)$ be an argumentation system, let $\mathcal{K} \subseteq \mathcal{L}$ be a knowledge base and let A, B be arguments. A attacks B iff*

1. *There exists a rule $r \in \mathcal{D}$ such that $Conc(A) = \neg n(r)$ and $r \in Rules(B)$ (undercut) or*
2. *There exists an argument $B' \in Sub(B)$ such that $Conc(A) = \neg Conc(B')$ or $\neg Conc(A) = Conc(B')$ (rebuttal)*

Example 3. Let $A_1 = [a; \neg k; a, \neg k \Rightarrow \neg d]$ and $A_2 = [d]$ be arguments. A_1 rebuts A_2 and A_2 rebuts A_1.

Now that attacks have been defined, an argument graph can be obtained by constructing all arguments and their attacks.

Definition 10 (Argument Graph from Structured Argumentation). *Let $AS = (\mathcal{L}, \mathcal{D}, n)$ be an argumentation system and let $\mathcal{K} \subseteq \mathcal{L}$ be a knowledge base. The argument graph of AS is defined as $(\mathcal{A}, \mathcal{R})$ with $\mathcal{A} = \{A \mid A$ is an argument in$(AS, \mathcal{K})\}$ and $\mathcal{R} = \{(A, B) \mid A, B$ are arguments in $(AS, \mathcal{K})$ and A attacks $B\}$.*

Example 4. Let $AS = (\mathcal{L}, \mathcal{D}, n)$ be an argumentation system and let $\mathcal{K} \subseteq \mathcal{L}$ be a knowledge base with $\mathcal{D} = \{\neg k \Rightarrow \neg d\}$ and $\mathcal{K} = \{a, \neg k, d\}$. The argument graph of $(AS, \mathcal{K})$ is $(\mathcal{A}, \mathcal{R})$ and it is defined as $\mathcal{A} = \{[a], [\neg k], A_1, A_2\}$ with A_1, A_2 as in Ex. 3 and a set of attacks $\mathcal{R} = \{(A_1, A_2), (A_2, A_1)\}$.

---

[1] ASPIC+ defines a third kind of attack, the undermining, from attacks on premises. In our case underminings would be a subset of rebuttals since every premise is an argument for itself.

5   Bimodal Graphs

Bimodal graphs capture arguments both on the object-level and on the meta-level. Every object-level argument and every object-level attack is supported by at least one meta-level argument. Meta-level arguments can only attack meta-level arguments, and object-level arguments can only attack object-level arguments. A bimodal graph therefore has two components, one argument graph for the meta-level and another argument graph for the object-level, alongside a "supports"-relation that originates in the meta-level and targets attacks and arguments on the object-level.

Definition 11 (Bimodal Argument Graph). *A bimodal argument graph is a tuple $(\mathcal{A}_O, \mathcal{A}_M, \mathcal{S}_A, \mathcal{S}_R, \mathcal{R}_O, \mathcal{R}_M)$ with*

1. *$\mathcal{A}_O, \mathcal{A}_M$ are sets such that $\mathcal{A}_O \cap \mathcal{A}_M = \emptyset$, object- and meta-level arguments*
2. *$\mathcal{R}_O \subseteq \mathcal{A}_O \times \mathcal{A}_O$, for object-level attacks*
3. *$\mathcal{R}_M \subseteq \mathcal{A}_M \times \mathcal{A}_M$, for meta-level attacks*
4. *$\mathcal{S}_A \subseteq \mathcal{A}_M \times \mathcal{A}_O$, meta-level arguments supporting object-arguments*
5. *$\mathcal{S}_R \subseteq \mathcal{A}_M \times \mathcal{A}_O \times \mathcal{A}_O$, meta-level arguments supporting object-level attacks*
6. *For all $A \in \mathcal{A}_O$ there exists a $B \in \mathcal{A}_M$ such that $(A, B) \in \mathcal{S}_A$*
7. *For all $(A_1, A_2) \in \mathcal{R}_O$ there exists a $B \in \mathcal{A}_M$ such that $(B, A_1, A_2) \in \mathcal{S}_R$*

The object-level argument graph is $(\mathcal{A}_O, \mathcal{R}_O)$, and the meta-level argument graph is $(\mathcal{A}_M, \mathcal{R}_M)$. These two components are connected by the "supports"-relations $\mathcal{S}_R$ and $\mathcal{S}_A$. This support is the only structural interaction between meta- and object-level. Definition 11 Cond. 6 ensures that every object-level argument is supported by at least one meta-level argument, and Def. 11 Cond. 7 ensures that every object-level attack is supported by at least one meta-level argument.

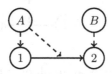

Fig. 2. Bimodal graph for the object-level graph $(\{1, 2\}, \{(1, 2)\})$. The support relation is indicated by the dashed arrows. There are two meta-level arguments $(\mathcal{A}_M = \{A, B\})$, two object-arguments $\mathcal{A}_O = \{1, 2\}$ with no meta-attacks $\mathcal{R}_M = \emptyset$ and a single object-level attack $\mathcal{R}_O = \{(1, 2)\}$. The supports are $\mathcal{S}_A = \{(A, 1), (B, 2)\}$ and $\mathcal{S}_R = \{(A, 1, 2)\}$.

Every extension of the meta-level induces a subgraph ("perspective") of the object-level graph with potentially many object-level extensions, as defined next.

Definition 12 (Perspective). *Let $\mathcal{G} = (\mathcal{A}_O, \mathcal{A}_M, \mathcal{S}_A, \mathcal{S}_R, \mathcal{R}_O, \mathcal{R}_M)$ be a bimodal argument graph and let $x \in \{\mathsf{grounded}, \mathsf{preferred}\}$. An x-perspective of $\mathcal{G}$ is a tuple $(\mathcal{A}'_O, \mathcal{R}'_O)$ if there exists an extension $\mathcal{E} \in \Sigma_x(\mathcal{A}_M, \mathcal{R}_M)$ with*

1. $\mathcal{A}'_O = \{A \mid \exists B \in \mathcal{E} \text{ such that } (B, A) \in \mathcal{S}_A\}$
2. $\mathcal{R}'_O = \{(A_1, A_2) \mid \exists B \in \mathcal{E} \text{ such that } (B, A_1, A_2) \in \mathcal{S}_R\}$

The function $P_x(\mathcal{G})$ returns all x-perspectives of a bimodal argument graph $\mathcal{G}$. For all object-level arguments B, if the meta-level argument for B has no attackers, then B is in every extension.

Proposition 1. *Let* $\mathcal{G} = (\mathcal{A}_O, \mathcal{A}_M, \mathcal{S}_A, \mathcal{S}_R, \mathcal{R}_O, \mathcal{R}_M)$ *be a bimodal argument graph and let* $(A, B) \in \mathcal{S}_A$ *such that* $\nexists C \in \mathcal{A}_M$ *such that* $(C, A) \in \mathcal{R}_M$. *Then, the object-level argument* B *is in every perspective of* $\mathcal{G}$.

Since meta-level arguments reason about object-level arguments, an object-argument may be present (acceptable) in one perspective and missing from another perspective. If there is no conflict on the meta-level, the bimodal argument graph simply yields the same results as the object-level graph on its own.

Definition 13 (Controversial Graph). *Let* $\mathcal{G} = (\mathcal{A}_O, \mathcal{A}_M, \mathcal{S}_A, \mathcal{S}_R, \mathcal{R}_O, \mathcal{R}_M)$ *be a bimodal argument graph.* $\mathcal{G}$ *is* controversial *iff* $\mathcal{R}_M \neq \emptyset$. $\mathcal{G}$ *is* uncontroversial *iff* $\mathcal{G}$ *is not controversial.*

Example 5. The graph shown in Fig. 5 has a disputed argument, (1). For the grounded meta-extension $\{C, D\}$, there is a unique perspective $(\{2\}, \emptyset)$. The two preferred meta-extensions $\mathcal{E}_1 = \{A, C, D\}$ and $\mathcal{E}_2 = \{B, C, D\}$ induce two perspectives $(\{1, 2\}, \{(1, 2)\})$ and $(\{2\}, \emptyset)$ (coinciding with the grounded perspective).

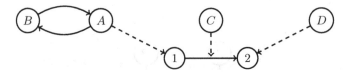

The simplest uncontroversial bimodal graph only has one meta-level argument which supports all object-level arguments and their attacks.

Proposition 2. *Let* $\mathcal{G} = (\mathcal{A}_O, \mathcal{A}_M, \mathcal{S}_A, \mathcal{S}_R, \mathcal{R}_O, \mathcal{R}_M)$ *be an uncontroversial bimodal graph. Then,* $P_{\text{grounded}}(\mathcal{G}) = P_{\text{preferred}}(\mathcal{G}) = \{(\mathcal{A}_O, \mathcal{R}_O)\}$.

The following result shows that multiple argument graphs can be combined in a single bimodal graph in a way that each one of the original graphs is represented by an admissible perspective of the combination.

Theorem 1. *Let* $\mathcal{G}_O$ *be an argument graph. For every nonempty set of subgraphs* $\mathcal{G}^* \subseteq Subgraphs(\mathcal{G}_O)$, *there exists a bimodal graph* $\mathcal{G}_B = (\mathcal{A}_O, \mathcal{A}_M, \mathcal{S}_A, \mathcal{S}_R,$ $\mathcal{R}_O, \mathcal{R}_M)$ *such that* $P_{\text{admissible}}(\mathcal{G}_B) = \mathcal{G}^*$ *and* $(\mathcal{A}_O, \mathcal{R}_O) = \mathcal{G}_O$.

If the graphs in $\mathcal{G}^*$ represent a range of argument graphs and it is uncertain what the actual argument graph looks like (unlike a classical abstract argumentation system, where the graph is defined with certainty and the question is which arguments to accept). The number of admissible perspectives (ie the number of admissible extensions of the meta-graph) can then be interpreted as a indicator of the uncertainty inherent in $\mathcal{G}^*$, the original set of graphs.

Another interpretation of Theorem 1 is that of merging multiple sets of knowledge. If the graphs in $\mathcal{G}^*$ represent knowledge bases – perhaps parts of the same global graph – then creating the bimodal graph $\mathcal{G}_B$ is a merge operation that leaves the original sources intact.

6   Meta-ASPIC

Meta-ASPIC uses the language and reasoning of structured argumentation to capture meta-level argumentation. In terms of the meta-level argument hierarchy presented by Wooldridge [18], meta-ASPIC is located on level Δ_2, the first meta-tier. Arguments in meta-ASPIC can refer to object-level arguments (ie to arguments on Δ_1), but not vice versa. Self-reference is therefore not an issue. It is, however, possible to argue about attacks on the object-level and about the applicability of rules, the "constituents" of arguments. Such an argumentative model of structured argumentation is useful when the original definitions need to be extended, for example to incorporate argument schemes (Section 7), preferences, or attacks on attacks.

Before we introduce meta-ASPIC, we will describe the language and notation used in the definitions. The language $\mathcal{L}_m$ of meta-ASPIC is that of grounded predicates $p(t_1, \ldots, t_n)$ applied to terms t_i. A term is either an object-level symbol or a grounded predicate. The rules of meta-ASPIC will be grounded using elements of an object-level knowledge base $\mathcal{K}_O$ and a set of object-level rules $\mathcal{D}_O$. $\mathcal{K}_m$ consists of meta-level facts about defeasible object-level rules $(\text{Rule}(a_1, \ldots, a_n \Rightarrow a))$ and about object-level facts $\text{Fact}(a)$.

Example 6. $\text{Rule}(\text{a} \Rightarrow \text{b})$ is a predicate where $\text{a} \Rightarrow \text{b}$ is a defeasible rule.

The definition of a structured argument (Def. 6) is captured on the meta-level by creating a set of meta-level rules for object-level facts $(\mathcal{D}_k)$ and another set of meta-level rules $(\mathcal{D}_d)$ that represent the object-level argument structure. The first set contains exactly one grounded predicate for each of the rules and facts in $\mathcal{K}_O$, as defined next. The predicate $\text{Arg}(A, C)$ plays a central role as it denotes an argument A with conclusion C.

Definition 14 (Standard meta-system). *Let $AS_O = (\mathcal{L}_O, \mathcal{D}_O, n_O)$ be an argumentation system (Def. 4) and let $\mathcal{K}_O$ be a knowledge base. Let $\mathcal{G} = (\mathcal{A}, \mathcal{R})$ be the argument graph (Def. 10) of $(AS_O, \mathcal{K}_O)$. Let $\lceil \cdot \rceil$ denote the function that maps defeasible rules to their textual representation.*

The standard meta-system of $(AS_O, \mathcal{K}_O)$, $AS_m = (\mathcal{L}_m, \mathcal{D}_m, n_m)$ with knowledge base $\mathcal{K}_m$ where $\mathcal{D}_m = D_k \cup D_d \cup D_{att}$ is defined as

$$\mathcal{K}_m = \{\text{Fact}(\varphi) \mid \varphi \in \mathcal{K}_O\} \cup \{\text{Rule}(r) \mid r \in \mathcal{D}_O\}$$
$$D_k = \{\text{Fact}(\varphi) \Rightarrow \text{Arg}([\varphi], \varphi) \mid [\varphi] \in \mathcal{A}\}$$
$$D_d = \{\text{Arg}(A_1, \varphi_1), \dots, \text{Arg}(A_n, \varphi_n), \text{Rule}(r) \Rightarrow \text{Arg}([A_1; \dots; A_n; r], \varphi) \mid$$
$$\exists B = [A_1; \dots; A_n; r](B \in \mathcal{A} \wedge \forall 1 \leq i \leq n.\varphi_i = Conc(A_i))\}$$
$$D_{att} = \{\text{Arg}(A_1, c_1), \text{Arg}(A_2, c_2) \Rightarrow \text{Attacks}(A_1, A_2) \mid$$
$$(A_1, A_2) \in \mathcal{R} \wedge Conc(A_1) = c_1 \wedge Conc(A_2) = c_2\}$$
$$\mathcal{L}_m = \{Q \mid Q \text{ is a grounded predicate in } \mathcal{D}_m\} \cup \{\lceil r \rceil \mid r \in \mathcal{D}_O\}$$
$$n_m = \lceil \cdot \rceil$$

Object-level arguments, object-level attacks and the "sub-argument"-relation over object-level arguments are represented in the standard meta-system as the following result shows.

Proposition 3. *Let $AS = (\mathcal{L}, \mathcal{D}, n)$ be an argumentation system, let $\mathcal{K}$ be a knowledge base and let AS_m be the standard meta-system of AS with knowledge base $\mathcal{K}_m$. Let $(\mathcal{A}, \mathcal{R})$ be the argument graph of $(AS, \mathcal{K})$ and let $(\mathcal{A}_m, \mathcal{R}_m)$ be the argument graph of $(AS_m, \mathcal{K}_m)$. Then,*

1. *$\forall A \in \mathcal{A}, \exists A_m \in \mathcal{A}_m$ such that $Conc(A_m) = \text{Arg}(A, Conc(A))$*
2. *$\forall (A, B) \in \mathcal{R}, \exists A_m \in \mathcal{A}_m$ such that $Conc(A_m) = \text{Att}(A, B)$*
3. *$\forall A, B \in \mathcal{A}$. If $A \sqsubseteq B$ then $\exists A_m, B_m \in \mathcal{A}_m$ such that $A_m \sqsubseteq B_m$, $Conc(A_m) = \text{Arg}(A, Conc(A))$ and $Conc(B_m) = \text{Arg}(B, Conc(B))$*

A meta-ASPIC system is a structured argumentation system that contains knowledge of an argumentation system in the form of meta-predicates.

Definition 15 (meta-ASPIC). *Let $AS' = (\mathcal{L}', \mathcal{D}', n')$ be an argumentation system (Def. 4) and let $\mathcal{K}'$ be a knowledge base. $(AS', \mathcal{K}')$ is a meta-ASPIC system iff there exists an argumentation system AS_O with knowledge base $\mathcal{K}_O \neq \emptyset$ such that $AS_m = (\mathcal{L}_m, \mathcal{D}_m, n_m)$ is the standard meta-system (Def. 14) of $(AS_O, \mathcal{K}_O)$ such that $\mathcal{L}_m \subseteq \mathcal{L}'$, $\mathcal{D}_m \subseteq \mathcal{D}'$, and $n_m \subseteq n'$.*

Example 7. Let $AS, \mathcal{K}$ be an object-level argumentation system and knowledge base as in Ex. 4. The rules to represent facts in the corresponding standard meta system are $D_k = \{\text{Fact}(a) \Rightarrow \text{Arg}([a], a), \text{Fact}(\neg k) \Rightarrow \text{Arg}([\neg k], \neg k), \dots\}$. The single defeasible rule results in $\{\text{Arg}([a], a), \text{Arg}([\neg k], \neg k), \text{Rule}(a, \neg k \Rightarrow \neg d) \Rightarrow \text{Arg}([a; \neg k; a, \neg k \Rightarrow \neg d], \neg d)\} = D_d$

The definitions of meta-ASPIC ensure that the resulting arguments conform with bimodal argument graphs. A meta-ASPIC system can thus be transformed into a bimodal argument graph which can be used to evaluate arguments using extension semantics, such as those of Dung [8]. In Def. 16, we separate meta-level arguments from those on the object-level, by classifying them according to their conclusions. Essentially, arguments whose conclusion is $\text{Arg}(X, Y)$ act as meta-support for an object-level argument X with conclusion Y. Attacks are determined likewise, using the predicate $\text{Attacks}(X_1, X_2)$.

Definition 16 (Bimodal Graph from meta-ASPIC). *Let AS be a meta-ASPIC system with knowledge base $\mathcal{K}$ and let $(\mathcal{A}, \mathcal{R})$ be the argument graph of $(AS, \mathcal{K})$ (Def. 10). Let $\mathcal{G}_{meta} = (\mathcal{A}_O, \mathcal{A}_M, \mathcal{S}_A, \mathcal{S}_R, \mathcal{R}_O, \mathcal{R}_M)$ be a bimodal argument graph. $\mathcal{G}_{meta}$ is the bimodal graph of $(AS, \mathcal{K})$ iff*

1. $\mathcal{A}_O = \{X \mid \exists A \in \mathcal{A}.\exists Y.Conc(A) = \texttt{Arg}(X,Y)\}$
2. $\mathcal{A}_M = \mathcal{A}$
3. $\mathcal{S}_A = \{(A,X) \mid \exists A \in \mathcal{A}.\exists Y.Conc(A) = \texttt{Arg}(X,Y)\}$
4. $\mathcal{S}_R = \{(A,X,Y) \mid \exists A \in \mathcal{A}.Conc(A) = \texttt{Attacks}(X,Y)\}$
5. $\mathcal{R}_O = \{(X,Y) \mid \exists A \in \mathcal{A}.Conc(A) = \texttt{Attacks}(X,Y)\}$
6. $\mathcal{R}_M = \mathcal{R}$

A standard meta-system contains one meta-level argument for each object-level argument and for each object-level attack. The bimodal graph of a standard meta-system does not have any attacks on the meta-level.

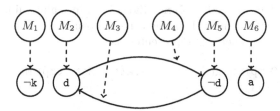

Fig. 3. A bimodal argument graph reflecting Ex. 7. The M_i are meta-level arguments, e.g. $M_1 = [\texttt{Fact}(\neg k); \texttt{Fact}(\neg k) \Rightarrow \texttt{Arg}(\neg k, \neg k)]$ and $M_4 = [\texttt{Arg}(A_2, d); \texttt{Arg}(A_5, \neg d); \texttt{Arg}(A_2, d); \texttt{Arg}(A_5, \neg d) \Rightarrow \texttt{Attacks}(A_2, A_5)]$ where A_i is the object-level argument of M_i.

Proposition 4. *Let $AT = (AS, \mathcal{K})$ be a standard meta-system with knowledge base $\mathcal{K}$ and let $\mathcal{B} = (\mathcal{A}_O, \mathcal{A}_M, \mathcal{S}_A, \mathcal{S}_R, \mathcal{R}_O, \mathcal{R}_M)$ be its bimodal graph (Def. 16). Then, $\mathcal{B}$ is uncontroversial.*

Meta-ASPIC as presented so far provides a baseline for argument schemes, as we explain in the next section.

7   Argument Schemes

Argument schemes are patterns of informal reasoning [17]. An argument scheme consists of a set of conditions and a conclusion. If the conditions are met, then the conclusion holds. Each argument scheme is associated with set of critical questions. Each critical question identifies possible attacks on arguments derived from argument schemes, by pointing out either a condition that must hold for an argument scheme to be applied, or an exception that renders an argument scheme invalid for a specific instance.

   The classification of informal arguments by argument schemes helps to identify similar kinds of argumentation. Argument schemes can also be used to identify weaknesses in argumentation, by making explicit the underlying assumptions of an argument and by providing a list of typical attacks on arguments from argument schemes, in the form of critical questions.

Throughout this section, we develop the notion of argument schemes based on a use case from our industry partner. This use case is presented in the examples, starting with Ex. 8.

Example 8. Several engineers are designing a rib that is part of a wing. They are currently trying to decide on a material (Mat). While in reality there is a choice of a large number of alloys and composites, we assume here that the principal decision is only that of aluminium (Al) or composite materials (Comp). The choice will be represented as Mat(Comp) or Mat(Al).

The two options are mutually exclusive. This constitutes the first argument scheme used in this example: Argument from Alternative. This alternative is expressed by the fact Alter(Mat(Comp), Mat(Al)). Since aluminium and composites cannot both be chosen at the same time, choosing one means excluding the other. The argument scheme is represented as

$$\texttt{Arg}(A_1, \texttt{Mat(Comp)}), \texttt{Arg}(A_2, \texttt{Mat(Al)}), \texttt{Alter(Mat(Comp), Mat(Al))} \atop \Rightarrow \texttt{Attacks}(A_1, A_2) \qquad (MR_1)$$

And an analogous rule MR_2 with the conclusion Attacks(A_2, A_1). We use labels MR_i for defeasible rules which result in meta-level arguments. It is important to note that, even though MR_1 and MR_2 seem to be based on the logical axiom *tertium non datur*, it behaves differently, because the assumption that a third option does not exist can be attacked (and indeed there are more than two possible materials for the component).

Having established the external constraints of the solution, we now turn to the actual debate about the materials. Engineer E is recognised by her peers as an expert on metallurgy (abbreviated Mly) and suggests to use aluminium.

$$\texttt{Expert(E, Mly)}, \texttt{Domain(Mat, Mly)}, \texttt{Asserts(E, Mat(Al))} \atop \Rightarrow \texttt{Arg([Mat(Al)], Mat(Al))} \qquad (MR_3)$$

In reality, arguments from expert opinion usually do not just state a conclusion without backing it up with further evidence. Instead, expert arguments summarise the expert's reasoning as well as the conclusion [16]. For example, E might recommend aluminium based on her experience with similar designs. Due to the limited space we omit these details here.

It turns out that the knowledge of E, the expert, may be outdated because E has not published any work on metallurgy recently (NoPub(E, Mly)). This opens the argument from expert opinion to an attack on one of its premises (conditions):

$$\texttt{NoPub(E, Mly)} \Rightarrow \neg\texttt{Expert(E, Mly)} \qquad (MR_4)$$

This attack is an example of a common pattern. Every argument scheme is associated with a list of "Critical Questions", questions which point to potential weaknesses of the argument. Some critical questions, such as the one expressed in MR_4, target the conditions of an argument scheme. Another type of critical questions is aimed at exceptions to the applicability of a scheme.

Another common pattern of argumentation is to argue from (positive or negative) consequences. In our example, heavy components increase the fuel consumption of airplanes. Minimising weight is therefore very important in aerospace design. The relatively high weight of aluminium is a reason to avoid it. This argument scheme is known as Argument from Negative Consequences (bringing about A will result in C, C is negative, therefore A should not be brought about). Conversely, using composites will have positive consequences, since it is lighter. The following two rules represent this reasoning about consequences.

$$\texttt{BadCons(Mat(Al))} \Rightarrow \texttt{Arg([BadCons(Mat(Al))]}, \neg\texttt{Mat(Al))} \qquad (MR_5)$$

$$\texttt{GoodCons(Mat(Comp))} \Rightarrow \texttt{Arg([GoodCons(Mat(Comp))]}, \texttt{Mat(Comp))} \qquad (MR_6)$$

With the meta-level rules MR_1 to MR_6, several arguments from argument schemes can be formed. The bimodal graph of this example is shown below. Since the meta-level extensions conincide, there is only one perspective which results in the acceptance O_C and $O_{\overline{Al}}$ on the object-level.

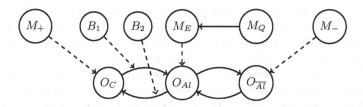

Fig. 4. Bimodal graph of Ex. 8. The arguments are $M_+ = [\texttt{GoodCons(Mat(Comp))}; MR_6]$ (Arg. from Positive Consequences), $B_1 = [\texttt{Alter(Mat(Comp),Mat(Al))}; MR_+]$ (Arg. from Alternative), $M_E = [\texttt{Expert(E,Mly),Domain(D,Mly),Asserts(E,Mat(Al))}; MR_3]$ (Arg. from Expert Opinion), $M_Q = [\texttt{NoPub}; MR_4]$ and $M_- = [\texttt{BadCons(Mat(Al))}; MR_5]$

Argument schemes affect the object-level arguments and object-level attacks. They can therefore be defined using the appropriate predicates of meta-ASPIC.

Definition 17 (Argument Scheme). *Let* $\varphi_1, \ldots, \varphi_n \Rightarrow \varphi$ *be defeasible rule.* $\varphi_1, \ldots, \varphi_n \Rightarrow \varphi$ *is an argument scheme iff* $\varphi \in \{\texttt{Arg}(A, X), \texttt{Attacks}(A_1, A_2)\}$ *for any arguments* A, A_1, A_2.

Critical questions are defined similarly. Definition 18 ensures that arguments from argument schemes can be attacked by arguments whose last rule is a critical question.

Definition 18 (Critical Question). *Let* $r = \varphi_1, \ldots, \varphi_n \Rightarrow \varphi$ *be an argument scheme and let* $c = \psi_1, \ldots, \psi_n \Rightarrow \psi$ *be a defeasible rule.* c *is a* critical question *for* r *iff* $\psi = \neg\varphi_n$ *(Condition) or* $\psi = \neg n(r)$ *(Exception)*

This section demonstrated how the language of meta-ASPIC may be used to model the use of argument schemes as meta-level arguments. The case study illustrates the advantages of modeling argument schemes using meta-argumentation for both arguments and attacks on the object-level.

8   Discussion

We presented an approach to meta-level argumentation with argument schemes based on three lines of work: Structured argumentation, meta-level argumentation and bimodal graphs. The framework for structured argumentation is a lightweight model that allows one to express object-level and meta-level arguments. The interactions of arguments are evaluated using bimodal argument graphs. We developed a set of argument schemes based on a case study from the aerospace industry.

The version of meta-ASPIC presented in this paper only uses a subset of the original ASPIC+ system. However, we formally prove elsewhere that the full meta-ASPIC gives exactly the same results as ASPIC+. Argumentation systems that already use ASPIC+ can thus easily be transformed into using meta-ASPIC to gain extensibility as demonstrated for example in Section 7.

Our model of argument schemes goes beyond a recent proposal by Sklar *et al.* [15] in two ways. Firstly, the argument schemes we consider include not only social argumentation patterns (authority, *ad hominem*, etc.) but also factual patterns determined by context (Argument from Alternative in Ex. 8 and Argument from Analogy), which are employed frequently in engineering. Secondly, the meta-argumentation system presented above handles arguments about attacks (object-level argument A attacks object-level argument B) using meta-arguments, whereas in the approach by Sklar *et al.*, the notion of the status of an argument separates successful attacks from unsuccessful ones. In our system, an unsuccessful object-level attack would be represented by a meta-level attack on the argument that supports the object-level attack.

We are considering two avenues for future work. The first one is to extend the theoretical foundation of bimodal graphs, in particular to explore their relation to abstract argumentation in a similar fashion to [11], in order to deepen the understanding of argument schemes. The second direction is to implement the framework in order to measure its performance and usefulness, particularly regarding the case study.

Acknowledgement. This work is supported by SAP AG and the Invest NI Collaborative Grant for R&D - RD1208002.

References

1. Atkinson, K., Bench-Capon, T., Modgil, S.: Argumentation for decision support. In: Bressan, S., Küng, J., Wagner, R. (eds.) DEXA 2006. LNCS, vol. 4080, pp. 822–831. Springer, Heidelberg (2006)

2. Atkinson, K., Bench-Capon, T., Mcburney, P.: A dialogue game protocol for multi-agent argument over proposals for action. Autonomous Agents and Multi-Agent Systems 11(2), 153–171 (2005)

3. Bex, F., Prakken, H., Reed, C., Walton, D.: Towards a formal account of reasoning about evidence: Argumentation schemes and generalisations. Artificial Intelligence and Law 11, 125–165 (2003)

4. Boella, G., van der Torre, L., Villata, S.: On the acceptability of meta-arguments. In: IEEE/WIC/ACM International Joint Conferences on Web Intelligence and Intelligent Agent Technologies, WI-IAT 2009, vol. 2, pp. 259–262 (September 2009)

5. Caminada, M., Amgoud, L.: On the evaluation of argumentation formalisms. Artificial Intelligence 171, 286–310 (2007)

6. Cayrol, C., Lagasquie-Schiex, M.C.: On the acceptability of arguments in bipolar argumentation frameworks. In: Godo, L. (ed.) ECSQARU 2005. LNCS (LNAI), vol. 3571, pp. 378–389. Springer, Heidelberg (2005)

7. Cayrol, C., Lagasquie-Schiex, M.C.: Bipolarity in argumentation graphs: Towards a better understanding. In: Benferhat, S., Grant, J. (eds.) SUM 2011. LNCS, vol. 6929, pp. 137–148. Springer, Heidelberg (2011)

8. Dung, P.: On the acceptability of arguments and its fundamental role in non-monotonic reasoning, logic programming and n-person games. Artificial Intelligence 77(2), 321–357 (1995)

9. Gordon, T.F., Walton, D.: Legal reasoning with argumentation schemes. In: Proceedings of the 12th International Conference on Artificial Intelligence and Law, pp. 137–146 (2009)

10. Hunter, A.: Reasoning about the appropriateness of proponents for arguments. In: AAAI 2008, pp. 89–94 (2008)

11. Modgil, S., Bench-Capon, T.: Metalevel argumentation. Journal of Logic and Computation (2010)

12. Nouioua, F., Risch, V.: Bipolar argumentation frameworks with specialized supports. In: 2010 22nd IEEE International Conference on Tools with Artificial Intelligence, ICTAI, vol. 1, pp. 215–218 (2010)

13. Prakken, H., Wyner, A., Bench-Capon, T., Atkinson, K.: A formalisation of argumentation schemes for legal case-based reasoning in aspic+. Journal of Logic and Computation in press (2013)

14. Prakken, H.: An abstract framework for argumentation with structured arguments. Argument & Computation (2010)

15. Sklar, E., Parsons, S., Singh, M.P.: Towards an argumentation-based model of social interaction. In: Proceedings of the Tenth International Workshop on Argumentation in Multi-Agent Systems, ArgMAS 2013 (2013)

16. Walton, D.: Appeal to Expert Opinion. Pennsylvania State University Press, University Park (1997)

17. Walton, D., Reed, C., Macagno, F.: Argumentation Schemes. Cambridge University Press (2008)

18. Wooldridge, M., McBurney, P., Parsons, S.: On the meta-logic of arguments. Argumentation in Multi-agent Systems 4049, 42–56 (2006)

19. Wyner, A.Z., Bench-Capon, T.J.M., Atkinson, K.M.: Towards formalising argumentation about legal cases. In: Proceedings of the 13th International Conference on Artificial Intelligence and Law, pp. 1–10. ACM (2011)

Efficiently Estimating the Probability of Extensions in Abstract Argumentation⋆

Bettina Fazzinga, Sergio Flesca, and Francesco Parisi

DIMES - Università della Calabria, 87036 Rende (CS), Italy
{bfazzinga,flesca,fparisi}@dimes.unical.it

Abstract. Probabilistic abstract argumentation combines Dung's abstract argumentation framework with probability theory to model uncertainty in argumentation. In this setting, we deal with the fundamental problem of computing the probability $Pr^{sem}(S)$ that a set S of arguments is an *extension* according to a semantics *sem*. We focus on three popular semantics (i.e., *complete*, *grounded*, and *preferred*) for which the state-of-the-art approach is that of estimating $Pr^{sem}(S)$ by using a Monte-Carlo simulation technique, as computing $Pr^{sem}(S)$ has been proved to be intractable. In this paper, we detect and exploit some properties of these semantics to devise a new Monte-Carlo simulation approach which is able to estimate $Pr^{sem}(S)$ using much fewer samples than the state-of-the-art approach, resulting in a significantly more efficient estimation technique.

1 Introduction

Argumentation allows disputes to be modeled, which arise between two or more parties, each of them providing arguments to assert her reasons. Although argumentation is strongly related to philosophy and law, it has gained remarkable interest in AI as a reasoning model for representing dialogues, making decisions and handling inconsistency/uncertainty [9,10,24].

The *abstract argumentation framework* (AAF) introduced in the seminal paper [12] is a simple but powerful argumentation framework. An AAF is a pair $\langle A, D \rangle$ consisting of a set A of *arguments*, and of a binary relation D over A, called *defeat* (or, equivalently, *attack*) relation. Basically, an argument is an abstract entity that may attack and/or be attacked by other arguments. For instance, consider the following scenario (inspired by an example in [20]), where we are interested in deciding whether to organize or not a BBQ party in our garden on Saturday. Assume that our decision should be taken considering the argument a, which is *"Our friends will have great fun at the party"*, and the argument b, which is *"Saturday will rain"* (according to the BBC weather forecasting service). This scenario can be modeled by the AAF $\mathcal{A}$, whose set of arguments is $\{a, b\}$, and whose defeat relation consists of the defeat $\delta = (b, a)$, meaning that the fun at the party is jeopardized if it rains.

Several semantics for AAFs, such as *admissible*, *complete*, *grounded*, and *preferred*, have been proposed [12,13,7] to identify "reasonable" sets of arguments, called *extensions*. Basically, each of these semantics corresponds to some properties which "certify"

⋆ The first two authors were supported by EJRM project.

W. Liu, V.S. Subrahmanian, and J. Wijsen (Eds.): SUM 2013, LNAI 8078, pp. 106–119, 2013.

whether a set of arguments can be profitably used to support a point of view in a discussion. For instance, a set S of arguments is an extension according to the admissible semantics if it has two properties: it is *conflict-free* (that is, there is no defeat between arguments in S), and every argument (outside S) attacking an argument in S is counterattacked by an argument in S. Intuitively enough, the fact that a set S is an extension according to the admissible semantics means that, using the arguments in S, you do not contradict yourself, and you can rebut to anyone who uses any of the arguments outside S to contradict yours. The other semantics correspond to other ways of determining whether a set of arguments would be a "good point" in a dispute, and will be described in the core of the paper. The fundamental problem of verifying whether a set of arguments is an extension according to one of the above-mentioned semantics has been studied in [17,15].

As a matter of fact, in the real world, arguments and defeats are often uncertain, thus, several proposals have been made to model uncertainty in AAFs, by considering weights, preferences, or probabilities associated with arguments and/or defeats. In this regard, [14,21,26,25] have recently extended the original Dung framework in order to achieve probabilistic abstract argumentation frameworks (PrAFs), where uncertainty of arguments and defeats is modeled by exploiting the probability theory. In particular, [21] proposed a PrAF where both arguments and defeats are associated with probabilities. For instance, a PrAF $\mathcal{F}_{\mathcal{A}}$ can be obtained from the AAF $\mathcal{A}$ by considering the arguments a, b, and the defeat δ as probabilistic events, having probabilities $Pr(a) = .9$, $Pr(b) = .7$, and $Pr(\delta) = .9$. Basically, this means that there is some uncertainty about the fact that our friends will have fun at the party, about the truthfulness of the BBC weather forecasting service, and about the fact that the bad weather forecast actually entails that the party will be disliked by our friends.

The issue of how to assign probabilities to arguments and defeats in the PrAF proposed in [21], has been deeply investigated in [19,20], where the *justification* and the *premise* perspectives have been introduced. In this paper, we do not address this issue, but we assume that the probabilities of arguments and defeats are given. We deal with the probabilistic counterpart of the problem of verifying whether a set of arguments is an extension according to a semantics, that is, the problem of *determining the probability $Pr_{\mathcal{F}}^{sem}(S)$ that a set S of arguments is an extension according to a given semantics sem*. To this end, we consider the PrAF proposed in [21], which is based on the notion of *possible world*. Basically, given a PrAF $\mathcal{F}$, a possible world represents a (deterministic) scenario consisting of some subset of the arguments and defeats in $\mathcal{F}$. Hence, a possible world can be viewed as an AAF containing exactly the arguments and the defeats occurring in the represented scenario. For instance, for the above-introduced PrAF $\mathcal{F}_{\mathcal{A}}$, the possible world $\langle \{a\}, \emptyset \rangle$ is the AAF representing the scenario where only a occurs, while the possible world $\langle \{a, b\}, \{\delta\} \rangle$ is the AAF representing the scenario where all the arguments and defeats occur.

In [21] it was shown that a PrAF admits a unique probability distribution over the set of possible worlds, which assigns a probability value to each possible world coherently with the probabilities of arguments and defeats. This follows from the assumption that arguments are viewed as pairwise independent probabilistic events, while each defeat is viewed as a probabilistic event conditioned by the occurrence of the arguments it relates, but independent from any other event. Once shown that a PrAF admits a unique probability distribution over the set of possible worlds, the probability $Pr_{\mathcal{F}}^{sem}(S)$ is naturally defined as the sum of the probabilities of the possible worlds where the set S of arguments is an extension according to the semantics *sem*. Unfortunately, as pointed out in [18], computing $Pr_{\mathcal{F}}^{sem}(S)$ is intractable (actually, $FP^{\#P}$-complete) for the three popular semantics *complete*, *grounded*, and *preferred*. Indeed, for these semantics, the state-of-the-art approach is that of estimating $Pr_{\mathcal{F}}^{sem}(S)$ by a Monte-Carlo simulation approach, as proposed in [21], since the complexity of computing $Pr_{\mathcal{F}}^{sem}(S)$ is prohibitive.

Main Contributions. In this paper, we propose a new Monte-Carlo-based simulation technique for estimating the probability $Pr_{\mathcal{F}}^{sem}(S)$, where *sem* is one of the following semantics: *complete*, *grounded*, *preferred*. In more detail, our strategy relies on the fact that a set S of arguments is an extension according to a semantics in {*complete*, *grounded*, *preferred*} only if it is *conflict-free* (resp., an *admissible* extension), and on the fact that computing the probability that S is conflict-free (resp., an admissible extension) is in *PTIME* [18]. Starting from this, we devise a strategy for estimating $Pr_{\mathcal{F}}^{sem}(S)$ which consists in:

i) first, computing an estimate of $Pr_{\mathcal{F}}^{sem|CF}(S)$ (resp., $Pr_{\mathcal{F}}^{sem|AD}(S)$), that is the conditional probability that S is an extension according to *sem* given that S is conflict-free (resp., an admissible extension) by adopting a Monte-Carlo simulation approach, and

ii) then, yielding an estimate of $Pr_{\mathcal{F}}^{sem}(S)$ by multiplying the estimate of $Pr_{\mathcal{F}}^{sem|CF}(S)$ (resp., $Pr_{\mathcal{F}}^{sem|AD}(S)$) with the probability that S is is conflict-free (resp., an admissible extension).

This strategy is implemented in two algorithms: the first one yields an estimate of $Pr_{\mathcal{F}}^{sem}(S)$ by estimating $Pr_{\mathcal{F}}^{sem|CF}(S)$, the second one provides an estimate of $Pr_{\mathcal{F}}^{sem}(S)$ by estimating $Pr_{\mathcal{F}}^{sem|AD}(S)$. Hence, differently from the approach proposed in [21], where the aim of the Monte-Carlo simulation is that of estimating $Pr_{\mathcal{F}}^{sem}(S)$, in our approach the Monte-Carlo simulation is exploited to estimate either $Pr_{\mathcal{F}}^{sem|CF}(S)$ or $Pr_{\mathcal{F}}^{sem|AD}(S)$. This implies that, instead of considering the whole set of possible worlds of $\mathcal{F}$ as sample space (as done in [21]), we work over a reduced sample space, that is, either the subset of the possible worlds wherein S is conflict-free, or the subset of possible worlds wherein S is an admissible extension (depending whether $Pr_{\mathcal{F}}^{sem|CF}(S)$ or $Pr_{\mathcal{F}}^{sem|AD}(S)$ is being estimated). Finally, we experimentally validate our approach showing that both algorithms outperform the approach in [21], and that the second algorithm is faster than the first one in most practical cases.

2   Preliminaries

We now overview Dung's framework and its probabilistic extension introducesd in [21].

2.1   Abstract Argumentation

An *abstract argumentation framework* [12] (*AAF*) is a pair $\langle A, D \rangle$, where A is a finite set, whose elements are referred to as *arguments*, and $D \subseteq A \times A$ is a binary relation over A, whose elements are referred to as *defeats* (or *attacks*). An argument is an abstract entity whose role is determined by its relationships with other arguments. Given an AAF $\mathcal{A}$, we also refer to the set of its arguments and the set of its defeats as $Arg(\mathcal{A})$ and $Def(\mathcal{A})$, respectively. Given arguments $a, b \in A$, we say that a *defeats* b iff there is $(a, b) \in D$. Similarly, a set $S \subseteq A$ *defeats* an argument $b \in A$ iff there is $a \in S$ such that a *defeats* b; and argument a *defeats* S iff there is $b \in S$ such that a *defeats* b. A set $S \subseteq A$ is said to be *conflict-free* if there are no $a, b \in S$ such that a *defeats* b. An argument a is said to be *acceptable* w.r.t. $S \subseteq A$ iff $\forall b \in A$ such that b *defeats* a, there is $c \in S$ such that c *defeats* b.

Several semantics for AAFs have been proposed to identify "reasonable" sets of arguments, called *extensions*. We consider the following well-known semantics [12]: *admissible* (ad), *complete* (co), *grounded* (gr), and *preferred* (pr). A set $S \subseteq A$ is

- an *admissible* extension iff S is conflict-free and all its arguments are acceptable w.r.t. S;
- a *complete* extension iff S is admissible and S contains all the arguments that are acceptable w.r.t. S;
- a *grounded* extension iff S is a minimal (w.r.t. $\subseteq$) complete set of arguments;
- a *preferred* extension iff S is a maximal (w.r.t. $\subseteq$) admissible set of arguments;

Example 1. Consider the AAF $\langle A, D \rangle$ obtained by extending the AAF $\mathcal{A} = \langle \{a, b\}, \{\delta_1 = (b, a)\} \rangle$ presented in the introduction as follows. The set A of arguments is $\{a, b, c\}$, where c is the new argument *"Saturday will be sunny"* (according to the Telegraph weather forecasting service). The set D of defeats is $\{\delta_1 = (b, a), \delta_2 = (b, c), \delta_3 = (c, b)\}$, where δ_2 and δ_3 encode the fact that arguments b and c attack each other. As $S = \{a, c\}$ is conflict-free and every argument in S is acceptable w.r.t. S, it holds that S is admissible. As S is maximally admissible, it a preferred extension. It is easy to check that S is complete, while it is not grounded since it is not minimally complete.

Given an AAF $\mathcal{A}$, a set $S \subseteq Arg(\mathcal{A})$ of arguments, and a semantics *sem* $\in \{$ad, co, gr, pr$\}$, we define the function *ext*$(\mathcal{A}, sem, S)$ which returns *true* if S is an extension according to *sem*, *false* otherwise.

2.2 Probabilistic Abstract Argumentation

We now review the *probabilistic* abstract argumentation framework (*PrAF*) proposed in [21].

Definition 1 (PrAF). *A PrAF is a tuple* $\langle A, P_A, D, P_D \rangle$ *where* $\langle A, D \rangle$ *is an AAF, and* P_A *and* P_D *are, respectively, functions assigning a non-zero[1] probability value to each argument in A and defeat in D, that is,* $P_A : A \to (0, 1]$ *and* $P_D : D \to (0, 1]$.

Basically, the value assigned by P_A to an argument a represents the probability that a actually occurs, whereas the value assigned by P_D to a defeat (a, b) represents the conditional probability that a defeats b given that both a and b occur.

---

[1] Assigning probability equal to 0 to arguments/defeats is redundant.

The meaning of a PrAF is given in terms of possible worlds, each of them representing a scenario that may occur in the reality. Given a PrAF $\mathcal{F}$, a possible world is modeled by an AAF which is derived from $\mathcal{F}$ by considering only a subset of its arguments and defeats. More formally, given a PrAF $\mathcal{F} = \langle A, P_A, D, P_D \rangle$, a possible world w of $\mathcal{F}$ is an AAF $\langle A', D' \rangle$ such that $A' \subseteq A$ and $D' \subseteq D \cap (A' \times A')$. The set of the possible worlds of $\mathcal{F}$ will be denoted as $pw(\mathcal{F})$.

Example 2. As a running example, consider the PrAF $\mathcal{F} = \langle A, P_A, D, P_D \rangle$ where A and D are those of Example 1, and assume that $P_D(\delta_2) = P_D(\delta_3) = 1$ (meaning that arguments b and c attack each other in all the possible scenarios), and that $P_A(c) = .2$ (this corresponds to the assumption that the Telegraph weather forecasting service has a low reliability). Furthermore, recall that $P_A(a) = .9$, $P_A(b) = .7$, $P_D(\delta_1) = .9$, as defined in the introduction. The set $pw(\mathcal{F})$ consists of the following possible worlds:

$w_1 = \langle \emptyset, \emptyset \rangle$ $w_2 = \langle \{a\}, \emptyset \rangle$ $w_3 = \langle \{b\}, \emptyset \rangle$ $w_4 = \langle \{c\}, \emptyset \rangle$ $w_5 = \langle \{a,b\}, \emptyset \rangle$ $w_6 = \langle \{a,c\}, \emptyset \rangle$ $w_7 = \langle \{b,c\}, \emptyset \rangle$ $w_8 = \langle A, \emptyset \rangle$ $w_9 = \langle \{a,b\}, \{\delta_1\} \rangle$ $w_{10} = \langle \{b,c\}, \{\delta_3\} \rangle$ $w_{11} = \langle \{b,c\}, \{\delta_2\} \rangle$ $w_{12} = \langle \{b,c\}, \{\delta_2, \delta_3\} \rangle$ $w_{13} = \langle A, \{\delta_1\} \rangle$ $w_{14} = \langle A, \{\delta_1, \delta_3\} \rangle$ $w_{15} = \langle A, \{\delta_1, \delta_2\} \rangle$ $w_{16} = \langle A, D \rangle$ $w_{17} = \langle A, \{\delta_2\} \rangle$ $w_{18} = \langle A, \{\delta_3\} \rangle$ $w_{19} = \langle A, \{\delta_2, \delta_3\} \rangle$

An interpretation for a PrAF $\mathcal{F} = \langle A, P_A, D, P_D \rangle$ is a probability distribution function I over the set $pw(\mathcal{F})$ of the possible worlds. Assuming that arguments represent pairwise independent events, and that each defeat represents an event conditioned by the occurrence of its argument events but independent from any other event, the interpretation for the PrAF $\mathcal{F} = \langle A, P_A, D, P_D \rangle$ is as follows. Each $w \in pw(\mathcal{F})$ is assigned by I the probability:

$$I(w) = \prod_{a \in Arg(w)} P_A(a) \cdot \prod_{a \in A \setminus Arg(w)} (1 - P_A(a)) \cdot \prod_{\delta \in Def(w)} P_D(\delta) \cdot \prod_{\delta \in \overline{D}(w) \setminus Def(w)} (1 - P_D(\delta))$$

where $\overline{D}(w)$ is the set of defeats that may appear in the possible world w, that is $\overline{D}(w) = D \cap (Arg(w) \times Arg(w))$.

Example 3. Continuing our running example, the interpretation I for $\mathcal{F}$ is as follows. First of all, observe that, for each possible world $w \in pw(\mathcal{F})$, if both arguments b and c belong to $Arg(w)$ and $\delta_2 \notin Def(w)$ or $\delta_3 \notin Def(w)$, then $I(w) = 0$. The probabilities of the other possible worlds are the following:
$I(w_1) = (1 - P_A(a)) \cdot (1 - P_A(b)) \cdot (1 - P_A(c)) = .024$; $I(w_2) = .216$; $I(w_3) = .056$;
$I(w_4) = .006$; $I(w_5) = P_A(a) \cdot P_A(b) \cdot (1 - P_A(c)) \cdot (1 - P_D(\delta_1)) = .0504$;
$I(w_6) = .054$; $I(w_9) = .4536$; $I(w_{12}) = .014$; $I(w_{16}) = .1134$; $I(w_{19}) = .0126$.

The probability $Pr_{\mathcal{F}}^{sem}(S)$ that a set S of arguments is an extension according to a given semantics *sem* is defined as the sum of the probabilities of the possible worlds w for which S is an extension according to *sem* (i.e., $ext(w, sem, S) = \texttt{true}$).

Definition 2 ($Pr_{\mathcal{F}}^{sem}(S)$). *Given a PrAF $\mathcal{F}$, a set S, and a semantics sem, the probability $Pr_{\mathcal{F}}^{sem}(S)$ that S is an extension according to sem is* $Pr_{\mathcal{F}}^{sem}(S) = \sum_{\substack{w \in pw(\mathcal{F}) \\ \wedge ext(w, sem, S)}} I(w)$.

Example 4. In our running example, the probabilities that the sets $S_1 = \{b\}$, and $S_2 = \{ac\}$ are admissible are as follows: $Pr_{\mathcal{F}}^{ad}(S_1) = I(w_3) + I(w_5) + I(w_9) + I(w_{12}) + I(w_{16}) + I(w_{19}) = .7$ $Pr_{\mathcal{F}}^{ad}(S_2) = I(w_6) + I(w_{16}) + I(w_{19}) = .18$

In the following we will also refer to the probability that a set S of arguments is conflict-free, that is the sum of probabilities of the possible worlds w wherein S is conflict-free. Let $cf(w, S)$ be a function returning *true* iff S is conflict-free in w. Though cf is not a semantics, with a little abuse of notation, we denote as $Pr_{\mathcal{F}}^{\mathsf{cf}}(S)$ the probability that S is conflict-free, that is $Pr_{\mathcal{F}}^{\mathsf{cf}}(S) = \sum_{w \in pw(\mathcal{F}) \wedge cf(w,S)} I(w)$.

3 Estimating Extension Probability Using Monte-Carlo Simulation

We first describe the state-of-the-art approach for estimating the probability $Pr_{\mathcal{F}}^{sem}(S)$ that a set S of arguments is an extension according to a semantics *sem*, then we introduce two algorithms that, as we show in Section 4, significantly speed up the estimation of $Pr_{\mathcal{F}}^{sem}(S)$.

Throughout this section, as well as in the rest of the paper, we assume that a PrAF $\mathcal{F} = \langle A, P_A, D, P_D \rangle$, a set $S \subseteq A$ of arguments, and a semantics sem are given.

3.1 The State-of-the-Art Approach

In this section, we briefly review the Monte-Carlo simulation approach proposed in [21], which is implemented by Algorithm 1. Algorithm 1 estimates the probability $Pr_{\mathcal{F}}^{sem}(S)$ by repeatedly sampling the set $pw(\mathcal{F})$ of possible worlds (i.e., the set of AAFs that can be induced by $\mathcal{F}$). It takes as input $\mathcal{F}$, S, *sem*, an error level ϵ, and a confidence level $z_{1-\alpha/2}$, and it returns an estimate $\widehat{Pr}_{\mathcal{F}}^{sem}(S)$ of the probability $Pr_{\mathcal{F}}^{sem}(S)$ such that $Pr_{\mathcal{F}}^{sem}(S)$ lies in the interval $\widehat{Pr}_{\mathcal{F}}^{sem}(S) \pm \epsilon$ with confidence level $z_{1-\alpha/2}$. Algorithm 1 works as follows. It samples n AAFs from the set $pw(\mathcal{F})$ (Lines 2-12), and, for each of them, it checks whether S is an extension according to *sem* (Line 10); if it is the case, variable x which keeps track of the number of AAFs for which S is an extension according to *sem* is incremented by 1. At each iteration, an AAF is generated by randomly selecting an argument $a \in A$ according to its probability $P_A(a)$ (Lines 4-6), and then randomly selecting a defeat such that both of its arguments are present in the set of generated arguments according to its probability (Lines 7-9). The number n of AAFs to be sampled to achieve the required error level ϵ with confidence level $z_{1-\alpha/2}$ is determined by exploiting the Agresti-Coull interval [1]. In particular, according to [1] the estimated value p of $Pr_{\mathcal{F}}^{sem}(S)$ after x successes in n samples is $p = \frac{x + (z_{1-\alpha/2}^2)/2}{n + z_{1-\alpha/2}^2}$ (Line 11), and the number of samples ensuring that the error level is ϵ with confidence level $z_{1-\alpha/2}$ is $\overline{n} = \frac{z_{1-\alpha/2}^2 \cdot p \cdot (1-p)}{\epsilon^2} - z_{1-\alpha/2}^2$ (Line 11). Thus, Algorithm 1 stops after that $\overline{n}$ samples have been generated (Line 12), and returns the proportion $\frac{x}{n}$ of successes in the number of n generated samples.

Algorithm 1 *State-of-the-art algorithm for approximating* $Pr_{\mathcal{F}}^{sem}(S)$
Input: $\mathcal{F} = \langle A, P_A, D, P_D \rangle$; $S \subseteq A$; sem; An error level ϵ; A confidence level $z_{1-\alpha/2}$
Output: $\widehat{Pr}_{\mathcal{F}}^{sem}(S)$ s.t. $Pr_{\mathcal{F}}^{sem}(S) \in [\widehat{Pr}_{\mathcal{F}}^{sem}(S) - \epsilon, \ \widehat{Pr}_{\mathcal{F}}^{sem}(S) + \epsilon]$ with confidence $z_{1-\alpha/2}$
01: $x = n = 0$;
02: **do**

```
03:     Arg = Def = ∅
04:     for each a ∈ A do
05:         Generate a random number r ∈ [0, 1]
06:         if r ≤ P_A(a) then Arg = Arg ∪ {a}
07:     for each ⟨a, b⟩ ∈ D s.t. a, b ∈ Arg do
08:         Generate a random number r ∈ [0, 1]
09:         if r ≤ P_D(⟨a, b⟩) then Def = Def ∪ {⟨a, b⟩}
10:     if ext(⟨Arg, Def⟩, sem, S) then x=x+1;
```

11:     $n=n+1; \quad p = \dfrac{x+(z_{1-\alpha/2}^2)/2}{n+z_{1-\alpha/2}^2}; \quad \overline{n} = \dfrac{z_{1-\alpha/2}^2 \cdot p \cdot (1-p)}{\epsilon^2} - z_{1-\alpha/2}^2$

```
12: while n ≤ n̄
13: return x/n
```

3.2   Estimating $Pr_{\mathcal{F}}^{sem}(S)$ by Sampling AAFs Wherein S Is Conflict-Free

In this section, we introduce a Monte-Carlo approach for estimating $Pr_{\mathcal{F}}^{sem}(S)$, whose main idea is that of sampling only the AAFs of $pw(\mathcal{F})$ wherein S is conflict-free. In fact, our algorithm estimates $Pr_{\mathcal{F}}^{sem}(S)$ by first computing the probability $Pr_{\mathcal{F}}^{cf}(S)$ that S is conflict-free and then estimating the conditional probability $Pr_{\mathcal{F}}^{sem|CF}(S)$ that S is an extension according to *sem* given that S is conflict-free.

Let CF be the event *"S is conflict-free"*. To compute the probability $Pr_{\mathcal{F}}^{cf}(S)$ of CF, our algorithm exploits the following fact proved in [18], which entails that $Pr_{\mathcal{F}}^{cf}(S)$ can be computed in $O(|S|^2)$.

Fact 1 ($\mathbf{Pr_{\mathcal{F}}^{cf}(S)}$). $Pr_{\mathcal{F}}^{cf}(S) = \prod_{a \in S} P_A(a) \cdot \prod_{\substack{⟨a, b⟩ \in D \\ \wedge a \in S \wedge b \in S}} \left(1 - P_D(⟨a, b⟩)\right)$.

Since for the considered semantics (i.e., complete, grounded, preferred), S is an extension according to *sem* only if S is conflict-free, it holds that $Pr_{\mathcal{F}}^{sem}(S) = \dfrac{Pr_{\mathcal{F}}^{sem|CF}(S)}{Pr_{\mathcal{F}}^{cf}(S)}$.

Hence, $Pr_{\mathcal{F}}^{sem}(S)$ can be estimated by first determining the exact value of $Pr_{\mathcal{F}}^{cf}(S)$ in polynomial time (Fact 1), and then estimating $Pr_{\mathcal{F}}^{sem|CF}(S)$ by sampling the AAFs wherein CF occurs (that is, AAFs wherein S is conflict-free). However, to accomplish this, we need to know the value of the probabilities of the argument events and defeat events given that CF occurs. Given an argument $a \in A$ and a defeat $⟨a, b⟩ \in D$, we denote as $Pr(a|CF)$ (resp., $Pr(⟨a, b⟩|CF)$) the probability that argument a (resp., defeat $⟨a, b⟩$) occurs given that CF occurs. The following lemma states that $Pr(a|CF)$ coincides with $P_A(a)$ if a is not in S, otherwise $Pr(a|CF) = 1$; and that $Pr(⟨a, b⟩|CF)$ coincides with $P_D(⟨a, b⟩)$ if $a \notin S$ or $b \notin S$, otherwise $Pr(⟨a, b⟩|CF)$ is zero.

Lemma 1. *Given a PrAF $\mathcal{F} = ⟨A, P_A, D, P_D⟩$ and a set $S \subseteq A$ of arguments, then*

- $\forall a \in S, Pr(a|CF)=1; \quad \forall a \in A \setminus S, Pr(a|CF)=P_A(a);$
- $\forall ⟨a, b⟩ \in D$ *such that* $a, b \in S, Pr(⟨a, b⟩|CF) = 0;$
- $\forall ⟨a, b⟩ \in D \setminus \{⟨a, b⟩ \in D \text{ s.t. } a, b \in S\}, Pr(⟨a, b⟩|CF) = P_D(⟨a, b⟩).$

Hence, our algorithm samples the AAFs wherein CF occurs by randomly selecting arguments and defeats according to the probabilities given in Lemma 1.

Algorithm 2 estimates $Pr_{\mathcal{F}}^{sem}(S)$, with error level ϵ and confidence level $z_{1-\alpha/2}$, by first computing $Pr_{\mathcal{F}}^{cf}(S)$ (Line 1), next, computing the estimate $\widehat{Pr}_{\mathcal{F}}^{sem|CF}(S)$ of $Pr_{\mathcal{F}}^{sem|CF}(S)$ (Lines 2-14); and, finally, returning $\widehat{Pr}_{\mathcal{F}}^{sem}(S) = \widehat{Pr}_{\mathcal{F}}^{sem|CF}(S) \cdot Pr_{\mathcal{F}}^{cf}(S)$ (Line 15).

Algorithm 2 *Estimating* $Pr_{\mathcal{F}}^{sem}(S)$ *by sampling AAFs wherein S is conflict-free*
Input *and* **Output** *as in Algorithm 1*
01: *Compute* $Pr_{\mathcal{F}}^{cf}(S)$ *as indicated in Fact 1*
02: $x = n = 0$;
03: **do**
04:     $Arg = S;\ Def = \emptyset$;
05:     **for each** $a \in A \setminus S$ **do**
06:         *Generate a random number* $r \in [0, 1]$
07:             *if* $r \le Pr(a|CF)$ *then* $Arg = Arg \cup \{a\}$
08:     **for each** $\langle a, b \rangle \in D$ such that $a, b \in Arg$ **do**
09:         **if** $a \notin S \vee b \notin S$ **do**
10:             *Generate a random number* $r \in [0, 1]$
11:                 *if* $r \le Pr(\langle a, b \rangle |CF)$ *then* $Def = Def \cup \{\langle a, b \rangle\}$
12:     **if** $ext(\langle Arg, Def \rangle, sem, S)$ **then** $x = x + 1$;
13:     $n = n + 1;\ p = \dfrac{x + z_{1-\alpha/2}^2/2}{n + z_{1-\alpha/2}^2};\ \overline{n} = \dfrac{z_{1-\alpha/2}^2 \cdot p \cdot (1-p)}{\epsilon^2} \cdot (Pr_{\mathcal{F}}^{cf}(S))^2 - z_{1-\alpha/2}^2$
14: **while** $n \le \overline{n}$
15: **return** $x/n \cdot Pr_{\mathcal{F}}^{cf}(S)$

The core of Algorithm 2 computes $\widehat{Pr}_{\mathcal{F}}^{sem|CF}(S)$ by exploiting the results of Lemma 1 as follows. At each iteration, it generates an AAF $\langle Arg, Def \rangle$ by first adding all the $a \in S$ to Arg, since $Pr(a|CF) = 1$ (Line 4). Then, the arguments $a \in A \setminus S$ are randomly added to Arg according to their probability $Pr(a|CF) = P_A(a)$ (Lines 5-7). Moreover, as every generated AAF has to not contain any defeat $\langle a, b \rangle$ where $a, b \in S$ (due to $Pr(\langle a, b \rangle |CF) = 0$), Algorithm 2 randomly adds to Def only the defeats $\langle a, b \rangle$ such that a or b is in the set $Arg \setminus S$ according to probability $Pr(\langle a, b \rangle |CF) = P_D(\langle a, b \rangle)$ (Lines 8-11). After such an AAF has been generated, variable x is incremented by 1 if S is an extension according to sem (Line 12).

Algorithm 2 takes as input the error level ϵ and the confidence level $z_{1-\alpha/2}$ to get $Pr_{\mathcal{F}}^{sem}(S)$ lying in the interval $\widehat{Pr}_{\mathcal{F}}^{sem}(S) \pm \epsilon$ with confidence level $z_{1-\alpha/2}$. However, in the core of Algorithm 2 we do not compute $\widehat{Pr}_{\mathcal{F}}^{sem}(S)$, but we compute $\widehat{Pr}_{\mathcal{F}}^{sem|CF}(S)$, that is an estimate of $Pr_{\mathcal{F}}^{sem|CF}(S)$. Hence, we need to determine the error level $\overline{\epsilon}$ to be taken into account to get $Pr_{\mathcal{F}}^{sem|CF}(S)$ lying in the interval $\widehat{Pr}_{\mathcal{F}}^{sem|CF}(S) \pm \overline{\epsilon}$, which in turn entails $Pr_{\mathcal{F}}^{sem}(S)$ lying in the interval $\widehat{Pr}_{\mathcal{F}}^{sem}(S) \pm \epsilon$. Since $\widehat{Pr}_{\mathcal{F}}^{sem}(S) = \widehat{Pr}_{\mathcal{F}}^{sem|CF}(S) \cdot Pr_{\mathcal{F}}^{cf}(S)$, an estimate $\widehat{Pr}_{\mathcal{F}}^{sem|CF}(S)$ such that $Pr_{\mathcal{F}}^{sem|CF}(S)$ lies in the interval $\widehat{Pr}_{\mathcal{F}}^{sem|CF}(S) \pm \overline{\epsilon}$ corresponds to an estimate $\widehat{Pr}_{\mathcal{F}}^{sem}(S)$ such that $Pr_{\mathcal{F}}^{sem}(S)$ lies in the interval $[\widehat{Pr}_{\mathcal{F}}^{sem|CF}(S) \pm \overline{\epsilon}] \cdot Pr_{\mathcal{F}}^{cf}(S)$. Thus, $\overline{\epsilon} = \epsilon / Pr_{\mathcal{F}}^{cf}(S)$.

Furthermore, we need to determine the number $\overline{n}$ of AAFs to be sampled to ensure that the error level of $\widehat{Pr}_{\mathcal{F}}^{sem|CF}(S)$ is $\overline{\epsilon}$ with confidence level $z_{1-\alpha/2}$. According to the

Agresti-Coull interval [1], $\overline{n} = \frac{z_{1-\alpha/2}^2 \cdot p \cdot (1-p)}{(\overline{\epsilon})^2} - z_{1-\alpha/2}^2$. Since $\overline{\epsilon} = \epsilon / Pr_{\mathcal{F}}^{cf}(S)$, we

obtain $\overline{n} = \frac{z_{1-\alpha/2}^2 \cdot p \cdot (1-p) \cdot (Pr_{\mathcal{F}}^{cf}(S))^2}{\epsilon^2} - z_{1-\alpha/2}^2$ (Line 13).

3.3   Estimating $Pr_{\mathcal{F}}^{sem}(S)$ by Sampling AAFs Wherein S Is Admissible

We now introduce a Monte-Carlo approach for estimating $Pr_{\mathcal{F}}^{sem}(S)$ where only the AAFs wherein S is an extension according to the admissible semantics are sampled.

Let AD be the event *"S is an admissible extension"*. We will show that $Pr_{\mathcal{F}}^{sem}(S)$ can be estimated by first computing (in polynomial time) the probability $Pr_{\mathcal{F}}^{ad}(S)$ that AD occurs and then estimating the probability $Pr_{\mathcal{F}}^{sem|AD}(S)$ that S is an extension according to *sem* given that AD occurs.

The probability $Pr_{\mathcal{F}}^{ad}(S)$ can be computed in polynomial time by exploiting the following fact [18], which entails that $Pr_{\mathcal{F}}^{ad}(S)$ can be computed in time $O(|S| \cdot |A|)$.

Fact 2 ($Pr_{\mathcal{F}}^{ad}(S)$). $Pr_{\mathcal{F}}^{ad}(S) = Pr_{\mathcal{F}}^{cf}(S) \cdot \prod_{d \in A \setminus S} (P_1(S,d) + P_2(S,d) + P_3(S,d))$, where:
$P_1(S,d) = 1 - P_A(d)$, and $P_2(S,d) = P_A(d) \cdot \prod_{\substack{\langle d,b \rangle \in D \\ \wedge b \in S}} (1 - P_D(\langle d,b \rangle))$, and

$$P_3(S,d) = P_A(d) \cdot \left(1 - \prod_{\substack{\langle d,b \rangle \in D \\ \wedge b \in S}} (1 - P_D(\langle d,b \rangle))\right) \cdot \left(1 - \prod_{\substack{\langle a,d \rangle \in D \\ \wedge a \in S}} (1 - P_D(\langle a,d \rangle))\right).$$

As for the case of sampling only AAFs wherein S is conflict-free, since for all the semantics *sem* we consider, S is an extension according to *sem* only if S is an admissible extension, it holds that $Pr_{\mathcal{F}}^{sem}(S) = \frac{Pr_{\mathcal{F}}^{sem|AD}(S)}{Pr_{\mathcal{F}}^{ad}(S)}$.

To sample only AAFs wherein S is an admissible extension, we need to know the probability of the argument events given that AD occurs. This probability $Pr(a|AD)$ is given by the first two items of Lemma 2, which also states the following probabilities that, as it will be clearer in the following, are exploited to estimate $Pr_{\mathcal{F}}^{sem|AD}(S)$:

(*i*) the probability $Pr(\langle a,b \rangle | AD)$ that $\langle a,b \rangle$ occurs, given that AD occurs, for each defeat $\langle a,b \rangle$ such that both a and b belong to S;

(*ii*) the probability $Pr(a \to S | AD \wedge a)$ that a defeats at least one argument in S, given that both AD and the argument event a occur, for each argument a not in S;

(*ii*) the probability $Pr(\langle a,b \rangle | AD \wedge b \not\to S)$ that $\langle a,b \rangle$ occurs, given that AD occurs and that the event "b does not defeat any argument in S" occurs, for each defeat $\langle a,b \rangle$ such that a belongs to S while b does not, or both a and b are not in S.

Lemma 2. *Given a PrAF $\mathcal{F} = \langle A, P_A, D, P_D \rangle$ and a set $S \subseteq A$ of arguments, then*

- $\forall a \in S, Pr(a|AD) = 1$;
- $\forall a \in A \setminus S, Pr(a|AD) = \frac{P_2(S,a) + P_3(S,a)}{P_1(S,a) + P_2(S,a) + P_3(S,a)}$;
- $\forall \langle a,b \rangle \in D$ s.t. $a,b \in S, Pr(\langle a,b \rangle | AD) = 0$;
- $\forall \langle a,b \rangle \in D$ s.t. $a,b \in A \setminus S, Pr(\langle a,b \rangle | AD \wedge b \not\to S) = P_D(\langle a,b \rangle)$;
- $\forall a \in A \setminus S, Pr(a \to S | AD \wedge a) = \frac{P_3(S,a)}{P_2(S,a) + P_3(S,a)}$;
- $\forall \langle a,b \rangle \in D$ s.t. $a \in S \wedge b \in A \setminus S, Pr(\langle a,b \rangle | AD \wedge b \not\to S) = P_D(\langle a,b \rangle)$.

where $P_1(S,a)$, $P_2(S,a)$, and $P_3(S,a)$ are defined as in Fact 2.

Algorithm 3 *Estimating $Pr_{\mathcal{F}}^{sem}(S)$ by sampling AAFs wherein S is admissible*
Input *and* **Output** *as in Algorithm 1*
01: *Compute $Pr_{\mathcal{F}}^{ad}(S)$ as in Fact 2*
02: $x = n = 0$;
03: *do*
04:     $Arg = S$; $Def = \emptyset$; $defeatS = \emptyset$;
05:     *for each* $a \in A \setminus S$ *do*
06:         *Generate a random number* $r \in [0, 1]$
07:         *if* $r \leq Pr(a|\mathrm{AD})$ *then*
08:             $Arg = Arg \cup \{a\}$
09:             *if* $r \leq Pr(a \rightarrow S|\mathrm{AD} \wedge a)$ *then*
10:                 $Def = Def \cup generateAtLeastOneDefeatAndDefend(\mathcal{F}, \langle Arg, Def \rangle, S, a)$
11:                 $defeatS = defeatS \cup \{a\}$
12:     *for each* $\langle a, b \rangle \in D$ *s.t.* $(a, b \in Arg \setminus S) \vee (a \in S \wedge b \in Arg \setminus S \wedge b \notin defeatS)$ *do*
13:         *Generate a random number* $r \in [0, 1]$
14:         *if* $r \leq Pr(\langle a, b \rangle|\mathrm{AD} \wedge b \nrightarrow S)$ *then* $Def = Def \cup \{\langle a, b \rangle\}$
15:     *if* $ext(\langle Arg, Def \rangle, sem, S)$ *then* $x=x+1$;
16:     $n=n+1$; $p = \dfrac{x+z_{1-\alpha/2}^2/2}{n+z_{1-\alpha/2}^2}$; $\overline{n} = \dfrac{z_{1-\alpha/2}^2 \cdot p \cdot (1-p)}{\epsilon^2} \cdot Pr_{\mathcal{F}}^{ad}(S) - z_{1-\alpha/2}^2$
17: *while* $n \leq \overline{n}$
18: *return* $x/n \cdot Pr_{\mathcal{F}}^{ad}(S)$

Algorithm 3 estimates $Pr_{\mathcal{F}}^{sem}(S)$ by sampling AAFs wherein S is an admissible extension. Analogously to Algorithm 2, Algorithm 3 first determines the (exact) probability $Pr_{\mathcal{F}}^{ad}(S)$ that S is an admissible extension, as specified in Fact 2 (Line 1); next, it computes the estimate $\widehat{Pr}_{\mathcal{F}}^{sem|\mathrm{AD}}(S)$ of $Pr_{\mathcal{F}}^{sem|\mathrm{AD}}(S)$ (Lines 2-17); and, finally, it determines the estimate $\widehat{Pr}_{\mathcal{F}}^{sem}(S)$ of $Pr_{\mathcal{F}}^{sem}(S)$ as $\widehat{Pr}_{\mathcal{F}}^{sem}(S) = \widehat{Pr}_{\mathcal{F}}^{sem|\mathrm{AD}}(S) \cdot Pr_{\mathcal{F}}^{ad}(S)$ (Line 18).

The core of Algorithm 3 determines $\widehat{Pr}_{\mathcal{F}}^{sem|\mathrm{AD}}(S)$ by exploiting the results of Lemma 2 as follows. At each iteration, it generates an AAF $\langle Arg, Def \rangle$ by first adding all the $a \in S$ to Arg (Line 4), since $Pr(a|\mathrm{AD}) = 1$. Next, to guarantee that S is an admissible extension in the AAF being generated, each argument $a \in A \setminus S$, is randomly added to Arg according to its probability $Pr(a|\mathrm{AD})$ (Lines 5-8).

Next, for each argument $a \in A \setminus S$ which has been added to Arg, we distinguish two cases: (i) the event $a \rightarrow S$ occurs (i.e., the random number r generated for deciding whether a should be added to Arg is less than or equal to $Pr(a \rightarrow S|\mathrm{AD} \wedge a)$ – Line 9) (ii) the event $a \rightarrow S$ does not occur.

In case (i), to guarantee that S is an admissible extension in the AAF being generated, Algorithm 3 randomly generates a non-empty set $\Delta(a)$ of defeats, and it adds the defeats in $\Delta(a)$ to the set of defeats Def of the AAF being generated (Line 10). Specifically, $\Delta(a) = \Delta'(a) \cup \Delta''(a)$ is such that $|\Delta'(a)|, |\Delta''(a)| \geq 1$, all the defeats in $\Delta'(a)$ are of the form $\langle a, b \rangle$ with $b \in S$, and all the defeats in $\Delta''(a)$ are of the form $\langle c, a \rangle$ with $c \in S$. That is, $\Delta'(a)$ consists of defeats from a toward S and, vice versa, $\Delta''(a)$ consists of defeats from S toward a. The fact that $|\Delta''(a)| \geq 1$ (which means that S defeats a) ensures that S remains an admissible extension even adding the defeats in $\Delta(a)$ to Def. The generation of the defeats in $\Delta(a)$ is accomplished by function *generateAtLeastOneDefeatAndDefend* (Line 10) in linear time w.r.t. the size of S, and set *defeatS* is used to keep track of the arguments a' for which $\Delta(a)$ has been generated (Line 11).

In case (ii), to guarantee that S is an admissible extension in the AAF being generated, no defeat from a towards S is generated and defeats from S to a are generated according to the probability $Pr(\langle b, a \rangle | \text{AD} \wedge a \not\rightarrow S)$. This is done at Line 14, where Algorithm 3 also randomly adds to Def the defeats $\langle a, b \rangle$ such that both a and b belong to $A \setminus S$, according to their probability. Observe that, since every AAF generated must not contain any defeat $\langle a, b \rangle$ where $a, b \in S$, Algorithm 3 does not add any of these defeats to Def.

After that such an AAF has been generated, analogously to Algorithm 2, Algorithm 3 checks whether S is an extension according to *sem*, and, if this is the case, variable x is incremented by 1 (Line 15). Moreover, reasoning as in the case of Algorithm 2, it can be shown that (*i*) the error level $\bar{\epsilon}$ to be taken into account when computing $\widehat{Pr}_{\mathcal{F}}^{sem|\text{AD}}(S)$ is $\bar{\epsilon} = \epsilon / Pr_{\mathcal{F}}^{\text{ad}}(S)$, where ϵ is the error level for estimating $Pr_{\mathcal{F}}^{sem}(S)$, and that (*ii*) the number of AAFs to be sampled to ensure that the error level of $\widehat{Pr}_{\mathcal{F}}^{sem|\text{AD}}(S)$ is $\bar{\epsilon}$ with confidence level $z_{1-\alpha/2}$ is $\overline{n} = \frac{z_{1-\alpha/2}^2 \cdot p \cdot (1-p) \cdot (Pr_{\mathcal{F}}^{\text{ad}}(S))^2}{\epsilon^2} - z_{1-\alpha/2}^2$ (Line 16).

4   Experimental Results

We now show the results of the experiments we carried out to compare the efficiency of the three algorithms presented in the previous section. Specifically, we compare Algorithm 1 (*A1*) with Algorithm 2 (*A2*) and Algorithm 3 (*A3*) in terms of number of generated samples and evaluation times needed to compute the probability $\widehat{Pr}_{\mathcal{F}}^{sem}(S)$ for the complete, grounded and preferred semantics. To this end, we denote as $samples(Ak)$ and $time(Ak)$, with $k \in \{1, 2, 3\}$, the average number of samples and the average execution time of the runs of algorithm Ak, respectively, and we use the following performance measures:

- $ImpS(A2) = \frac{samples(A2)}{samples(A1)}$ and $ImpS(A3) = \frac{samples(A3)}{samples(A1)}$, for measuring the improvement of *A2* and *A3* w.r.t. *A1*, in terms of number of generated samples;
- $ImpT(A2) = \frac{time(A2)}{time(A1)}$ and $ImpT(A3) = \frac{time(A3)}{time(A1)}$, for measuring the improvement of *A2* and *A3* w.r.t. *A1*, in terms of execution time.

To perform a thorough experimental evaluation of the performances of the three algorithms, we varied the size of the set A of arguments of the input PrAF, the composition of its set of defeats and the size of S as follows. We varied the size of A considering every even number from 12 to 40 and, for each $|A|$, we considered 5 PrAFs having different sets of defeats. For each of the so obtained PrAFs, we considered 5 sets S of arguments, whose size was chosen in the interval $[20\%, 40\%] \cdot |A|$, and such that $Pr_{\mathcal{F}}^{\text{cf}}(S)$ and $Pr_{\mathcal{F}}^{\text{ad}}(S)$ ranged in the interval $[.5, .8]$ and $[.4, .7]$, respectively. For each of these combinations of PrAF and S, we made 200 runs of each algorithm for each of the three semantics, and we averaged the number of samples and the execution times.

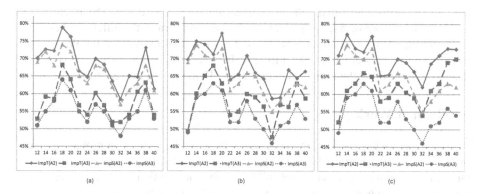

Fig. 1. Improvements of A2 and A3 vs A1 for (a) complete, (b) grounded, (c) preferred semantics

In all the experiments, we considered as error level $\epsilon = 0.005$ and as confidence level $z_{1-\alpha/2} = 95\%$. All experiments have been carried out on an Intel i7 CPU with 6GB RAM running Windows 7.

The results of our experiments are shown in Fig. 1. Specifically, Fig. 1(a) (resp., Fig. 1(b)) reports the results for the complete (resp., grounded) semantics, whereas Fig. 1(c) reports the results for the preferred semantics. On the one hand, the experiments show that, on average, $ImpT(A2)$ is equal to 70%, meaning that the time required by A2 to estimate $Pr_{\mathcal{F}}^{sem}(S)$ is 70% of the time required by A1. As regards the number of generated samples, $ImpS(A2)$ is 65% on average. On the other hand, the experiments show that, on average, $ImpT(A3)$ is equal to 60%, and $ImpS(A3)$ is equal to 55%. Hence, for every semantics, A2 and A3 perform much better than A1.

Moreover, the experiments show that, on average, the time improvement is worse than the sample improvement for both A2 and A3 and all the three considered semantics. Intuitively, this derives from the fact that a single Monte-Carlo iteration of A2 (as well as the single iteration of A3) is on average more time consuming than a single Monte-Carlo iteration of A1. As regards the evaluation times, considering the PrAFs consisting of 40 arguments, A_1 needs on average 3 (resp., 5, 73) ms for estimating $Pr_{\mathcal{F}}^{sem}(S)$ in the case of the complete (resp., grounded, preferred) semantics [2]. We also implemented an algorithm performing the exact computation of $Pr_{\mathcal{F}}^{sem}(S)$ by directly applying the Definition 2, obtaining that it requires prohibitive evaluation times as expected. In particular, considering PrAFs consisting of 18, 19 and 20 arguments, on average, the exact computation required 11 and 72 minutes on PrAFs consisting of 18 and 19 arguments, respectively, while it was halted without giving results after 9 hours on the PrAFs consisting of 20 arguments.

Summing up, since on average $ImpT(A3)$ is about 10% better than $ImpT(A2)$, we can conclude that A3 performs better on our dataset. However, this does not preclude

---

[2] Observe that the higher evaluation time for the preferred semantics is basically due to the fact that the verification problem of checking whether a set of arguments is an extension according to this semantics is $CoNP$-complete, while it is polynomial for the other two semantics [17,15].

that $A2$ performs better than $A3$ on different datasets, where $Pr_{\mathcal{F}}^{cf}(S)$ is closer to $Pr_{\mathcal{F}}^{ad}(S)$ than in our dataset. In fact, in our dataset, on average, the difference between $Pr_{\mathcal{F}}^{cf}(S)$ and $Pr_{\mathcal{F}}^{ad}(S)$ is about 10%. Since it is likely that for most of the PrAFs representing real-world situations, the difference between $Pr_{\mathcal{F}}^{cf}(S)$ and $Pr_{\mathcal{F}}^{ad}(S)$ is higher than 10%, it is fair to claim that $A3$ outperforms $A2$ in most of the real-life contexts.

5   Related Work

Recently approaches for handling uncertainty in AAFs by relying on probability theory have been proposed in [14,25,21,26]. [14] proposed a PrAF where uncertainty is taken into account by specifying probability distribution functions (PDFs) over possible worlds and shown how an instance of the proposed PrAF can be obtained by specifying a probabilistic assumption-based argumentation framework (introduced by themselves). In the same spirit, [25] defined a PrAF as a PDF over the set of possible worlds, and introduced a probabilistic version of a fragment of ASPIC framework [23] that can be used to instantiate the proposed PrAF. [21] proposed a PrAF where probabilities are directly associated with arguments and defeats, instead of being associated with possible worlds, and then proposed a Monte-Carlo simulation approach to estimate the probability $Pr^{sem}(S)$ that a set S is an extension according to semantics sem. In [21], as well as in [14,25], $Pr^{sem}(S)$ is defined as the sum of the probabilities of the possible worlds where S is an extension, according to sem. [26] did not define a probabilist version of a classical semantics, but introduced a new probabilistic semantics based on specifying a class of PDFs, called *p-justifiable* PDFs, over sets of possible AAFs, and shown that this probabilistic semantics generalizes the complete semantics.

Though in the above-cited works probability theory is recognized as a fundamental tool to model uncertainty, a deeper understanding of the role of probability theory in abstract argumentation was developed only later in [19,20], where the connection among argumentation theory, classical logic, and probability theory was investigated.

Besides the approaches that model uncertainty by relying on probability theory, other proposals represent uncertainty by weights or preferences on arguments and/or defeats [8,6,4,22,16,11], or by relying on the possibility theory [5,2,3].

Although the approaches based on weights, preferences, possibilities, or probabilities to model uncertainty have been proved to be effective in different contexts, there is no common agreement on what kind of approach should be used in general. In this regard, [19,20] observed that the probability-based approaches may take advantage from relying on a well-established and well-founded theory, whereas the approaches based on weights or preferences do not conform to well-established theories yet.

6   Conclusions

In this paper, we focused on estimating the probability $Pr_{\mathcal{F}}^{sem}(S)$ that a set S of arguments is an extension for $\mathcal{F}$ according to a semantics sem, where sem is the *complete*, the *grounded*, or the *preferred* semantics. In particular, we proposed two algorithms for estimating $Pr_{\mathcal{F}}^{sem}(S)$, which outperform the state-of-the-art algorithm proposed in [21], both in terms of number of generated samples and evaluation time.

References

1. Agresti, A., Coull, B.A.: Approximate is better than "exact" for interval estimation of binomial proportions. The American Statistician 52(2), 119–126 (1998)
2. Alsinet, T., Chesñevar, C.I., Godo, L., Sandri, S., Simari, G.R.: Formalizing argumentative reasoning in a possibilistic logic programming setting with fuzzy unification. Int. J. Approx. Reasoning 48(3) (2008)
3. Alsinet, T., Chesñevar, C.I., Godo, L., Simari, G.R.: A logic programming framework for possibilistic argumentation: Formalization and logical properties. Fuzzy Sets and Systems 159(10), 1208–1228 (2008)
4. Amgoud, L., Cayrol, C.: A reasoning model based on the production of acceptable arguments. Ann. Math. Artif. Intell. 34(1-3), 197–215 (2002)
5. Amgoud, L., Prade, H.: Reaching agreement through argumentation: A possibilistic approach. In: KR, pp. 175–182 (2004)
6. Amgoud, L., Vesic, S.: A new approach for preference-based argumentation frameworks. Ann. Math. Artif. Intell. 63(2), 149–183 (2011)
7. Baroni, P., Giacomin, M.: Semantics of abstract argument systems. In: Argumentation in Artificial Intelligence, pp. 25–44 (2009)
8. Bench-Capon, T.J.M.: Persuasion in practical argument using value-based argumentation frameworks. J. Log. Comput. 13(3), 429–448 (2003)
9. Bench-Capon, T.J.M., Dunne, P.E.: Argumentation in artificial intelligence. Artif. Intell. 171(10-15), 619–641 (2007)
10. Besnard, P., Hunter, A. (eds.): Elements Of Argumentation. The MIT Press (2008)
11. Coste-Marquis, S., Konieczny, S., Marquis, P., Ouali, M.A.: Weighted attacks in argumentation frameworks. In: KR (2012)
12. Dung, P.M.: On the acceptability of arguments and its fundamental role in nonmonotonic reasoning, logic programming and n-person games. Artif. Intell. 77(2), 321–358 (1995)
13. Dung, P.M., Mancarella, P., Toni, F.: Computing ideal sceptical argumentation. Artif. Intell. 171(10-15), 642–674 (2007)
14. Dung, P.M., Thang, P.M.: Towards (probabilistic) argumentation for jury-based dispute resolution. In: COMMA, pp. 171–182 (2010)
15. Dunne, P.E.: The computational complexity of ideal semantics. Artif. Intell. 173(18) (2009)
16. Dunne, P.E., Hunter, A., McBurney, P., Parsons, S., Wooldridge, M.: Weighted argument systems: Basic definitions, algorithms, and complexity results. Artif. Intell. 175(2) (2011)
17. Dunne, P.E., Wooldridge, M.: Complexity of abstract argumentation. In: Argumentation in Artificial Intelligence, pp. 85–104 (2009)
18. Fazzinga, B., Flesca, S., Parisi, F.: On the complexity of probabilistic abstract argumentation. In: IJCAI (2013)
19. Hunter, A.: Some foundations for probabilistic abstract argumentation. In: COMMA, pp. 117–128 (2012)
20. Hunter, A.: A probabilistic approach to modelling uncertain logical arguments. Int. J. Approx. Reasoning 54(1), 47–81 (2013)
21. Li, H., Oren, N., Norman, T.J.: Probabilistic argumentation frameworks. In: Modgil, S., Oren, N., Toni, F. (eds.) TAFA 2011. LNCS, vol. 7132, pp. 1–16. Springer, Heidelberg (2012)
22. Modgil, S.: Reasoning about preferences in argumentation frameworks. Artif. Intell. 173(9-10), 901–934 (2009)
23. Prakken, H.: An abstract framework for argumentation with structured arguments. Argument & Computation 1(2), 93–124 (2010)
24. Rahwan, I., Simari, G.R. (eds.): Argumentation in Artificial Intelligence. Springer (2009)
25. Rienstra, T.: Towards a probabilistic dung-style argumentation system. In: AT, pp. 138–152 (2012)
26. Thimm, M.: A probabilistic semantics for abstract argumentation. In: ECAI, pp. 750–755 (2012)

AFs with Necessities: Further Semantics and Labelling Characterization

Farid Nouioua

LSIS, Aix-Marseille Univ.,
Avenue Escadrille Normandie Niemen. 13397 Marseille Cedex 20
farid.nouioua@lsis.org

Abstract. The Argumentation Frameworks with Necessities (AFNs) proposed in [17] are a kind of bipolar AFs extending Dung AFs with a support relation having the particular meaning of necessity. This paper is a continuation of this work in two respects. First, we complete the acceptability semantics picture by defining the well-founded, the complete and the semi-stable semantics for AFNs. We show that the proposed semantics keep the same properties as those given for Dung AFs and represent proper generalizations of them (in absence of the necessity relation, the classical semantics are recovered). Then, we show how to generalize Caminada's labelling algorithms in presence of a necessity relation to compute the extensions under the studied semantics for AFNs.

Keywords: abstract argumentation, necessity relation, acceptability semantics, computing extensions, labelling algorithms.

1 Introduction

Dung's abstract argumentation theory is today one of the most influential theories in artificial intelligence approaches for argumentation. One of the numerous extensions of Dung AFs tries to take into account positive interactions between arguments in addition to attacks that represent negative interactions. An interesting question is then to know how to represent this kind of positive interaction and how to handle it without necessarily use an equivalent Dung AF ?

The main approaches to handle supports represent them explicitly. These approaches include the bipolar argumentation frameworks (BAFs) [10] [11], the deductive supports approach [5] and the abstract dialectical frameworks [6] among others. We have discussed in [17] the advantages and the limits of these different proposals and introduced the Argumentation Frameworks with Necessities (AFNs) as a kind of bipolar AFs where the support relation has the particular meaning of "necessity". It has been shown that thanks to this specification of the meaning of support, it was possible to generalize some acceptability semantics (stable and preferred) in a natural way without borrowing techniques from logic programs (LPs) or necessarily making use of a Meta Dung model.

To complete the picture, we continue this research line by exploring further acceptability semantics for AFNs including complete, grounded and semi-stable

W. Liu, V.S. Subrahmanian, and J. Wijsen (Eds.): SUM 2013, LNAI 8078, pp. 120–133, 2013.

semantics. We show in particular how to keep the same "high level" definitions and properties of the different acceptability semantics of AFs in the case of AFNs by just a suitable incorporation of the necessity relation into some "basic level" concepts. Thus, we generalize the fact that unlike other bipolar formalisms, to draw conclusions, it is not necessary to first translate the AFN into a Dung AF or to use techniques from LPs. Moreover, we show that in case where the necessity relation is empty, we recover exactly the original versions of these semantics.

In [9] [8][16], a nice characterization of acceptability semantics using the idea of labelling has been proposed. Extensions under a given semantics are characterized by labellings satisfying some conditions that depend on this semantics. This approach gave rise to the *labelling algorithms* that compute the extensions of an AF under a given semantics. Another contribution of this paper is to generalize Caminada's labelling characterization to the case of AFNs and to propose new labelling algorithms for the acceptability semantics in AFNs.

In section 2. we recall some basics of Dung AFs. Section 3. presents different acceptability semantics for AFNs. In section 4, the labelling approach for acceptability semantics for Dung AFs is generalized to AFNs and section 5 describes adapted labelling algorithms for AFNs able to compute different kinds of extensions (grounded, admissible, preferred, stable and semi-stable). In section 6, we give some concluding remarks and some perspectives for future work.

2   Brief Reminder of Dung AFs

An AF [12] is a pair $\mathcal{A} = \langle AR, att \rangle$ where AR is a set of arguments and att is a binary attack relation on AR. We abuse notation and write S att a (resp. a att S) for $S \subseteq AR$ and $a \in AR$ to denote that b att a (resp. a att b) for some $b \in S$. A subset $S \subseteq \mathcal{A}$ is conflict-free if there are no $a, b \in S$ such that a att b, S defends a if for each $b \in AR$, if b att a then S att b and S is an admissible set if it is conflict-free and defends all its elements. The characteristic function of an AF $\mathcal{A}$ is a function $\mathcal{F}_\mathcal{A} : 2^\mathcal{A} \to 2^\mathcal{A}$ such that : $\mathcal{F}_\mathcal{A}(S) = \{a | S \text{ } defends \text{ } a\}$. We denote by S^+ the set of arguments attacked by $S : S^+ = \{a | S \text{ } att \text{ } a\}$.

Several acceptability semantics have been defined to capture sets of arguments that may be collectively accepted from an AF. Let $S \subseteq AR$. S is a complete extension iff it is admissible and contains any argument it defends. S is a grounded extension iff it is the least fixed point of $\mathcal{F}_\mathcal{A}$. S is a preferred extension iff it is a $\subseteq$-maximal admissible set. S is a stable extension iff S is an admissible set that attacks any argument outside it (i.e., $S^+ = AR \setminus S$) and S is a semi-stable extension iff S is is a complete extension that maximizes $S \cup S^+$.

3   AFNs and Their Acceptability Semantics

An AFN [17] is defined by $\mathcal{G} = \langle AR, att, \mathcal{N} \rangle$ where AR is a set of arguments, att is an attack relation and $\mathcal{N}$ is a necessity relation. att is interpreted exactly as in Dung AFs : a att b means that if a is accepted then b is not accepted. The new relation $\mathcal{N}$ is interpreted in a dual way as follows : For $E \subseteq AR$ and $b \in AR$,

$E \mathcal{N} b$ means that if no argument of E is accepted then b is not accepted (the acceptance of b requires the acceptance of at least one argument of E).

Notice that $E \mathcal{N} b$ does not imply the existence of a "primitive" necessity relation between element(s) of E and b. For example if $E = \{a, c\}$, the expression $E \mathcal{N} b$ may not be replaced neither by the two expressions $\{a\} N b$ and $\{c\} N b$ nor by only one of them. Indeed, using the two expressions together means that to accept b the two arguments a and c must both be accepted whereas $E \mathcal{N} b$ means that the acceptance of b requires at least the acceptance of one of the two argements a or c. By using only the expression $\{a\} N b$ (resp. $\{c\} N b$) the acceptance of b will be totally independent from c (resp. a) which does not correspond no more to the intended meaning of $E \mathcal{N} b$. As an example, suppose that we have the following three rules in a LP : $r_1 : a \leftarrow b.$, $r_2 : a \leftarrow c.$ and $r_3 : d \leftarrow a$. The application of r_3 requires at least r_1 or r_2 to be applied (either only one of them or the two together). This is captured by the expression : $\{r_1, r_2\} \mathcal{N} r_3$. This meaning cannot be expressed neither by writing only $\{r_1\} \mathcal{N} r_3$ or only $\{r_2\} \mathcal{N} r_3$ nor by writing both $\{r_1\} \mathcal{N} r_3$ and $\{r_2\} \mathcal{N} r_3$.

As for the attack relation, we do not suppose any particular property on the necessity relation $\mathcal{N}$.

In this section, we generalize the acceptability semantics to AFNs [1]. The new semantics are defined very similarly to the case of Dung AFs except that instead of using conflict-freeness as a minimal requirement, this notion is strengthened by an additional notion of *coherence* that takes into account the relation $\mathcal{N}$. The notions of defense, characteristic function and the set of attacked arguments by a given set are also adapted to take into account the relation $\mathcal{N}$.

Definition 1. (from [17]). Let $S \subseteq AR$, S is *closed under* $\mathcal{N}^{-1}$ iff for each $a \in S$, if there is $E \subseteq AR$ such that $E \mathcal{N} a$ then $E \cap S \neq \emptyset$. An argument $a \in S$ is *N-Cycle-Feee in* S iff for each $E \subseteq AR$ s.t. $E \mathcal{N} a$, either $E \cap S = \emptyset$ or there is $b \in E \cap S$ s.t. b is N-Cycle-Free in S. S is N-Cycle-Free iff each $a \in S$ is N-Cycle-Free in S. S is *coherent* iff S is N-Cycle-Free and closed under $\mathcal{N}^{-1}$. Finally, S is *strongly coherent* iff S is coherent and conflict-free.

Intuitively, in a coherent set S, each argument is provided by at least one of its necessary arguments and no risk of a deadlock due to necessity cycles is present. We introduce the notion of a *powerful* argument to capture this meaning of coherence at the individual (i.e., the argument) level.

Definition 2. Let $\mathcal{G} = \langle AR, att, \mathcal{N} \rangle$ be an AFN and $S \subseteq AR$. An argument a is powerful in S iff $a \in S$ and there is a sequence $a_0, \ldots, a_k$ of elements of S such that $a_k = a$, there is no $E \subseteq AR$ s.t. $E \mathcal{N} a_0$ and for $1 \leq i \leq k$: for each $E \subseteq AR$, if $E \mathcal{N} a_i$ then $E \cap \{a_0, \ldots, a_{i-1}\} \neq \emptyset$.

Coherent sets are characterized in terms of powerful arguments as follows :

---

[1] The work in [17] has treated this generalization for admissible sets, stable and preferred extensions. Here we consider further acceptability semantics.

Proposition 1. Let $\mathcal{G} = \langle AR,\ att,\ \mathcal{N} \rangle$ be an AFN and $S \subseteq AR$. S is coherent iff each $a \in S$ is powerful in S.

The second ingredient in the generalization of acceptability semantics to AFNs is to redefine the notions of defense, characteristic function and arguments attacked (here we call them deactivated) by a given set of arguments.

Definition 3. (Defense / characteristic function / deactivated arguments). Let $S \subseteq AR$ and $a \in AR$. We say that S defends a iff $S \cup \{a\}$ is coherent and for each $b \in AR$, if b att a then for each coherent subset C of AR that contains b, S att C. Based on this new notion of defense, the characteristic function of an AFN is defined as for classical AFs by $F : 2^{AR} \to 2^{AR}$ where $F(S) = \{a \mid S\ defends\ a\}$. The set of arguments deactivated by S is defined by $S^+ = \{a \mid S\ att\ a$ or there is $E \subseteq AR\ s.t.\ E\ \mathcal{N}\ a\ and\ S \cap E = \emptyset\}$.

Now, we are ready to define the different acceptability semantics of AFNs in a very similar way as in AFs as follows :

Definition 4. (Acceptability semantics for AFNs). Let $S \subseteq AR$. S is : an *admissible set* iff S is strongly coherent and defends all its arguments; a *complete* extension iff S is admissible and contains any argument it defends; a *grounded* extension iff S is the least fixed-point of F; a *preferred* extension iff S is a maximal admissible set; a *stable* extension iff S is a complete extension and $S^+ = AR \setminus S$ and a *semi-stable* extension iff S is a complete extension that maximizes $S \cup S^+$.

It turns out that the main properties of the different acceptability semantics for AFs continue to hold in the case of AFNs.

Proposition 2. Let $S \subseteq AR$. S is :

- an admissible set iff $S \subseteq F(S)$; a complete extension iff $S = F(S)$; a grounded extension iff S is a minimal complete extension; a preferred extension iff S is a maximal complete extension.
- there is exactly one grounded extension; zero, one or several stable extensions and at least one preferred extension.
- each stable extension is semi-stable and each semi-stable extension is preferred. The inverses are not true.
- if stable extensions exist, then they coincide with the semi-stable extensions.

The new semantics for AFNs are proper generalizations of the the corresponding semantics for Dung AFs :

Proposition 3. . Let $\mathcal{G} = \langle AR,\ att,\ \mathcal{N} \rangle$ with $\mathcal{N} = \emptyset$ and $S \subseteq AR$. S is an admissible set (resp. complete, grounded, preferred, stable, semi-stable extension) of $\mathcal{G}$ iff S is an admissible set (resp. complete, grounded, preferred, stable, semi-stable extension) of the AF $\mathcal{F} = \langle AR,\ att \rangle$.

Example 1. . Let us consider the three AFNs depicted in figure 1.
$\mathcal{G}_1 = \langle AR = \{a, b, c\}, att = \{(b, c)\}, \mathcal{N} = \{(\{a\}, b), (\{b\}, a)\}\rangle$,
$\mathcal{G}_2 = \langle AR = \{a, b, c, d\}, att = \{(a, b), (d, c)\}, \mathcal{N} = \{(\{b, c\}, a)\}\rangle$ and
$\mathcal{G}_3 = \langle AR = \{a, b, c, d, e, f, g\}, att = \{(a, b), (b, a), (c, d), (d, e), (e, c), (e, f),$
$(g, g)\}, \mathcal{N} = \{(\{b\}, c), (\{g\}, f)\}\rangle$;

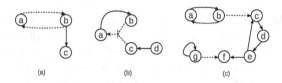

Fig. 1. Figure 1. (a) the AFN $\mathcal{G}_1$, (b) the AFN $\mathcal{G}_2$, (c) the AFN $\mathcal{G}_3$

$\mathcal{G}_1$ has two admissible sets : $\emptyset$ and $\{c\}$. Indeed, c is attacked by b but there is no coherent subset containing b (the only set containing b and closed under $\mathcal{N}$ is $\{a, b\}$ which is not coherent). This means that $\{c\}$ defends its unique element c. We have $\mathcal{F}(\emptyset) = \{c\}$ and $\mathcal{F}(\{c\}) = \{c\}$ and we can check that $\{c\}$ is the unique fixed point of $\mathcal{F}$. Thus, $\{c\}$ is the unique complete extension of $\mathcal{G}_1$ which is its grounded extension and its unique preferred extension. It is also its unique stable and semi-stable extension since $\{c\}^+ = \{a, b\} = AR \setminus \{c\}$.

The strongly coherent sets of $\mathcal{G}_2$ are : $\emptyset$, $\{b\}$, $\{c\}$, $\{d\}$, $\{a, c\}$, $\{b, c\}$, $\{b, d\}$. Only $\emptyset$ and $\{d\}$ defend their elements and are admissible. We have $\mathcal{F}(\emptyset) = \{d\}$ and $\mathcal{F}(\{d\}) = \{d\}$ and we can check that $\{d\}$ is the unique fixed point of $\mathcal{F}$. Thus, $\{d\}$ is the unique complete extension of $\mathcal{G}_2$ which is its grounded extension and its unique preferred extension. We have $\{d\}^+ = \{a, c\} \neq AR \setminus \{d\}$. $\mathcal{G}_2$ has no stable extension and admits $\{d\}$ as a unique semi-stable extension.

$\mathcal{G}_3$ has four admissible sets : $\emptyset$, $\{a\}$, $\{b\}$ and $\{a, d\}$. We have $\mathcal{F}(\emptyset) = \emptyset$, $\mathcal{F}(\{a\}) = \{a, d\}$, $\mathcal{F}(\{b\}) = \{b\}$ and $\mathcal{F}(\{a, d\}) = \{a, d\}$. Thus, $\emptyset$, $\{b\}$ and $\{a, d\}$ are the complete extensions $\mathcal{G}_3$. Among them $\emptyset$ is its grounded extension. $\{b\}$ and $\{a, d\}$ are its preferred extensions. We have $\{b\}^+ = \{a, f\} \neq AR \setminus \{b\}$ and $\{a, d\}^+ = \{b, c, e, f\} \neq AR \setminus \{a, d\}$. $\mathcal{G}_3$ has no stable extension and since $\{b\} \cup \{b\}^+ \subseteq \{a, d\} \cup \{a, d\}^+$, only $\{a, d\}$ is a semi-stable extension of $\mathcal{G}_3$.

3.1 AFNs and AFs

Given an AFN $\mathcal{G} = \langle AR, att, \mathcal{N}\rangle$, a first question we are interested in is to know if it is always possible to find a Dung AF with exactly the same arguments and which contains all the information encoded in $\mathcal{G}$. It has been shown in [17] that the answer is positive when the necessity relation is defined between single arguments (for AFNs where if $E \mathcal{N} a$ then E is a singleton). The idea is to add the implicit attacks that result from the interaction between attacks and necessities as follows : if a attacks b and b is necessary for c then a attacks indirectly c and if a requires b and b attacks c then a attacks indirectly c.

We show here that the answer is negative in the general case and one may need a greater number of arguments to encode all the information of an AFN in an AF. To show this, let us take the AFN $\mathcal{G}_2$ of example 1 and let us suppose that $\mathcal{F} = \langle AR, att' \rangle$ is an AF encoding the same information as $\mathcal{G}_2$. It is clear that $(a, b), (d, c)$ are in att'. The AF $\langle AR, \{(a, b), (d, c)\}\rangle$ does not have the same extensions for all the considered semantics. Apart from these two attacks, any other possible attack from an argument x to an argument y $(x, y \in AR)$ is not present directly or indirectly in $\mathcal{G}_2$. In particular we cannot say that d attacks a because a may be obtained either by having c or b and d attacks only b. The solution is to represent separately the two different ways to obtain a (by providing b and by providing c) as two meta arguments, say A_1 and A_2. Only the second meta argument, involving a and c, is attacked by d.

More generally, given an AFN $\mathcal{G} = \langle AR, att, \mathcal{N} \rangle$ and $a \in AR$, each coherent set $\mathcal{C} \subseteq AR$ containing a and minimal (no subset of $\mathcal{C}$ containing a is coherent) is a meta argument in the AF encoding $\mathcal{G}$. Let AR' denote the set of all such meta arguments. Notice that any non powerful argument in AR will not be represented in AR' since it does not belong to any coherent set. For $\mathcal{C}_1, \mathcal{C}_2 \in AR'$, $\mathcal{C}_1$ attacks $\mathcal{C}_2$ iff x att y for some $x \in \mathcal{C}_1$ and $y \in \mathcal{C}_2$. Let att' be the resulting attack relation.

Theorem 1. . Let $\mathcal{G} = \langle AR, att, \mathcal{N} \rangle$ be an AFN, $\mathcal{F} = \langle AR', att' \rangle$ be the corresponding AF and $S \subseteq AR$. S is a grounded (resp. complete, preferred, stable, semi-stable) extension of $\mathcal{G}$ iff $\mathcal{F}$ admits a set $Y = \{\mathcal{C}_1, ..., \mathcal{C}_n\} \subseteq AR'$ as an extension under the same semantics such that $S = \mathcal{C}_1 \cup ... \cup \mathcal{C}_n$.

Now, there are AFNs whose corresponding AFs contain a number of arguments that is exponential with respect to the number of arguments in the initial AFN. To show this, let us take the example of the AFN $\mathcal{G} = \langle AR, att, \mathcal{N} \rangle$ where $AR = \{a\} \cup AR_1 \cup ... AR_n$, each AR_i contains p arguments $(p > 1)$ and AR_i $mathcalN$ a (for $1 \leq i \leq n$). Let $\mathcal{F}$ be the corresponding AF. The number of arguments in $\mathcal{G}$ is $1 + p \times n$. Each set $\{a, b_1, ... b_n\}$ such that $b_i \in AR_i$ for $1 \leq i \leq n$ is a minimal coherent set containing a, i.e., is a meta argument in $\mathcal{F}$. a gives rize to p^n meta arguments and each $x \in AR \setminus \{a\}$ gives rise to one meta argument $(\{x\})$. The total number of the meta arguments is then $p^n + p \times n$. Thus, even if the information present in an AFN may always be encoded by a Dung AF, the use of an AFN in general, may allow a representation that is significantly more concise than that obtained by using the corresponding AF.

4   A Labelling Characterization

In this section, the labelling approach for acceptability semantics in Dung AFs [16] is generalized to AFNs. Let $\mathcal{G} = \langle AR, att, \mathcal{N} \rangle$ be an AFN. A labelling is a function $\mathcal{L} : AR \longrightarrow \{in, out, undec\}$. We let $in(\mathcal{L}) = \{a \in AR | \mathcal{L}(a) = in\}$, $out(\mathcal{L}) = \{a \in AR | \mathcal{L}(a) = out\}$ and $undec(\mathcal{L}) = \{a \in AR | \mathcal{L}(a) = undec\}$ and we write a labelling $\mathcal{L}$ as a triplet $(in(\mathcal{L}), out(\mathcal{L}), undec(\mathcal{L}))$. The notion of a legal label is generalized in AFNs as follows.

Definition 5. Let a be an argument and $\mathcal{L}$ be a labelling.

- a is legally *in* iff a is labelled *in* and the two following conditions hold : (1) for each argument b, if b *att* a then $b \in out(\mathcal{L})$ (all attackers of a are *out*) and (2) for each set of arguments E, if $E \; \mathcal{N} \; a$ then $E \cap in(\mathcal{L}) \neq \emptyset$ (at least one argument from each necessary set for a is *in*).
- a is legally *out* iff a is labelled *out* and at least one of the two following conditions holds : either (1) there is an argument b such that b *att* a and $b \in in(\mathcal{L})$ (at least one attacker of a is *in*) or (2) there is a set of arguments E, such that $E \; \mathcal{N} \; a$ and $E \subseteq out(\mathcal{L})$ (all the arguments of at least one necessary set for a are *out*).
- a is legally *undec* iff a is labelled *undec* and the three following conditions hold : (1) for each argument b, if b *att* a then $b \notin in(\mathcal{L})$ (no attacker of a is *in*), (2) for each set $E \subseteq AR$, if $E \; \mathcal{N} \; a$ then $E \nsubseteq out(\mathcal{L})$ (not all the arguments of any necessary set for a are *out*) and (3) either there is an argument b such that b *att* a and $b \notin out(\mathcal{L})$ or there is $E \subseteq AR$ such that $E \; \mathcal{N} \; a$ and $E \cap in(\mathcal{L}) = \emptyset$ (either at least one attacker of a is not *out* or at least one necessary set for a does not contain any argument that is *in*).

Notice that for $\mathcal{N} = \emptyset$, we find exactly the original definitions of legal labels given in [16]. In addition to legality of labels, the presence of necessity relation imposes two further constraints. Any argument which is not powerful in AR does not belong to any extension and must be labelled *out* and since each extension E under any semantics must be coherent, the set of *in* arguments of any labelling characterizing any acceptability semantics for an AFN must be coherent. Labellings that satisfy these constraints are called *safe* labellings.

Definition 6. We say that a labelling $\mathcal{L}$ is *safe* iff the set $in(\mathcal{L})$ is coherent and for each $a \in AR$: if a is not powerful in A then $a \in out(\mathcal{L})$.

Notice that in a safe labelling, the relation $\mathcal{N}$ does not play any role in determining whether or not an argument is legally *in*, *out* or *undec*. The decision depends only on the attack relation like in classical AFs. Once the notion of labelling is extended to take into account the necessity relation, the different kinds of labellings are defined as usual except that they must always be safe.

Definition 7. A labelling $\mathcal{L}$ is : *admissible* iff $\mathcal{L}$ is safe and without any illegally *in* or illegally *out* arguments; *complete* iff $\mathcal{L}$ is admissible and without any illegally *undec* arguments; *grounded* iff $\mathcal{L}$ is complete and $in(\mathcal{L})$ is $\subseteq$-minimal; *preferred* iff $\mathcal{L}$ is complete and $in(\mathcal{L})$ is $\subseteq$-maximal; *stable* iff $\mathcal{L}$ is complete and $undec(\mathcal{L}) = \emptyset$ and *semi-stable* iff $\mathcal{L}$ is complete and $undec(\mathcal{L})$ is $\subseteq$-minimal.

For Dung AFs (i.e. an AFN where $\mathcal{N} = \emptyset$), any set of arguments is safe. In this case, we obtain exactly the classical definitions for legally *in*, *out* and *undec* arguments and for the different kinds of labellings. The relationship between labellings and acceptability semantics for AFNs is given as follows.

Theorem 2. S is an admissible set (resp. complete, grounded, preferred, stable, semi-stable extension) iff there is an admissible (resp. complete, grounded, preferred, stable, semi-stable) labelling $\mathcal{L}$ such that $S = in(\mathcal{L})$.

Example 1 (cont.). Let us take again the AFNs of example 1.

Consider the labellings: $\mathcal{L}_1 = (\{c\}, \{b\}, \{a\})$, $\mathcal{L}_2 = (\emptyset, \emptyset, \{a, b, c\})$, $\mathcal{L}_3 = (\emptyset, \{a, b\}, \{c\})$, and $\mathcal{L}_4 = (\{c\}, \{a, b\}, \emptyset)$ for $\mathcal{G}_1$. In $\mathcal{L}_1$ c is legally in because $\mathcal{L}_1(b) = out$ but b is illegally out. Moreover, $\mathcal{L}_1$ is not safe because b is not powerful in AR but $a \notin out(\mathcal{L}_1)$. Thus, $\mathcal{L}_1$ is not admissible. $\mathcal{L}_2$ is not safe for the same reason and thus, is not admissible. In $\mathcal{L}_3$, a and b are legally out but c is illegally $undec$. $\mathcal{L}_3$ is admissible but not complete. In $\mathcal{L}_4$ c is legally in and a and b are legally out. Moreover, $\mathcal{L}_4$ is safe and thus it is admissible and complete (no argument is illegally $undec$ in $\mathcal{L}_4$). In summary, we can verify that $\mathcal{L}_3$ and $\mathcal{L}_4$ are the admissible labellings of $\mathcal{G}_1$ and $\mathcal{L}_4$ is its unique complete labelling which is also its unique grounded and preferred labelling. Moreover, since $undec(\mathcal{L}_4) = \emptyset$, $\mathcal{L}_4$ is also the unique stable and semi-stable labelling of $\mathcal{G}_1$.

$\mathcal{L}_1 = (\emptyset, \emptyset, \{a, b, c, d\})$ and $\mathcal{L}_2 = (\{d\}, \{c\}, \{a, b\})$ are the admissible labellings of $\mathcal{G}_2$. $\mathcal{L}_2$ is the only complete labelling of $\mathcal{G}_2$ (d is illegally $undec$ in $\mathcal{L}_1$) which is also its unique grounded and preferred labelling. Moreover, since $undec(\mathcal{L}_2) \neq \emptyset$, $\mathcal{L}_4$ is not stable but it is a semi-stable labelling of $\mathcal{G}_1$.

$\mathcal{L}_1 = (\emptyset, \emptyset, \{a, b, c, d, e, f, g\})$, $\mathcal{L}_2 = (\{a\}, \{b, c\}, \{d, e, f, g\})$, $\mathcal{L}_3 = (\{b\}, \{a\}, \{c, d, e, f, g\})$ and $\mathcal{L}_4 = (\{a, d\}, \{b, c, e\}, \{f, g\})$ are the admissible labellings of $\mathcal{G}_3$. Among them, only $\mathcal{L}_2$ is not complete (d is illegally $undec$ in $\mathcal{L}_2$). The grounded labelling (which minimizes the in arguments) is $\mathcal{L}_1$, the preferred labellings (which maximize the in arguments) are $\mathcal{L}_3$ and $\mathcal{L}_4$. No complete labelling has an empty set of $undec$ arguments, thus no labelling is stable. The only semi-stable stable labelling (which minimizes $undec$ arguments) is $\mathcal{L}_4$.

5   Labeling Algorithms for AFNs

In this section we adapt the labelling algorithms given in [16] for grounded, preferred, stable and semi-stable extensions to the case of AFNs. We show that, a suitable handling of the necessity relations allows us to keep the main form of the existing algorithms proposed originally for Dung AFs in the case of AFNs.

5.1   Grounded Semantics

For Dung AFs, the grounded labelling is constructed by a successive application of the characteristic function starting from the empty set and stopping when a fixed point is found. First, the algorithm labels by in all the arguments that are not attacked. Then each argument attacked by these arguments is labelled out. After that, any argument whose all attackers are labelled out is labelled in, and so on. When no more arguments my be labelled in or out, the algorithm gives to the remaining unlabelled arguments the $undec$ label and stops.

We propose a similar algorithm for the grounded semantics in AFNs with two additional considerations : at the very beginning of the algorithm, all the non powerful arguments in AR are labelled out[2]; and the addition of new in or out arguments takes account of both the attack and the necessity relations.

The first point is required since the non powerful arguments must not be present in the grounded extension and their attacks must not be taken into account. Once this first operation is done, it is straightforward to verify that the set of in arguments obtained after each iteration is coherent. For the second point, the condition of adding a new in argument is strengthened : an argument becomes in not only if all of its attackers are out but in addition if at least an argument of each set that is necessary for it is in. The condition of adding a new out argument is weakened : an argument becomes out either if one of its attackers is in or if all the arguments of one of its necessary sets are out. Here is the algorithm that computes the grounded labelling of an AFN :

```
01. NB-ARGs := ∅;
02. L₀ := (∅, NB-ARGs, ∅)
03. repeat
04.     in(Lᵢ₊₁) := in(Lᵢ) ∪ {x|x is not labelled in Lᵢ and ∀y ∈ AR :
05.         if y att x then y ∈ out(Lᵢ) and ∀E ⊆ AR : if E N x then
06.         E ∩ in(Lᵢ) ≠ ∅};
07.     out(Lᵢ₊₁) := out(Lᵢ) ∪ {x|x is not labelled in Lᵢ and either
08.         ∃y ∈ AR : y att x and y ∈ in(Lᵢ) or ∃E ⊆ AR : E N x and
09.         E ⊆ out(Lᵢ₊₁)};
10. until (Lᵢ₊₁ = Lᵢ)
11. L_G := (in(Lᵢ), out(Lᵢ), AR \ (in(Lᵢ) ∪ out(Lᵢ)))
```

5.2   Preferred, Stable and Semi-stable Semantics

The original labelling algorithms for preferred, stable and semi-stable semantics are adapted to the case of AFNs by injecting two main modifications : introducing a new operation (*the cleaning operation*) at the beginning of each iteration of the algorithm and modifying slightly the so-called transition step. The operation of cleaning a labelling $\mathcal{L}$ filters out the arguments that are not powerful in $in(\mathcal{L})$ in order to ensure coherence in the extensions[3].

---

[2] The detection of arguments of a set S of arguments that are not powerful is performed by a simple iterative algorithm. This algorithm starts by taking up each argument $a \in in(\mathcal{L})$ for which there is no $E \subseteq AR$ such that $E \, \mathcal{N} \, a$. Then at the iteration i, if the set of all the arguments obtained so far is H_{i-1} then $H_i = H_{i-1} \cup \{a | if \, E \, \mathcal{N} \, a \, then \, E \cap H_{i-1} \neq \emptyset\}$. The algorithm stops as soon as no more argument can be added ($H_i = H_{i-1}$). The remaining arguments : $in(\mathcal{L}) \setminus H_i$ are all the non powerful arguments.

[3] The application of the cleaning operation at the beginning of the algorithm filters out the non powerful arguments in AR as in the case of the algorithm for the grounded extension. We need to use this operation again at the beginning of each subsequent iteration because it is possible that a subset of a coherent set is not coherent.

Definition 8. Let $\mathcal{L}$ be a labelling without illegally *out* arguments. The labelling resulting from *cleaning* $\mathcal{L}$, denoted $\mathcal{C}(\mathcal{L})$, is obtained by: (1) for each argument a of $in(\mathcal{L})$ which is not powerful in $in(\mathcal{L})$, change the label of a from *in* to *out*, (2) then, while there remain illegally *out* arguments, change their label to *undec*.

The first step changes simultaneously to *out* all the *in* arguments that are not powerful in $\mathcal{L}$. In fact, these arguments form collectively a part of the *in* arguments that must not be kept. The second step is a propagation step that ensures that it does not remain any argument that is illegally *out* in the resulting labelling. It is obvious that the following holds :

Proposition 4. If $\mathcal{L}$ is a labelling then $\mathcal{C}(\mathcal{L})$ is safe and does not contain any argument that is illegally *out*.

In a classical transition step, the label of an illegally *in* argument a is changed to *out*. Then, only a or some of the arguments attacked by a may become illegally *out* and must receive the *undec* label. In AFNs, because of the necessity relation, further arguments may become illegally *out* and must be labelled *undec*. The process must be repeated until no more such argument remains. Of course, if $\mathcal{N} = \emptyset$ then we find the classical definition of a transition step.

Definition 9. Let $\mathcal{L}$ be a safe labelling and a be an argument illegally *in* in $\mathcal{L}$. A transition step on a in $\mathcal{L}$ consists of the following : (1) change the label of a from *in* to *out*; then (2) while there remain elements in $out(\mathcal{L})$ that are illegally *out*, change their labels to *undec*.

Based on the previous notions, a transition sequence is defined as a list $[\mathcal{C}(\mathcal{L}_0), a_1, \mathcal{C}(\mathcal{L}_1), \ldots, a_n, \mathcal{C}(\mathcal{L}_n)]$ ($n \geq 0$) where each argument a_i ($0 \leq i \leq n$) is illegally *in* in $\mathcal{C}(\mathcal{L}_{i-1})$ and $\mathcal{L}_n$ is the resulting labelling of the transition step on a_i in $\mathcal{C}(\mathcal{L}_{i-1})$. A transition sequence is terminated iff $\mathcal{C}(\mathcal{L}_n)$ does not contain any argument that is illegally *in*. As in the classical case, throughout a transition sequence, the number of arguments labelled *in* is not increasing while the number of arguments labelled *undec* is not decreasing.

Proposition 5. Let $[\mathcal{C}(\mathcal{L}_0), a_1, \mathcal{C}(\mathcal{L}_1), \ldots, a_n, \mathcal{C}(\mathcal{L}_n)]$ be a transition sequence. For each $i \geq 0 : in(\mathcal{C}(\mathcal{L}_{i+1})) \subseteq in(\mathcal{C}(\mathcal{L}_i))$ and $undec(\mathcal{C}(\mathcal{L}_i)) \subseteq undec(\mathcal{C}(\mathcal{L}_{i+1}))$.

An immediate consequence of the previous proposition is that if the number of arguments is finite then, each terminated sequence is also finite. Now, the following result shows that all the admissible sets and only them are found from all the possible terminated transition sequences starting from the *all-in labelling* (the labelling where all the arguments are *in*).

Theorem 3. If $[\mathcal{C}(\mathcal{L}_0), a_1, \mathcal{C}(\mathcal{L}_1), \ldots, a_n, \mathcal{C}(\mathcal{L}_n)]$ $(n \geq 0)$ be a terminated transition sequence where $\mathcal{L}_0$ is the *all-in labelling* then $\mathcal{C}(\mathcal{L}_n)$ is an admissible labelling. Inversely, for each admissible labelling $\mathcal{L}$ there exists a terminated transition sequence $[\mathcal{C}(\mathcal{L}_0), a_1, \mathcal{C}(\mathcal{L}_1), \ldots, a_n, \mathcal{C}(\mathcal{L}_n)]$ $(n \geq 0)$ where $\mathcal{L}_0$ is the all-in labelling and $\mathcal{C}(\mathcal{L}_n) = \mathcal{L}$.

Admissible sets are computed by constructing a tree with $\mathcal{C}$ *(all-in labelling)* as a root and for each node $\mathcal{C}(\mathcal{L})$ and each a that is illegally *in* in $\mathcal{C}(\mathcal{L})$, if $\mathcal{L}'$ is the result of a transition step on a in $\mathcal{C}(\mathcal{L})$ then $\mathcal{C}(\mathcal{L}')$ is a child for $\mathcal{C}(\mathcal{L})$. The leaves of the tree (the set of the final elements of all the terminated transition sequences) correspond to the admissible sets.

To compute the extensions under a given semantics, some optimization techniques are proposed in [16] for Dung AFs. Fortunately, these techniques continue to be valid for AFNs and so for LPs. Let us recall them briefly. At a given point of the algorithm, let Σ be the set of candidate extensions found so far.

For preferred (resp. semi-stable) semantics, we keep the admissible labellings that maximize (resp. minimize) the set of *in* (resp. *undec*) arguments. Thus, if the current labelling $\mathcal{L}$ is such that $in(\mathcal{L}) \subseteq in(\mathcal{L}')$ (resp. $undec(\mathcal{L}') \subseteq undec(\mathcal{L})$) for some $\mathcal{L}' \in \Sigma$ then we stop developing the current branch. Indeed, from proposition 4., any descendant $\mathcal{L}''$ of $\mathcal{L}$ is worse than $\mathcal{L}'$ because $in(\mathcal{L}'') \subseteq in(\mathcal{L}')$ (resp. $undec(\mathcal{L}') \subseteq undec(\mathcal{L}'')$). Moreover, if a new labelling $\mathcal{L}$ is found, $\mathcal{L}$ is added to the set of candidate labellings and any candidate labelling that is worse than $\mathcal{L}$ is removed. For stable semantics, we keep only admissible labellings where the set of *undec* arguments is empty. Thus, if the current labelling $\mathcal{L}$ is such that $undec(\mathcal{L}) \neq \emptyset$ we stop developing the current branch. Indeed, from proposition 4., for any descendant $\mathcal{L}''$ of $\mathcal{L}$ we will have $undec(\mathcal{L}'') \neq \emptyset$. So, the current branch may not yield a stable labelling.

A third optimization technique uses the so-called *superillegally in* arguments. An illegally *in* argument is also superillegally *in* iff it is attacked by an argument which is either legally *in* or legally *undec*. In presence of a superillegally *in* arguments, it suffices to perform a transition step on one of them instead of performing it on all the illegally *in* arguments. The reason is that if a is superillagally *in* in a labelling $\mathcal{L}$ then, any transition sequence that starts from $\mathcal{L}$ and do not perform any transition step on a yields a labelling in which a stays illegally *in*. Fortunately, this reason continue to be valid for AFNs :

Proposition 6. Let $\mathcal{L}_0$ be a safe labelling where an argument a is superillegally *in* and $[\mathcal{L}_0, a_1, \mathcal{C}(\mathcal{L}_1), \ldots, a_n, \mathcal{C}(\mathcal{L}_n)]$ is a transition sequence where $a \notin \{a_1, \ldots, a_n\}$ then a is illegally *in* in $\mathcal{C}(\mathcal{L}_n)$.

In addition, since a transition step is always performed on a safe labelling, the relation $\mathcal{N}$ is not used in determining illegally *in* or superillegally *in* arguments. Now, let us give the algorithm to compute preferred labellings and describe how to modify it in order to compute stable and semi-stable labellings.

```
01. set-labels = ∅; find-labels(all-in);
02. print set-labels; end;
03. procedure find-labels(Lab)
04.    #cleaning the laballing Lab
05.    L = clean(Lab);
06.    #if L is worse than an existing labelling, then stop
07.    if ∃L' ∈ set-labels : in(L) ⊂ in(L') then
08.       return
09.    #if a transition sequence has terminated, then remove
10.    #the worse candidates, add the new labelling and stop
11.    if no argument is illegally in in L
12.       for each L' ∈ set-labels do
13.          #if L''s in arguments are a subset of
14.          #L's in arguments then remove L'
15.          if in(L') ⊂ in(L) then
16.             set-labels := set-labels \ {L'}
17.          end if
18.       end for
19.       #add L as a new candidate
20.       set-labels := set-labels ∪ {L}
21.       return;
22.    else
23.       if ∃ superillegally in arguments in L  then
24.          x := some superillegally argument in L;
25.          find-labels(transition-step(L,x));
26.       else
27.          for each x that is illegally in in L do
28.             find-labels(transition-step(L,x));
29.          end for
30.       end if
31.    end if
32. end.
```

For the semi-stable semantics, the tree is pruned when the set of *undec* arguments in the current labelling is a superset of the *undec* arguments in an existing candidate. Thus, line 7 must be replaced by : " if ∃L' ∈ set-labels : $undec(L') \subset undec(L)$ then". For stable semantics the tree is pruned as soon as the current labelling has a non empty set of *undec* arguments. Thus, we have to replace the line 7 by : " if $undec(L) \neq \emptyset$ then". Moreover, since a stable extension is never strictly included in another one, lines 12-18 can be removed.

6 Concluding Remarks and Perspectives

In this paper, we continue the work presented in [17] about the AFNs in two directions. First, we introduce new acceptability semantics for AFNs, namely the complete, the grounded and the semi-stable semantics. We define all these semantics in a very similar way to the corresponding semantics defined for classical AFs by injecting the necessity relation in defining some elementary notions. The semantics proposed for AFNs keep the same properties and relationships between each others as the corresponding classical semantics and they represent proper generalizations of them (they coincide with them if the necessity relation is empty). We show also that any AFN may be represented as an equivalent Dung AF. However, in the general case, an AFN may express information that cannot be encoded in the equivalent AF without using an exponential number of arguments with respect to the number of arguments in the original AFN. This means that, from a representational point of view, there are situations where AFNs are significantly more concise in representing knowledge than classical Dung AFs. The second contribution of this work is a generalization of Caminada's labelling characterization and algorithms proposed for Dung AFs to the case of AFNs. we show that by suitable handling of the necessity relation, the modifications required in this generalization are not so significant and the main form of the original algorithm is kept. Moreover, if the necessity relation is empty, we recover exactly the original labelling algorithms used for Dung AFs.

Like Dung AFs, AFNs remain very abstract and do not suppose any internal structures of arguments. Their practical use requires an instantiation step that allows to build arguments as well as attacks and necessities from "concrete" knowledge bases. Several works have been proposed in this domain and most of them use knowledge bases expressed either in classical logic or as logic programs. The first class of works (see for example [1], [15]) follows the logical-based argumentation approach [4] and constructs arguments from a classical logic knowledge base $\mathcal{K}$ as couples of the form (*support, claim*) where *support* is a minimal set of formulas of $\mathcal{K}$ that entails *claim* which is a formula of $\mathcal{K}$. Recent work (see [2] [3]) has seriously challenged the adequacy of the acceptability semantics of Dung AFs in systems obtained by such instantiation.

Other works construct arguments from knowledge bases expressed as LPs. In [12] arguments are build from LPs as couples (K, c) where c is a defeasible consequence of K and in [7] arguments are build as tree like structures of rules. In [17] each rule is an argument. This kind of instantiation gives concrete account of AFs and sheds more light on the possible links between argumentation and LPs. Some links between argumentation acceptability semantics and LP semantics have been established as the link between : stable models in LPs and stable semantics in AFs [12] or in AFNs [17], well-founded semantics in LPs and grounded semantics in AFs [12] and partial stable models in LPs and complete semantics in AFs [18]. A natural perspective is then to continue the investigation for further possible links between the other semantics proposed both in logic programming (like L-stable semantics [14]) and in argumentation theory (like ideal semantics [13]), by using our instantiation method. Another perspective is to use the labelling algorithms to propose new solvers for both argumentation theory and logic programming under other semantics than the stable semantics for which very efficient solvers are available today. Last but not least, we want to generalize the dialectical proof procedures proposed for Dung AFs (see [16]) to AFNs and also for LPs. This is beneficial in practice when the goal is to know if an argument (resp. atom) is accepted or not under a given semantics wrt some acceptability criteria (skeptical, credulous, ...) without computing all the extensions (resp. models) under the considered semantics.

Acknowledgments. This work is funded by the ANR (Agence Nationale de la Recherche), ASPIQ Project. Ref: ANR-12-BS02-0003.

References

1. Amgoud, L., Besnard, P.: Bridging the gap between abstract argumentation systems and logic. In: Godo, L., Pugliese, A. (eds.) SUM 2009. LNCS, vol. 5785, pp. 12–27. Springer, Heidelberg (2009)
2. Amgoud, L.: Stable Semantics in Logic-Based Argumentation. In: Hüllermeier, E., Link, S., Fober, T., Seeger, B. (eds.) SUM 2012. LNCS, vol. 7520, pp. 58–71. Springer, Heidelberg (2012)
3. Amgoud, L.: The Outcomes of Logic-Based Argumentation Systems under Preferred Semantics. In: Hüllermeier, E., Link, S., Fober, T., Seeger, B. (eds.) SUM 2012. LNCS, vol. 7520, pp. 72–84. Springer, Heidelberg (2012)
4. Besnard, P., Hunter, A.: Elements of Argumentation. The MIT Press (2008)
5. Boella, G., Gabbay, D.M., Van Der Torre, L., Villata, S.: Support in Abstract Argumentation. In: Proceedings of COMMA 2010, pp. 40–51 (2010)
6. Brewka, G., Woltran, S.: Abstract Dialectical Frameworks. In: Proceedings of KR 2012, Toronto, Canada, pp. 102–111 (2010)
7. Caminada, M.W.A., Carnielli, W.A., Dunne, P.E.: Semi-Stable Semantics. J. Log. Comp. 22(5), 1207–1254 (2012)
8. Caminada, M.W.A.: Semi-Stable Semantics. In: Proceedings of COMMA 2006, Liverpool, UK, pp. 121–130 (2006)
9. Caminada, M.W.A.: An Algorithm for Computing Semi-Stable Semantics. In: Mellouli, K. (ed.) ECSQARU 2007. LNCS (LNAI), vol. 4724, pp. 222–234. Springer, Heidelberg (2007)
10. Cayrol, C., Lagasquie-Schiex, M.C.: On the acceptability of arguments in bipolar argumentation frameworks. In: Godo, L. (ed.) ECSQARU 2005. LNCS (LNAI), vol. 3571, pp. 378–389. Springer, Heidelberg (2005)
11. Cayrol, C., Lagasquie-Schiex, M.C.: Coalitions of arguments: A tool for handling bipolar argumentation frameworks. Int. J. Intell. Syst. 25(1), 83–109 (2010)
12. Dung, P.M.: On the acceptability of arguments and its fundamental role in nonmonotonic reasoning, logic programming and n-person games. Artif. Intel. 77, 321–357 (1995)
13. Dung, P.M., Mancarella, P., Toni, F.: Computing ideal sceptical argumentation. Artif. Intel. 171(10-15), 642–674 (2007)
14. Eiter, T., Leone, N., Sacca, D.: On the partial semantics for disjunctive deductive a databases. Ann. Math. Artif. Intell. 19(1-2), 59–96 (1997)
15. Gorogiannis, N., Hunter, A.: Instantiating abstract, argumentation with classical logic arguments: Postulates and properties. Artif. Intel. 175, 1479–1497 (2011)
16. Modgil, S., Caminada, M.W.A.: Proof Theories and Algorithms for Abstract Argumentation Frameworks. In: Rahwan, I., Simari, G. (eds.) Argumentation in Artificial Intelligence, pp. 105–129 (2009)
17. Nouioua, F., Risch, V.: Argumentation Frameworks with Necessities. In: Benferhat, S., Grant, J. (eds.) SUM 2011. LNCS, vol. 6929, pp. 163–176. Springer, Heidelberg (2011)
18. Wu, Y., Caminada, M., Gabbay, D.: Complete Extensions in Argumentation Coincide with 3-Valued Stable Models in Logic Programming. Studia logica 93(2-3), 383–403 (2009)

Ranking-Based Semantics for Argumentation Frameworks

Leila Amgoud and Jonathan Ben-Naim

IRIT – CNRS,
118, route de Narbonne, 31062, Toulouse Cedex 09, France

Abstract. An argumentation system consists of a set of interacting arguments and a semantics for evaluating them. This paper proposes a new family of semantics which rank-orders arguments from the most acceptable to the weakest one(s). The new semantics enjoy two other main features: i) an attack weakens its target but does not kill it, ii) the number of attackers has a great impact on the acceptability of an argument. We start by proposing a set of rational postulates that such semantics could satisfy, then construct various semantics that enjoy them.

1 Introduction

Argumentation is a reasoning model based on the construction and evaluation of interacting arguments. The most popular semantics were proposed by Dung in his seminal paper [6]. Those semantics as well as their refinements (e.g. in [3,5]) partition the powerset of the set of arguments into two classes: *extensions* and *non-extensions*. Every extension represents a coherent point of view. An *absolute* status is assigned to each argument: *accepted* (if it belongs to every extension), *rejected* (if it does not belong to any extension), and *undecided* if it is in some extensions and not in others. Those semantics are based in particular on the following considerations:

- *Killing:* The impact of an attack from an argument b to an argument a is drastic, that is, if b belongs to an extension, then a is automatically excluded from that extension (i.e., a is killed).
- *Existence:* One successful attack against an argument a has the same effect on a as any number of successful attacks. Indeed, one such attack is sufficient to kill a, several attacks cannot kill a to a greater extent.
- *Absoluteness:* The three possible status of the arguments are absolute, that is, they make sense even without comparing them with each other.
- *Flatness:* All the accepted arguments have the same level of acceptability.

These four considerations seem rational in applications like paraconsistent reasoning. For example, the killing consideration makes sense in this application, because arguments are formulas and attacks correspond to contradictions, and it is natural to consider that one contradiction is lethal.

However, in other applications, e.g. decision-making, some of these considerations are debatable. First, the killing principle is problematic in decision-making, because an attack does not necessarily kill its target, but just weakens it. Suppose for instance that the two following arguments a and b are exchanged by two doctors:

W. Liu, V.S. Subrahmanian, and J. Wijsen (Eds.): SUM 2013, LNAI 8078, pp. 134–147, 2013.

a: The patient should have a surgery since he has cancer.

b: The statistics show that the probability that a surgery will improve the state of the patient is low.

In this case, the attack from *b* only weakens *a*, it does not kill *a*. The doctor may still choose to do the surgery since it gives (a small) chance for the patient to survive.

Next, the existence consideration is also debatable. Suppose a seller provides the following argument *a* in favor of a given car:

a: This car is certainly powerful since it is made by Peugeot.

b_1: The engines of Peugeot cars break down before 300000km.

b_2: The airbags of Peugeot cars are not reliable.

b_3: The spare part is very expensive.

If the buyer receives the argument b_1 against Peugeot (thus against *a*), then he accepts less *a*. The situation becomes worse if he receives b_2 and b_3. Indeed, the more arguments he receives against *a*, the less his confidence in *a*.

The flatness consideration is also debatable in decision-making. Suppose for example that *a* is not attacked, *b* is attacked only by *a*, and *c* is attacked only by *b*. Then, *a* and *c* are both accepted and have the same level of acceptability. But, in applications like decision-making, it is reasonable to consider that an attack from a non-attacked node (or any number of non-attacked nodes) does not kill the destination node. So, *b* is only weakened, which means that its attack against *c* should have some effect, that is, the level of acceptability of *c* should be lower than that of *a*.

To sum up, existing semantics may be well-suited for reasoning but not for applications like decision-making. In the present paper, we propose a new family of semantics that are based on the following *graded* considerations:

- *Weakening:* Arguments cannot be killed (however, they can be weakened to an extreme extent). As a consequence, an attack from an argument *b* to an argument *a* always decreases the degree of acceptability of *a* (possibly only by an infinitesimal amount). The greater the acceptability of *b*, the greater the decrease in the acceptability of *a*.
- *Counting:* The more numerous the attacks against *a*, the greater the decrease in the acceptability of *a*.
- *Relativity:* The degrees of acceptability of the arguments are relative, that is, they do not make sense when they are not compared with each other.
- *Graduality:* There is an arbitrarily large number of degrees of acceptability.

In our approach, a semantics is a function that transforms any *argumentation graph* into a *ranking* on its set of arguments: from the most accepted to the weakest one(s). Our first step consists in proposing formal postulates, each of which is an intuitive and desirable property that a semantics may enjoy. Our postulates are based on the four informal graded considerations described earlier: weakening, counting, relativity, and graduality. Such an axiomatic approach allows a better understanding of semantics and a more precise comparison between different proposals. We investigate dependencies and compatibilities between postulates. In a second step, we construct two ranking-based semantics satisfying certain postulates.

2   Ranking-Based Semantics

An argumentation framework consists of a set of arguments and a set of attacks between them. Arguments represent reasons to believe in statements, doing actions, etc. Attacks express conflicts between pairs of arguments. In what follows, both components are assumed to be abstract entities.

Definition 1 (Argumentation framework). *An* argumentation framework *is an ordered pair* $\mathbf{A} = \langle \mathcal{A}, \mathcal{R} \rangle$, *where* $\mathcal{A}$ *is a finite set of* arguments *and* $\mathcal{R}$ *a binary relation on* $\mathcal{A}$ *(i.e.,* $\mathcal{R} \subseteq \mathcal{A} \times \mathcal{A}$*). We call* $\mathcal{R}$ *an* attack relation *and* $a\mathcal{R}b$ *means that* a *attacks* b.

We turn to the notion of attacker:

Notation. *Let* $\mathbf{A} = \langle \mathcal{A}, \mathcal{R} \rangle$ *be an argumentation framework and* $a \in \mathcal{A}$. *We define that* $\mathrm{Arg}(\mathbf{A}) = \mathcal{A}$ *and* $\mathrm{Att}_{\mathbf{A}}(a) = \{b \in \mathcal{A} \mid b\mathcal{R}a\}$. *When the context is clear, we write* $\mathrm{Att}(a)$ *for short. The same goes for all notations.*

As in classical approaches to argumentation [6], since arguments may be conflicting, it is important to evaluate them and to identify the ones to rely on for inferring conclusions (in case of handling inconsistency in knowledge bases) or making decisions, etc. For that purpose, we propose ranking-based semantics which rank-order the set of arguments from the most acceptable to the weakest one(s). Thus, unlike existing semantics which assign an *absolute* status (accepted, rejected, undecided) to each argument, the new approach compares pairs of arguments.

Definition 2 (Ranking). *A* ranking *on a set* $\mathcal{A}$ *is a binary relation* $\preceq$ *on* $\mathcal{A}$ *such that:* $\preceq$ *is total (i.e.,* $\forall a, b \in \mathcal{A}$, $a \preceq b$ *or* $b \preceq a$*) and transitive (i.e.,* $\forall a, b, c \in \mathcal{A}$, *if* $a \preceq b$ *and* $b \preceq c$, *then* $a \preceq c$*). Intuitively,* $a \preceq b$ *means that* a *is* at least as acceptable *as* b. *So,* $b \not\preceq a$ *means that* a *is* strictly more acceptable *than* b.

We emphasize that, unlike in certain other works, the equal-or-more acceptable argument in an expression of the form $a \preceq b$ is on the *left-hand side* (i.e., a takes precedence over b; the rank of a is above that of b; etc.).

Definition 3 (Ranking-based semantics). *A* ranking-based semantics *is a function* **S** *that transforms any argumentation framework* $\mathbf{A} = \langle \mathcal{A}, \mathcal{R} \rangle$ *into a ranking on* $\mathcal{A}$.

A ranking should not be arbitrary, but should obey some postulates. By postulate, we mean any reasonable principle, be it very general or very specific.

3   Postulates for Semantics

First of all, a ranking on a set of arguments should be defined only on the basis of the attacks between arguments, it should not depend on the identity of the arguments (at least when the data only consist of nodes and arrows). So, our first postulate says that two equivalent argumentation frameworks should give rise to two equivalent rankings. Let us first define the notion of equivalence between two argumentation frameworks.

Definition 4 (Isomorphism). *Let* $\mathbf{A} = \langle \mathcal{A}, \mathcal{R} \rangle$ *and* $\mathbf{A}' = \langle \mathcal{A}', \mathcal{R}' \rangle$ *be two argumentation frameworks. An* isomorphism *from* $\mathbf{A}$ *to* $\mathbf{A}'$ *is a bijective function* f *from* $\mathcal{A}$ *to* $\mathcal{A}'$ *such that* $\forall\, a, b \in \mathcal{A}, a\mathcal{R}b$ *iff* $f(a)\mathcal{R}'f(b)$.

We define formally our first postulate and then exemplify it.

Postulate 1 (Abstraction). *A ranking-based semantics* $\mathbf{S}$ *satisfies* abstraction (Ab) *iff for any two frameworks* $\mathbf{A} = \langle \mathcal{A}, \mathcal{R} \rangle$ *and* $\mathbf{A}' = \langle \mathcal{A}', \mathcal{R}' \rangle$, *for any isomorphism* f *from* $\mathbf{A}$ *to* $\mathbf{A}'$, *we have that* $\forall\, a, b \in \mathcal{A}$, $\langle a, b \rangle \in \mathbf{S}(\mathbf{A})$ *iff* $\langle f(a), f(b) \rangle \in \mathbf{S}(\mathbf{A}')$.

Example 1. *Consider the two argumentation frameworks depicted in the figure below.*

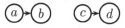

The postulate (Ab) *ensures that the ranking relation between a and b is the same as the one between c and d.*

It is worth pointing out that extension-based semantics (i.e., Dung's semantics) obey in some sense this postulate. For instance, both argumentation frameworks of Example 1 have one preferred extension containing the non-attacked argument (a, resp. c).

The second postulate states the following: the question whether an argument a is at least as acceptable as an argument b should be independent of any argument c that is neither connected to a nor to b, that is, there is no path from c to a or b (ignoring the direction of the edges). Let us first define the independent parts of an argumentation framework.

Definition 5 (Weak connected component). *A* weak connected component *of an argumentation framework* $\mathbf{A}$ *is a maximal subgraph of* $\mathbf{A}$ *in which any two vertices are connected to each other by a path (ignoring the direction of the edges). We denote by* Com($\mathbf{A}$) *the set of every argumentation framework* $\mathbf{B}$ *such that* $\mathbf{B}$ *is a weak connected component of* $\mathbf{A}$ *or the graph union of several weak connected components of* $\mathbf{A}$.

We turn to our second postulate and to an example.

Postulate 2 (Independence). *A ranking-based semantics* $\mathbf{S}$ *satisfies* independence (In) *iff for every argumentation framework* $\mathbf{A}$, $\forall\, \mathbf{B} \in$ Com($\mathbf{A}$), $\forall\, a, b \in$ Arg($\mathbf{B}$), $\langle a, b \rangle \in \mathbf{S}(\mathbf{A})$ *iff* $\langle a, b \rangle \in \mathbf{S}(\mathbf{B})$.

Example 1 (Cont). Assume that the two graphs of Example 1 constitute a single argumentation framework. Then, (In) ensures that the ranking relation between a and b (and the one between c and d) remains the same after the fusion of the two frameworks.

Given our weakening principle (detailed in the introduction), it is natural to consider that a non-attacked argument is more acceptable (and thus ranked higher) than an attacked argument. In other words, there is no full reinstatement for arguments. The third postulate reflects this idea.

Postulate 3 (Void Precedence). *A ranking-based semantics* **S** *satisfies* void precedence (VP) *iff for every argumentation framework* $\mathbf{A} = \langle \mathcal{A}, \mathcal{R} \rangle$, $\forall a, b \in \mathcal{A}$, *if* $\mathrm{Att}(a) = \emptyset$ *and* $\mathrm{Att}(b) \neq \emptyset$, *then* $\langle b, a \rangle \notin \mathbf{S}(\mathbf{A})$.

Example 1 (Cont). (VP) ensures that a is ranked higher than b, and c higher than d.

Non-attacked arguments are also favored by extension-based semantics. They belong to any extension under grounded, complete, stable, and preferred semantics. Thus, they are accepted. However, they may have the same status (accepted) as attacked arguments (which are defended). Let us consider the following example.

Example 2. *Assume the argumentation framework depicted in the figure below.*

The grounded extension of this framework is $\{a, c\}$. *The arguments* a *and* c *are both accepted whereas* b *is rejected. Our approach ranks* a *higher than* c *since* c *is attacked, thus weakened. Thus, it ensures a more refined treatment of arguments.*

Since an attack always weakens its target, the next postulate states that having attacked attackers is better than having non-attacked attackers (assuming the number of attackers is the same). In other words, being *defended* is better than not being defended. First, we formally introduce the notion of defender:

Notation. *Let* $\mathbf{A} = \langle \mathcal{A}, \mathcal{R} \rangle$ *be an argumentation framework and* $a \in \mathcal{A}$. *We denote by* $\mathrm{Def}_\mathbf{A}(a)$ *the set of all defenders of* a *in* $\mathbf{A}$, *that is,* $\mathrm{Def}_\mathbf{A}(a) = \{b \in \mathcal{A} \mid \exists c \in \mathcal{A}, c\mathcal{R}a \text{ and } b\mathcal{R}c\}$.

Next, we turn to the postulate and to an example.

Postulate 4 (Defense Precedence). *A ranking-based semantics* **S** *satisfies* defense precedence (DP) *iff for every argumentation framework* $\mathbf{A} = \langle \mathcal{A}, \mathcal{R} \rangle$, $\forall a, b \in \mathcal{A}$, *if* $|\mathrm{Att}(a)| = |\mathrm{Att}(b)|$, $\mathrm{Def}(a) \neq \emptyset$, *and* $\mathrm{Def}(b) = \emptyset$, *then* $\langle b, a \rangle \notin \mathbf{S}(\mathbf{A})$.

Example 3. *Consider the argumentation framework depicted in the figure below.*

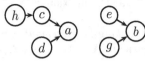

Both arguments a *and* b *have two attackers. The two attackers of* b *are not attacked, thus they are strong. However,* a *is defended by* h, *thus the attacker* c *is weakened. To sum up,* a *has one strong and one weak attacker, while* b *has two strong attackers. So,* (DP) *ensures that* a *is ranked higher than* b.

The two next postulates are based on both the weakening and the counting principles: the more the attackers of an argument a are numerous and acceptable, the less a is acceptable. The first postulate, called *counter-transitivity*, corresponds to a large version of this combined principle, the second one, called *strict counter-transitivity*, corresponds to a strict version.

More precisely, counter-transitivity says that an argument a should be ranked at least as high as an argument b, if the attackers of b are at least as numerous and acceptable as those of a. Let us first introduce a relation that compares sets of arguments on the basis of a ranking on the arguments.

Definition 6 (Group comparison). *Let $\preceq$ be a ranking on a set $\mathcal{A}$ of arguments. For all $A, B \subseteq \mathcal{A}$, $\langle A, B \rangle \in \mathtt{Gr}(\preceq)$ iff there exists an injective function f from B to A such that $\forall a \in B, f(a) \preceq a$. Intuitively, $\langle A, B \rangle \in \mathtt{Gr}(\preceq)$ iff the elements of the group A are at least as numerous and acceptable as those of B.*

To put the emphasize on the meaning of $\mathtt{Gr}(\preceq)$, we derive the following fact:

Proposition 1. *Let $\preceq$ be a ranking on a set $\mathcal{A}$ of arguments and $A, B \subseteq \mathcal{A}$. If $\langle A, B \rangle \in \mathtt{Gr}(\preceq)$, then:*

- $|A| \geq |B|$;
- *for all $b \in B$, $\exists a \in A$ such that $a \preceq b$.*

We are ready to formally state the postulate based on argument-group comparisons:

Postulate 5 (Counter-Transitivity). *A ranking-based semantics $\mathbf{S}$ satisfies the postulate* counter-transitivity (CT) *iff for every argumentation framework $\mathbf{A} = \langle \mathcal{A}, \mathcal{R} \rangle$, $\forall a, b \in \mathcal{A}$, if $\langle \mathtt{Att}(b), \mathtt{Att}(a) \rangle \in \mathtt{Gr}[\mathbf{S}(\mathbf{A})]$, then $\langle a, b \rangle \in \mathbf{S}(\mathbf{A})$.*

Example 3 (Cont). (CT) ensures that a is ranked at least as high as b.

Strict counter-transitivity is another mandatory postulate in our approach. Loosely speaking, it says that an argument a should be ranked strictly higher than an argument b, if the attackers of b are more numerous or more acceptable than those of a.

Definition 7 (Strict group comparison). *Let $\preceq$ be a ranking on a set $\mathcal{A}$ of arguments. For all $A, B \subseteq \mathcal{A}$, $\langle A, B \rangle \in \mathtt{Sgr}(\preceq)$ iff there exists an injective function f from B to A such that the two following conditions hold:*

- $\forall a \in B, f(a) \preceq a$;
- $|B| < |A|$ *or* $\exists a \in B, a \not\preceq f(a)$.

Intuitively, $\langle A, B \rangle \in \mathtt{Sgr}(\preceq)$ iff the elements of A are strictly better than those of B from a global point of view based on both cardinality and acceptability.

Postulate 6 (Strict Counter-Transitivity). *A ranking-based semantics $\mathbf{S}$ satisfies* strict counter-transitivity (SCT) *iff for every argumentation framework $\mathbf{A} = \langle \mathcal{A}, \mathcal{R} \rangle$, $\forall a, b \in \mathcal{A}$, if $\langle \mathtt{Att}(b), \mathtt{Att}(a) \rangle \in \mathtt{Sgr}[\mathbf{S}(\mathbf{A})]$, then $\langle b, a \rangle \notin \mathbf{S}(\mathbf{A})$.*

Example 3 (Cont). (SCT) ensures that a is strictly more acceptable than b.

We turn to situations where the cardinality of the attackers and their quality (i.e., acceptability) are opposed. Here is an example.

Example 4. *Consider the argumentation framework depicted in the figure below.*

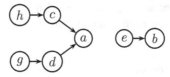

If one non-attacked attacker is sufficient to kill an argument (which is the case in most approaches to argumentation), then the argument a should naturally be ranked higher than b. But, in our approach, as explained in the introduction, no number of attacked or non-attacked attackers can kill an argument. They can just weaken it. Consequently, in this example, a is attacked by two weakened arguments, while b is attacked by one strong argument. As usual, we have to make a choice: give precedence to cardinality over quality (i.e. two weakened attackers are worse for the target than one strong attacker), or on the contrary give precedence to quality over cardinality.

In certain applications such as decision-making, both options are reasonable. For example, suppose we have to buy a car and we are considering a red one and a blue one. In addition, the arguments of Example 4 correspond to the following statements:

b = The red car has got 5 stars out of 5 in our favorite car magazine;
e = The magazine does not take into account the fact that the red car is 1000 euros more expensive than the blue one;
a = The blue car has got 5 stars out of 5 in our favorite car magazine;
c = The magazine does not take into account the fact that there is a probability of 0.5 that the blue car engine breaks down before 300000km. The reparations would cost 2000 euros;
h = A friend of ours is a mechanic. He would offer us a 10% discount on engine reparation;
d = The magazine does not take into account the fact that there is a probability of 0.5 that the blue car will be stolen from us before 10 years. The insurance will pay for another blue car, but there is a deductibility provision of 2000 euros;
g = In our neighborhood, the rate of motor vehicle theft is 10% lower than the average.

In this example, it is intuitive to consider that b is more acceptable than a. Indeed, it is obvious that the group $\{c, d\}$ is stronger than the singleton $\{e\}$, despite the fact that the former is slightly weakened by h and g. Now, suppose that the argument e is replaced by the following one:

e = The magazine does not take into account the fact that the red car is 4000 euros more expensive than the blue one.

This time it is intuitive to consider that a is more acceptable than b.

To summarize, with abstract nodes and arrows as arguments and attacks, the outcome of Example 4 is debatable. We can give precedence to cardinality over quality (i.e. b is more acceptable than a) or on the contrary give precedence to quality over cardinality (i.e. a is more acceptable than b). Both options are rational. We turn to two axioms representing these two choices.

First, *cardinality precedence* says that an argument a should be ranked higher than an argument b, if the attackers of a are less numerous than those of b.

Postulate 7 (Cardinality Precedence). *A ranking-based semantics* **S** *satisfies* cardinality preference (CP) *iff for every argumentation framework* $\mathbf{A} = \langle \mathcal{A}, \mathcal{R} \rangle$, $\forall a, b \in \mathcal{A}$, *if* $|\mathtt{Att}(a)| < |\mathtt{Att}(b)|$, *then* $\langle b, a \rangle \notin \mathbf{S}(\mathbf{A})$.

Next, *quality precedence* says that an argument a should be ranked higher than an argument b, if at least one attacker of b is ranked higher than any attacker of a.

Postulate 8 (Quality Precedence). *A ranking-based semantics* **S** *satisfies* quality precedence (QP) *iff for every argumentation framework* $\mathbf{A} = \langle \mathcal{A}, \mathcal{R} \rangle$, $\forall a, b \in \mathcal{A}$, *if there exists* $c \in \mathtt{Att}(b)$ *such that* $\forall d \in \mathtt{Att}(a)$, $\langle d, c \rangle \notin \mathbf{S}(\mathbf{A})$, *then* $\langle b, a \rangle \notin \mathbf{S}(\mathbf{A})$.

The last postulate says that, all other things remaining equal, a distributed defense is better than a focused one. This postulate is not at all mandatory. It simply represents a reasonable choice that one can make in very specific situations. More precisely, the idea is to compare two arguments having the same number of attackers and the same number of defenders. In addition, each defender attacks exactly one attacker. The postulate says that, in this case, the best kind of defense is the totally distributed one, i.e. each defender attacks a distinct attacker. In some sense, there is no "overkill".

First, we formally define what is a simple and distributed defense.

Definition 8 (Simple/distributed defense). *Let* $\mathbf{A} = \langle \mathcal{A}, \mathcal{R} \rangle$ *be an argumentation framework and* $a \in \mathcal{A}$.

- *The defense of* a *in* **A** *is* simple *iff every defender of* a *attacks exactly one attacker of* a.
- *The defense of* a *in* **A** *is* distributed *iff every attacker of* a *is attacked by at most one argument.*

We are ready to define our last postulate:

Postulate 9 (Distributed-Defense Precedence). *A ranking-based semantics* **S** *satisfies* distributed-defense precedence (DDP) *iff for any argumentation framework* $\mathbf{A} = \langle \mathcal{A}, \mathcal{R} \rangle$, $\forall a, b \in \mathcal{A}$ *such that* $|\mathtt{Att}(a)| = |\mathtt{Att}(b)|$ *and* $|\mathtt{Def}(a)| = |\mathtt{Def}(b)|$, *if the defense of* a *is simple and distributed and the defense of* b *is simple but not distributed, then* $\langle b, a \rangle \notin \mathbf{S}(\mathbf{A})$.

Let us illustrate these concepts on the following example.

Example 5. *Consider the argumentation framework depicted in the figure below.*

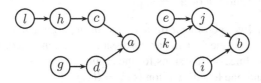

The two arguments a *and* b *have the same number of defenders:* $\mathtt{Def}(a) = \{h, g\}$ *and* $\mathtt{Def}(b) = \{e, k\}$. *However, the defense of* a *is simple and distributed while the defense of* b *is simple but not distributed. The postulate* (DDP) *ensures that* a *is more acceptable than* b, *despite the fact that the defenders of* a *are weaker than those of* b.

4   Relationships between Postulates

So far we have proposed a set of postulates that are suitable for defining a ranking-based semantics in argumentation theory. In the present section, we briefly study their dependencies, as well as their compatibilities (i.e., whether they can be satisfied together by a semantics). We start by showing that the postulates (CT), (SCT), (VP) and (DP) are not independent.

Proposition 2. *Let* S *be a ranking-based semantics:*

- *if* S *satisfies* (SCT), *then it satisfies* (VP);
- *if* S *satisfies both* (CT) *and* (SCT), *then it satisfies* (DP).

Let us now check the compatibility of the postulates. Unsurprisingly, (CP) and (QP) cannot be satisfied together. Example 4 already illustrates this issue. Indeed, (QP) prefers a to b, while (CP) prefers the converse.

Proposition 3. *No ranking-based semantics can satisfy both* (CP) *and* (QP).

In the next section, we construct a ranking-based semantics showing the following compatibility result:

Proposition 4. *The postulates* (Ab), (In), (CT), (SCT), (CP), *and* (DDP) *are compatible.*

5   Discussion-Based and Burden-Based Semantics

This section introduces two semantics satisfying most of our postulates, namely those that are compatible with (CP).

The first semantics, called *discussion-based semantics*, is centered on a notion of linear discussion similar to 'argumentation line' in [8]. A linear discussion is a sequence of arguments such that each argument attacks the argument preceding it in the sequence.

Definition 9 (Linear discussions). *Let* $\mathbf{A} = \langle \mathcal{A}, \mathcal{R} \rangle$ *be an argumentation framework and* $a \in \mathcal{A}$. *A linear discussion for* a *in* $\mathbf{A}$ *is a sequence* $s = \langle a_1, \ldots, a_n \rangle$ *of elements of* $\mathcal{A}$ *(where* n *is a positive integer) such that* $a_1 = a$ *and* $\forall\, i \in \{2, 3, \ldots, n\}$ $a_i \mathcal{R} a_{i-1}$. *The length of* s *is* n. *We say that:* s *is* won *iff* n *is odd;* s *is* lost *iff* n *is even.*

Let us illustrate this notion on an example.

Example 5 (Cont). Two won linear discussions for the argument a are e.g., $s_1 = \langle a \rangle$ and $s_2 = \langle a, d, g \rangle$ and one lost linear discussion is, for instance, $s_3 = \langle a, c, h, l \rangle$. Similarly, three won linear discussions for the argument b are $s'_1 = \langle b \rangle$, $s'_2 = \langle b, j, e \rangle$ and $s'_3 = \langle b, j, k \rangle$ and one lost discussion is $s'_4 = \langle b, i \rangle$.

The basic idea behind the semantics is the following: for every argument a, for every positive integer i, we count the number of linear discussions for a of length i. We positively count the lost discussions and negatively count the won discussions. So, in any case, the smaller the number calculated, the better the situation for a.

Definition 10 (Discussion count). *Let* $\mathbf{A} = \langle \mathcal{A}, \mathcal{R} \rangle$ *be an argumentation framework,* $a \in \mathcal{A}$, *and* i *a positive integer. We define that:*

$$\mathtt{Dis}_{\mathbf{A}i}(a) = \begin{cases} -N & \text{if } i \text{ is odd;} \\ N & \text{if } i \text{ is even;} \end{cases}$$

where N *is the number of linear discussions for* a *in* $\mathbf{A}$ *of length* i.

Example 5 (Cont). The following table provides the discussion counts $\mathtt{Dis}_{\mathbf{A}i}$ of the two arguments a and b.

i	a	b
1	-1	-1
2	2	2
3	-2	-2
4	1	0

Our strategy is to lexicographically rank the arguments on the basis of their won and lost linear discussions.

Definition 11 (Discussion-based semantics). *The ranking-based semantics* Dbs *transforms any argumentation framework* $\mathbf{A} = \langle \mathcal{A}, \mathcal{R} \rangle$ *into the ranking* $\mathtt{Dbs}(\mathbf{A})$ *on* $\mathcal{A}$ *such that* $\forall a, b \in \mathcal{A}$, $\langle a, b \rangle \in \mathtt{Dbs}(\mathbf{A})$ *iff one of the two following cases holds:*

- $\forall i \in \{1, 2, \ldots\}$, $\mathtt{Dis}_i(a) = \mathtt{Dis}_i(b)$;
- $\exists i \in \{1, 2, \ldots\}$, $\mathtt{Dis}_i(a) < \mathtt{Dis}_i(b)$ *and* $\forall j \in \{1, 2, \ldots, i-1\}$, $\mathtt{Dis}_j(a) = \mathtt{Dis}_j(b)$.

Example 5 (Cont). For every $i \in \{1, 2, 3\}$, $\mathtt{Dis}_i(a) = \mathtt{Dis}_i(b)$. However, $\mathtt{Dis}_4(a) > \mathtt{Dis}_4(b)$. Thus, $\langle a, b \rangle \notin \mathtt{Dbs}(\mathbf{A})$, i.e., b is strictly more acceptable than a.

At first sight, the infinite character of the set $\{1, 2, \ldots\}$ of all positive integers may look like an issue from a computational point of view. Indeed, $\mathtt{Dis}_i(a)$ may never stop evolving. This is due to the possible presence of cycles in the argumentation framework. But, if $\mathtt{Dis}_i(a)$ never stops evolving, it evolves cyclically. So, we strongly conjecture that there exists a threshold t such that if $\forall i \leq t$, $\mathtt{Dis}_i(a) = \mathtt{Dis}_i(b)$, then $\forall i > t$, $\mathtt{Dis}_i(a) = \mathtt{Dis}_i(b)$. Such an equality-ensuring threshold would be dependent on the length of the longest elementary cycle in the argumentation framework. This threshold would be useful to write a program implementing our discussion-based semantics.

Note also that the computation can simply be done up to a fixed step t. The greater t, the closer the ranking obtained to the actual discussion-based ranking.

Next, the postulates represent theoretical validations for our semantics:

Theorem 1. Dbs *satisfies* (Ab), (In), (CT), (SCT), *and* (CP).

From Proposition 2, it is immediate that Dbs satisfies additional postulates:

Corollary 1. Dbs *satisfies* (VP) *and* (DP).

Theorem 2. Dbs *does not satisfy* (DDP).

Next, we show that Dbs treats odd and even length cycles in a similar way:

Proposition 5. *Let* $\mathbf{A} = \langle \mathcal{A}, \mathcal{R} \rangle$ *be an argumentation framework. Suppose that* $\mathbf{A}$ *takes the form of a unique cycle, i.e. there exists an enumeration* $\langle a_1, \ldots, a_n \rangle$ *of* $\mathcal{A}$ *(without repetition and where* n *is a positive integer) such that* $\forall i \in \{1, 2, \ldots, n-1\}$, $\mathrm{Att}(a_i) = \{a_{i+1}\}$, *and* $\mathrm{Att}(a_n) = \{a_1\}$. *Then,* $\forall a, b \in \mathcal{A}$, $\langle a, b \rangle \in \mathrm{Dbs}(\mathbf{A})$.

The second semantics, called *burden-based semantics*, satisfies (DDP). It follows a multiple steps process. At each step, it assigns a *burden number* to every argument. In the initial step, this number is 1 for all arguments. Then, in each step, all the burden numbers are simultaneously recomputed on the basis of the number of attackers and their burden numbers in the previous step. More precisely, for every argument a, its burden number is set back to 1, then, for every argument b attacking a, the burden number of a is increased by a quantity *inversely* proportional to the burden number of b in the previous step. More formally:

Definition 12 (Burden numbers). *Let* $\mathbf{A} = \langle \mathcal{A}, \mathcal{R} \rangle$ *be an argumentation framework,* $i \in \{0, 1, \ldots\}$, *and* $a \in \mathcal{A}$. *We denote by* $\mathrm{Bur}_{\mathbf{A}i}(a)$ *the burden number of* a *in the* i^{th} *step, i.e.:*

$$\mathrm{Bur}_i(a) = \begin{cases} 1 & \text{if } i = 0; \\ 1 + \Sigma_{b \in \mathrm{Att}(a)} 1/\mathrm{Bur}_{i-1}(b) & \text{otherwise.} \end{cases}$$

By convention, if $\mathrm{Att}(a) = \emptyset$, then $\Sigma_{b \in \mathrm{Att}(a)} 1/\mathrm{Bur}_{i-1}(b) = 0$.
Let us illustrate this function on the following example.

Example 2 (Cont). The burden numbers of each argument are summarized in the table below. Note that these numbers will not change beyond step 2.

Step i	a	b	c
0	1	1	1
1	1	2	2
2	1	2	1.5
⋮	⋮	⋮	⋮

We lexicographically compare two arguments on the basis of their burden numbers.

Definition 13 (Burden-based semantics). *The ranking-based semantics* Bbs *transforms any argumentation framework* $\mathbf{A} = \langle \mathcal{A}, \mathcal{R} \rangle$ *into the ranking* $\mathrm{Bbs}(\mathbf{A})$ *on* $\mathcal{A}$ *such that* $\forall a, b \in \mathcal{A}$, $\langle a, b \rangle \in \mathrm{Bbs}(\mathbf{A})$ *iff one of the two following cases holds:*

- $\forall i \in \{0, 1, \ldots\}$, $\mathrm{Bur}_i(a) = \mathrm{Bur}_i(b)$;
- $\exists i \in \{0, 1, \ldots\}$, $\mathrm{Bur}_i(a) < \mathrm{Bur}_i(b)$ *and* $\forall j \in \{0, 1, \ldots, i-1\}$, $\mathrm{Bur}_j(a) = \mathrm{Bur}_j(b)$.

As for the discussion-based semantics, an equality-ensuring threshold probably exists for the burden-based semantics. Such a threshold would make possible an exact computation, despite the fact that $\{0, 1, \ldots\}$ is infinite.

Note that both semantics (Dbs and Bbs) do not take into account possible dependencies between an argument and one of its attackers, nor the dependencies between two attackers. Actually, Dbs and Bbs rank the arguments only on the basis of the structure obtained by "unrolling" the cycles. For example, our semantics do not distinguish between a loop (e.g. $a\mathcal{R}a$) and a cycle (e.g. $a\mathcal{R}b$, $b\mathcal{R}a$). The notion of dependence is hard to capture and beyond the scope of this paper. Our goal in the present paper is essentially to introduce a new kind of semantics, basic postulates for it, and instances satisfying those postulates.

We turn to the postulate-based analysis of Bbs:

Theorem 3. Bbs *satisfies* (Ab), (In), (CT), (SCT), (CP), *and* (DDP).

From Proposition 2, it satisfies more postulates:

Corollary 2. Bbs *satisfies* (VP) *and* (DP).

Let us see on examples how the semantics works.

Example 2 (Cont). According to Bbs, the argument a is strictly more acceptable than c which is itself strictly more acceptable than b.

Note that Bbs returns a more refined result than Dung's semantics. Indeed, the set $\{a, c\}$ is a (preferred, grounded, stable) extension according to [6]. Our approach refines the result by ranking a higher than c since it is not attacked. This does not mean that Bbs semantics coincides with Dung's ones. The following example shows that the two approaches may return different results since they are grounded on different principles.

Example 4 (Cont). The argumentation framework has a unique extension $\{h, g, a, e\}$ which is grounded, preferred and stable. Thus, the argument b is rejected. Let us now apply the Bbs semantics on the same framework. The table below provides the burden numbers of the arguments.

Step i	h	g	c	d	a	e	b
0	1	1	1	1	1	1	1
1	1	1	2	2	3	1	2
2	1	1	2	2	2	1	2
⋮	⋮	⋮	⋮	⋮	⋮	⋮	⋮

Bbs provides the following ranking: $h, g, e \preceq c, d, b \preceq a$. Thus, b is more acceptable than a. The reason is that b has less attackers and Bbs give precedence to the cardinality of the attackers over their quality.

Example 5 (Cont). According to Bbs, a is strictly more acceptable than b.

Note that in this example, the semantics Dbs returns the converse. This shows that the two semantics may return very different results. This difference comes from the postulate DDP which is satisfied by Bbs but violated by Dbs.

As with Dbs, we show next that the Bbs semantics treats odd and even length cycles in a similar way.

Proposition 6. *Let* $\mathbf{A} = \langle \mathcal{A}, \mathcal{R} \rangle$ *be an argumentation framework. Suppose that* $\mathbf{A}$ *takes the form of a unique cycle, i.e. there exists an enumeration* $\langle a_1, \ldots, a_n \rangle$ *of* $\mathcal{A}$ *(without repetition and where* n *is a positive integer) such that* $\forall i \in \{1, 2, \ldots, n-1\}$, $\mathtt{Att}(a_i) = \{a_{i+1}\}$, *and* $\mathtt{Att}(a_n) = \{a_1\}$. *Then,* $\forall\, a, b \in \mathcal{A}$, $\langle a, b \rangle \in \mathtt{Bbs}(\mathbf{A})$.

6 Related Work

There are three works in the literature which are somehow related to our contribution. The first attempts were done in [1,2] where the authors identified different principles and compared existing semantics wrt them. The principles are tailored for extension-based semantics, and do not apply for ranking-based ones.

The work in [4] is closer to ours. The authors defined a notion of gradual acceptability. The idea is to assign a numerical value to each argument on the basis of its attackers. The properties of the valuation function are unclear. Our approach defines, through a set of formal postulates, the desirable properties of our semantics.

In [7], Dung's abstract framework was extended by considering weighted attacks. The basic idea is to remove *some* attacks up to a certain degree representing the tolerated incoherence, and then apply existing semantics to the new graph(s) by ignoring completely the weights. This leads to extensions which are not conflict-free in the sense of the attack relation. Consider the following weighted framework. If one tolerates incoherence up to degree 1 ($\beta = 1$), then the attack from a to b is ignored. Consequently, $\emptyset$ and $\{a, b\}$ are two β-grounded extensions.

This approach is different from ours for several reasons.

First, it does not obey the four graded considerations at the basis of our postulates and semantics (i.e., weakening, counting, relativity, and graduality), it rather obeys the four traditional non-graded considerations described in the introduction (i.e., killing, existence, absoluteness, and flatness). Indeed, weights are only used for deciding which attacks can be ignored when computing the extensions.

The second main difference stems from the fact that weights of attacks are *inputs* of the argumentation system of [7]. In our approach, degrees are located in the *output*, i.e. we compute the relative degree of acceptability of each argument. Note that the more an argument is acceptable, the more the attacks emanating from it are important. However, this does not mean that weights of attacks are generated. In our approach, the three arguments a, b and c are equivalent with regard to Bbs and Dbs. Finally, our semantics can be extended to deal with weighted attacks as input.

7 Conclusion

The paper develops an axiomatic approach for defining semantics for argumentation frameworks. It proposes postulates (each of which represents a criterion) that a semantics may satisfy. The approach offers thus a theoretical framework for comparing semantics. It is worth emphasizing that only some of the postulates (e.g. abstraction) are satisfied by Dung's semantics (when the arguments are ranked on the basis of their status, i.e. accepted arguments are ranked above undecided ones, which are ranked above rejected ones). The other postulates are based on graded considerations which may be natural in applications like decision-making.

Another novelty of our approach is that it computes the acceptability of arguments without passing through multiple points of view. Its basic idea is to compute a complete ranking on the set of arguments. The paper proposes two novel semantics that satisfy the postulates but that do not necessarily return the same results. An important future work is to find sufficiently many postulates to characterize our semantics.

References

1. Amgoud, L., Vesic, S.: A new approach for preference-based argumentation frameworks. Annals of Mathematics and Artificial Intelligence 63(2), 149–183 (2011)
2. Baroni, P., Giacomin, M.: On principle-based evaluation of extension-based argumentation semantics. Artificial Intelligence 171(10-15), 675–700 (2007)
3. Baroni, P., Giacomin, M., Guida, G.: Scc-recursiveness: a general schema for argumentation semantics. Artificial Intelligence Journal 168, 162–210 (2005)
4. Cayrol, C., Lagasquie-Schiex, M.-C.: Graduality in Argumentation. Journal of Artificial Intelligence Research (JAIR) 23, 245–297 (2005)
5. Dung, P., Mancarella, P., Toni, F.: Computing ideal skeptical argumentation. Artificial Intelligence Journal 171, 642–674 (2007)
6. Dung, P.M.: On the Acceptability of Arguments and its Fundamental Role in Non-Monotonic Reasoning, Logic Programming and n-Person Games. AIJ 77, 321–357 (1995)
7. Dunne, P.E., Hunter, A., McBurney, P., Parsons, S., Wooldridge, M.: Weighted argument systems: Basic definitions, algorithms, and complexity results. Artificial Intelligence 175(2), 457–486 (2011)
8. García, A., Simari, G.: Defeasible logic programming: an argumentative approach. Theory and Practice of Logic Programming 4(1-2), 95–138 (2004)

A Logical Theory about Dynamics in Abstract Argumentation

Richard Booth[1], Souhila Kaci[2], Tjitze Rienstra[1,2], and Leendert van der Torre[1]

[1] Université du Luxembourg
6 rue Richard Coudenhove-Kalergi, Luxembourg
{richard.booth,tjitze.rienstra,leon.vandertorre}@uni.lu
[2] LIRMM (CNRS/Université Montpellier 2)
161 rue Ada, Montpellier, France
souhila.kaci@lirmm.fr

Abstract. We address dynamics in abstract argumentation using a logical theory where an agent's belief state consists of an argumentation framework (AF, for short) and a constraint that encodes the outcome the agent believes the AF *should* have. Dynamics enters in two ways: (1) the constraint is strengthened upon learning that the AF should have a certain outcome and (2) the AF is expanded upon learning about new arguments/attacks. A problem faced in this setting is that a constraint may be inconsistent with the AF's outcome. We discuss two ways to address this problem: First, it is still possible to form consistent *fallback beliefs*, i.e., beliefs that are most plausible given the agent's AF and constraint. Second, we show that it is always possible to find AF expansions to restore consistency. Our work combines various individual approaches in the literature on argumentation dynamics in a general setting.

Keywords: Argumentation, Dynamics, Knowledge Representation.

1 Introduction

In Dung-style argumentation [1] the argumentation framework (AF for short) is usually assumed to be static. There are, however, many scenarios where argumentation is a dynamic process: Agents may learn that an AF must have a certain outcome and may learn about new arguments/attacks. These are two basic issues that a theory about argumentation dynamics should address.

Some of these aspects have received attention in recent years. For example, the so called *enforcing problem* [2] is concerned with the question of whether and how an AF can be modified to make a certain set of arguments accepted. Other work studies the impact on the outcome of an AF when a new argument comes into play [3] or studies the issue of reasoning with incomplete AFs [4].

We address the problem by answering the following research questions: *How can we model an agent's belief about the outcome of an AF?* and *How can we characterize the effects of an agent learning that the AF should have a certain outcome, or learning about new arguments/attacks?*

W. Liu, V.S. Subrahmanian, and J. Wijsen (Eds.): SUM 2013, LNAI 8078, pp. 148–161, 2013.
© Springer-Verlag Berlin Heidelberg 2013

The basis of our approach is a logical *labeling language*, interpreted by labelings that assign to each argument a label indicating that it is *accepted, rejected* or *undecided* [5]. Formulas in this language are statements about the acceptance of the arguments of an AF. This allows us to reason about the outcome of an AF in terms of beliefs, rather than extensions or labelings.

We take an agent's belief state to consist of an AF and a formula encoding a constraint on the outcome of the AF. The constraint is strengthened upon learning that the AF should have a certain outcome. Furthermore, the agent's AF is expanded upon learning about new arguments and attacks. These two operations are modeled by a *constraint expansion* and *AF expansion* operator.

A problem faced in this setting is that the constraint on the AF's outcome may be inconsistent with its actual outcome, preventing the agent from forming consistent beliefs. We call such a state *incoherent*. We appeal to the intuition that an AF provides the agent with the ability to argue for the plausibility of the beliefs that it induces. Incoherence thus means that the agent is unable to argue for the plausibility of her beliefs using the AF.

We show that there are two ways to deal with this. First, we show that, given an incoherent belief state, it is always possible to come up with an expansion of the AF that restores coherence. Such AF expansions can be thought of as providing the missing arguments necessary to argue for her beliefs. Second, we show that it is always possible to form consistent *fallback beliefs*, which represent the "most rational" outcome of the agent's AF, given the constraint. Finally, we present an answer-set program for computing fallback belief, i.e., for determining whether or not some formula is a fallback belief in a particular belief state.

Our theory about argumentation dynamics combines several individual approaches in the literature in a general setting. For example, the issue of restoring coherence is related to the enforcing problem [2]; other ways to characterize the effect of an AF expansion have been studied in [3] and our notion of fallback belief is related to principles developed in [4].

A brief outline of this paper: In section 2 we introduce our labeling logic, together with the necessary basics of argumentation theory. Next, we present our belief state model and associated expansion operators in section 3. We then discuss in sections 4 and 5 how to deal with incoherent belief states, i.e., by restoring coherence via AF expansion and by using fallback belief. In section 6 we present an ASP encoding for computing fallback belief. Having focused in these sections on the complete semantics, we turn in section 7 to a discussion of a number of additional semantics. In section 8 we discuss related work and we conclude and discuss future work in section 9.

2 Preliminaries

We start out with some preliminaries concerning Dung-style abstract argumentation theory [1]. According to this theory, argumentation can be modeled using an *argumentation framework*, which captures two basic notions, namely arguments and attacks among arguments. We limit ourselves to the abstract setting, meaning that we do not specify the content of arguments in a formal way. Nevertheless, arguments should be understood to consist of a *claim* and a *reason*, i.e., some consideration that counts in favor of believing the claim to be true, while attacks among arguments stem from conflicts between different claims and reasons. We assume in this paper that argumentation frameworks are finite.

Definition 1. *An* argumentation framework *(AF for short) is a pair* (A, R) *where A is a finite set of* arguments *and* $R \subseteq A \times A$ *is an* attack *relation.*

Given an AF (A, R) we say that x is an *attacker* of y, whenever $(x, y) \in R$. The outcome of an AF consists of possible points of view on the acceptability of its arguments. In the literature, these points of view are represented either by sets of acceptable arguments, called *extensions* or by *argument labelings*, which are functions assigning to each argument a label *in*, *out* or *undecided*, indicating that the argument is respectively accepted, rejected or neither [5]. The two representations are essentially reformulations of the same idea as they can be mapped 1-to-1 such that extensions correspond to sets of in-labeled arguments [5]. For the current purpose we choose to adopt the labeling-based approach.

Definition 2. *A* labeling *of an AF* $F = (A, R)$ *is a function* $L : A \to \{I, O, U\}$. *We denote by* $I(L), O(L)$ *and* $U(L)$ *the set of all arguments* $x \in A$ *such that* $L(x) = I$, $L(x) = O$ *or* $L(x) = U$, *respectively, and by* $\mathcal{M}_F$ *the set of all labelings of* F.

Various conditions are used to single out labelings that represent rational points of view. The following gives rise to what is called the *complete* semantics:

Definition 3. *Let* $F = (A, R)$ *be an AF and* $L \in \mathcal{M}_F$ *a labeling. We say that* L *is* complete *iff for each* $x \in A$ *it holds that:*
  - $L(x) = I$ *iff* $\forall y \in A$ *s.t.* $(y, x) \in R$, $L(y) = O$,
  - $L(x) = O$ *iff* $\exists y \in A$ *s.t.* $(y, x) \in R$ *and* $L(y) = I$,

Thus, under the complete semantics, the outcome of an AF consists of labelings in which an argument is in iff its attackers are out and is out iff it has an attacker that is in. Many of the other semantics proposed in the literature, such as the *grounded*, *preferred* and *stable* semantics [1] are based on selecting particular subsets of the set of complete labelings:

Definition 4. *Let* L *be a complete labeling of the AF* F. L *is called:*
  - grounded *iff there is no complete labeling* L' *of* F *s.t.* $I(L') \subset I(L)$,
  - preferred *iff there is no complete labeling* L' *of* F *s.t.* $I(L) \subset I(L')$,
  - stable *iff* $U(L) = \emptyset$.

We focus on the complete semantics but briefly discuss the others in section 7.

Example 1. Consider the AF shown in figure 1, which has three complete labelings, namely IOOI, OIOI and UUUU. (We denote labelings by strings of the form ABC... where A, B, C, ... are the labels of the arguments $a, b, c, \ldots$)

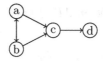

Fig. 1. An argumentation framework

A flexible way to reason about the outcome of an AF is by using a logical *labeling language*. Formulas in this language assign a label to an argument or are boolean combinations of such assignments. The language, given an AF $F = (A, R)$, is denoted by $\mathcal{L}_F$ and is generated by the following BNF, where $x \in A$:

$$\phi := \mathbf{in}_x \mid \mathbf{out}_x \mid \mathbf{u}_x \mid \neg\phi \mid \phi \vee \phi \mid \top \mid \bot.$$

We also use the connectives $\wedge, \rightarrow, \leftrightarrow$, defined as usual in terms of $\neg$ and $\vee$. Next, we define a *satisfaction* relation between labelings and formulas:

Definition 5. *Let F be an AF. The satisfaction relation $\models_F \subseteq \mathcal{M}_F \times \mathcal{L}_F$ is defined by:*

- $L \models_F \mathbf{in}_x$ *iff* $L(x) = I$,
- $L \models_F \mathbf{out}_x$ *iff* $L(x) = O$,
- $L \models_F \mathbf{u}_x$ *iff* $L(x) = U$,
- $L \models_F \phi \vee \psi$ *iff* $L \models_F \phi$ *or* $L \models_F \psi$,
- $L \models_F \neg\phi$ *iff* $L \not\models_F \phi$,
- $L \models_F \top$ *and* $L \not\models_F \bot$.

A model of a formula ϕ is a labeling $L \in \mathcal{M}_F$ such that $L \models_F \phi$. We denote by $[\phi]_F$ the set of labelings satisfying ϕ, defined by $[\phi]_F = \{L \in \mathcal{M}_F \mid L \models_F \phi\}$. We write $\phi \models_F \psi$ iff $[\phi]_F \subseteq [\psi]_F$ and $\phi \equiv_F \psi$ iff $[\phi]_F = [\psi]_F$.

Whenever the AF we talk about is clear from the context, we drop the subscript F from $\models_F$, $[\ldots]_F$ and $\equiv_F$.

Using this labeling language, we can reason about the outcome of an AF by talking about *beliefs* induced by the AF. These beliefs can be represented by a formula ϕ such that $[\phi]$ is exactly the set of complete labelings of F. It is worth noting that ϕ can be formulated in a straightforward way:

Proposition 1. *Let $F = (A, R)$ be an AF. It holds that a labeling L is a complete labeling of F iff L is a model of the formula*

$$\bigwedge_{x \in A}((\mathbf{in}_x \leftrightarrow (\bigwedge_{(y,x) \in R}\mathbf{out}_y)) \wedge (\mathbf{out}_x \leftrightarrow (\bigvee_{(y,x) \in R}\mathbf{in}_y))).$$

Example 2. Among the beliefs induced by AF in figure 1 are $\neg\mathbf{out}_d$ and $(\mathbf{in}_a \vee \mathbf{in}_b) \leftrightarrow \mathbf{in}_d$ and $\neg(\mathbf{in}_a \wedge \mathbf{in}_b)$.

Finally, *conflict-freeness* is considered to be a necessary (but not sufficient) condition for any labeling to be considered rational. We will make use of the following definition:

Definition 6. *Let $F = (A, R)$ be an AF. A labeling L of F is said to be* conflict-free *iff L is a model of the formula*

$$\wedge_{x \in A}(\mathtt{in}_x \rightarrow ((\wedge_{(y,x) \in R}\mathtt{out}_y) \wedge (\wedge_{(x,y) \in R}\mathtt{out}_y))).$$

We denote this formula by Cf_F. We say that ϕ is is conflict-free iff $Cf_F \not\models \neg\phi$.

Thus, in a conflict-free labeling any neighbor of an in-labeled argument is out. Note that we deviate from the usual definition (see e.g. [6]), which allows neighbors of an in-labeled argument to be undecided. The reason is that, given our definition, conflict-freeness can be seen to generalize completeness in a dynamic setting, in the sense that a conflict-free labeling of an AF is always (part of) a complete labeling of some expansion of the AF. The benefit of this will become clear in the following sections.

Example 3. Some examples of conflict-free labelings of the AF in figure 1 are IOOO, UUOI and OOOO. Examples of labelings that are not are IIOO and UUIO.

3   Belief States

On the one hand, AFs interpreted under the complete semantics induce beliefs about the status of arguments (and consequently about argument's claims and reasons) that are rational in the sense that the arguments and attacks in the AF can be used to argue for the plausibility of these beliefs. For example, given the AF $(\{b, a\}, \{(b, a)\})$, the belief $\mathtt{out}_a$ can, informally speaking, be argued for by pointing out that a is attacked by b which, in turn, is not attacked and should thus be accepted. Furthermore, these beliefs are defeasible, because learning about new arguments and attacks may cause old beliefs to be retracted.

On the other hand, an agent may learn or come to desire some claim to be true or false, without being aware of arguments to argue for the plausibility of it. This bears on the outcome that the AF *should* have, according to the agent. To model scenarios like these, we define an agent's belief state to consist not only of an AF, but also a constraint that the agent puts on its outcome.

Definition 7. *A* belief state *is a pair $S = (F, K)$, where $F = (A, R)$ is an AF and $K \in \mathcal{L}_F$ the agent's* constraint. *We define $K(S)$ by $K(S) = K$ and $Bel(S)$ by $[Bel(S)] = \{L \in [K] \mid L$ is a complete labeling of $F\}$. We say that the agent* believes ψ *iff $Bel(S) \models \psi$ and that S is* coherent *iff $Bel(S) \not\models \bot$.*

Thus, the belief $Bel(S)$ of an agent is formed by the outcome of the AF in conjunction with the constraint. Intuitively, the plausibility of the agent's belief can be argued for only if it is consistent, i.e., only if the belief state is coherent. An incoherent state is thus a state in which the agent is prevented from forming beliefs that can be shown to be plausible via the AF.

We turn again to incoherence in the following section. We first define two expansion operators: one that strengthens the agent's constraint and one that expands the AF. The *constraint expansion operator* takes as input a belief state and a formula ϕ representing a constraint that is to be incorporated into the new belief state. It is defined as follows.

Definition 8. *Let F be an AF, $S = (F, K)$ a belief state and $\phi \in \mathcal{L}_F$. The constraint expansion of S by ϕ, denoted $S \oplus \phi$ is defined by $S \oplus \phi = (F, K \wedge \phi)$.*

Example 4. Let $S_1 = (F, \top)$ where F is the AF shown in figure 1. We do not have $Bel(S_1) \models \text{in}_d$. That is, the agent does not believe that d is in. Consider the constraint expansion $S_2 = S_1 \oplus (\text{in}_a \vee \text{in}_b)$. Now we have $Bel(S_2) \models \text{in}_d$. That is, after learning that either a or b is in, the agent believes that d is in.

As to expanding the AF, we make two assumptions: First, we assume that arguments and attacks are not "forgotten". This means that elements can be added to an AF but not removed. Second, we assume that attacks between arguments are determined once the arguments are known. This means that no new attacks can be added between arguments already present in the agent's AF. Such expansions are called *normal expansions* by Baumann and Brewka [2]. We call a set of new arguments and attacks an *AF update*:

Definition 9. *Let $F = (A, R)$ be an AF. An AF update for F is a pair $F^* = (A^*, R^*)$ where A^* is a set of added arguments, such that $A \cap A^* = \emptyset$ and $R^* \subseteq ((A \cup A^*) \times (A \cup A^*)) \setminus (A \times A)$ a set of added attacks.*

The *AF expansion operator* is defined as follows:

Definition 10. *Let $F = (A, R)$ be an AF, $S = (F, K)$ a belief state and $F^* = (A^*, R^*)$ an AF update for F. The AF expansion of S by F^*, denoted by $S \otimes F^*$ is defined by $S \otimes F^* = ((A \cup A^*, R \cup R^*), K)$.*

Example 5. Consider the belief state $S_1 = (F, \text{out}_a \vee \text{out}_b)$ where F is the AF shown in figure 1. Note that we do not have, e.g., $Bel(S_1) \models \text{in}_b$. Now consider the AF expansion $S_2 = (S_1 \otimes (\{e\}, \{(e, a)\})$. Now we do have $Bel(S_2) \models \text{in}_b$.

The two operators just defined allow us to study our belief state model in a dynamic setting, where an agent's belief state changes after new constraints on the AF's outcome are acquired or after adding new arguments and attacks.

4   Restoring Coherence through AF Expansion

In the previous section we presented a belief state model which includes, besides the agent's AF, a constraint on its outcome. We also explained that incoherence (i.e., the belief induced by the AF being inconsistent with the constraint) prevents the agent from forming beliefs that can be shown to be plausible via the agent's AF. The question is then: can the AF be expanded in such a way that the beliefs induced by it *are* consistent with the agent's constraints? In other words: can we restore coherence by expanding the AF in some way? Consider the following example.

Example 6. Let $S_1 = (F, \top)$ where F is the AF shown in figure 1. Suppose the agent learns that both a and b are out. The resulting state $S_2 = S_1 \oplus (\text{out}_a \wedge \text{out}_b)$ is incoherent, i.e., we have $Bel(S_2) \models \bot$. Now suppose the agent learns about

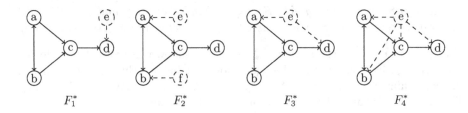

Fig. 2. Four argumentation framework updates

arguments e and f, attacking a and b. The corresponding AF update is shown as F_2^* in figure 2. The resulting state is $S_3 = S_2 \otimes (\{e, f\}, \{(e, a), (f, b)\})$. Coherence is now restored: $Bel(S_3) \not\models \bot$. In S_3 the agent believes, e.g., that c is in and d is out: $Bel(S_3) \models \mathrm{in}_c \wedge \mathrm{out}_d$. Notice that F_4^*, too, restores coherence in state S_2, whereas F_1^* and F_3^* do not.

This example shows that it is indeed possible to expand an AF such that coherence is restored. Note, also, that the AF updates F_2^* and F_4^* can be understood to provide the "missing explanation" for the agent's constraint $\mathrm{out}_a \wedge \mathrm{out}_b$. That is, a and b are out *because* there are arguments attacking (among possibly other arguments) a and b. We can show that, as long as the agent's constraint is conflict-free, there always exists some AF expansion that restores coherence. That the agent's constraint is required to be conflict-free follows from the fact that attacks between existing arguments cannot be removed. Proofs are omitted due to space constraints.

Theorem 1. *Let (F, K) be an incoherent belief state where K is conflict-free. There exists an AF update F^* for F such that $(F, K) \otimes F^*$ is coherent.*

This result essentially says that incoherence of a belief state can be understood to mean that the agent's AF is incomplete and needs to be expanded with additional arguments and attacks. A related result, called the *conservative strong enforcing* result, was presented by Baumann and Brewka [2]. However, this result deals only with the possibility of making some set of arguments accepted. By contrast, we deal with arbitrary formulas expressible in the logical labeling language.

5   Fallback Belief

In example 6, the agent learns that a and b are out, resulting in the belief state becoming incoherent and beliefs becoming inconsistent. Nevertheless, it is still possible to form reasonable, consistent beliefs given this constraint, even without performing a coherence restoring AF expansion. To see what we mean, it is enough to just look at the AF in figure 1 and see that, once a and b are out, c should be in and d should be out. However, there are no complete labelings satisfying these assignments of labels. Thus to form such beliefs, which we call *fallback* beliefs, we must adopt a different method.

The starting point is to define a *rationality order* over conflict-free labelings, used to determine their relative rationality. Consider an assignment, to each AF F, of a total pre-order (i.e., a complete, transitive and reflexive order) $\preceq_F$ over conflict-free labelings of F. Given a set $M \subseteq [Cf_F]$ we define $min_{\preceq_F}(M)$ by $min_{\preceq_F}(M) = \{L \in M \mid \forall L' \in M, L \preceq_F L'\}$. Following terminology used in belief revision, we call such an assignment *faithful* if the minimal labelings according to $\preceq_F$ are exactly the complete labelings of F.

Definition 11. *A faithful assignment* assigns to each AF F a total pre-order $\preceq_F \subseteq [Cf_F] \times [Cf_F]$ s.t. $L \in min_{\preceq_F}([Cf_F])$ iff L is a complete labeling of F. If $L \preceq_F L'$, we say that L is at least as rational as L'.

In an incoherent state, i.e., when all fully rational labelings of the AF F are ruled out, the agent can fall back on the remaining labelings that are most rational according to the ordering $\preceq_F$. These labelings can be used to form fallback beliefs, the idea being that they represent the best outcome of the AF given the agent's constraint. Given a belief state S, we denote the fallback belief in S by $Bel^*(S)$. The type of belief we end up with can be characterized by an appropriate adaptation of the well known KM postulates [7]:

Theorem 2. *The following are equivalent:*

1. *There exists a faithful assignment mapping each F to a total pre-order $\preceq_F$ such that for each K, $[Bel^*((F,K))] = min_{\preceq_F}([K] \cap [Cf_F])$.*
2. *For each $S = (F,K)$, Bel^* satisfies:*
   P1: $Bel^*(S) \models K(S) \wedge Cf_F$.
   P2: If S is coherent then $Bel^*(S) \equiv Bel(S)$.
   P3: If $K(S)$ is conflict-free then $Bel^*(S)$ is conflict-free.
   P4: If $F_1 = F_2$ and $K_1 \equiv K_2$ then $Bel^*((F_1,K_1)) \equiv Bel^*((F_2,K_2))$.
   P5: $Bel^*(S) \wedge \psi \models Bel^*(S \oplus \psi)$.
   P6: If $Bel^*(S) \wedge \psi$ is conflict-free then $Bel^*(S \oplus \psi) \models Bel^*(S) \wedge \psi$.

Thus, if we define Bel^* by $[Bel^*((F,K))] = min_{\preceq_F}([K] \cap [Cf_F])$ then fallback belief behaves like the one-shot revision, by the constraint K, of the outcome of F under the complete semantics. The postulates in proposition 2 now embody conditions of minimal change w.r.t. the fully rational outcome of the AF, rather than an arbitrary KB. The original postulates were discussed by Katsuno and Mendelzon [7], who built on the AGM approach to belief revision [8]. Here we content ourselves with pointing out how our postulates differ from the original ones. First of all, P1, P2, P3 and P6 are changed to account for the fact that only conflict-free labelings are considered possible. Second, in P5 and P6 conjunction is substituted with $\oplus$, Finally, P4 requires the AFs (and thus orderings) in the two belief states to be equivalent, as well as the constraint.

The question we need to answer now is: when is one conflict-free labeling of an AF F to be more rational than another? That is, how should $\preceq_F$ order arbitrary conflict-free labelings of F? A natural way to do this is by looking at the arguments that are *illegally* labeled [6]. This is defined as follows:

Definition 12. *Let $F = (A, R)$ be an AF and $L \in \mathcal{M}_F$ a labeling of F. An argument $x \in A$ is said to be:*

- *Illegally in iff $L(x) = I$ and $\exists y \in A, (y, x) \in R$ and $L(y) \neq O$,*
- *Illegally out iff $L(x) = O$ and $\nexists y \in A, (y, x) \in R$ such that $L(y) = I$,*
- *Illegally undecided iff $L(x) = U$ and $\exists y \in A, (y, x) \in R$ and $L(y) = I$ or $\nexists y \in A, (y, x) \in R$ such that $L(y) = U$.*

We denote by $Z_F^I(L), Z_F^O(L)$ and $Z_F^U(L)$ the sets of arguments that are, respectively, illegally in, out and undecided in L.

Intuitively, an illegally labeled argument indicates a local violation of the condition imposed on the argument's label according to the complete semantics. It can be checked, for example, that a labeling L is a complete labeling iff it has no arguments illegally labeled. It can also be checked that, in a conflict-free labeling, arguments are never illegally in. Thus in judging the relative rationality of a conflict-free labeling L, we only have to look at the sets $Z_F^O(L)$ and $Z_F^U(L)$.

What, exactly, do the sets $Z_F^O(L)$ and $Z_F^U(L)$ tell us about how rational L is? To answer this we have to look at what it takes to turn L into a complete labeling. We say that an AF update that turns L into (part of) a complete labeling of the (expanded) AF is an AF update that *completes L*. Formally:

Definition 13. *Let $F^* = (A^*, R^*)$ be an AF update for $F = (A, R)$ and L a conflict-free labeling of F. We say that F^* completes L iff there is a complete labeling L' of the AF $(A \cup A^*, R \cup R^*)$ such that $(L' \downarrow A) = L$, where $(L \downarrow A)$ is a function defined by $(L \downarrow A)(x) = L(x)$, for all $x \in A$.*

As a measure for the "impact" of an AF update, Baumann looked at the number of added attacks [9]. In our setting it is more appropriate to look at the number of arguments in the existing AF that are attacked by the AF update. We call this this the *attack degree* of the AF update.

Definition 14. *Let $F^* = (A^*, R^*)$ be an AF update for $F = (A, R)$. We denote by $\delta_F(F^*)$ the attack degree of F^*, defined by $\delta_F(F^*) = |\{x \in A \mid \exists y \in A^*, (y, x) \in R^*\}|$.*

The key is that the sets $Z_F^O(L)$ and $Z_F^U(L)$ inform us about the minimal impact it would take to complete L, or to turn L into a fully rational point of view. That is, it informs us about the minimal attack degree of an AF update that completes L:

Proposition 2. *Let L be a conflict-free labeling of an AF F. If F^* completes L then $\delta_F(F^*) \geq |Z_F^O(L) \cup Z_F^U(L)|$.*

We use the cardinality of the sets $Z_F^O(L)$ and $Z_F^U(L)$ as the criterion to define the rationality order $\preceq_F$, making the assumption that the agent believes that conflict-free labelings that require less impact to be turned into a complete labeling are more rational. We now define a faithful assignment as follows: Let F be an AF and $L, L' \in [Cf_F]$,

$$L \preceq_F L' \text{ iff } |Z_F^O(L) \cup Z_F^U(L)| \leq |Z_F^O(L') \cup Z_F^U(L')|$$

Now, the outcome of the AF according to the agent's fallback belief is the outcome that would hold if some minimal impact, coherence restoring AF update would be performed.

Example 7. The table below represents $\preceq_F$ for the AF F shown in figure 1.

0	1	2	3	4
OIOI	OIOO UUUO	OOIO UUOO	OOOI OUUO	OOOO OUOU OUOO
UUUU	OIOU IOOO	OUUU UUOU	OOUU UOOI	OOOU UOOO
IOOI	UUOI IOOU	UOUU UOUU	OUOI UOUO	OOUO UOOU

The table groups labelings by to the number of arguments illegally labeled. These arguments are underlined and the numbers are shown in the column headers. This determines the ordering $\preceq_F$ as follows: $L \prec_F L'$ iff L is in another column to the left of L'. We have the following fallback beliefs:

- $Bel^*(F, \mathsf{out}_a \wedge \mathsf{out}_b) \models \mathsf{in}_c$ (if a and b are out then c is in).
- $Bel^*(F, \mathsf{in}_c) \models \mathsf{out}_a \wedge \mathsf{out}_b$ (if c is in then a and b must be out).
- $Bel^*(F, \mathsf{out}_d) \models \neg(\mathsf{in}_a \wedge \mathsf{in}_b)$ (even if d is out, a and b cannot both be in).
- $Bel^*(F, \mathsf{out}_d) \models \mathsf{u}_a \rightarrow \mathsf{u}_c$ (even if d is out, if a is undecided then so is c).

Note that none of these inferences can be made by looking only at the complete labelings of F.

As the following theorem states more formally, and as we pointed out above, fallback belief is formed by assuming the most rational outcome of an AF in an incoherent state to be the outcome that would hold after a coherence restoring AF update with minimal impact. That is, if coherence is restored using an AF update with a minimal attack degree, then the agent's regular belief in the updated state includes the agent's fallback belief in the old state.

Theorem 3. *Let S be an incoherent belief state and F_1^* a minimal coherence restoring update (i.e., $S \otimes F_1^*$ is coherent and there is no F_2^* such that $S \otimes F_2^*$ is coherent and $\delta_F(F_2^*) < \delta_F(F_1^*)$). It holds that $Bel(S \otimes F_1^*) \models Bel^*(S)$.*

Example 8. Let $S = (F, \mathsf{out}_c)$ be a belief state with $F = (\{a, b, c, d, e\}, \{(a, b), (b, c), (d, e), (e, c)\})$. We have $[Bel^*(S)] = \{\mathtt{IOOIO}, \underline{\mathtt{O}}\mathtt{IOIO}, \mathtt{IOO}\underline{\mathtt{O}}\mathtt{I}\}$. Three minimal coherence restoring AF updates are: $F_1^* = (\{f\}, \{(f, c)\})$, $F_2^* = (\{f\}, \{(f, a)\})$ and $F_3^* = (\{f\}, \{(f, d)\})$. We have that $Bel(S \otimes F_n^*) = \psi$, where $[\psi] = \{\mathtt{IOOIOI}\}$ if $n = 1$; $[\psi] = \{\mathtt{OIOIO}\}$, if $n = 2$ and $[\psi] = \{\mathtt{IOOOII}\}$, if $n = 3$. It can be checked that, for all $n \in \{1, 2, 3\}$, $Bel(S \otimes F_n^*) \models \phi$ and thus $Bel(S \otimes F_n^*) \models Bel^*(S)$.

6 Computing Fallback Beliefs with ASP

Answer-set programming has proven to be a useful mechanism to compute extensions of AFs under various semantics [10,11,12]. The idea is to encode both the AF and a so called *encoding* of the semantics in a single program of which the stable models correspond to the extensions of the AF.

In this section we show that the problem of deciding whether a formula ϕ is a fallback belief in a state (F, K) can be solved, too, using an answer-set program. The encoding, shown in listing 6, turns out to be surprisingly simple, and works as follows. The AF is assumed to be encoded (line 1) using the predicates `arg/1` and `att/2`. For example, the AF of figure 1 is encoded by the facts `arg(a)`, `arg(b)`, `arg(c)`, `arg(d)`, `att(a,b)`, `att(a,c)`, `att(b,a)`, `att(b,c)` and `att(c,d)`. The choice rule on line 2 ensures that each argument $x \in A$ gets one of three labels, expressed by the predicates `in/1`, `out/1` and `undec/1`. On lines 3 and 4 conflict-freeness is ensured. Given just these constraints, stable models correspond to conflict-free labelings of F. Lines 5-10 are used to establish whether an argument $x \in A$ is illegally labeled, expressed by the predicate `illegal(x)`. The cardinality of this predicate is minimized on line 12. Finally, the agent's constraint is assumed to be encoded (line 11) using statements restricting the possible labels assigned to arguments. For example, the constraint $out_a \vee out_b$ is encoded by the choice rule 1 `{out(a), out(b)}` 2, and the constraint $out_a \wedge out_b$ by the two facts `out(a)` and `out(b)`. The (optimal) stable models now correspond to maximally rational conflict-free labelings that satisfy the constraint.

```
 1   % <-- Framework encoding here -->
 2   1 { in(X), out(X), undec(X) } 1 :- arg(X).
 3   out(Y) :- att(X, Y), in(X).
 4   out(X) :- att(X, Y), in(Y).
 5   legally_out(X)   :- out(X), att(Y, X), in(Y).
 6   legally_undec(X) :- undec(X), att(Y, X), undec(Y).
 7   illegally_out(X)   :- out(X), not legally_out(X).
 8   illegally_undec(X) :- undec(X), not legally_undec(X).
 9   illegal(X) :- illegally_out(X).
10   illegal(X) :- illegally_undec(X).
11   % <-- Constraint encoding here -->
12   #minimize { illegal(X) }.
```

Program 1. An answer set program to compute fallback belief

The program is compatible with the *Gringo* grounder (version 3.0.5) and *Clasp* answer set solver (version 2.1.2) [13]. The optimal stable models can be obtained by running the solver with the option `--opt-all`. The final step of the complete procedure amounts to checking whether the formula ϕ is true in every optimal stable model. Alternatively, the set of stable models of the program can be converted into a formula in disjunctive normal form that represents the agent's whole fallback belief.

7   Additional Semantics

We have focused in this paper on the complete semantics. Some of the notions we introduced can be adapted to other semantics in a straightforward way. For example, we can define a family of types of *s-belief* for a semantics $s \in \{Co, St, Pr, Gr\}$ (for Complete, Stable, Preferred, Grounded) as follows:

Definition 15. *Let $F = (A, R)$ be an AF, $S = (F, K)$ be the agent's belief state and $s \in \{Co, St, Pr, Gr\}$. We define $Bel_s(S)$ by $[Bel_s(S)] = \{L \in [K] \mid L$ is an s-labeling of $F\}$. We say that the agent s-believes ϕ iff $Bel_s(S) \models \phi$.*

It can be checked that we have $Bel_{Gr}(S) \models Bel_{Co}(S)$ and $Bel_{St}(S) \models Bel_{Pr}(S) \models Bel_{Co}(S)$. This follows directly from the fact that grounded labelings are also complete, stable also preferred, and so on. Now consider e.g. the following notion of *s-coherence*:

Definition 16. *Let S be a belief state and $s \in \{Co, St, Pr, Gr\}$. We say that S is s-coherent iff $Bel_s(S) \not\models \bot$.*

Given the notions of s-belief and s-coherence we can generalize theorem 1:

Theorem 4. *Let $s \in \{Co, St, Pr, Gr\}$ and let (F, K) be an s-incoherent belief state where K is conflict-free. There exists an AF update F^* for F such that $(F, K) \otimes F^*$ is s-coherent.*

Fallback belief, however, is less straightforward to adapt, as the corresponding rationality orderings would have to combine different criteria, i.e. minimizing/-maximizing in-labeled arguments w.r.t. set-inclusion and minimizing illegally labeled arguments, meaning we have to deal with partial pre-orders.

8   Related Work

In this section we give a short overview of related work. We already mentioned the relation of our work with the *enforcing problem* [2]. The authors present a result stating that every conflict-free extension can be enforced (i.e., made accepted under a semantics) with an appropriate AF expansion. In our setting we consider more general types of enforcing, not limited only to acceptance of sets of arguments. Our theorem 4 thus strengthens their possibility result.

Next, different ways to characterize the impact of AF expansions have been studied. This includes minimality w.r.t. the number of added attacks, studied in the context of the enforcing problem [9]. Further criteria were defined in the study of the impact on the outcome of an AF of adding an argument [3]. A limitation in that work is that it considers only additions of a single argument. A slightly different perspective is taken in the work of Liao, Lin and Koons [14], where the impact of adding arguments and attacks plays a role in the efficient recomputation of the extensions of an AF.

The ordering presented in section 5 is related to a preferential model semantics for argumentation [15] and a study of nonmonotonic inference relations to reason with AFs [4]. Also related are *open labelings* [16], which have a purpose similar to ours, i.e., to identify arguments to attack in order to make a labeling consistent with an AF, and an approach where arguments are labeled with formulas expressing instructions on what to attack in order to change the argument's status under the grounded semantics [17]. We should also mention other work in which (parts of) argumentation theory are formalized using logics. They include models using modal logics [18,19]; translations of the problem of computing extensions to problems in classical logic or ASP [20,12]; and a study of a logical language consisting of attack and defense connectives [21].

Finally, our model is related to the concept of a *constrained AF*, where an AF is combined with a constraint on the status of the arguments [22]. However, these constraints must be consistent with the AF's outcome under the admissible semantics, limiting the types of constraints that can be dealt with. Furthermore, this work does not explore the relation between constraints and AF expansions.

9   Conclusion and Future Work

We believe that theories about dynamics in abstract argumentation should address two issues: First, agents may learn or come to desire that an AF must have a certain outcome and second, agents may expand their AF. Our solution centers on the issue of dealing with incoherence after constraining the AF's outcome. Two ways to deal with this are AF expansion and by using fallback belief.

We plan to extend our work in a number of directions. First, our model allows iterated updates only under the assumption that new observations never contradict old ones. In order to allow this we have to look at revising the agent's constraint in the light of conflicting observations. Second, a number of generalizations are possible. For example, we may drop the requirement that observations are conflict-free and we can allow removal of arguments and attacks.

Finally, we plan to investigate connections between the areas of abstract argumentation and belief revision beyond those presented in this paper. We believe that the approach of using a logical labeling language to reason about the outcome of an AF is an essential step towards establishing such connections.

Acknowledgements. Richard Booth is supported by the National Research Fund, Luxembourg (DYNGBaT project).

References

1. Dung, P.M.: On the acceptability of arguments and its fundamental role in non-monotonic reasoning, logic programming and n-person games. Artif. Intell. 77(2), 321–358 (1995)
2. Baumann, R., Brewka, G.: Expanding argumentation frameworks: Enforcing and monotonicity results. In: Baroni, P., Cerutti, F., Giacomin, M., Simari, G.R. (eds.) COMMA. Frontiers in Artificial Intelligence and Applications, vol. 216, pp. 75–86. IOS Press (2010)
3. Cayrol, C., de Saint-Cyr, F., Lagasquie-Schiex, M.: Change in abstract argumentation frameworks: Adding an argument. Journal of Artificial Intelligence Research 38(1), 49–84 (2010)
4. Booth, R., Kaci, S., Rienstra, T., van der Torre, L.: Monotonic and non-monotonic inference for abstract argumentation. In: FLAIRS (2013)

5. Caminada, M.: On the issue of reinstatement in argumentation. In: Fisher, M., van der Hoek, W., Konev, B., Lisitsa, A. (eds.) JELIA 2006. LNCS (LNAI), vol. 4160, pp. 111–123. Springer, Heidelberg (2006)

6. Baroni, P., Caminada, M., Giacomin, M.: An introduction to argumentation semantics. Knowledge Eng. Review 26(4), 365–410 (2011)

7. Katsuno, H., Mendelzon, A.O.: Propositional knowledge base revision and minimal change. Artificial Intelligence 52(3), 263–294 (1991)

8. Alchourrón, C.E., Gärdenfors, P., Makinson, D.: On the logic of theory change: Partial meet contraction and revision functions. Journal of symbolic logic, 510–530 (1985)

9. Baumann, R.: What does it take to enforce an argument? Minimal change in abstract argumentation. In: Raedt, L.D., Bessière, C., Dubois, D., Doherty, P., Frasconi, P., Heintz, F., Lucas, P.J.F. (eds.) ECAI. Frontiers in Artificial Intelligence and Applications, vol. 242, pp. 127–132. IOS Press (2012)

10. Toni, F., Sergot, M.: Argumentation and answer set programming. In: Balduccini, M., Son, T.C. (eds.) Gelfond Festschrift. LNCS (LNAI), vol. 6565, pp. 164–180. Springer, Heidelberg (2011)

11. de la Banda, M.G., Pontelli, E. (eds.): ICLP 2008. LNCS, vol. 5366. Springer, Heidelberg (2008)

12. Egly, U., Gaggl, S.A., Woltran, S.: Answer-set programming encodings for argumentation frameworks. Argument and Computation 1(2), 147–177 (2010)

13. Gebser, M., Kaufmann, B., Kaminski, R., Ostrowski, M., Schaub, T., Schneider, M.: Potassco: The potsdam answer set solving collection. AI Communications 24(2), 107–124 (2011)

14. Liao, B.S., Jin, L., Koons, R.C.: Dynamics of argumentation systems: A division-based method. Artif. Intell. 175(11), 1790–1814 (2011)

15. Roos, N.: Preferential model and argumentation semantics. In: Proceedings of the 13th International Workshop on Non-Monotonic Reasoning, NMR 2010 (2010)

16. Gratie, C., Florea, A.M.: Argumentation semantics for agents. In: Cossentino, M., Kaisers, M., Tuyls, K., Weiss, G. (eds.) EUMAS 2011. LNCS, vol. 7541, pp. 129–144. Springer, Heidelberg (2012)

17. Boella, G., Gabbay, D.M., Perotti, A., van der Torre, L., Villata, S.: Conditional labelling for abstract argumentation. In: Modgil, S., Oren, N., Toni, F. (eds.) TAFA 2011. LNCS, vol. 7132, pp. 232–248. Springer, Heidelberg (2012)

18. Grossi, D.: On the logic of argumentation theory. In: van der Hoek, W., Kaminka, G.A., Lespérance, Y., Luck, M., Sen, S. (eds.) AAMAS, pp. 409–416. IFAAMAS (2010)

19. Schwarzentruber, F., Vesic, S., Rienstra, T.: Building an epistemic logic for argumentation. In: del Cerro, L.F., Herzig, A., Mengin, J. (eds.) JELIA 2012. LNCS, vol. 7519, pp. 359–371. Springer, Heidelberg (2012)

20. Besnard, P., Doutre, S.: Checking the acceptability of a set of arguments. In: Delgrande, J.P., Schaub, T. (eds.) NMR, pp. 59–64 (2004)

21. Boella, G., Hulstijn, J., van der Torre, L.W.N.: A logic of abstract argumentation. In: Parsons, S., Maudet, N., Moraitis, P., Rahwan, I. (eds.) ArgMAS 2005. LNCS (LNAI), vol. 4049, pp. 29–41. Springer, Heidelberg (2006)

22. Coste-Marquis, S., Devred, C., Marquis, P.: Constrained argumentation frameworks. In: Doherty, P., Mylopoulos, J., Welty, C.A. (eds.) KR, pp. 112–122. AAAI Press (2006)

Sound Source Localization from Uncertain Information Using the Evidential EM Algorithm

Xun Wang[1,2,*], Benjamin Quost[1], Jean-Daniel Chazot[2], and Jérôme Antoni[3]

[1] Heudiasyc, UMR CNRS 7253, Université de Technologie
de Compiègne, Compiègne, France
`xun.wang@hds.utc.fr`
[2] Roberval, UMR CNRS 7337, Université de Technologie
de Compiègne, Compiègne, France
[3] LVA, INSA Lyon, Lyon, France

Abstract. We consider the problem of sound sources localization from acoustical measurements obtained from a set of microphones. We formalize the problem within a statistical framework: the pressure measured by a microphone is interpreted as a mixture of the signals emitted by the sources, pervaded by a Gaussian noise. Maximum-likelihood estimates of the parameters of the model (locations and strengths of the sources) may then be computed via the EM algorithm. In this work, we introduce two sources of uncertainties: the location of the microphones and the wavenumber. First, we show how these uncertainties may be transposed to the data using belief functions. Then, we detail how the localization problem may be studied using a variant of the EM algorithm, known as Evidential EM algorithm. Eventually, we present simulation experiments which illustrate the advantage of using the Evidential EM algorithm when uncertain data are available.

Keywords: Localization of sound sources, Inverse problem, EM algorithm, Belief function, Evidential EM algorithm, Uncertain data.

1 Introduction

In this paper, we consider the problem of sound sources localization. We assume there exist N sound sources on the plane, and our aim is to determine their position using measurements made by an array of microphones. The sound pressure measured by each microphone is interpreted as a superimposed signal composed of N components, each of which has been emitted by a sound source. Our purpose is to estimate the locations and strengths of the sound sources.

Feder and Weinstein [9] investigated the parameter estimation problem of superimposed signals using the EM algorithm. This approach makes it possible to compute iteratively maximum-likelihood (ML) estimates of the parameters of a model which depends on unobserved (or "complete") variables. In the case of sound source estimation, each observed variable (the pressure measured by a microphone) is the sum of several complete ones (the signals propagated by the

---

* Corresponding author.

W. Liu, V.S. Subrahmanian, and J. Wijsen (Eds.): SUM 2013, LNAI 8078, pp. 162–175, 2013.
© Springer-Verlag Berlin Heidelberg 2013

sources towards this microphone). The EM algorithm proceeds with the complete likelihood (that is, the likelihood of the complete variables), the expectation of which is maximized at each iteration. Cirpan and Cekli [2,3] and Kabaoglu et al. [10] studied the localization of near-field sources using the EM algorithm. This work is a particular case of [9], in which the sources are located using polar coordinates. In our work, we investigated the problem of sound source localization. However, unlike in [2,3,10], we use general coordinates, and we explicitly describe the propagation process from the sources to the microphones using a specific operator.

It should be stressed out that uncertainties may pervade the measurement process, so that the sound pressures received by the microphones are not exactly known. For instance, it may be difficult to perfectly assess the positions of the microphones, for example due to the vibration of the antenna on which they are set. The medium may also be the cause to some uncertainties: in particular, the wavenumber can vary, due to a significant variation of the temperature between the sound sources and the microphones. In this paper, we present how both these sources of uncertainty may be taken into account in the localization process.

The theory of belief functions, also known as Dempster-Shafer theory, is a powerful tool for managing and mining uncertain data. The theory was developed by Dempster and Shafer [5,6,13]. The problem of statistical inference was addressed in [13], and developed by Denoeux [7]. In this latter work, the author proposed a framework in which data uncertainty is represented using belief functions. He then introduced an extension of the EM algorithm, the Evidential EM (E2M) algorithm, which makes it possible to estimate the parameters of the model from such uncertain data. In this paper, we will adopt this approach for representing the uncertainties arising from ill-known microphone locations and wavenumber, and for estimating the locations and strengths of the sound sources.

The organization of this paper is as follows. In Section 2, we give the basic description of the sound sources localization model in the case of precise data and we show how the EM algorithm may be used to solve the parameter estimation problem. In Section 3, we show how imprecise microphone locations and wavenumber may induce an uncertainty on the sound pressures measured by the microphones. Then, in Section 4, we detail our parameter estimation method for such uncertain data using the E2M algorithm. Finally, Section 5 presents simulation experiments which show the advantage of the E2M algorithm in coping with uncertain data, and Section 6 concludes the paper.

2   Sound Source Localization via the EM Algorithm

2.1   Basic Description of the Model

We assume that there are M microphones on a line, with known locations $(\theta_m, 0), m = 1, \ldots, M$. We consider N sound sources, the coordinates of the n-th source being $(\xi_n, \eta_n), n = 1, \ldots, N$. The signal received by the m-th microphone in each snapshot is the sum of components from different sources, altered by a complex-valued Gaussian distributed noise:

$$x_p = G(\xi, \eta)A + \epsilon_p. \tag{1}$$

Here $x_p = (x_{1p}, \ldots, x_{Mp})^T$ is the vector of pressures measured by the M microphones in the p-th snapshot. The M-by-N matrix $G(\xi, \eta) = (G(\xi_n, \eta_n, \theta_m))_{m=1,n=1}^{M,N}$ describes the sound propagation process: the (m, n)-th entry of this matrix is the Green function $G(\xi_n, \eta_n, \theta_m) = \frac{e^{jk\sqrt{(\xi_n - \theta_m)^2 + \eta_n^2}}}{2\pi\sqrt{(\xi_n - \theta_m)^2 + \eta_n^2}}$, that is the operator which transforms the signal emitted by the n-th source into the signal received by the m-th microphone. The vector $A = (A_1, \ldots, A_N)^T$ contains the strengths of the sound sources. Eventually, $\epsilon_p = (\epsilon_{1p}, \ldots, \epsilon_{Mp})^T$ is a complex Gaussian-distributed noise. Note that $Re(\epsilon_p)$ and $Im(\epsilon_p)$ are independent $N(\mathbf{0}, \frac{\sigma^2}{2}I_M)$ M-dimensional random variables (here, I_M stands for the M-by-M identity matrix).

2.2   The EM Algorithm

In this section, we give a brief introduction of the EM algorithm [4]. Let us denote by x the incomplete or observed data, with probability density function (pdf) $g(x|\Phi)$. Similarly, y stands for the complete (unknown) data, with pdf $f(y|\Phi)$. Both f and g depend on the parameter vector Φ, which is to be estimated by maximizing the log-likelihood of the observed data (observed log-likelihood)

$$L(\Phi) = \log g(x|\Phi),$$

over Φ. For this purpose, the EM algorithm proceeds with the complete log-likelihood by iterating back and forth between two steps:

- the E-step, where the expectation $Q(\Phi|\Phi') = E(\log f(y|\Phi)|x, \Phi')$ over the unknown variables is computed, knowing the parameter vector Φ' estimated at the previous iteration;
- the M-step, where ML estimates of the parameters are determined by maximizing $Q(\Phi|\Phi')$ with respect to Φ.

It may be shown that, under regularity conditions, the EM algorithm converges towards a local maximum of the observed log-likelihood [4,17].

2.3   Model Estimation Using the EM Algorithm

We now present how the problem of sound source localization may be addressed using the EM algorithm. Let $x = (x_1, \ldots, x_P)$ be the vector of observed data. For each observed vector of pressures x_p, the complete data are the contributions $y_p = (y_{1p}, \ldots, y_{Np})$ of the sound sources to these measured pressures: each vector y_{np} represents the set of pressures emitted by the n-th source and received by the microphones. Thus, x_p is related to the y_{np} by

$$x_p = \sum_{n=1}^{N} y_{np}.$$

We have $y_{np} \propto N(G_n(\xi_n, \eta_n)A_n, \frac{\sigma^2}{N}I_M)$, where $G_n(\xi_n, \eta_n)$ is the n-th column of the matrix $G(\xi, \eta)$. Since the joint pdf of the complete data over all snapshots is

$$f(y; \xi, \eta, A, \sigma) = \prod_{p=1}^{P} \prod_{n=1}^{N} f(y_{np}; \xi_n, \eta_n, A_n, \sigma),$$

we may write the complete log-likelihood:

$$\log L(y; \xi, \eta, A, \sigma) = -2MNP \log \sigma - \frac{N}{\sigma^2} \sum_{p=1}^{P} \sum_{n=1}^{N} |y_{np} - G_n(\xi_n, \eta_n)A_n|^2. \quad (2)$$

Then, given the parameters $\Phi^l = (\xi^l, \eta^l, A^l, \sigma^l)$ estimated at iteration l, the $(l+1)$-th iteration of EM algorithm consists in the E- and M-steps.

E-step: compute $Q(\Phi|\Phi^l) = E(\log L(y; \xi, \eta, A, \sigma)|x, \xi^l, \eta^l, A^l, \sigma^l)$. For this purpose, let us remind the following theorem ([15]):

Theorem 1. *Let X and Y be n-dimensional Gaussian random vectors with expectation m_X and m_Y and with covariance matrix Σ_{XX} and Σ_{YY}. Let $\Sigma_{XY} = Cov(X, Y)$ and $\Sigma_{YX} = Cov(Y, X)$, then the conditional pdf of Y given X is*

$$N(m_Y + \Sigma_{YX}\Sigma_{XX}^{-1}(x - m_X), \Sigma_{YY} - \Sigma_{YX}\Sigma_{XX}^{-1}\Sigma_{XY}).$$

Since $(x_p, y_{np})^T$ are jointly Gaussian with expectation $(G(\xi, \eta)A \ G_n(\xi_n, \eta_n)A_n)^T$, assuming that y_{n_1p} and y_{n_2p} are uncorrelated for $n_1 \neq n_2$, the covariance matrix of $(x_p, y_{np})^T$ is $\begin{pmatrix} \sigma^2 I_M & \frac{\sigma^2}{N}I_M \\ \frac{\sigma^2}{N}I_M & \frac{\sigma^2}{N}I_M \end{pmatrix}$. Then by Theorem 1, we obtain the conditional expectation of y_{np}:

$$E(y_{np}|x, \xi^l, \eta^l, A^l, \sigma^l) = G_n(\xi_n^l, \eta_n^l)A_n^l + \frac{1}{N}(x_p - G(\xi^l, \eta^l)A^l); \quad (3)$$

in the following, we will denote this expectation by v_{np}^l. For each n, we have

$$\arg \max_{\xi_n, \eta_n, A_n} Q(\Phi|\Phi^l)$$

$$= \arg \min_{\xi_n, \eta_n, A_n} \sum_{p=1}^{P} E\left(|y_{np} - G_n(\xi_n, \eta_n)A_n|^2 |x, \xi^l, \eta^l, A^l, \sigma^l\right)$$

$$= \arg \min_{\xi_n, \eta_n, A_n} \sum_{p=1}^{P} |v_{np}^l - G_n(\xi_n, \eta_n)A_n|^2. \quad (4)$$

M-step: compute $\Phi^{l+1} = (\xi^{l+1}, \eta^{l+1}, A^{l+1}, \sigma^{l+1})$ so as to maximize $Q(\Phi; \Phi^l)$.

The update equation for the strength A_n of the sources is obtained from (4), for $n = 1 \ldots, N$:

$$A_n^{l+1} = \arg\min_{A_n} \left| e_n^l - G_n(\xi_n, \eta_n) A_n \right|^2 = \frac{G_n(\xi_n, \eta_n)^H e_n^l}{G_n(\xi_n, \eta_n)^H G_n(\xi_n, \eta_n)}, \quad (5)$$

in which we write

$$e_n^l = \frac{1}{P} \sum_{p=1}^{P} v_{np}^l. \quad (6)$$

Therefore, the estimates for the source location are obtained by minimizing

$$\frac{1}{P} \sum_{p=1}^{P} \left| v_{np}^l - G_n(\xi_n, \eta_n) \frac{G_n(\xi_n, \eta_n)^H e_n^l}{G_n(\xi_n, \eta_n)^H G_n(\xi_n, \eta_n)} \right|^2,$$

which gives

$$\begin{aligned} \xi_n^{l+1}, \eta_n^{l+1} &= \arg\min_{\xi_n, \eta_n} \left| e_n^l - G_n(\xi_n, \eta_n) \frac{G_n(\xi_n, \eta_n)^H e_n^l}{G_n(\xi_n, \eta_n)^H G_n(\xi_n, \eta_n)} \right|^2 \\ &= \arg\max_{\xi_n, \eta_n} (e_n^l)^H \frac{G_n(\xi_n, \eta_n) G_n(\xi_n, \eta_n)^H}{G_n(\xi_n, \eta_n)^H G_n(\xi_n, \eta_n)} e_n^l. \quad (7) \end{aligned}$$

Finally, by computing and maximizing $E(\log L(y; \xi^{l+1}, \eta^{l+1}, A^{l+1}, \sigma) | x, \xi^l, \eta^l, A^l, \sigma^l)$ with respect to σ, we obtain the estimate of the variance σ^2:

$$(\sigma^2)^{l+1} = \frac{1}{MP} \sum_{p=1}^{P} \sum_{n=1}^{N} \left[\frac{M(N-1)(\sigma^l)^2}{N^2} + \left| v_{np}^l - G_n(\xi_n^{l+1}, \eta_n^{l+1}) A_n^{l+1} \right|^2 \right]. \quad (8)$$

We can see that (7) is a 2-parameter optimization problem. If we assume further that all noise sources are on a line ($\eta_n = z$), it boils down to a single-parameter optimization problem, which is easy to solve.

Eventually, the strategy for estimating the parameters of the model using the EM algorithm may be summarized as follows:

1. For $l = 0$, pick starting values for the parameters $\xi^0, \eta^0, A^0, \sigma^0$.
2. For $l \geq 1$:
   - obtain e_n^l from (6),
   - obtain $\xi_n^{l+1}, \eta_n^{l+1}$, for $n = 1, \ldots, N$ from (7);
   - obtain $A_n^{l+1}, n = 1, \ldots, N$ by substituting $\xi_n^{l+1}, \eta_n^{l+1}$ back into (5);
   - obtain $(\sigma^2)^{l+1}$ from (8).
3. Continue this process until convergence: stop when the relative increase of the observed data (incomplete data) log-likelihood is less than a given threshold κ:

$$\frac{\log L(x; \Phi^{l+1}) - \log L(x; \Phi^l)}{\log L(x; \Phi^l)} < \kappa. \quad (9)$$

We remark that the computation of $(\sigma^2)^{l+1}$ from (8) may be skipped if we just care the situation of the sound sources, since the estimates of the location and the strength do not depend on σ^2.

3   Uncertainty Representation Using Belief Functions

3.1   Uncertain Measurements

In practice, uncertainty may pervade the measurement process, so that the sound pressures measured by microphones are not precise. For example, it may be difficult to give an exact location for the microphones due to the vibration of the antenna. The medium may also be the cause to some uncertainties: e.g., the wavenumber may be ill-known, due to a significant variation of the temperature and thus of the sound velocity between the sound sources and the microphones.

In the next subsection, we will give a short introduction of the belief functions theory, which is a powerful tool for representing and managing uncertain information. Then we will describe how the uncertainties on the microphone locations and wavenumber may be transferred to the observed data in the belief function framework.

3.2   Belief Functions

Let X be a variable taking values in a finite domain Ω. Uncertain information about X may be represented by a *mass function* $m^{\Omega} : 2^{\Omega} \to [0, 1]$, where 2^{Ω} stands for the power set of Ω, such that $\sum_{A \subseteq \Omega} m(A) = 1$. Any subset A of Ω such that $m(A) > 0$ is called *focal element* of m. A mass function m may also be represented by its associated *belief* and *plausibility functions*. Both are defined for all $A \subseteq \Omega$, by:

$$Bel(A) = \sum_{B \subseteq A} m(B), \quad Pl(A) = \sum_{B \cap A \neq \phi} m(B).$$

We can interpret $Bel(A)$ as the degree to which the evidence supports A, while $Pl(A)$ can be interpreted as an upper bound on the degree of support that could be assigned to A if further evidence was available. Eventually, note that the function $pl : \Omega \to [0, 1]$ such that $pl(\omega) = Pl(\{\omega\})$ is the *contour function* associated to m^{Ω}.

Belief Functions on the Real Line. Here we consider the case in which the domain $\Omega_X = \mathcal{R}$. In this case, a mass density can be defined as a function m from the set of closed real intervals to $[0, +\infty)$ such that $m([u, v]) = f(u, v)$ for all $u \leq v$, where f is a two-dimensional probability density function with support in $\{(u, v) \in \mathcal{R}^2 : u \leq v\}$. The intervals $[u, v]$ such that $m([u, v]) > 0$ are called *focal intervals* of m. The contour function pl corresponding to m is defined by the integral:

$$pl(x) = \int_{-\infty}^{x} \int_{x}^{+\infty} f(u, v) dv du.$$

One important special case of continuous belief functions are Bayesian belief functions, for which focal intervals are reduced to points. Then the two-dimensional pdf has the following form: $f(u, v) = p(u)\delta(u - v)$, where p is a univariate pdf and δ is the Dirac delta function. If we assume further that p is a Gaussian pdf, then $pl(x)$ is Gaussian contour function.

3.3   Uncertain Data Model Using Belief Functions

We consider here the same model as described in Section 2. We have M microphones with coordinates $(\theta_m, 0), m = 1, \ldots, M$, and N sound sources, with coordinates $(\xi_n, \eta_n), n = 1, \ldots, N$. The complete data are the signals emitted by the sources: $y_p = (y_{1p}, y_{2p}, \ldots, y_{Np})$, in which $y_{np} = G_n(\xi_n, \eta_n)A_n + \epsilon_{np}$, such that $x_p = \sum\limits_{n=1}^{N} y_{np}$ for all $p = 1, \ldots, P$. In this section, we present how the uncertainties on the microphone locations and on the wavenumber may be transposed to the data. A variance estimation technique via first-order Taylor expansion [1] is used here.

Assume X_1 and X_2 are two independent real-valued random variables and $E(X_1) = \mu_1$, $E(X_2) = \mu_2$, $\text{Var}(X_1) = \sigma_1^2$, $\text{Var}(X_2) = \sigma_2^2$. Furthermore, we assume $f(x_1, x_2)$ is second-order differentiable and has real-valued inputs and complex outputs. By first-order Taylor expansion in (μ_1, μ_2) we have

$$f(x_1, x_2) \approx f(\mu_1, \mu_2) + \frac{\partial f}{\partial x_1}(\mu_1, \mu_2)(x_1 - \mu_1) + \frac{\partial f}{\partial x_2}(\mu_1, \mu_2)(x_2 - \mu_2),$$

and by the independence of X_1 and X_2, we obtain

$$E\left(|f(X_1, X_2) - f(\mu_1, \mu_2)|^2\right) \approx \left|\frac{\partial f}{\partial X_1}(\mu_1, \mu_2)\right|^2 \sigma_1^2 + \left|\frac{\partial f}{\partial X_2}(\mu_1, \mu_2)\right|^2 \sigma_2^2.$$

Let us remind that the p-th pressure measured by the m-th microphone is

$$x_{mp} = \sum_{n=1}^{N} y_{mnp} = \sum_{n=1}^{N} G(\xi_n, \eta_n, \theta_m)A_n = \sum_{n=1}^{N} \frac{e^{jk\sqrt{(\xi_n-\theta_m)^2+\eta_n^2}}}{2\pi\sqrt{(\xi_n-\theta_m)^2+\eta_n^2}} A_n,$$

for $m = 1, \ldots, M, p = 1, \ldots, P$. Assume that the imprecise knowledge of the microphone locations and of the wavenumber is expressed by variances σ_θ^2 (assumed to be the same for all the microphones) and σ_k^2. First, we remark that

$$\frac{\partial x_{mp}}{\partial k} = \sum_{n=1}^{N} \frac{jA_n}{2\pi} e^{jkr_{mn}}, \tag{10}$$

$$\frac{\partial x_{mp}}{\partial \theta_m} = \sum_{n=1}^{N} \frac{A_n}{2\pi}\left[\frac{jk}{r_{mn}}e^{jkr_{mn}} - \frac{1}{r_{mn}^2}e^{-jkr_{mn}}\right]\left(-\frac{\xi_n - \theta_m}{r_{mn}}\right), \tag{11}$$

$m = 1, \ldots, M$, where $r_{mn} = \sqrt{(\xi_n - \theta_m)^2 + \eta_n^2}$ is the distance from the n-th source to the m-th microphone.

Then, we can transfer the uncertainties on the microphone locations and on the wavenumber to the measured pressure. We assume that our imprecise knowledge on the actual measured pressure may be represented using a Gaussian contour function $N(\mu_{mp}^*, \sigma_{mp}^2)$ with expectation $\mu_{mp}^* = x_{mp}$ and variance

$$\sigma_{mp}^2 \approx \left|\frac{\partial x_{mp}}{\partial k}\right|^2 \sigma_k^2 + \left|\frac{\partial x_{mp}}{\partial \theta_m}\right|^2 \sigma_\theta^2. \tag{12}$$

Note that this variance is a function of ξ_n, η_n and A_n for $n = 1, \ldots, N$.

4   Sound Source Localization from Credal Data

4.1   Likelihood Function of a Credal Sample

Let Y be a discrete random vector taking values in Ω_Y, with probability function $p_Y(y|\Phi)$. Let y denote a realization of Y, referred to as the *complete* data. In some cases, y is not precisely observed, but it is known for sure that $y \in A$ for some $A \subseteq \Omega_Y$. The likelihood function given such imprecise data is:

$$L(\Phi; A) = p_Y(A; \Phi) = \sum_{y \in A} p_Y(y|\Phi).$$

More generally, our knowledge of y may be not only imprecise, but also uncertain; it can be described by a mass function m on Ω_Y with focal elements $A_1, \ldots, A_r$ and corresponding masses $m(A_1), \ldots, m(A_r)$. To extend the likelihood function, Denoeux [7] proposes to compute the weighted sum of the terms $L(\Phi; A_i)$ with coefficients $m(A_i)$, which leads to the following expression:

$$L(\Phi; m) = \sum_{i=1}^{r} m(A_i) L(\Phi; A_i) = \sum_{y \in \Omega_Y} p_Y(y|\Phi) pl(y). \tag{13}$$

The likelihood function $L(\Phi; m)$ thus only depends on m through its associated contour function pl. Therefore, we will write indifferently $L(\Phi; m)$ or $L(\Phi; pl)$.

The above definitions can be straightforwardly transposed to continuous case. Assume that Y is a continuous random vector with probability density function $p_Y(y|\Phi)$ and let $pl : \Omega_X \rightarrow [0,1]$ be the contour function of a continuous mass function m on Ω_X. The likelihood function given pl can be defined as:

$$L(\Phi; pl) = \int_{\Omega_Y} p_Y(y; \Phi) pl(y) dy, \tag{14}$$

assuming this integral exists and is nonzero.

4.2   The Evidential EM Algorithm

Here we remind how the classical EM algorithm may be extended so as to estimate the parameters of the model when the data at hand are imprecise. An extensive presentation of this approach may be found in [7]. The new method is called Evidential EM (E2M) algorithm, which maximizes the generalized criterion introduced in the previous section.

At iteration l, the E-step of the E2M algorithm consists in computing

$$Q(\Phi; \Phi^l) = E_{\Phi^l} \left[\log L(\Phi; y) | pl(x) \right]. \tag{15}$$

Note that this expectation is now computed with respect to the imprecise sample known through the contour function $pl(x)$. The M-step is unchanged and requires the maximization of $Q(\Phi; \Phi^l)$ with respect to Φ. As in the EM algorithm, the E2M algorithm alternately repeats the E- and M-steps defined above until the relative increase of the observed-data likelihood becomes smaller than a given threshold.

4.3   Sound Source Localization via the E2M Algorithm

In this section, we present the main results which lead to the update equations of the parameter estimates ξ, η, A and σ^2. For this purpose, we give a lemma which will be used later. Due to page limitation, we omit the proof.

Lemma 1. *If X follows M-dimensional complex Gaussian distribution with mean μ_1 and covariance matrix $\sigma_1^2 I_M$, that is $f(x) = \phi(x; \mu_1, \sigma_1^2 I_M)$, and if X is known through the Gaussian contour function $pl(x) = \phi(x; \mu_2, \sigma_2^2 I_M)$, then*

$$f(x)pl(x) = \phi(x; \frac{\sigma_1^2 \mu_2 + \sigma_2^2 \mu_1}{\sigma_1^2 + \sigma_2^2}, \frac{\sigma_1^2 \sigma_2^2}{\sigma_1^2 + \sigma_2^2} I_M)\phi(\mu_1; \mu_2, (\sigma_1^2 + \sigma_2^2)I_M)$$

and

$$f(x|pl(x)) = \phi\left(x; \frac{\sigma_2^2 \mu_1 + \sigma_1^2 \mu_2}{\sigma_1^2 + \sigma_2^2}, \frac{\sigma_1^2 \sigma_2^2 I_M}{\sigma_1^2 + \sigma_2^2}\right).$$

The log-likelihood of the complete data y is

$$\log L(y; \xi, \eta, A, \sigma) = -2MNP\log\sigma - \frac{N}{\sigma^2}\sum_{p=1}^{P}\sum_{n=1}^{N}|y_{np} - G_n(\xi_n, \eta_n)A_n|^2.$$

Since the partial knowledge of the actual pressure measured by the m-th microphone is represented by a Gaussian contour function $pl(x_{mp}) = N(x_{mp}; \mu_{mp}^*, \sigma_{mp}^2)$, by the second equation of Lemma 1,

$$f(x_{mp}|pl(x_{mp})) = \phi\left(x_{mp}; \frac{\sigma_{mp}^2[G(\xi, \eta)A]_m + \sigma^2\mu_{mp}^*}{\sigma_{mp}^2 + \sigma^2}, \frac{\sigma_{mp}^2 \sigma^2}{\sigma_{mp}^2 + \sigma^2}\right),$$

where $[v]_m$ stands for the m-th element of the vector v. Then, by Theorem 1,

$$f(y_{mnp}|x_{mp}) = \phi\left(y_{mnp}; [G_n(\xi_n, \eta_n)A_n]_m + \frac{1}{N}(x_{mp} - [G(\xi, \eta)A]_m), \frac{N-1}{N^2}\sigma^2\right),$$

from which we can deduce

$$\begin{aligned}
&f(y_{mnp}|x_{mp})f(x_{mp}|pl(x_{mp}))\\
&= N^2\phi\left(x_{mp}; [G(\xi, \eta)A - N(y_{np} - G_n(\xi_n, \eta_n)A_n)]_m, (N-1)\sigma^2\right)\\
&\quad \phi\left(x_{mp}; \frac{\sigma_{mp}^2[G(\xi, \eta)A]_m + \sigma^2\mu_{mp}^*}{\sigma_{mp}^2 + \sigma^2}, \frac{\sigma_{mp}^2 \sigma^2}{\sigma_{mp}^2 + \sigma^2}\right).
\end{aligned}$$

Then, the conditional probability of y_{mnp} given $pl(x_{mp})$ is

$$f(y_{mnp}|pl(x_{mp})) = \int f(y_{mnp}|x_{mp})f(x_{mp}|pl(x_{mp}))dx_{mp}, \tag{16}$$

and by the first equation in Lemma 1, we can write

$$f(y_{mnp}|pl(x_{mp})) = \phi\left(y_{mnp}; v_{mnp}, \frac{1}{N^2}\frac{N\sigma^2\mu_{mp}^* + (N-1)\sigma^4}{\sigma_{mp}^2 + \sigma^2}\right),$$

where $v_{mnp} = [G_n(\xi_n, \eta_n)A_n]_m + \frac{1}{N}\frac{\sigma^2\left(\mu_{mp}^* - [G(\xi,\eta)A]_m\right)}{\sigma_{mp}^2 + \sigma^2}$. Then we obtain

$$E(y_{mnp}|pl(x_{mp}); \xi^l, \eta^l, A^l, \sigma^l) = v_{mnp}^l,$$

where v_{mnp}^l is obtained by replacing the parameters ξ, η, A, σ by $\xi^l, \eta^l, A^l, \sigma^l$ in v_{mnp}. Eventually, for each n, for obtaining the estimates of ξ_n, η_n, A_n, the E2M algorithm amounts to solve, at iteration l,

$$\arg\max_{\xi_n, \eta_n, A_n} E\left(\log L(y; \xi, \eta, A, \sigma)|pl(x); \xi^l, \eta^l, A^l, \sigma^l\right)$$

$$= \arg\min_{\xi_n, \eta_n, A_n} \sum_{p=1}^{P}\sum_{m=1}^{M}\left|v_{mnp}^l - [G_n(\xi_n, \eta_n)A_n]_m\right|^2. \tag{17}$$

The M-step of the E2M algorithm thus corresponds to the following computations:

1. From (17), we obtain the estimate of A_n at $(l+1)$-th step

$$A_n^{l+1} = \frac{G_n(\xi_n, \eta_n)^H e_n^l}{G_n(\xi_n, \eta_n)^H G_n(\xi_n, \eta_n)}, \tag{18}$$

$n = 1, \ldots, N$, where

$$e_n^l = (e_{1n}^l, e_{2n}^l, \ldots, e_{Mn}^l)^T \tag{19}$$

and $e_{mn}^l = \frac{1}{P}\sum_{p=1}^{P} v_{mnp}^l$.

2. Therefore, the update equation for the source location are obtained by

$$\arg\min_{\xi_n, \eta_n} \sum_{m=1}^{M}\left|e_{mn}^l - \frac{[G_n(\xi_n, \eta_n)]_m G_n(\xi_n, \eta_n)^H}{|G_n(\xi_n, \eta_n)|^2} e_n^l\right|^2$$

$$= \arg\min_{\xi_n, \eta_n} (e_n^l)^H \frac{G_n(\xi_n, \eta_n)G_n(\xi_n, \eta_n)^H}{|G_n(\xi_n, \eta_n)|^2} e_n^l. \tag{20}$$

3. Finally, we compute $E\left(\log L(y; \xi^{l+1}, \eta^{l+1}, A^{l+1}, \sigma)|pl(x); \xi^l, \eta^l, A^l, \sigma^l\right)$ and maximize it with respect to σ, then the estimate of variance σ^2 is given by

$$(\sigma^2)^{l+1} = \frac{1}{MP}\sum_{p=1}^{P}\sum_{n=1}^{N}\sum_{m=1}^{M}\left[\frac{N(\sigma^l)^2\sigma_{mp}^2 + (N-1)(\sigma^l)^4}{N^2((\sigma^l)^2 + \sigma_{mp}^2)}\right.$$

$$\left. + \left|v_{mnp}^l - [G_n(\xi_n^{l+1}, \eta_n^{l+1})A_n^{l+1}]_m\right|^2\right]. \tag{21}$$

Finally, we could summarize the algorithm as the EM algorithm at the end of Section 2.3, but Equations (5-8) are replaced by Equations (18-21). The stopping criterion in this E2M algorithm is the same as in Equation (9), but the observed-data log-likelihood is replaced by $\log L(\Phi; pl)$, which is computed by:

$$\log L(\Phi; pl) = -MP \log \pi - \sum_{m=1}^{M} \sum_{p=1}^{P} \left[\log(\sigma^2 + \sigma_{mp}^2) + \frac{\left| \mu_{mp}^* - [G(\xi, \eta)A]_m \right|^2}{\sigma^2 + \sigma_{mp}^2} \right].$$

5   Experiments

5.1   Data Generation

We consider $M = 11$ microphones situated on the x-axis with locations $\theta_m = 0 :$ $0.1 : 1$, and $N = 2$ sound sources with actual coordinates $\xi_1 = 0.3$, $\xi_2 = 0.7$, and $\eta_1 = \eta_2 = 0.3$ (the locations of microphones and sound sources are displayed in Figure 1 using pink stars and black crosses respectively.). The strengths of both sources are $A_1 = 0.6$ and $A_2 = 0.8$. The theoretical wavenumber value is $k = 2\pi f/c$, where the sound frequency is $f = 500$Hz and the sound velocity is assumed to be $c = 340m/s$.

The number of snapshots (the amount of pressures measured by each microphone) is set to $P = 100$. Then, we introduce noise in the microphone locations and wavenumber, as follows. For a given value of σ_k, we generate a wavenumber value according to a Gaussian distribution with mean k and standard deviation σ_k. Similarly, we introduce noise in the microphone locations using means $\theta_1, \ldots, \theta_{11}$ and standard deviation σ_θ. Then, we obtain simulated pressures using Equation (1), in which we set $\sigma = 0.05$. The contour functions modeling the uncertainty on the pressures may be obtained using the method detailed in Section 3.3. The level of noise in the data, and consequently the amount of uncertainty, may thus be controlled through the parameters σ_θ and σ_k.

Note that the quality of the estimates obtained via the EM and E2M algorithms depend on the starting values for the parameters. Therefore, for a given dataset, we let both algorithms run using 5 different sets of starting values, retaining the solution with highest log-likelihood. Remark that the variances of the contour functions σ_{mp}^2 used in the E2M algorithm are computed using the parameters ξ_n, η_n and A_n estimated via the EM algorithm. Since the data are randomly generated, the above procedure (from data generation to model estimation) is repeated 30 times, so that mean square errors (MSE) on the parameter values and associated 95% confidence intervals may be computed.

5.2   Results

First, we set $\sigma_k = 0.5$ and $\sigma_\theta = 0.05$, and we estimate the sound source locations (ξ_n, η_n) using EM and E2M. Figure 1 shows the 30 estimation results of the source locations computed by EM (left figure) and E2M (right figure).

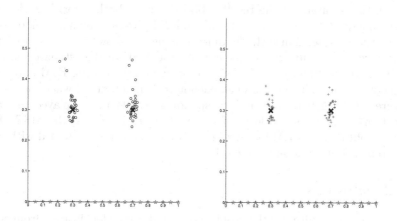

Fig. 1. Locations of the sources estimated using the EM (left) and E2M (right) algorithms, with $\sigma_k = 0.5$ and $\sigma_\theta = 0.05$

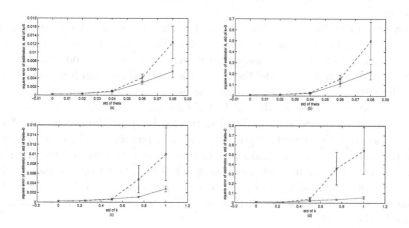

Fig. 2. MSE and 95% confidence intervals for the estimates of sound source locations (ξ, η) (left) and strengths A (right) using EM (blue dash lines) and E2M (red full lines), according to the level of uncertainty in the microphone locations (top) and wavenumber (bottom)

The estimates obtained using E2M clearly exhibit a smaller spread around the actual locations of the sources, which demonstrates its interest in terms of dealing with uncertain data.

To corroborate these observations, we now study the MSE of the parameters (ξ_n, η_n) and A_n estimated via both algorithms. We set $\sigma_k = 0$ and increase σ_θ from 0 to 0.08. The estimated MSE of the sound source locations and strengths are displayed in Figure 2 (top) along with 95% confidence intervals. Without surprise, the accuracy of the results obtained using both algorithms decreases as the amount of noise increases. However, E2M proves to be much more robust to the level of noise than EM: the difference clearly shows the interest of using E2M. More specifically, when σ_θ increases, the MSE of the estimates obtained via EM increase dramatically, while the MSE obtained using E2M stay under an acceptable level. The same phenomenon may be observed when $\sigma_\theta = 0$ and σ_k increases from 0 to 1. The corresponding results are displayed in Figure 2 (bottom). Again, as the level of uncertainty σ_k increases, the MSE of the estimates obtained via EM increases much more than those obtained with E2M, which remains at an acceptable level.

6   Conclusions

In this paper, we addressed the problem of sound source localization from acoustical pressures measured by a set of microphones. The problem may be solved in a statistical setting, by assuming that each pressure measured by a microphone is the sum of contributions of the various sources, pervaded by a Gaussian noise. The EM algorithm may then be used to compute maximum-likelihood estimates of the model.

However, in many applications, some parameters of the model, such as the microphone locations or the wavenumber, may be pervaded with uncertainty. In this work, we show how this uncertainty may be transposed on the measured pressures. We propose to model these uncertainties using belief functions. In this case, the parameters of the model may be estimated using a variant of the EM algorithm, known as the Evidential EM algorithm.

The results obtained on simulated data clearly show the advantage of taking into account the uncertainty on the data, in particular when the degree of noise due to the ill-known parameters (microphone locations and wavenumber) is high. Then, the results obtained using the E2M algorithm are more robust than those obtained using the EM algorithm. The generalization of our method to the general case of spatial sources, as well as its validation on real data, are left for further work.

Acknowledgments. This work has been partially funded by the European Union. Europe is committed in Picardy with the FEDER.

References

1. Bevington, P.R., Keith, D.: Data Reduction and Error Analysis for the Physical Sciences, 3rd edn. McGraw-Hill, New York (2003)
2. Cirpan, H.A., Cekli, E.: Deterministic Maximun Likelihood Approach for Localization of Near-Field Sources. International Journal of Electronics and Communications 56(1), 1–10 (2002)
3. Cirpan, H.A., Cekli, E.: Unconditional Maximum Likelihood Approach for Localization of Near-Field Sources: Algorithm and Performance Analysis. International Journal of Electronics and Communications 57(1), 9–15 (2003)

 4. Dempster, A.P., Laird, N.M., Rubin, D.B.: Maximum Likelihood from Incomplete Data via the EM Algorithm. Journal of the Royal Statistical Society, Series B (Methodological) 39(1), 1–38 (1977)
 5. Dempster, A.P.: Upper and Lower Probabilities Induced by a Multivalued Mapping. Annals of Mathematical Statistics 38(2), 325–339 (1967)
 6. Dempster, A.P.: Upper and Lower Probabilities Generated by a Random Closed Interval. Annals of Mathematical Statistics 39(3), 957–966 (1968)
 7. Denoeux, T.: Maximum Likelihood Estimation from Uncertain Data in the Belief Function Framework. IEEE Transactions on Knowledge and Data Engineering 25(1), 119–130 (2013)
 8. Denoeux, T.: Maximum Likelihood Estimation from Fuzzy Data Using the EM Algorithm. Fuzzy Sets and Systems 183(1), 72–91 (2011)
 9. Feder, M., Weinstein, E.: Parameter Estimation of Superimposed Signals Using EM Algorithm. IEEE Transactions on Acoustics, Speech and Signal Processing 36(4) (2005)
10. Kabaoglu, N., Cirpan, H.A., Cekli, E., Paker, S.: Deterministic Maximum Likelihood Approach for 3-D Near-Field Source Localization. International Journal of Electronics and Communications 57(5), 345–350 (2003)
11. Quost, B., Denœux, T.: Clustering Fuzzy Data Using the Fuzzy EM algorithm. In: Deshpande, A., Hunter, A. (eds.) SUM 2010. LNCS (LNAI), vol. 6379, pp. 333–346. Springer, Heidelberg (2010)
12. Render, R.A., Walker, H.F.: Mixture Densities, Maximum Likelihood and the EM Algorithm. SIAM Review 26(2), 195–239 (1984)
13. Shafer, G.: A Mathematical Theory of Evidence. Princeton University Press, Princeton (1976)
14. Smets, P.: Belief Functions on Real Numbers. International Journal of Approximate Reasoning 40(3), 181–223 (2005)
15. Rhodes, I.B.: A Tutorial Introduction to Estimation and Filtering. IEEE Transaction on Automatic Control AC-16(6), 688–706 (1971)
16. Williams, E.G.: Fourier Acoustic: Sound Radiation and Nearfield Acoustical Holography. Academic Press (1999)
17. Wu, J.C.F.: On the Convergence Properties of the EM Algorithm. Annals of Statistics 11(1), 95–103 (1983)

An Improvement of Subject Reacquisition
by Reasoning and Revision

Jianbing Ma[1,2], Weiru Liu[2], Paul Miller[2], and Fabian Campbell-West[2]

[1] School of Electronics, Electrical Engineering and Computer Science, Queen's University
Belfast, Belfast BT7 1NN, UK
[2] School of Design, Engineering and Computing, Bournemouth University, BH12 5BB, UK
{w.liu,p.miller,f.h.campbell-west}@qub.ac.uk,
jma@bournemouth.ac.uk

Abstract. CCTV systems are broadly deployed in the present world. Despite
this, the impact on anti-social and criminal behaviour has been minimal. Subject reacquisition is a fundamental task to ensure in-time reaction for intelligent
surveillance. However, traditional reacquisition based on face recognition is not
scalable, hence in this paper we use reasoning techniques to reduce the computational effort which deploys the time-of-flight information between interested
zones such as airport security corridors. Also, to improve accuracy of reacquisition, we introduce the idea of revision as a method of post-processing. We demonstrate the significance and usefulness of our framework with an experiment which
shows much less computational effort and better accuracy.

Keywords: Subject Reacquisition, Time-of-Flight, CCTV Surveillance, Event
Reasoning, Revision.

1 Introduction

During the last decade, there has been massive investment in Closed-Circuit TeleVision
(CCTV) technology in the UK. Currently, there are approximately four million CCTV
cameras operationally deployed. Despite this, the impact on anti-social and criminal
behaviour has been minimal. Although most incidents, also called events, are captured
on video, there is no response because very little of the data is actively analysed in real-
time. Consequently, CCTV operates in a passive mode, simply collecting enormous
volumes of video data. For this technology to be effective, CCTV has to become active
by alerting security analysts in real-time so that they can stop or prevent the undesirable
behaviour. Such a quantum leap in capability will greatly increase the likelihood of
offenders being caught, a major factor in crime prevention.

To ensure in-time reaction for intelligent surveillance, one fundamental task to utilize
CCTV videos is subject reacquisition[1] [1, 2, 28]. Subject reacquisition is the process
of identifying a particular subject at a specific point in space and time given knowledge of a previous observation. There are many methods of performing reacquisition,

---

[1] It is also called object reacquisition, here we use the term subject reacquisition since we focus
on reacquiring people.

W. Liu, V.S. Subrahmanian, and J. Wijsen (Eds.): SUM 2013, LNAI 8078, pp. 176–189, 2013.

but the general case is illustrated in Fig. 1. At each checkpoint (a camera/sensor) the face information is captured and stored in a central repository. A comparison of these features will reveal the most suitable match from a previous checkpoint for a subject at the current checkpoint.

In the context of this paper and the concept demonstrator it is based on, the sensors are standard CCTV cameras. A video analytic sub-system performs face detection and recognition. As the number of subjects in the system increases, the reacquisition problem becomes more difficult and mismatch becomes more likely. In addition, since comparisons are made between the live subject and all previously observed subjects the system is not scalable with large numbers of subjects. It is necessary to keep the potential number of comparisons at a manageable number, which is a primary goal of the event reasoning part.

To study subject reacquisition, in this paper, we consider a scenario which involves a simple secure corridor. At each end of the corridor is video camera. One agent is responsible for managing the events generated by these sensors. The aim of the system is to perform subject reacquisition. When presented with live data of subject X at the second sensor, the subject reacquisition problem is to identify which subject previously seen at the first sensor matches subject X (Same person always presents slightly different images in different sensors). In this context, subject reacquisition is also known as closed-set matching. This is one of the simplest scenarios that can benefit from artificial intelligence techniques. Also, note that this kind of secure corridor does exist in a set of airports, e.g., the Gatwick Airport[2].

In this scenario, two simplification assumptions are introduced and listed as follows:

- Only one subject shall be considered at one time(no occlusions etc.)
- The subject will be co-operative at key points (i.e. looking at the camera for a few frames)

The first assumption removes the need for tracking the subjects spatially and the second removes the risk that no face is present. These two assumptions are reasonable. Actually they are fully achieved in the secure corridor case in the Gatwick airport and other airports.

Currently, common approaches for subject reacquisition are simply by face recognition. Although face provides a rich source of information, its recognition comes with the price of uncertainty. A great deal of research effort has been applied to this area aiming for increasing robustness. In all non-trivial scenarios, there will be sources of errors in a pure video-analysis system and, if unchecked, these errors will propagate and cause further mistakes.

In this paper, an alternative approach is proposed where classic imperfect subject reacquisition (by face recognition algorithms) results are enhanced by using artificial intelligence approaches including event reasoning[3] and belief revision[4]. A real-world experiment is introduced and the baseline results are presented. More precisely, in this

---

[2] Currently they check passengers manually.

[3] As in [24, 18, 20, 25, 5, 16, 21–23, 7], event reasoning uses domain information to combine basic events to get high-level information.

[4] Belief revision depicts the process that an agent revises its belief state upon receiving new information [13, 19, 8, 9, 15, 12, 14, 11, 10, 17].

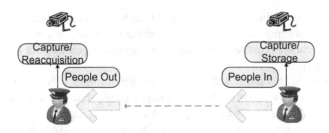

Fig. 1. Classic Subject Reacquisition

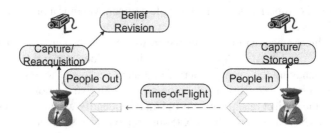

Fig. 2. Subject Reacquisition Enhanced by Reasoning and Revision

paper, event reasoning is used as pre-processing to reduce the size of comparison sets while belief revision, or the spirit of revision, is used as post-processing to enhance the recognition accuracy, as shown in Fig. 2. Here by comparison set we mean when a subject X is detected at sensor 2, the set of face models from sensor 1 that should be extracted to be compared with the face model of X.

The remainder of this paper is organized as follows. In Section 2, we briefly introduce our face recognition method. In Section 3, we state our event reasoning component using the time-of-flight information to reduce sizes of comparison sets. Section 4 describes the experimental scenario. In Section 5, we discuss how to use revision to improve accuracy of classification and reduce reacquisition failure. In Section 6, we conclude the paper.

2 Face Recognition

In this section, we only briefly introduce the method we used for face recognition since it is not the main focus of the paper.

The face recognition component is designed to be modular, so that different algorithms can be substituted for any component provided it has an appropriate interface.

With each new frame provided by the sensor it is necessary to search for faces. Face detection is a mature research area and there are many solutions available, among which we have tried two famous detectors: the Viola-Jones detector [27] and the Luxand detector [6]. The Viola-Jones detector is well-known, robust and good at finding faces in images in terms of precision and recall, but is not consistent in the bounding box returned of the detected face. So eventually we use the Luxand detector which provides more consistent results. Again, the Face Detection sub-system is modular, so any algorithm that can return a bounding box of a face in an image can be used.

We have also applied illumination compensation which is the process of adjusting the pixel intensities of an input image to match the illumination profile of a reference image. The algorithm used in the system is adapted from [4].

Faces of the same subject within a series of frames are registered to one face model and this model is stored in a buffer.

When a subject X is detected at sensor 2 reacquisition is performed by comparing every available face image to the set of known models (and corresponding subjects). And the subject in the set whose face model matches the most is chosen as the reacquired subject.

For each detected subject, a SubjectDetected event is created which contains four pieces of information needed. Let event i be denoted by the tuple

$$e^i = (o, d, t, M)$$

where o is the source identifier, d is the subject identifier, t is the timestamp and M is the face model. The source identifier is simply an integer that uniquely identifies the sensor. The subject identifier is also an integer, but is relative to the source. For example the subject with ID 1 at sensor 1 is not necessarily the same as the subject with ID 1 at sensor 2. The timestamp is the frame number, but it is assumed that this is synchronised across all sensors. The face model is a MACE filter produced from the face image. In implementation, M can be recorded by a pointer to the face model.

In addition, it is useful to discuss which methods can be used to integrate the reliabilites of face recognition generated by different frames. Now we use a simple maximum coverage method. That is, if a face model is considered the most plausible by face recognition in one frame, then its weight is added by one. After considering all frames, we normalize the weights of all possible face models. We are trying to use probabilistic and evidential [3, 26] methods to compare with the current one.

3   Reasoning by Time-of-Flight

When a subject is detected at sensor 2 the agent chooses face models from sensor 1 to match to the live data at sensor 2. Without limiting the comparisons the system is not scalable. In addition, the greater the size of comparison set the more likely it is a mismatch will be made. However, it is possible that by limiting the comparisons that the real subject is not included.

Our system uses domain knowledge, in the form of an average time-of-flight between the sensors, to determine appropriate models. For instance, a person cannot appear within the detection zone A, and then appear in zone B within 10 seconds. But if the time interval is changed to 10-20 seconds, then it could be possible; if the time-of-flight is 20-40 seconds, it is highly possible; if the time is greater than 40 seconds, it is less likely, etc.. Empirical probabilities for such information can be found by tests. However, by now, we will use simple settings that in some special range of time-of-flight, the certainty is 1 instead of probabilities.

Let E be the set of all events an agent has received so far. The most recent event is e^N (the N-th event). If $e^N.o = 2$, then the subject in this event needs to be reacquired. Let C be the comparison set of events that contain face models relative to e^N. The set C is composed according to the following rule:

$$C = \{e^i : e^i.o = 1, \tau_1 < e^N.t - e^i.t < \tau_2\}$$

where τ_1, τ_2 are thresholds on time-of-flight determined through domain knowledge indicating the lower and upper time limits for selecting subjects for comparison. Note that since e^N is the most recent event, $e^N.t \geq e^i.t$ is always true. This rule states that the time difference between the event of the current subject being detected at sensor 2 and the subject in event e^i being detected at sensor 1 is within the range (τ_1, τ_2). This rule gives event correlation between the SubjectDetected events at sensors 1 and 2. This is directly analogous to the data association problem in conventional tracking in which observations, i.e. SubjectDetected events, are associated with predictions, i.e. the time difference (τ_1, τ_2).

The average time-of-flight between sensors 1 and 2 is denoted by λ_F and is part of the domain knowledge of the application. This domain knowledge is used to calculate the time thresholds:

$$\tau_1 = \lambda_F - s\sqrt{\lambda_F}, \ \tau_2 = \lambda_F + s\sqrt{\lambda_F}$$

where s in the two above equations is a scalar which determines how many standard deviations from the mean the threshold should be set.

4 Example Scenario

In the example scenario, 30 subjects pass between two checkpoints (sensors), as shown in Fig. 3. The time of arrival of each subject at a checkpoint is determined by a random

Fig. 3. All 30 subjects in the experiment

variable with a Poisson distribution. The Poisson distribution is used in queuing models for this purpose. It is characterized by a mean value, λ, which defines the probability distribution according to the standard Poisson distribution such that:

$$p(t; \lambda) = \frac{\lambda^t e^{-\lambda}}{t!}$$

where t is a non-negative integer.

Let the time of arrival of subject i at checkpoint 1 be denoted by t_i. Without loss of generality we set $t_1 = 1$. The difference in arrival time, $a_{i,j}$, between two consecutive subjects i and j can be modeled as a random variable with an Exponential distribution of mean denoted by λ_A. The arrival time of subject i at checkpoint 1 can be written as

$$t_i = t_{i-1} + a_{i-1,i}$$

for $i > 1$. For this scenario the values of $a_{i,j}$ were determined by a random number generator with $\lambda_A = 3$.

The time-of-flight between checkpoints 1 and 2 can also be modeled as a random variable with a Poisson distribution of mean λ_F. Let the time-of-flight for subject i be denoted by F_i. Then the time of arrival of subject i at checkpoint 2, denoted by u_i, can be written as

$$u_i = t_i + F_i.$$

For this scenario the values of F_i were determined by a random number generator with $\lambda_F = 11$.

The video data for the example scenario was created by manually editing existing video footage to meet the requirements of the timing model above.

The method for choosing the comparison set, C, is described in the previous section. Given the events produced for the example scenario, it is necessary to set $\lambda_F = 11$ and $s = 3.5$ to ensure the comparison set always includes the correct subject. At $s \leq 3$ the comparison set for some subject at sensor 2 will be empty.

For illustration simplicity, here we list the system behavior on nine subjects, which generates nine events at each sensor for a total of 18 input events. Reacquisition is performed for every input event at the second sensor for a total of nine output events.

Fig. 4. The subject order at each sensor and the comparison sets for each subject

Fig. 4[5] shows the comparison set for each subject. Fig. 4 provides the orders each subject is present at each camera and the comparison set for each subject. To avoid confusion, the subjects are numbered according to their order at sensor 1. Between sensors 1 and 2 subjects 3 and 5 changed places. This means subject 5 moved quickly between the sensors and subject 3 moved slowly. Subjects 6 and 7 also changed places while moving between sensors.

---

[5] For the sake of privacy, we have blurred the faces used.

Also from Fig. 4, we can see that our event reasoning method indeed reduces the size of comparison set a lot. Classically each subject at sensor 2 should compare to 9 nine candidates, but in our system, it only needs to compare to 5.3 candidates on average. That is, we have a 41% decrease in computational effort. Actually, in our 30 people experiment, the average size of comparison sets is 8, which saves 72% computational effort. Not surprisingly, if the scale of scenario becomes larger, it will save even more computational effort.

5   Revision

In this section, we discuss how to revise the face recognition results provided that there are obvious errors in the results. First, we describe the experimental result.

5.1   Comparison Reasoning

The subject reacquisition result based on the comparison set is shown in the following Fig. 5. In Fig. 5, we can see that in terms of classification results three out of nine

Fig. 5. Reacquisition result based on the comparison set

reacquisitions are incorrect, i.e., reacquisitions of subjects 1, 6, and 7, and one subject was not reacquired at all.

Classification accuracy is defined as:

$$\text{Accuracy} = \frac{\text{correct classifications}}{\text{total classifications}}.$$

The results above yield an accuracy of 67%.

Reacquisition failure can occur for the following reasons:

1. A subject is falsely detected and therefore the reacquired model is incorrect.
2. A subject is not detected when present at the second sensor, so reacquisition is not performed.
3. The subject is detected but wrongly reacquired. This can occur for the following reasons:
   (a) The comparison set C is empty.
   (b) The comparison set C does not contain the true subject.
   (c) The true subject is not chosen from the comparison set C.

Reasons 1, 2 and 3c on the list are failures of the video analytics, rather than the event reasoning part. Reasons 3a and 3b are failures of the reasoning.

Reacquisition failure is clearly shown in Fig. 5.1. Until a better description of system

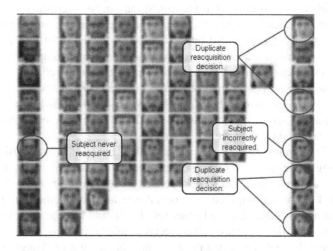

Fig. 6. Types of Errors in Reacquisition

failure is developed, the Reacquisition Failure measure will be calculated as

$$\text{Reacq.Failure} = \frac{\text{incorrect reacquisitions+missed reacquisitions}}{\text{possible correct reacquisitions}} = 4/9 = 44\%.$$

This is different to 1-Accuracy, as it considers models that are not reacquired at all. The failure measure is a worst-case view of the system, relative to the subjects rather than just the system outcome. It can be described as the proportion of 'problematic' subjects. Note that it is possible for the failure measure to be greater than 1, which reflects the imbalance between the number of ways the system can be correct versus the number of ways the system can fail.

Note that Subject 9 is incorrectly reacquired as the seventh subject at sensor 2. This raises the important issue of what to do once reacquisition has occurred. If the reacquisition is assumed correct, then subject 9 would be removed from subsequent comparisons and the final comparison set would be empty.

Note that if the system makes an incorrect classification that implies the system has, or will, either make a duplicate reacquisition decision or will switch subjects. A duplicate reacquisition implies that one of the subjects has not been reacquired. Subject switching is when two subjects are confused with each other.

In addition, in our 30 people experiment, the accuracy rate is 87% (26/30).

5.2   Revision of Reacquisition Results

When we have conflict reacquisition results, i.e., duplicate reacquisition, missing someone out, etc., we then use revision as a post-processing method to improve accuracy.

First, let us note that due to the closed world assumption, mistakes cannot happen in isolation, one mistake should infer another. That is, if we have missed out on someone, then there should be duplicate reacquisition, and vice versa.

The basic idea of revision in subject reacquisition can be described as follows: when we have conflict classification results, we can first determine the more reliable classification result (which is provided by the degree of certainties of face classification results), then we remove that candidate in the other conflicting comparison set. For instance, if both subjects X and Y are classified as person A while reliability (i.e., degree of certainty) shows the classification of X to A is more plausible, then we will remove candidate A in the comparison set of Y, and then we can choose the best match for Y in the new comparison set. This process is not strictly belief revision but deploys the idea of revision in artificial intelligence.

Here we should note that mistakes can cascade. That is, if in the revision process, we have wrongly removed a correct candidate from a comparison set, e.g., removing A from the comparison set of Y, and we choose B as the reacquisition result of Y, then it is possible that B has been reacquired by subject Z which makes another conflict between the reacquisitions of Y and Z and a further revision should be taken, and so on. In this sense, a wrong revision can destroy all correct reacquisition results.

To overcome this deficiency, we need to limit the amount of changes that can be made. Here we propose two kinds of revision:

– One step revision
– Limited revision based on threshold

By one step revision, we mean we do not proceed if a revision result induces a further conflict. For instance, suppose X and Y are classified as A, and Z is classified as B. Assume $r(X \to A) > r(Y \to A)$ (here $r(X \to A)$ is the degree of certainty for X classified as A), Y should be revised as a second choice, and suppose it is B which leads to further conflict with the reacquisition of Z. Now if $r(X \to A) < r(Z \to B)$, we just keep the original result, i.e., X, Y are classified as A and Z as B, else we keep the revised result, i.e., X is classified as A and Y, Z as B. And we finish here.

In Algorithm 1, notation C_Y means the comparison set of Y. By applying Algorithm 1, we have the following result showing in Fig. 7: From Fig. 7, we find that two incorrect reacquisitions have been corrected, which improves the Accuracy from 67% to 89%, a 22% increase in Accuracy, and the reacquisition failure rate is reduced from 44% to 22%. Also, from Fig. 7, we know that revision cannot always achieve consistency.

Algorithm 1. One Step Revision

Require: All subjects, their comparison sets and their reacquired results.
Ensure: Revised reacquired results.
1: **for** each set S of conflict subjects **do**
2:    Revision = 1;
3:    A = classified result of each subject in S;
4:    $X = max_S\{r(X) : X \in S\}$;
5:    **for** each subject Y in $S \setminus \{X\}$ **do**
6:      $TempReacq(Y) = max_{C_Y}\{r(Y \to B) : B \in C_Y \setminus \{A\}\}$;
7:      B = TempReacq(Y);
8:      **if** exist Z, Reacq(Z)=B and $r(X \to A) < r(Z \to B)$ **then**
9:        Revision = 0;
10:       BREAK;
11:     **end if**
12:    **end for**
13:    **if** Revision = 1 **then**
14:      **for** each subject Y in $S \setminus \{X\}$ **do**
15:        $Reacq(Y) = max_{C_Y}\{r(Y \to B) : B \in C_Y \setminus \{A\}\}$;
16:      **end for**
17:    **end if**
18: **end for**
19: **return** Revised reacquired results.

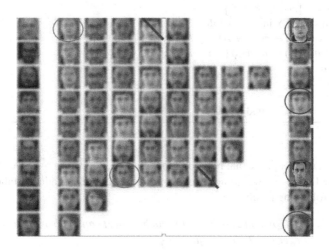

Fig. 7. Revised Reacquisition Result by One Step Revision

In addition, for the 30 people experiment, the revised result achieves a remarkable 97% accuracy (29/30).

For limited revision, we should first introduce a threshold value t to indicate the difference between reliabilities (degrees of certainty) of classification results. As mentioned above, again we suppose X and Y are classified as A and Z classified as B.

If $r(X \to A) - r(Y \to A) >= t$, then Y should be revised to a second choice, else we do nothing. Now suppose Y is revised to B. Now

If $r(Y \to B) > r(Z \to B)$

  if $r(Y \to B) - r(Z \to B) > t$, then we go on to revise the classification of Z,
and so on.

  else we do nothing.

If $r(Y \to B) <= r(Z \to B)$

  if $r(Z \to B) - r(Y \to B) >= t$, then we go on to revise the classification of
Y, and so on.

  else we do nothing.

Each time reacquisition of some subject is changed, it may cause further revision. Since limited revision does not *rewind* as done in one step revision (e.g., the revision of Y from A to B can be rewound (Y changes back to A) by a further conflict reacquisition Z), the algorithm of limited revision looks very simple.

Algorithm 2. Limited Revision

Require: All subjects, their comparison sets and their reacquired results, a threshold value t,
  $0 < t < 1$.
Ensure: Revised reacquired results.
1: **while** exist a set S of conflict subjects **do**
2: A = classified result of each subject in S;
3: $X = max_S\{r(X) : X \in S\}$;
4: **for** each subject Y in $S \setminus \{X\}$ **do**
5:   **if** $r(X \to A) - r(Y \to A) >= t$ **then**
6:    $Reacq(Y) = max_{C_Y}\{r(Y \to B) : B \in C_Y \setminus \{A\}\}$;
7:   **end if**
8: **end for**
9: **end while**
10: **return** Revised reacquired results.

In this way, we can prove that revision (change of reacquisition result) can only happen for limited times, so we do not need to worry about the cascade mistake problem. We have the following result.

Proposition 1. *In Algorithm 2, revision for each subject can happen at most $\lfloor \frac{1}{t} \rfloor$ times. Here $\lfloor \frac{1}{t} \rfloor$ is an integer less than or equal to $\frac{1}{t}$.*

Proof of Proposition 1: We only need to note that each time the reacquisition for a subject is revised, its reliability to its reacquired result is reduced by at least t. So if it is revised by l times, then we have $lt <= 1$, which leads to $l \leq \lfloor \frac{1}{t} \rfloor$ (since l is an integer).

Of course, this $\lfloor \frac{1}{t} \rfloor$ is a theoretic limit. In practice, revision will happen much less. In our experiment, limited revision happens to get the same result as one-step revision, i.e., 97% (29/30). However, it could be expected that the results will be different when the experiment scale gets larger.

6 Conclusion

In this paper, we proposed a system for monitoring subjects passing between two sensors in a secure corridor. A hybrid real-synthetic scenario was created to model an authentic flow of subjects through the corridor. The sensors use video analytics to detect and learn face-based appearance models. Event reasoning using time-of-flight domain knowledge is proved helpful in reducing the comparison set and saves much computational effort. In addition, revision is deployed in this system to improve accuracy. Experimental results from the sequence show that revision does play an important role in increasing accuracy and decreasing reacquisition failure.

Applying artificial intelligence ideas to video surveillance is not a new idea, e.g., [24, 20], but it is seldom to see AI applied to subject reacquisition. Our paper hence can reminder researchers from the vision community and the AI community to be aware of the advantages of each area..

For future work, since the video analytics system is modular, there is potential for modifications to all aspects of the system. An evaluation of the different modules needs to be completed to ensure the best parameters are being used. Other video features, like clothing models and hair colour analysis can be incorporated.

For reasoning, currently generating the comparison is based on the time-of-flight of subjects through the corridor. The choice of the thresholds τ_1, τ_2 is most important. If the thresholds are too far apart the comparison set will include unnecessary samples. If the thresholds are too close the true subject may not be in the comparison set and the system will fail.

It may be more appropriate to use different values of s when calculating τ_1 and τ_2. A better scheme might be to start with strict bounds and relax them if the set C is empty. A problem with this, however, is that a non-empty set C that doesn't contain the true subject will definitely produce an incorrect reacquisition result. As a future work, we will investigate how to choose these two thresholds in general, and, if not well chosen, how does the system behave.

Detecting missed and/or duplicate reacquisitions is straightforward. Missed reacquisitions can be triggered by setting a maximum value on the time-of-flight between sensors. If a subject from sensor 1 has not been reacquired before that time then the agent must re-evaluate previous decisions with belief revision. Duplicate reacquisition occurs when two reacquisitions link to the same subject ID at the previous sensor.

Also, we are performing a much bigger scale of experiment to validate our system, and considering using other features (clothing color, Radio Frequency IDentification (RFID), etc.) to further improve accuracy.

References

1. Arth, C., Leistner, C., Bischof, H.: Object reacquisition and tracking in large-scale smart camera networks. In: Procs. of 1st ACM/IEEE Distributed Smart Cameras, ICDSC 2007, pp. 156–163 (2007)

2. Campbell-West, F., Wang, H., Miller, P.: Where is it? object reacquisition in surveillance video. In: Procs. of Machine Vision and Image Processing, IMVIP 2008, pp. 182–187 (2008)

3. Dempster, A.P.: A generalization of bayesian inference. J. Roy. Statist. Soc. Series B 30, 205–247 (1968)

4. Jiang, X., Fan, P., Ravyse, I., Sahli, H., Huang, J., Zhao, R., Zhang, Y.: Perception-based lighting adjustment of image sequences. In: Zha, H., Taniguchi, R.-i., Maybank, S. (eds.) ACCV 2009, Part III. LNCS, vol. 5996, pp. 118–129. Springer, Heidelberg (2010)

5. Liu, W., Miller, P., Ma, J., Yan, W.: Challenges of distributed intelligent surveillance system with heterogenous information. In: Procs. of QRASA, Pasadena, California, pp. 69–74 (2009)

6. Luxand.com. face recognition

7. Ma, J.: Qualitative approach to bayesian networks with multiple causes. IEEE Transactions on Systems, Man, and Cybernetics, Part A 42(2), 382–391 (2012)

8. Ma, J., Benferhat, S., Liu, W.: Revising partial pre-orders with partial pre-orders: A unit-based revision framework (2012)

9. Ma, J., Benferhat, S., Liu, W.: Revision over partial pre-orders: A postulational study. In: Hüllermeier, E., Link, S., Fober, T., Seeger, B. (eds.) SUM 2012. LNCS, vol. 7520, pp. 219–232. Springer, Heidelberg (2012)

10. Ma, J., Liu, W.: A general model for epistemic state revision using plausibility measures. In: Procs. of ECAI, pp. 356–360 (2008)

11. Ma, J., Liu, W.: Modeling belief change on epistemic states. In: Procs. of FLAIRS (2009)

12. Ma, J., Liu, W.: A framework for managing uncertain inputs: An axiomization of rewarding. Int. J. Approx. Reasoning 52(7), 917–934 (2011)

13. Ma, J., Liu, W., Benferhat, S.: A belief revision framework for revising epistemic states with partial epistemic states. In: Procs. of AAAI, pp. 333–338 (2010)

14. Ma, J., Liu, W., Dubois, D., Prade, H.: Revision rules in the theory of evidence. In: Procs. of ICTAI, pp. 295–302 (2010)

15. Ma, J., Liu, W., Dubois, D., Prade, H.: Bridging Jeffrey's rule, AGM revision and Dempster conditioning in the theory of evidence. International Journal on Artificial Intelligence Tools 20(4), 691–720 (2011)

16. Ma, J., Liu, W., Hunter, A.: Inducing probability distributions from knowledge bases with (in)dependence relations. In: Procs. of AAAI, pp. 339–344 (2010)

17. Ma, J., Liu, W., Hunter, A.: Modeling and reasoning with qualitative comparative clinical knowledge. Int. J. Intell. Syst. 26(1), 25–46 (2011)

18. Ma, J., Liu, W., Miller, P.: Event modelling and reasoning with uncertain information for distributed sensor networks. In: Deshpande, A., Hunter, A. (eds.) SUM 2010. LNCS, vol. 6379, pp. 236–249. Springer, Heidelberg (2010)

19. Ma, J., Liu, W., Miller, P.: Belief change with noisy sensing in the situation calculus. In: Procs. of UAI (2011)

20. Ma, J., Liu, W., Miller, P.: Handling sequential observations in intelligent surveillance. In: Benferhat, S., Grant, J. (eds.) SUM 2011. LNCS, vol. 6929, pp. 547–560. Springer, Heidelberg (2011)

21. Ma, J., Liu, W., Miller, P.: A characteristic function approach to inconsistency measures for knowledge bases. In: Hüllermeier, E., Link, S., Fober, T., Seeger, B. (eds.) SUM 2012. LNCS, vol. 7520, pp. 473–485. Springer, Heidelberg (2012)

22. Ma, J., Liu, W., Miller, P.: Evidential fusion for gender profiling. In: Hüllermeier, E., Link, S., Fober, T., Seeger, B. (eds.) SUM 2012. LNCS, vol. 7520, pp. 514–524. Springer, Heidelberg (2012)
23. Ma, J., Liu, W., Miller, P.: An evidential improvement for gender profiling. In: Denœux, T., Masson, M.-H. (eds.) Belief Functions: Theory & Appl. AISC, vol. 164, pp. 29–36. Springer, Heidelberg (2012)
24. Ma, J., Liu, W., Miller, P., Yan, W.: Event composition with imperfect information for bus surveillance. In: Procs. of AVSS, pp. 382–387. IEEE Press (2009)
25. Miller, P., Liu, W., Fowler, F., Zhou, H., Shen, J., Ma, J., Zhang, J., Yan, W., McLaughlin, K., Sezer, S.: Intelligent sensor information system for public transport: To safely go. In: Procs. of AVSS (2010)
26. Shafer, G.: A Mathematical Theory of Evidence. Princeton University Press (1976)
27. Viola, P., Jones, M.J.: Rapid object detection using a boosted cascade of simple features. In: Procs. of IEEE CVPR, pp. 511–518 (2001)
28. Walter, M.R., Friedman, Y., Antone, M., Teller, S.: Appearance-based object reacquisition for mobile manipulation. In: Procs. of IEEE Computer Vision and Pattern Recognition Workshops, CVPRW, pp. 1–8 (2010)

Belief Functions:
A Revision of Plausibility Conflict
and Pignistic Conflict

Milan Daniel

Institute of Computer Science, Academy of Sciences of the Czech Republic
Pod Vodárenskou věží 2, CZ – 182 07 Prague 8, Czech Republic
milan.daniel@cs.cas.cz

Abstract. Plausibility conflict of belief functions is based on decisional support / opposition of elements of a frame of discernment. It distinguishes conflict between belief functions from internal conflicts of individual functions.

This contribution presents a revision of plausibility conflict between belief functions. According to four types of conflicting sets, four variants of plausibility conflict are defined. Further, a new alternative approach — pignistic conflict — based on pignistic probability instead of on normalized plausibility of singletons is introduced. Its cautious version may be considered to be an improvement of Liu's degree of conflict cf.

Comparing the approaches, a relation of sum of conflicting belief masses $m_{\ominus}(\emptyset)$ and a relation of a distance of belief functions to conflict between belief functions are also discussed.

Keywords: belief functions, Dempster-Shafer theory, internal conflict, conflict between belief functions, plausibility conflict, pignistic conflict, degree of conflict, uncertainty.

1 Introduction

Belief functions are one of the widely used formalisms for uncertainty representation and processing that enable representation of incomplete and uncertain knowledge, belief updating, and combination of evidence. They present a principal notion of the Dempster-Shafer Theory or the Theory of Evidence [17].

When combining belief functions (BFs) by the conjunctive rules of combination, conflicts often appear which are assigned to $\emptyset$ by non-normalized conjunctive rule $\ominus$ or normalized by Dempster's rule of combination $\oplus$. Combination of conflicting BFs and interpretation of conflicts is often questionable in real applications, thus a series of alternative combination rules was suggested and a series of papers on conflicting belief functions was published, e.g. [2,4,14,20].

In [7], new ideas concerning interpretation, definition, and measurement of conflicts of BFs were introduced. We presented three new approaches to interpretation and computation of conflicts: combinational conflict, plausibility conflict, and comparative conflict. Important distinction of conflicts between BFs from

W. Liu, V.S. Subrahmanian, and J. Wijsen (Eds.): SUM 2013, LNAI 8078, pp. 190–203, 2013.
© Springer-Verlag Berlin Heidelberg 2013

internal conflicts of single BF, and distinction of a conflict between BFs from a difference between BFs was introduced there. Properties of the most elaborated and prospective of the three approaches — plausibility conflict of BFs — are studied in [10].

The presented study brings a revision and improvement of the plausibility conflict for a specific class of couples of BFs, where the original formulation does not work well; the improvement idea has motivated several variants of definition of conflicting set, from which the conflict is computed (Section 3). A new alternative approach based on pignistic probability instead of on normalized plausibility of singletons (normalized contour function) is also discussed and compared to the plausibility conflict (Sect. 4). Pignistic conflict is further compared to Liu's degree of conflict [14] in Section 4.2, which also discusses a relation of sum of conflicting masses and a relation of distance of BFs to conflict between the BFs.

2   State of the Art

2.1   General Primer on Belief Functions

We assume classic definitions of basic notions from theory of *belief functions* [17] on finite frames of discernment $\Omega_n = \{\omega_1, \omega_2, ..., \omega_n\}$, see also [4–9].

A *basic belief assignment (bba)* is a mapping $m : \mathcal{P}(\Omega) \longrightarrow [0,1]$ such that $\sum_{A \subseteq \Omega} m(A) = 1$; the values of the bba are called *basic belief masses (bbm)*. $m(\emptyset) = 0$ is usually assumed. A *belief function (BF)* is a mapping $Bel : \mathcal{P}(\Omega) \longrightarrow [0,1]$, $Bel(A) = \sum_{\emptyset \neq X \subseteq A} m(X)$. A *plausibility function* $Pl(A) = \sum_{\emptyset \neq A \cap X} m(X)$. There is a unique correspondence among m and corresponding Bel and Pl thus we often speak about m as of belief function.

A *focal element* is a subset X of the frame of discernment, such that $m(X) > 0$. If all the focal elements are *singletons* (i.e. one-element subsets of Ω), then we speak about a *Bayesian belief function (BBF)*; in fact, it is a probability distribution on Ω. In the case of $m(\Omega) = 1$ we speak about *vacuous BF (VBF)*.

Dempster's (conjunctive) rule of combination $\oplus$ is given as $(m_1 \oplus m_2)(A) = \sum_{X \cap Y = A} K m_1(X) m_2(Y)$ for $A \neq \emptyset$, where $K = \frac{1}{1-\kappa}$, $\kappa = \sum_{X \cap Y = \emptyset} m_1(X) m_2(Y)$, and $(m_1 \oplus m_2)(\emptyset) = 0$, see [17]; putting $K = 1$ and $(m_1 \oplus m_2)(\emptyset) = \kappa$ we obtain the *non-normalized conjunctive rule of combination* $\ominus$, see e. g. [18].

Normalized plausibility of singletons[1] of Bel is BBF such that $\frac{Pl(\{\omega_i\})}{\sum_{\omega \in \Omega} Pl(\{\omega\})}$; the formula is also used as definition of probability transformation Pl_P of BF Bel: $(Pl\_P(Bel))(\omega_i) = \frac{Pl(\{\omega_i\})}{\sum_{\omega \in \Omega} Pl(\{\omega\})}$ [3,5]. Smets' pignistic probability is given by $BetP(\omega) = \sum_{\omega \in X \subseteq \Omega} \frac{1}{|X|} \frac{m(X)}{1-m(\emptyset)}$ [18].

2.2   Conflict of Belief Functions

When combining two BFs Bel_1, Bel_2 given by m_1 and m_2 conflicting belief masses $m_1(X) > 0$, $m_2(Y) > 0$, where $X \cap Y = \emptyset$, often appear. The sum of

---

[1] Plausibility of singletons is called *contour function* by Shafer in [17], thus $Pl\_P(Bel)$ is a normalization of contour function in fact.

products of such conflicting masses corresponds to $m(\emptyset)$ when non-normalized conjunctive rule of combination $\circledcirc$ is applied and $m = m_1 \circledcirc m_2$. This sum is considered to be a conflict between belief functions Bel_1 and Bel_2 in $[17]^2$; this interpretation is commonly used when dealing with conflicting belief functions. Unfortunately, the name and interpretation of this notion does not correctly correspond to reality. We often obtain positive sum of conflicting belief masses even if two numerically same belief functions[3] are combined, see e.g. examples discussed by Almond [1] already in 1995 and by Liu [14], for another examples see [7].

Liu further correctly demonstrates [14] that neither distance nor difference are adequate measures of conflicts between BFs. (For relation of distance and conflict see Section 4.2.) Thus Liu uses a two-dimensional (composed) measure *degree of conflict* $cf(m_1, m_2) = (m_\circledcirc(\emptyset), difBet_{m_1}^{m_2})$ (see also Section 4.2). Liu puts together two previous non-adequate measures of conflict $m_\circledcirc(\emptyset)$ and a distance together as two components of a new measure of conflict between BFs cf; unfortunatelly this does not capture a nature of conflictness / non-conflictness between BFs. For detail see Example 5 in Section 4.2. Hence this progressive approach should be further developed.

Internal conflicts $IntC(m_i)$ which are included in particular individual BFs are distinguished from *conflict between BFs* $C(m_1, m_2)$ in [7]; the entire sum of conflicting masses is called *total conflict*; and three approaches to conflicts were introduced: combinational, plausibility and comparative. In this study, we will discuss the most elaborated and most prospective of the three approaches — the plausibility conflict.

An *internal conflict* of a BF is a conflict included inside an individual BF. BF is non-conflicting if it is consistent (it has no internal conflict) otherwise it is internally conflicting. A *conflict between BFs* is a conflict between opinions of believers which are expressed by the BFs (the individual attitudes of believers; particular BFs may be internally conflicting or non-conflicting). If there is a positive conflict between BFs, we simply say that the *BFs are mutually conflicting*; otherwise they are *mutually non-conflicting*, i.e., there is no conflict between them.

Analogously to the original $m_\circledcirc(\emptyset)$ and cf, the three approaches from [7], including the plausibility conflict (Def. 1 and 2), seem to be rather empirical. For introductive axiomatic studies of conflicts between BFs see [11] and [16], unfortunately these studies do not yet capture a real nature of conflict, as e.g. Martin adds a non-correctly presented or ad-hoc strong axiom of inclusion [16] and proposes an inclusion-weighted distance as a measure of conflict. Hence, this interesting and complex topic is still open for discussion and further development. The important ideas from [11] and [16] should be studied and elaborated together with those from [7].

---

[2] A *weight of conflict between BFs Bel_1 and Bel_2* is defined as $log\frac{1}{1-m_\circledcirc(\emptyset)}$ in [17].

[3] All BFs combined by $\oplus$ and $\circledcirc$ are assumed to be mutually independent, even if they are numerically same.
   $m_\circledcirc(\emptyset)$ is called *autoconflict* when numerically same BFs are combined [16].

Two BFs on a 2−element frame of discernment which both support/prefer the same element of the frame (i.e., both oppose the other element) are assumed to be mutually non-conflicting in [7] (there is no conflict between them); otherwise they are mutually conflicting. Unfortunately, the generalization of this assumption to general finite frame was not precise, thus the original definitions of plausibility conflict need correction (see Section 3). Conflict between BFs is distinguished from internal conflict in [7,10] thus two definitions were introduced in [7]:

Definition 1. *The* internal plausibility conflict *Pl-IntC of BF Bel is defined as*

$$Pl\text{-}IntC(Bel) = 1 - max_{\omega \in \Omega} Pl(\{\omega\}),$$

where Pl is the plausibility corresponding to Bel.

Definition 2. *The* conflicting set $\Omega_{PlC}(Bel_1, Bel_2)$ *is defined as the set of el-ements* $\omega \in \Omega_n$ *with conflicting Pl_P masses, i.e.,* $\Omega_{PlC}(Bel_1, Bel_2) = \{\omega \in \Omega_n \mid (Pl\_P(Bel_1)(\omega) - \frac{1}{n})(Pl\_P(Bel_2)(\omega) - \frac{1}{n}) < 0\}$.
Plausibility conflict between BFs Bel_1 *and* Bel_2 *is then defined by the formula*

$$Pl\text{-}C(Bel_1, Bel_2) = min(\ Pl\text{-}C_0(Bel_1, Bel_2), (m_1 \text{\textcircled{$\odot$}} m_2)(\emptyset) \ ),$$

where[4]

$$Pl\text{-}C_0(Bel_1, Bel_2) = \sum_{\omega \in \Omega_{PlC}(Bel_1, Bel_2)} \frac{1}{2} |Pl\_P(Bel_1)(\omega) - Pl\_P(Bel_2)(\omega)|.$$

If $\Omega_{PlC}(Bel_1, Bel_2) = \emptyset$ then BFs Bel_1 and Bel_2 on Ω_n are mutually non-conflicting. The reverse statement does not hold true for $n > 2$, see e.g. Example 1 (i.e. Example 5 from [7], Example 8 from [6]). Any two BFs $(m_1(\{\omega_1\}), m_1(\{\omega_2\})) = (a, b)$ and $(m_2(\{\omega_1\}), m_2(\{\omega_2\})) = (c, d)$ on Ω_2 are mutually non-conflicting iff $\Omega_{PlC}((a, b), (c, d)) = \emptyset$ iff $(a - b)(c - d) \geq 0$.

Using $m_{\text{\textcircled{$\odot$}}}(\emptyset)$, degree of conflict cf, or measures of conflict based on a distance, a misclassification of BFs as beeing in mutual conflict sometimes appear. This is not a problem of the plausibility conflict which detects conflict according to harmonious/disharmonious support/opposition of individual elements of the frame of discernment by the BFs in question. For more properties of plausibility conflict $Pl\text{-}C$ and its comparison with Liu's degree of conflict cf [14] see [10].

Example 1. Let us suppose Ω_6, now; and two intuitively non-conflicting BFs m_1 and m_2.

X :	$\{\omega_1\}$	$\{\omega_2\}$	$\{\omega_3\}$	$\{\omega_4\}$	$\{\omega_5\}$	$\{\omega_6\}$	$\{\omega_1, \omega_2, \omega_3, \omega_4\}$
$m_1(X)$:	1.00						
$m_2(X)$:							1.00

$Pl\_P(m_1) = (1.00, 0.00, 0.00, 0.00, 0.00, 0.00)$,
$Pl\_P(m_2) = (0.25, 0.25, 0.25, 0.25, 0.00, 0.00)$, (we mean $Pl\_P(Bel_i)$ for Bel_i cor-responding to m_i), $\Omega_{PlC}(m_i, m_j) = \{\omega_2, \omega_3, \omega_4\}$, as $Pl\_P(m_2)(\omega_i) = \frac{1}{4} > \frac{1}{6}$ for $i = 2, 3, 4$, whereas $Pl\_P(m_1)(\omega_i) = 0 < \frac{1}{6}$ for $i = 2, 3, 4$, (the other elements are non-conflicting: $Pl\_P(m_1)(\omega_1) = 1 > \frac{1}{6}$, $Pl\_P(m_2)(\omega_1) = \frac{1}{4} > \frac{1}{6}$, $Pl\_P(m_1)(\omega_i) = 0 = Pl\_P(m_2)(\omega_i)$ for $i = 5, 6$; $Pl\text{-}C(m_1, m_2) = min(0.375, 0.00) = 0.00$.

---

[4] $Pl\text{-}C_0$ is not a separate measure of conflict in general; it is just a component of $Pl\text{-}C$.

3 Revision and Improvement of Plausibility Conflict

3.1 A Set of Conflicting Belief Functions with Empty Ω_{PlC}

While developing a brand new approach to conflicts which is based on decomposition of belief functions to their conflicting and non-conflicting parts [8], and comparing the approach under development with plausibility conflict, a special class of conflicting pairs of BFs with $\Omega_{PlC} = \emptyset$ was observed: pairs of BFs which *support* and *oppose* same sets of *elements of Ω* (i.e., $m_i(\{\omega_j\})$ is always $\geq \frac{1}{n}$ or always $\leq \frac{1}{n}$ for $i = 1, 2$ and any fixed j), but where $\{\omega \mid Pl\_P_1(\omega) = max_i Pl\_P_1(\omega_i)\} \cap \{\omega \mid Pl\_P_2(\omega) = max_i Pl\_P_2(\omega_i)\} = \emptyset$, see e.g. Example 2.

Example 2. Let us suppose Ω_3, now; and two BFs m_1 and m_2 given as follows:

X :	$\{\omega_1\}$	$\{\omega_2\}$	$\{\omega_3\}$	$\{\omega_1,\omega_2\}$	$\{\omega_1,\omega_3\}$	$\{\omega_2,\omega_3\}$	Ω_3
$m_1(X)$:	0.3	0.2	0.1	0.3			0.1
$m_2(X)$:	0.3	0.4	0.1	0.1			0.1

$Pl\_P(m_1) = (0.466\overline{6}, 0.4000, 0.133\overline{3})$, $Pl\_P(m_2) = (0.3846, 0.4615, 0.1539)$,
$Pl\_P(m_1)(\omega_1), Pl\_P(m_1)(\omega_2), Pl\_P(m_2)(\omega_1), Pl\_P(m_2)(\omega_2) > 0.3\overline{3}$,
$Pl\_P(m_1)(\omega_3), Pl\_P(m_2)(\omega_3) < 0.3\overline{3}$, hence $\Omega_{PlC}(m_i, m_j) = \emptyset$ (as all the elements are non-conflicting according to definition from [7]);
$m(\emptyset) = 0.08 + 0.008 + 0.15 + 0.03 = 0.34$; $Pl\text{-}C(m_1, m_2) = min(0.00, 0.34) = 0.00$.

BFs m_1 and m_2 are classified as non-conflicting by $Pl\text{-}C$. There is really a relatively high consensus of BFs. Both m_1 and m_2 support the elements ω_1 and ω_2 (i.e., $m_i(\{\omega_j\}) > \frac{1}{3}$) and oppose ω_3 (i.e., $m_i(\{\omega_3\}) < \frac{1}{3}$). A majority of focal elements have the same belief: $Bel_1(\{\omega_1\}) = Bel_2(\{\omega_1\}) = 0.3$, $Bel_1(\{\omega_3\}) = Bel_2(\{\omega_3\}) = 0.1$, $Bel_1(\{\omega_1,\omega_2\}) = Bel_2(\{\omega_1,\omega_2\}) = 0.8$, $Bel_1(\{\omega_1,\omega_3\}) = Bel_2(\{\omega_1,\omega_3\}) = 0.4$; only beliefs of $\{\omega_2\}$ and of $\{\omega_2,\omega_3\}$ differ: $Bel_1(\{\omega_2\}) = 0.2, Bel_2(\{\omega_2\}) = 0.4$, $Bel_1(\{\omega_2,\omega_3\}) = 0.3, Bel_2(\{\omega_2,\omega_3\}) = 0.5$. Analogously, a majority of focal elements have the same plausibility: $Pl_1(\{\omega_2\}) = Pl_2(\{\omega_2\}) = 0.6$, $Pl_1(\{\omega_3\}) = Pl_2(\{\omega_3\}) = 0.2$, $Pl_1(\{\omega_1,\omega_2\}) = Pl_2(\{\omega_1,\omega_2\}) = 0.9$, $Pl_1(\{\omega_2,\omega_3\}) = Pl_2(\{\omega_2,\omega_3\}) = 0.7$; only plausibilities of $\{\omega_1\}$ and of $\{\omega_1,\omega_3\}$ differ: $Pl_1(\{\omega_1\}) = 0.7, Pl_2(\{\omega_1\}) = 0.5$, $Pl_1(\{\omega_1,\omega_3\}) = 0.8, Pl_2(\{\omega_1,\omega_3\}) = 0.6$.

The problem arises when we want to make a decision as ω_1 has highest belief, plausibility, Pl_P and also $BetP$ for Bel_1, whereas ω_2 has highest belief, plausibility, Pl_P and also $BetP$, in the case of Bel_2. (There is $BetP_1(\omega_1) = 0.48\overline{3}, BetP_1(\omega_2) = 0.38\overline{3}, BetP_1(\omega_3) = 0.13\overline{3}, BetP_2(\omega_1) = 0.38\overline{3}, BetP_2(\omega_2) = 0.48\overline{3}, BetP_2(\omega_3) = 0.13\overline{3}$.) Hence the belief functions produce different decision, thus it is not correct to classify them as non-conflicting.

The problem comes from the incorrect generalization of the idea of plausibility conflict from 2-element frame of discernment Ω_2 to a general Ω_n in [7]. When ω_i is supported (preferred) by a BF Bel, it automatically has higher values of Bel, Pl, Pl_P and $BetP$ than $\omega_j \neq \omega_i$, i.e. ω_i has the highest Pl_P of both two elements of Ω_2. A situation is much more complicated in the case of general frame of discernment, thus we have to add to conflicting set also elements with maximal Pl_P if they are different, independently of the fact whether both of these elements are preferred by both of the BFs or not.

Example 2 (cont.). Including ω_1 and ω_2 into conflicting set of m_1 and m_2 we obtain $\frac{1}{2}(0.4666-0.3846) = 0.082/2$ for ω_1 and $0.0615/2$ for ω_2, thus $Pl\text{-}C_0(m_1, m_2)$ $= (0.082+0,0615)/2 = 0.0718$, and $Pl\text{-}C(m_1, m_2) = min(0.0718, 0.34) = 0.0718$.

There is a small plausibility conflict $Pl\text{-}C$ between Bel_1 and Bel_2 which corresponds both to different decision made according to the beliefs (conflictness) and to high consensus of both the beliefs (the value of the conflict is small).

3.2   Four Variants of Correction of Conflicting Sets

According to the previous subsection it is obvious that we have to add singletons with the highest Pl (if they are different) into the conflicting set of the beliefs. In relation to this two questions arise:

A decision is made according to the highest (normalized) plausibility; is it necessary to consider singletons which have not the highest plausibility as elements of the conflicting set? Is it not enough to decide conflictness and compute a conflict between BFs just from maximal plausibilities if they arise at different singletons?

On the other side, there are many pairs of BFs which are not conflicting according to the plausibility conflict, but they are conflicting according to $m(\emptyset)$, conflict measures based on a distance [15,16] and according to Liu's cf [14]. Is it enough to add the singletons with the highest plausibility into the conflicting set or it is necessary to add something more, e.g., singletons with different ordering number when ordered with respect to their (normalized) plausibility?

Keeping these questions open for a future discussion, denoting the original conflicting set Ω_{PlC} from [7] (Section 2, Def. 2 here) as Ω_{PlC_0}, and defining a partial order Ord of the elements of Ω_n, we can define four different conflicting sets for any pair of BFs: Ω_{smPlC}, Ω_{spPlC}, Ω_{cpPlC} and Ω_{cbPlC}, according to the possible answers to the questions above.

Definition 3. *Let us denote* $\Omega_{PlC_0}(Bel_1, Bel_2) = \{\omega \in \Omega_n \mid (Pl\_P(Bel_1)(\omega) - \frac{1}{n})(Pl\_P(Bel_2)(\omega) - \frac{1}{n}) < 0\}$. *Let us define* $Ord : \Omega_n \longrightarrow \mathcal{P}(\{1, 2, ..., n\})$, *such that* $Ord(\omega) = \{k_1, k_1+1, ..., k_m\}$ *iff there exist just* k_1-1 *elements* ω_i *s.t.* $Pl\_P(\omega_i) > Pl\_P(\omega)$ *and there exist just* $n-k_m$ *elements* ω_j *s.t.* $Pl\_P(\omega_j) < Pl\_P(\omega)$.

Now, we can define simple conflicting set Ω_{smPlC} *by* $\Omega_{smPlC}(m_1, m_2) = \emptyset$ *if* $\{\omega \mid 1 \in Ord_1(\omega)\} \cap \{\omega \mid 1 \in Ord_2(\omega)\} \neq \emptyset$, $\Omega_{smPlC}(m_1, m_2) = \{\omega \mid 1 \in Ord_1(\omega)\} \cup \{\omega \mid 1 \in Ord_2(\omega)\}$ *if* $\{\omega \mid 1 \in Ord_1(\omega)\} \cap \{\omega \mid 1 \in Ord_2(\omega)\} = \emptyset$, *where* Ord_i *is partial order defined by* Pl_P_i *(i.e. defined by* Pl_i*).*

Let us further define support conflicting set Ω_{spPlC} *by* $\Omega_{spPlC}(m_1, m_2) = \Omega_{PlC_0}(m_1, m_2) \cup \Omega_{smPlC}(m_1, m_2)$; comparative conflicting set Ω_{cpPlC} *by* $\Omega_{cpPlC}(m_1, m_2) = \{\omega \mid Ord_1(\omega) \cap Ord_2(\omega) = \emptyset\}$ *and* combined conflicting set Ω_{cbPlC} *as* $\Omega_{cbPlC}(m_1, m_2) = \Omega_{PlC_0}(m_1, m_2) \cup \Omega_{cpPlC}(m_1, m_2)$.
Using these four types of conflicting sets we can define 4 different conflictness $smPl\text{-}C$, $spPl\text{-}C$, $cpPl\text{-}C$ *and* $cbPl\text{-}C$ *and related procedures for computation of conflict.*

Using any of these four variants of conflicting sets instead of the original Ω_{PlC} corrects the definition of the plausibility conflict between BFs (Def. 2).

3.3   Properties of Conflicting Sets Ω_{smPlC}, Ω_{spPlC}, Ω_{cpPlC}, Ω_{cbPlC}

Lemma 1. *The following inclusions hold true for any couple of belief functions* Bel_1, Bel_2:

(i) $\Omega_{smPlC}(Bel_1, Bel_2) \subseteq \Omega_{spPlC}(Bel_1, Bel_2) \subseteq \Omega_{cbPlC}(Bel_1, Bel_2)$.

(ii) $\Omega_{smPlC}(Bel_1, Bel_2) \subseteq \Omega_{cpPlC}(Bel_1, Bel_2) \subseteq \Omega_{cbPlC}(Bel_1, Bel_2)$.

(iii) *There exist couples of belief functions* Bel_1 *and* Bel_2 *such that the above inclusions are proper. Hence there are four different types of plausibility conflict in general.*

Proof. Proofs of (i) and (ii) follow the construction of conflicting sets. Proof of (iii) follows Examples 3 and 4.

Lemma 2. (i)   *The following holds true for any belief function* Bel *defined on n-element frame of discernment* Ω_n: $\Omega_{smPlC}(Bel, Bel) = \Omega_{spPlC}(Bel, Bel) = \Omega_{cpPlC}(Bel, Bel) = \Omega_{cbPlC}(Bel, Bel) = \Omega_{PlC_0}(Bel, Bel) = \emptyset$.

(ii) *The following holds true for any couple of belief functions* Bel_1, Bel_2 *defined on 2-element frame of discernment* Ω_2: $\Omega_{smPlC}(Bel_1, Bel_2) = \Omega_{spPlC}(Bel_1, Bel_2) = \Omega_{cpPlC}(Bel_1, Bel_2) = \Omega_{cbPlC}(Bel_1, Bel_2) = \Omega_{PlC_0}(Bel_1, Bel_2)$.

Corollary 1. *There is only one type of plausibility conflict (according to the previous definitions) on two-element frame of discernment* Ω_2.

Example 3. Let suppose Ω_3 now; and two BFs m_1 and m_2:

X :	$\{\omega_1\}$	$\{\omega_2\}$	$\{\omega_3\}$	$\{\omega_1,\omega_2\}$	$\{\omega_1,\omega_3\}$	$\{\omega_2,\omega_3\}$	Ω_3
$m_1(X)$:	0.3	0.3		0.1	0.3		
$m_2(X)$:	0.4	0.1		0.1	0.4		

$PlP(m_1) = (0.45, 0.35, 0.20)$, $PlP(m_2) = (0.45, 0.30, 0.25)$,
$Ord_1(\omega_1) = \{1\}$, $Ord_1(\omega_2) = \{2\}$, $Ord_1(\omega_3) = \{3\}$, $Ord_2(\omega_1) = \{1\}$, $Ord_2(\omega_2) = \{2\}$,
$Ord_2(\omega_3) = \{3\}$, thus $Ord_1 \equiv Ord_2$, and $\Omega_{smPlC}(m_1, m_2) = \Omega_{cpPlC}(m_1, m_2) = \emptyset$.
$PlP(m_1)(\omega_1), PlP(m_1)(\omega_2), PlP(m_2)(\omega_1) > 0.3\overline{3}$,
$PlP(m_1)(\omega_3), PlP(m_2)(\omega_2), PlP(m_2)(\omega_3) < 0.3\overline{3}$, hence $\Omega_{spPlC}(m_1, m_2) = \{\omega_2\}$ as $(PlP(m_1)(\omega_2) - \frac{1}{n})(PlP(m_2)(\omega_2) - \frac{1}{n}) = (0.35 - 0.3\overline{3})(0.30 - 0.3\overline{3}) < 0$,
hence $\omega_2 \in \Omega_{spPlC}(m_1, m_2)$ (both other elements are non-conflicting).
Thus we have $\emptyset = \Omega_{cpPlC}(m_1, m_2) \subset \Omega_{spPlC}(m_1, m_2) = \Omega_{cbPlC}(m_1, m_2) = \{\omega_2\}$.
Further $m(\emptyset) = 0.03 + 0.04 = 0.07$, hence $cpPl\text{-}C(m_1, m_2) = min(0.00, 0.07) = 0.00 \neq cbPl\text{-}C(m_1, m_2) = min(0.025, 0.07) = 0.025$.

Example 4. Let suppose Ω_3 now; and two BFs m_1 and m_2:

X :	$\{\omega_1\}$	$\{\omega_2\}$	$\{\omega_3\}$	$\{\omega_1,\omega_2\}$	$\{\omega_1,\omega_3\}$	$\{\omega_2,\omega_3\}$	Ω_3
$m_1(X)$:	0.6	0.2		0.1	0.1		
$m_2(X)$:	0.6		0.3	0.1			

$PlP(m_1) = (0.600, 0.26\overline{6}, 0.13\overline{3})$, $PlP(m_2) = (0.6\overline{6}, 0.06\overline{6}, 0.26\overline{6})$,
$Ord_1(\omega_1) = \{1\}$, $Ord_1(\omega_2) = \{2\}$, $Ord_1(\omega_3) = \{3\}$, $Ord_2(\omega_1) = \{1\}$, $Ord_2(\omega_2) = \{3\}$,
$Ord_2(\omega_3) = \{2\}$, thus $Ord_1(\omega_2) \cap Ord_2(\omega_2) = \emptyset = Ord_1(\omega_3) \cap Ord_2(\omega_3)$, and
$\Omega_{smPlC}(m_1, m_2) = \emptyset \neq \Omega_{cpPlC}(m_1, m_2) = \{\omega_2, \omega_3\}$. $PlP(m_1)(\omega_1), PlP(m_2)(\omega_1) > 0.3\overline{3}$, $PlP(m_1)(\omega_2), PlP(m_1)(\omega_3), PlP(m_2)(\omega_2), PlP(m_2)(\omega_3) < 0.3\overline{3}$, hence $\Omega_{spPlC}(m_1, m_2) = \emptyset$. Thus $\Omega_{spPlC}(m_1, m_2) \subset \Omega_{cpPlC}(m_1, m_2) = \Omega_{cbPlC}(m_1, m_2)$.

3.4   Summary of the Plausibility Conflict

There are four different generalizations of unique plausibility conflict from a 2-element frame of discernment to a general n-element frame.

According to inclusions from Lemma 1 we can see that the *simple plausibility conflict smPl-C* using the least conflicting set $\Omega_{smPlC}(Bel_1, Bel_2)$ is the weakest one of the four plausibility approaches; it tolerates more couples of BFs as mutually non-conflicting (there is no conflict between them). On the other side, the *combined plausibility conflict* using the greatest conflicting set $\Omega_{cbPlC}(Bel_1, Bel_2)$ is the most cautious (most strict) of the approaches. It classifies more conflicting couples of BFs and produces the greatest conflict *cbPl-C*. Both the *support plausibility conflict* and *comparative plausibility conflict* are mutually incomparable from this point of view; the values of both of them are between the values of the simple plausibility conflict and of the cautious combined plausibility conflict. We can formalize this by the following theorem.

Theorem 1. *The following holds true for four types of plausibility conflict and any belief functions Bel_1 and Bel_2.*
(i) $smPl\text{-}C(Bel_1, Bel_2) \leq spPl\text{-}C(Bel_1, Bel_2) \leq cbPl\text{-}C(Bel_1, Bel_2)$,
(ii) $smPl\text{-}C(Bel_1, Bel_2) \leq cpPl\text{-}C(Bel_1, Bel_2) \leq cbPl\text{-}C(Bel_1, Bel_2)$.

Simple plausibility conflict approach classifies two belief functions as mutually non-conflicting whenever they give the greatest decisional support (preference) to the same elements of the frame. The cautious combined plausibility conflict classifies two belief functions as mutually non-conflicting only when both beliefs provide the same sets of elements of Ω_n which are supported and opposed (to be more precise which are not opposed and which are not supported) and moreover when the orders of the elements according to the support/opposition are the same. Such BFs should be really mutually non-conflicting despite usually positive distance and frequent positive $(m_1 \ominus m_2)(\emptyset)$. The principle of all four variants of plausibility conflict is the same; the difference among them is only relatively small. When a positive simple plausibility conflict between BFs appears, some level of mutual conflictness of the BFs is hardly opposed. On the other hand, when cautious (combined) non-conflictness appears then mutual non-conflictnes of the BFs is hardly opposed.

The simple plausibility conflict is the simplest from the computational point of view. Anywhere, where just indetification of strong conflict between BFs is required this version is definitely sufficient.

Considering Martin's axioms [16], all four variants of the plausibility conflict satisfy axioms $1-4$ (non-negativity, identity, symmetry and normalization). The 5th axiom of inclusion is either non-correctly presented in [16], or it is simply too strong thus, it is not satisfied by any of four variants of the plausibility conflict.

4   Pignistic Conflict of Belief Functions

There are frequent questions: Why is normalized plausibility of singletons used in definition of conflict? Why not to use popular pignistic probability instead?

Yes, syntactically, pignistic probability can be used instead of normalized plausibility of singletons. We will obtain analogous definitions.

4.1 Definitions of Pignistic Conflict

Definition 4. *Let the* internal pignistic conflict *of BF Bel be defined as*

$$BetP\text{-}IntC(Bel) = 1 - max_{\omega \in \Omega}BetP(\omega),$$

where BetP is the pignistic probability corresponding to Bel.

This internal pignistic conflict is greater or equal to internal plausibility conflict as $BetP(\omega) \leq Pl(\omega)$. The internal plausibility conflict is defined by the plausibility values, not by the related probabilistic transformation PLP, thus it has a reasonable interpretation, which is hard to say about the internal pignistic conflict.

The *conflicting set* $\Omega_{BetC_0}(Bel_1, Bel_2)$ can be defined analogously to the plausibility version as the set of *elements* $\omega \in \Omega_n$ *with conflicting BetP values* $\Omega_{BetC_0}(Bel_1, Bel_2) = \{\omega \in \Omega_n \mid (BetP(Bel_1)(\omega) - \frac{1}{n})(BetP(Bel_2)(\omega) - \frac{1}{n}) < 0\}$. Further, there may by defined four variants of pignistic conflicting sets Ω_{smBetC}, Ω_{spBetC}, Ω_{cpBetC} and Ω_{cbBetC}, analogously to their plausibility patterns.

Definition 5. Pignistic conflict between BFs Bel_1 and Bel_2 is defined by the formula

$$Bet\text{-}C(Bel_1, Bel_2) = min(\ Bet\text{-}C_0(Bel_1, Bel_2), (m_1 \odot m_2)(\emptyset)\ ),$$

where

$$Bet\text{-}C_0(Bel_1, Bel_2) = \sum_{\omega \in \Omega_{BetC}(Bel_1, Bel_2)} \frac{1}{2}|BetP(Bel_1)(\omega) - BetP(Bel_2)(\omega)|,$$

substituting Ω_{smBetC}, Ω_{spBetC}, Ω_{cpBetC}, *or* Ω_{cbBetC} *for* Ω_{BetC}.

Inclusions analogous to those from Lemma 1 hold true here, and following the analogous definitions, the following theorem (an analogy of Theorem 1) also holds true.

Theorem 2. *The following holds true for four types of pignistic conflict and any belief functions Bel_1 and Bel_2.*
(i)   smBet-C$(Bel_1, Bel_2) \leq$ spBet-C$(Bel_1, Bel_2) \leq$ cbBet-C(Bel_1, Bel_2),
(ii) smBet-C$(Bel_1, Bel_2) \leq$ cpBet-C$(Bel_1, Bel_2) \leq$ cbBet-C(Bel_1, Bel_2).

Mutual comparison of four variants of the pignistic conflict is completely analogous to that of the plausibility conflict. Also Martin's axioms 1–4 are satisfied by all variants of the pignistic conflict, but the 5th one (inclusion) is not satisfied by any of them.

We have to note that in general there are two numerical differences between $BetP$ and PlP: i) different values of both probabilities, which make different values of related conflict between BFs, ii) even the order of the values can be different (it does not hold true $Ord\text{-}Pl(\omega) = Ord\text{-}Bet(\omega)$ in general) thus a couple of plausibility non-conflicting BFs may be pignisticly conflicting and vice-versa.

Thus the definition of pignistic conflict is analogous to that of plausibility conflict, however, these methods are not equivalent.

We have to further note the principal theoretical difference that Pl_P commutes with Dempster's rule of combination [3,5] whereas $BetP$ does not. Pl_P is the only probability transformation of BFs which is compatible with conjunctive combination rule. Conflict between belief functions is frequently used when belief functions are combined [2,13,14,20]. Here the compatibility plays very important role: it is an argument for using the plausibility conflict and for not-using its pignistic alternative.

While betting, we are interested in the most perspective elements only or in all the elements, thus from the betting point of view only simple pignistic conflict $smBet\text{-}C(Bel_1, Bel_2)$ and cautious (combined) pignistic conflict $cbBet\text{-}C(Bel_1, Bel_2)$ play some role (not $spBet\text{-}C$ and $cpBet\text{-}C$ which have not a reasonable interpretation from the betting point of view).

4.2 Relation of Pignistic Conflict $Bet\text{-}C$ and of Liu's Degree of Conflict cf

We can observe that pignistic conflict $Bet\text{-}C(Bel_1, Bel_2)$ has some similarities with Liu's degree of conflict $cf(Bel_1, Bel_2)$. Both the approaches use $(m_1 \ominus m_2)(\emptyset)$ and both use some pignistic probability, but there are also important differences. Let us recall the definition of degree of conflict cf and also Liu's general definition of conflict before analysing the relation of the approaches.

Definition 6 (Liu). *A conflict between two beliefs in DS theory can be interpreted as one source strongly supports one hypothesis and the other strongly supports another hypothesis, and the two hypothesis are not compatible ([14] Definition 8).*

Definition 7 (Liu). *Let m_1 and m_2 be two bbas on frame Ω and let $BetP_{m_1}$ and $BetP_{m_2}$ be the results of two pignistic transformations from them respectively then*
$$dif BetP_{m_i}^{m_j} = max_{A \subseteq \Omega}(|BetP_{m_i}(A) - BetP_{m_j}(A)|)$$
is called the distance between betting commitments of the two bbas ([14] Definition 9).

Definition 8 (Liu). *Let m_1 and m_2 be two bbas. Let $cf(m_1, m_2) = (m_\oplus(\emptyset), dif BetP)$ be a two-dimensional measure where $m_\oplus(\emptyset)$ is the mass of uncommitted belief when combining m_1 and m_2 with Dempster's rule and $dif BetP$ be the distance between betting commitments in Definition 9 (Def. 7 here), m_1 and m_2 are defined as in conflict iff both $dif BetP > \varepsilon$ and $m_\oplus(\emptyset) > \varepsilon$ hold true, where $\varepsilon \in [0, 1]$ is the threshold of conflict tolerance ([14] Definition 10).*

Note that there should be rather $m_\ominus(\emptyset)$ instead of $m_\oplus(\emptyset)$ (more precisely $(m_i\ominus m_j)(\emptyset)$) in fact. Further, we can simplify $difBetP_{m_i}^{m_j}$ using the following lemma from [10].

Lemma 3 (Daniel). *For any belief functions* Bel_i, Bel_j *given by bbas* m_i, m_j *holds true that:*

$$difBetP_{m_i}^{m_j} = Diff(BetP_{m_i}, BetP_{m_j}) = \frac{1}{2}\sum_{\omega\in\Omega}|BetP_{m_i}(\{\omega\}) - BetP_{m_j}(\{\omega\})|.$$

Liu correctly presents [14] that neither $m_\ominus(\emptyset)$ nor distance or difference are adequate measures of conflicts between BFs. Thus she uses the two-dimensional (composed) measure cf which just includes both the non-adequate components, which are really somehow related to a size of a conflict. What is this relation?

Interpretation of $m_\ominus(\emptyset)$ as a conflict between BFs is not easy as it somehow includes conflict between the belief functions, but also internal conflicts of both arguments of belief combination. Much clearer is the interpretation of the rest of belief masses: $1 - m_\ominus(\emptyset)$ is the sum of belief masses related to the conjunctive consensus of the arguments m_1 and m_2; this definitely does not include any conflict, thus $m_\ominus(\emptyset)$ is simply an upper bound of the conflict between belief functions.

What about a difference or a distance? It definitely does not hold true that the higher a distance is the higher is the conflict. It is sure that zero distance means zero conflict, as we suppose, that a belief function is non-conflicting with itself. What does a small distance mean? It is not necessary a small conflict, it could be just a small difference of two non-conflicting BFs (same elements supported, same elements opposed, same order of sizes of supported/opposed element) with numerically different values, for an example of BFs on Ω_2 see Fig. 1. What does a medium distance mean? Analogously, it can be no conflict, small conflict or medium conflict, (see the second triangle in Fig. 1). Large distance means conflicts (small, medium or large), just a very large distance means a very large conflict (see the last triangle in Fig. 1). We can simplify this by saying that analogously to $m_\ominus(\emptyset)$ a distance of BFs is also an upper bound of the conflict between them.

Degree of conflict cf successfully distinguishes conflict between BF both from their distance and from $m_\ominus(\emptyset)$. It correctly considers that there is no conflict between two identical BFs. Strongly mutually conflicting BFs are recognized as really conflicting, unfortunately there are still mutually non-conflicting BFs which are classified like conflicting by cf. A support of different non-compatible hypotheses is correctly classified as conflict according to Def. 6, on the other hand, support of the same or compatible hypotheses is not classified as non-conflict. See the following example:

Example 5. Let suppose BFs $m_1 = (0.55, 0.40; 0.05)$ and $m_2 = (0.95, 0.01; 0.04)$ on Ω_2 and $m_3 = (0.55, 0.20, 0.10, 0.10, 0, ..., 0; 0.05)$ and $m_4 = (0.85, 0.06, 0.02, 0.02, 0, ..., 0; 0.05)$ on Ω_4. There is $cf(m_1, m_2) = (0.3855, 0.40)$ and $cf(m_3, m_4) = (0.415, 0.30)$, thus some degree of conflict is classified by cf in both cases.

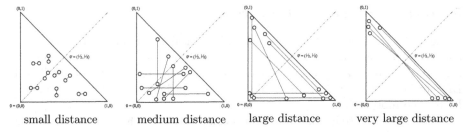

small distance   medium distance   large distance   very large distance

Fig. 1. Couples of BFs on Ω_2 with different sizes of a distance. BFs in the same half of a triangle are mutually non-conflicting; conflict appears if one BF is in the upper (left) and the other in the lower (right) part of the triangle; BFs from the axis are non-conflicting with any other. Pair (x, y) represents BF given by $m(\{\omega_1\}) = x$, $m(\{\omega_2\}) = y$ here; $m(\{\omega_1, \omega_2\}) = 1-x-y$.

Nevertheless, both believers 1 and 2 support (prefer) ω_1 and both of them oppose ω_2. Their beliefs differ only in a degree of support of ω_1 (opposition of ω_2). Similarly both believers 3 and 4 support (prefer) ω_1 and both of them oppose $\omega_2, \omega_3, \omega_4$, even orders of degrees of opposition of $\omega_2, \omega_3, \omega_4$ are the same for both believers. Thus also believers 3 and 4 are in accord and their beliefs differ only in a degree of support of ω_1 and degree of opposition of $\omega_2, \omega_3, \omega_4$. There is a difference between the beliefs but not a conflict between them.

Above, we have recognized $m_{\bigodot}(\emptyset)$ and $Diff(BetP_{m_i}, BetP_{m_j})$ as upper bounds of conflict. Thus, cf should rather be a number $\leq min(m_{\bigodot}(\emptyset), Diff(BetP_{m_i}, BetP_{m_j}))$ than a couple of these values, as cf may be limited by its upper bounds but not equal to them in general. The upper bound of cf looks very similarly to Bet-C; it differs only by the fact that differences $|BetP_{m_i}(\{\omega\}) - BetP_{m_j}(\{\omega\})|$ are summed over Ω_{BetC} by Bet-C whereas by cf over entire Ω.

This is the principal difference between the two approaches; Ω_{BetC} may be empty for mutually non-conflicting BFs which are in a positive distance, with positive $m_{\bigodot}(\emptyset)$. Classification of non-conflictness according to Ω_{BetC} better corresponds to Liu's Definition 8 from [14] (Def. 6 here), where conflict is given by strong support of incompatible hypothesis. (An idea of support (preference) or opposition of elements according to $BetP$ is analogous to that according to Pl_P [7,10]). Nevertheless non-conflictness classified according to Liu's Definition 10 is given by components of cf $m_{\bigodot}(\emptyset)$ and $Diff(BetP_{m_i}, BetP_{m_j})$ and by artificially determined threshold ε of a conflict tolerance, which has nothing to do with with a nature of the belief functions, which are or which are not in a conflict, for detail see [14]. The conflict tolerance should be used for decision how to handle beliefs which are in a conflict; not for classification whether the beliefs are or are not in a conflict.

Ω_{cbBetC} is the largest of the conflicting sets defined in the previous Section as variants of Ω_{BetC}, thus the cautious (combined) pignistic conflict is the closest one to cf. Thus $cbBet$-C can be considered as an real improvement of the degree of conflict cf. Nevertheless, for a utilization related to conflicting belief functions combination plausibility conflict is recommended, see the previous Section.

5   Ideas for a Future Research

A theoretical question, which BFs should be considered as mutually conflicting and which not, should be analysed in more detail together with ideas and results from [11,16]: to obtain a deeper understanding of the real nature of conflict of BFs; to be able to recommend which of the 4 variants of conflicting sets are better and more useful.

A new approach to conflict based on decomposition of BFs to their conflicting and non-conflicting parts is just under development.

Axiomatic approaches to conflict of BFs [11,16] should by studied and further elaborated together with the presented results and with conflicting and non-conflicting parts of BF.

6   Conclusion

An improvement of the plausibility conflict of belief functions was presented. A new alternative pignistic conflict was introduced and compared with the plausibility one. Higher importance and utility of the plausibility conflict, from the point of view of belief combination, was concluded.

Further, it was shown that pignistic conflict can be considered as an improvement of Liu's degree of conflict. Degree of conflict distinguishes conflict between BFs from $m_{\ominus}(\emptyset)$ and from distance/difference of BFs. Both plausibility and pignistic conflicts go further to the nature of a conflict, they capture also mutual non-conflictness of BFs which prefer/oppose same elements of the frame of discernment. They further distinguish internal conflict of single belief functions from conflict between them.

The plausibility conflict has better interpretation than pignistic one, as it does not depend on any extra additive assumption, just on plausibility equivalent to corresponding belief function and normalization, whereas the pignistic conflict depends on the pignistic transformation, thus on its linearity property assumption [19]. Normalized plausibility of singletons is further compatible with conjunctive rule of combination, it commutes with Dempster's rule whereas pignistic transformation does not comute with any rule for combination of belief functions [5].

We have compared two new approaches to conflict between BFs with Liu's degree of conflict. Unfortunately, any of the presented approaches to conflict between BFs still does not fully cover the complete nature of conflict between BFs as a difference is partially included in the definitions of all investigated approaches to conflict between belief functions.

Acknowledgments. This research is supported by the grant P202/10/1826 of the Grant Agency of the Czech Republic. The partial institutional support RVO: 67985807 is also acknowledged.

References

1. Almond, R.G.: Graphical Belief Modeling. Chapman & Hall, London (1995)
2. Ayoun, A., Smets, P.: Data association in multi-target detection using the transferable belief model. Int. Journal of Intelligent Systems 16(10), 1167–1182 (2001)

3. Cobb, B.R., Shenoy, P.P.: A Comparison of Methods for Transforming Belief Function Models to Probability Models. In: Nielsen, T.D., Zhang, N.L. (eds.) ECSQARU 2003. LNCS (LNAI), vol. 2711, pp. 255–266. Springer, Heidelberg (2003)
4. Daniel, M.: Distribution of Contradictive Belief Masses in Combination of Belief Functions. In: Bouchon-Meunier, B., Yager, R.R., Zadeh, L.A. (eds.) Information, Uncertainty and Fusion, pp. 431–446. Kluwer Academic Publishers, Boston (2000)
5. Daniel, M.: Probabilistic Transformations of Belief Functions. In: Godo, L. (ed.) ECSQARU 2005. LNCS (LNAI), vol. 3571, pp. 539–551. Springer, Heidelberg (2005)
6. Daniel, M.: New Approach to Conflicts within and between Belief Functions. Technical report V-1062, ICS AS CR, Prague (2009)
7. Daniel, M.: Conflicts within and between Belief Functions. In: Hüllermeier, E., Kruse, R., Hoffmann, F. (eds.) IPMU 2010. LNCS (LNAI), vol. 6178, pp. 696–705. Springer, Heidelberg (2010)
8. Daniel, M.: Non-conflicting and Conflicting Parts of Belief Functions. In: Coolen, F., de Cooman, G., Fetz, T., Oberguggenberger, M. (eds.) ISIPTA 2011: Proceedings of the 7th ISIPTA, pp. 149–158. Studia Universitätsverlag, Innsbruck (2011)
9. Daniel, M.: Introduction to an Algebra of Belief Functions on Three-element Frame of Discernment — A Quasi Bayesian Case. In: Greco, S., Bouchon-Meunier, B., Coletti, G., Fedrizzi, M., Matarazzo, B., Yager, R.R., et al. (eds.) IPMU 2012, Part III. CCIS, vol. 299, pp. 532–542. Springer, Heidelberg (2012)
10. Daniel, M.: Properties of Plausibility Conflict of Belief Functions. In: Rutkowski, L., Korytkowski, M., Scherer, R., Tadeusiewicz, R., Zadeh, L.A., Zurada, J.M. (eds.) ICAISC 2013, Part I. LNCS (LNAI), vol. 7894, pp. 235–246. Springer, Heidelberg (2013)
11. Destercke, S., Burger, T.: Revisiting the Notion of Conflicting Belief Functions. In: Denœux, T., Masson, M.-H. (eds.) Belief Functions: Theory & Appl. AISC, vol. 164, pp. 153–160. Springer, Heidelberg (2012)
12. Hájek, P., Havránek, T., Jiroušek, R.: Uncertain Information Processing in Expert Systems. CRC Press, Boca Raton (1992)
13. Lefèvre, É., Elouedi, Z., Mercier, D.: Towards an Alarm for Opposition Conflict in a Conjunctive Combination of Belief Functions. In: Liu, W. (ed.) ECSQARU 2011. LNCS, vol. 6717, pp. 314–325. Springer, Heidelberg (2011)
14. Liu, W.: Analysing the degree of conflict among belief functions. Artificial Intelligence 170, 909–924 (2006)
15. Martin, A., Jousselme, A.-L., Osswald, C.: Conflict measure for the discounting operation on belief functions. In: Proceedings of 11th International Conference on Information Fusion, Fusion 2008, Cologne, Germany (2008)
16. Martin, A.: About Conflict in the Theory of Belief Functions. In: Denœux, T., Masson, M.-H. (eds.) Belief Functions: Theory & Appl. AISC, vol. 164, pp. 161–168. Springer, Heidelberg (2012)
17. Shafer, G.: A Mathematical Theory of Evidence. Princeton University Press, Princeton (1976)
18. Smets, P.: The combination of evidence in the transferable belief model. IEEE-Pattern analysis and Machine Intelligence 12, 447–458 (1990)
19. Smets, P.: Decision Making in the TBM: the Necessity of the Pignistic Transformation. Int. Journal of Approximate Reasoning 38, 133–147 (2005)
20. Smets, P.: Analyzing the combination of conflicting belief functions. Information Fusion 8, 387–412 (2007)

Bipolar Possibility Theory
as a Basis for a Logic of Desires and Beliefs

Didier Dubois, Emiliano Lorini, and Henri Prade

IRIT-CNRS, Université Paul Sabatier, Toulouse, France

Abstract. Bipolar possibility theory relies on the use of four set functions. On the one hand, a weak possibility and a strong necessity measure are increasing set functions, which are respectively max-decomposable with respect to union and min-decomposable with respect to intersection. On the other hand, strong possibility and weak necessity measures are two decreasing set functions, which are respectively min-decomposable with respect to union and max-decomposable with respect to intersection. In the first part of the paper we advocate the use of the last two functions for modeling a notion of graded desire. Moreover, we show that the combination of weak possibility and strong possibility allows us to model a notion of realistic desire, i.e., a desire that does not only account for satisfactoriness but also for its epistemic possibility. In the second part of the paper we show that possibility theory offers a semantic basis for developing a modal logic of beliefs and desires.

1 Introduction

Possibility theory has been originally proposed as an alternative approach to probability for modeling epistemic uncertainty, independently by two authors. In economics, Shackle [26] advocated a new view of the idea of expectation in terms of degree of surprise (a substitute for a degree of impossibility). Later in computer sciences, Zadeh [28] introduced a setting for modeling the information originated from linguistic statements in terms of fuzzy sets (understood as possibility distributions). Zadeh's proposal for a possibility theory relies on the idea of possibility measure, a max-decomposable set function w.r.t. union with values in $[0, 1]$. However, in these works, the duality between possibility and necessity (captured by a min-decomposable set function with respect to intersection) was not exploited. Later, it has been recognized that two other set functions, which contrast with the two previous ones by their decreasingness, also make sense in this setting [10]. These two latter set functions, which are dual of each other, model an idea of strong (guaranteed) possibility and of weak necessity respectively, while the original possibility measure that evaluates the consistency between the considered event and the available information, corresponds to a weak potential possibility.

The framework of possibility theory with its four basic set functions exhibits a rich structure of oppositions, which can be also closely related to other structures of oppositions that exist in modal logics and other settings such that formal concept analysis for instance [11]. Moreover, possibility theory is graded since the four set functions can take values in the unit interval. This very general setting can not only be interpreted in terms of uncertainty. It makes sense for preference modeling as well [2]. But it is also of interest when modeling situations that require modal logic languages, and where

W. Liu, V.S. Subrahmanian, and J. Wijsen (Eds.): SUM 2013, LNAI 8078, pp. 204–218, 2013.

grading modalities is meaningful. For instance, when modeling uncertainty, necessity measures are useful for representing beliefs and their epistemic entrenchments [9].

We provide here an investigation of the potentials of possibility theory for modeling the concept of desire. Indeed, although this concept has been already investigated in the past in artificial intelligence [19,20],[1] up to now, no clear connection between a theory of desires and possibility theory has been built. The rest of the paper is organized as follows. Section 2 presents a background on possibility theory. Section 3 discusses the modeling of desires in terms of strong possibility, as well the dual notion of potential desire in terms of weak necessity. We conclude by defining a notion of realistic desire, in the sense of desiring something that one considers epistemically possible. In Section 4 we introduce a modal logic of beliefs and desires based on possibility theory, more precisely, of realistic desires. The extension of this logic to graded desires is outlined. Finally, Section 5 points out some lines for further research on the relationship between possibility theory and the logic of emotions. A first version of sections 1-3 is in [22].

2 Background on Possibility Theory

Let π be a mapping from a set of worlds W to $[0, 1]$ that rank-orders them. Note that this encompasses the particular case where π reduces to the characteristic function of a subset $E \subseteq W$. The possibility distribution π may represent a plausibility ordering (and E the available evidence) when modeling epistemic uncertainty, or a preference ordering (E is then the subset of satisfactory worlds) when modeling preferences. Let us recall the complete system of the 4 set functions underlying possibility theory [10] and their characteristic properties:

- i) The *(weak) possibility measure* (or potential possibility) $\Pi(A) = \max_{w \in A} \pi(w)$ evaluates to what extent there is a world in A that is possible. When π reduces to E, $\Pi(A) = 1$ if $A \cap E \neq \emptyset$, which expresses the consistency of the event A with E, and $\Pi(A) = 0$ otherwise. Possibility measures are characterized by the following decomposability property: $\Pi(A \cup B) = \max(\Pi(A), \Pi(B))$.
- ii) The dual *(strong or or actual) necessity measure* $N(A) = \min_{w \notin A} 1 - \pi(w) = 1 - \Pi(\overline{A})$ evaluates to what extent it is certain (necessarily true) that all possible worlds are in A. When π reduces to E, $N(A) = 1$ if $E \subseteq A$, which expresses that E entails event A (when E represents evidence), and $N(A) = 0$ otherwise. The duality of N w. r. t. Π expresses that A is all the more certain as the opposite event $\overline{A}$ is impossible. Necessity measures are characterized by the following decomposability property: $N(A \cap B) = \min(N(A), N(B))$.
- iii) The *strong* (or actual, or *"guaranteed"*) possibility measure $\Delta(A) = \min_{w \in A} \pi(w)$ evaluates to what extent *any* value in A is possible. When π reduces to E, $\Delta(A) = 1$ if $A \subseteq E$, and $\Delta(A) = 0$ otherwise. Strong possibility measures are characterized by the following property: $\Delta(A \cup B) = \min(\Delta(A), \Delta(B))$.

---

[1] For instance, in [19] Lang et al. propose a formal theory of desires based on Boutilier's logic QDT [4] in which two ordering relations representing preference and normality are given. The interpretation given to the statement "in context φ, I desire ψ" is "the best among the most normal $\varphi \wedge \psi$ worlds are preferred to the most normal $\varphi \wedge \neg\psi$ worlds" which is different from interpretation of desire given in this paper.

- iv) The dual *(weak)* (or potential) necessity measure $\nabla(A) = \max_{w \notin A} 1 - \pi(w) = 1 - \Delta(\overline{A})$ evaluates to what extent there is a value outside A that is impossible. When π reduces to E, $\nabla(A) = 1$ if $A \cup E \neq U$, and $\nabla(A) = 0$ otherwise. Weak necessity measures are characterized by property: $\nabla(A \cap B) = \max(\nabla(A), \nabla(B))$.

Δ, ∇ are decreasing set functions, while the (weak) possibility and (strong) necessity measures are increasing. A modal logic counterpart of these 4 modalities has been proposed in the *binary*-valued case (things are possible or impossible) [7]. There is a close link between Spohn functions and (weak) possibility / (strong) necessity measures [9].

3   Possibility Theory as Basis for a Logical Theory of Desires

The possibility and necessity operators Π and N have a clear epistemic meaning both in the frameworks of possibility theory, and of Spohn's uncertainty theory [27] (also referred to as 'κ calculus', or as 'rank-based system' and 'qualitative probabilities' [16]). Differently from the operators Π and N, the operators Δ and ∇ have not an intuitive interpretation in terms of epistemic attitudes. Indeed, although Δ and ∇ make sense from the point of view of possibility theory and also from a logical viewpoint, it is not fully clear which kind of mental attitudes these two operators aim at modeling. Here we defend the idea that Δ and ∇ can be viewed as operators modeling motivational mental attitudes such as goals or desires.[2] In particular, we claim that Δ can be used to model the notion of *desire*, whereas ∇ can be used to model the notion of *potential* desire. [3]

According to the philosophical theory of motivation based on Hume [18], a desire can be conceived as an agent's motivational attitude which consists in an anticipatory mental representation of a pleasant (or desirable) state of affairs (representational dimension of desires) that motivates the agent to achieve it (motivational dimension of desires). In this perspective, the motivational dimension of an agent's desire is realized through its representational dimension. For example when an agent desires to be at the Japanese restaurant eating sushi, he imagines himself eating sushi at the Japanese restaurant and this representation gives him pleasure. This pleasant representation motivates him to go to the Japanese restaurant in order to eat sushi.

Intuitively speaking, with the term *potential* desire, we refer to a weaker form of motivational attitude. We assume that an agent considers a given property φ potentially desirable if φ does not conflict with the agent's current desires. In this sense, φ is potentially desirable if it is not incompatible with the agent's current desires. Following ideas presented in [21], let us explain why the operator Δ is a good candidate for modeling the concept of desire and why ∇ is a good candidate for modeling the idea of desire compatibility. We define an agent's *mental state* as a tuple $M = (E, D)$ where:

- $E \subseteq W$ is a *non-empty* subset of the set of all worlds, and
- $D \subset W$ is a proper subset of the set of all worlds.

---

[2] We use the term 'motivational' mental attitude (e.g., a desire, a goal or an intention) in order to distinguish it from an 'epistemic' mental attitude such as knowledge or belief.

[3] Here, the word *potential* does not refer to the idea that φ would be desired by the agent as a consequence of his mental state, but the agent has not enough deductive power to become aware of it. It is more the idea that the agent has no reason not to desire φ. Another possible term is *desire admissibility* or *desire compatibility*.

The set E defines the set of worlds envisaged by the agent (i.e., the set of worlds that the agent considers possible), whereas D is the set of desirable worlds for the agent. Let $\mathcal{M}$ denote the set of all mental states. We here assume for every mental state M there exists a world with a minimal degree of desirability 0 (this is why $D \neq W$). This type of normality constraint for guaranteed possibility distributions is usually assumed in possibility theory. More generally, a *graded mental state* is a pair $M = (\pi, \delta)$ where:

- $\pi : W \to L$ is a normal possibility distribution over the set of all worlds, where 'normal' means that $\pi(w) = 1$ for some $w \in W$, and
- $\delta : W \to L$ is a function mapping every world w to its desirability (or pleasantness) degree in L, with $\delta(w) = 0$ for some $w \in W$.
- L is a bounded chain acting as a qualitative scale for possibility and desirability, that make these notions commensurate.

Note that while $\delta(w) = 1$ expresses complete desirability, $\delta(w) = 0$ expresses indifference, rather than repulsion. The condition $\delta(w) = 0$ for some $w \in W$ indicates that desire presupposes that not everything is desired.

3.1 Modeling Desire Using Δ Function

We here assume that in order to determine how much a proposition φ is desirable an agent takes into consideration the worst situation in which φ is true. Thus, denoting by $||\varphi||$ the set of situations where φ is true, for all graded mental states $M = (\pi, \delta)$ and for all propositions φ, we can interpret $\Delta(||\varphi||) = \min_{u \in ||\varphi||} D(u)$ as the extent to which the agent desires φ to be true. Let us justify the following two properties for desires: $\Delta(||\varphi \vee \psi||) = \min(\Delta(||\varphi||), \Delta(||\psi||))$ and $\Delta(||\varphi \wedge \psi||) \geq \max(\Delta(||\varphi||), \Delta(||\psi||))$.

According to the first property, an agent desires φ to be true with a given strength α and desires ψ to be true with a given strength β if and only if the agent desires φ or ψ to be true with strength equal to $\min(\alpha, \beta)$. Notice that in the case of epistemic states, this property would not make any sense because the plausibility of $\varphi \vee \psi$ should be clearly *at least* equal to the maximum of the plausibilities of φ and ψ. For the notion of desires, it seems intuitively satisfactory to have the opposite, namely the level of desire of $\varphi \vee \psi$ should be *at most* equal to the minimum of the desire levels of φ and ψ. Indeed, we only deal with here with "*positive*"[4] desires (i.e., desires to reach something with a given strength). Under this proviso, the level of desire of $\varphi \wedge \psi$ cannot be less than the maximum of the levels of desire of φ and ψ. According to the second property, the joint occurrence of two desired events φ and ψ is more desirable than the occurrence of one of the two events. This is the reason why in the right side of the equality we have the max. The latter property does not make any sense in the case of epistemic attitudes like beliefs, as the joint occurrence of two events φ and ψ is epistemically less plausible than the occurrence of a single event. On the contrary it makes perfect sense for motivational attitudes likes desires. By way of example, suppose Peter wishes to go to the cinema in the evening with strength α (i.e., $\Delta(||goToCinema||) = \alpha$)

---

[4] The distinction between positive and negative desires is a classical one in psychology. Negative desires correspond to state of affairs the agent wants to avoid with a given strength, and then desires the opposite to be true. However, we do not develop this bipolar view here.

and, at the same time, he wishes to spend the evening with his girlfriend with strength β (i.e., $\Delta(||stayWithGirlfriend||) = \beta$). Then, according to the preceding property, Peter wishes to to go the cinema with his girlfriend with strength at least $\max\{\alpha, \beta\}$ (i.e., $\Delta(||goToCinema \wedge stayWithGirlfriend||) \geq \max\{\alpha, \beta\}$). This is a reasonable conclusion because the situation in which Peter achieves his two desires is (for Peter) at least as pleasant as the situation in which he achieves only one desire. A similar intuition can be found in [5] about the min-decomposability of disjunctive desires, where however it is emphasized that it corresponds to a pessimistic view.

From the normality constraint of δ, we can deduce the following inference rule:

Proposition 1. *For every* $M \in \mathcal{M}$, *if* $\Delta(||\varphi||) > 0$ *then* $\Delta(||\neg\varphi||) = 0$.

This means that if an agent desires φ to be true — i.e., with some strength $\alpha > 0$ — then he does not desire φ to be false. In other words, an agent's desires must be consistent.

Note that the operator Δ satisfies the following additional property:

Proposition 2. *For every* $M \in \mathcal{M}$, *if* $||\varphi|| = \emptyset$ *then* $\Delta(||\varphi||) = 1$.

i.e., in absence of actual situations where φ is true, the property φ is desirable by default.

3.2 Modeling Potential Desire Using ∇

As pointed out above, we claim that the operator ∇ allows us to capture a concept of potential desire (or desire compatibility): $\nabla(||\varphi||)$ represents the extent to which an agent considers φ a potentially desirable property or, alternatively, the extent to which the property φ is not incompatible with the agent's desires. An interesting situation is when the property φ is *maximally* potentially desirable for the agent (i.e., $\nabla(||\varphi||) = 1$). This is the same thing as saying that the agent does not desire φ to be false (i.e., $\Delta(||\neg\varphi||) = 0$). Intuitively, this means that φ is totally potentially desirable in as much as the level of desire for $\neg\varphi$ is 0. In particular, given a graded mental state $M = (\pi, \delta)$, let $D = \{w \in W : \delta(w) > 0\}$ be the set of somewhat satisfactory or desirable worlds in M. Then, we have $\nabla(||\varphi||) = 1$ if and only if $\overline{D} \cap ||\neg\varphi|| \neq \emptyset$, i.e., $\neg\varphi$ is consistent with what is not desirable, represented by the set $\overline{D}$.

Another interesting situation is when the property φ is *maximally* desirable for the agent (i.e., $\Delta(||\varphi||) = 1$). This is the same thing as saying that $\neg\varphi$ is not at all potentially desirable for the agent (i.e., $\nabla(||\neg\varphi||) = 0$). It is worth noting that if an agent desires φ to be true, then φ should be *maximally* potentially desirable. This property is expressed by the following valid inference rule which follows straightforwardly from the previous one and from the definition of $\nabla(||\varphi||)$ as $1 - \Delta(||\neg\varphi||)$:

Proposition 3. *For every* M, *if* $\Delta(||\varphi||) > 0$ *then* $\nabla(||\varphi||) = 1$.

Let us now consider the case in which the agent does not desire φ (i.e., $\Delta(||\varphi||) = 0$). In this case two different situations are possible: either $\Delta(||\neg\varphi||) = 0$ and φ is *fully* compatible with the agent's desires (i.e., $\nabla(||\varphi||) = 1$), or $\Delta(||\neg\varphi||) > 0$ and then φ is not *fully* compatible with the agent's desires (i.e., $\nabla(||\varphi||) < 1$).

3.3 Some Valid Inference Rules for Desires

The following is a valid inference rule for Δ-based logic, see [7,12] for the proof:

Proposition 4. *For every $M \in \mathcal{M}$, if $\Delta(||\varphi \wedge \psi||) \geq \alpha$ and $\Delta(||\neg\varphi \wedge \chi||) \geq \beta$ then $\Delta(||\psi \wedge \chi||) \geq \min(\alpha, \beta)$.*

Therefore, if we interpret Δ as a desire operator, we have that if an agent desires $\varphi \wedge \psi$ with strength at least α and desires $\neg\varphi \wedge \chi$ with strength at least β, then he desires $\psi \wedge \chi$ with strength at least $\min(\alpha, \beta)$. This seems a reasonable property of desires. By way of example, suppose Peter desires to be in a situation in which he drinks red wine and eats a pizza with strength at least α and, at the same time, he desires to be in a situation in which he does not drink red wine and eats tiramisú as a dessert with strength at least β. Then, it is reasonable to conclude that Peter desires to be in a situation in which he eats both a pizza and tiramisú with strength at least $\min(\alpha, \beta)$.

Another rule, never studied, mixes Δ (*alias* actual desire) and ∇ (potential desire):

Proposition 5. *For every $M \in \mathcal{M}$, if $\Delta(||\varphi \wedge \psi||) \geq \alpha$ and $\nabla(||\neg\varphi \wedge \chi||) \geq \beta$ then $\nabla(||\psi \wedge \chi||) \geq \alpha * \beta$, where $\alpha * \beta = \alpha$ if $\alpha > 1 - \beta$ and $\alpha * \beta = 0$ if $1 - \beta \geq \alpha$.*

Proof. First, we have by duality $\Delta(||\varphi \wedge \psi||) \geq \alpha \Leftrightarrow \nabla(||\neg\varphi \vee \neg\psi||) \leq 1 - \alpha$. Then observe $\neg\varphi \wedge \chi \equiv (\neg\varphi \vee \neg\psi) \wedge (\neg\varphi \vee \psi) \wedge \chi$. Thus $\nabla(||\neg\varphi \wedge \chi||) = \max(\nabla(||\neg\varphi \vee \neg\psi||), \nabla(||(\neg\varphi \vee \psi) \wedge \chi||)) \geq \beta$ which leads to $\max(1 - \alpha, \nabla(||\psi \wedge \chi||)) \geq \beta$ from which the result follows. The last inequality is obtained by noticing that $\nabla(||(\neg\varphi \vee \psi) \wedge \chi||) \leq \nabla(||\psi \wedge \chi||)$ due to the decreasingness of ∇. It can be shown that $\alpha * \beta$ is the tightest lower bound that can be established for the above pattern. $\qquad \square$

Thus, in particular, if φ is *fully* potentially desirable ($\nabla(||\varphi||) = 1$), and $\neg\varphi \wedge \psi$ is *fully* desirable ($\Delta(||\neg\varphi \wedge \psi||) = 1$), then ψ is *fully* potentially desirable ($\nabla(||\psi||) = 1$). The two above inference rules are the counterparts of the following inference rule:

$$\text{if } N(||\varphi \vee \psi||) \geq \alpha \text{ and } N(||\neg\varphi \vee \chi||) \geq \beta \text{ then } N(||\psi \vee \chi||) \geq \min(\alpha, \beta)$$

(the basic inference rule in standard possibilistic logic), and of the following one [8]:

$$\text{if } N(||\varphi \vee \psi||) \geq \alpha \text{ and } \Pi(||\neg\varphi \vee \chi||) \geq \beta \text{ then } \Pi(||\psi \vee \chi||) \geq \alpha * \beta$$

with $\alpha * \beta = \alpha$ if $\alpha > 1 - \beta$ and $\alpha * \beta = 0$ if $1 - \beta \geq \alpha$. They are themselves the graded counterparts of two inference rules well-known in modal logic [13,8].

3.4   Realistic Desires

Besides, $\Delta(||\varphi||) = \alpha$ implies that for any ψ logically independent from φ, it holds that $\Delta(||\varphi \wedge \psi||) \geq \alpha$ and $\Delta(||\varphi \wedge \neg\psi||) \geq \alpha$, which may sound counterintuitive. Indeed, suppose you wish to choose a menu and you prefer to eat fish than not to a certain degree. It means that you should wish to eat fish with white wine, and fish with red wine to a degree at least as high. Yet, you may dislike very much to drink red wine with fish. Your desire for fish presupposes the restaurant offers white wine as well. So you express your desire for fish is conditioned to the possibility of having white wine as well. In other words, you believe that in fish restaurants it is more likely to find white wine than red wine. Modeling desire irrespectively of what you assume to be possible is liable to such kind of paradoxes when using the set-function δ.

This discussion suggests that a *realistic desire* can be defined as one whose realization is considered possible by the agent. The most natural representation consists in restricting crist mental states to pairs (E, D) such that $E \cap D \neq \emptyset$, called *realistic mental states*, the set of which is denoted by $\mathcal{M}^r$. Indeed, if $E \cap D = \emptyset$, then the agent knows that desirable states are impossible in his view. Then the agent with mental state (E, D) is said to *realistically desire* φ if and only if $\Delta(||\varphi||) = 1$ (that is, $||\varphi|| \subseteq D$) and $\Pi(||\varphi||) = 1$ (that is, $||\varphi|| \cap E \neq \emptyset$).

A more conservative notion of realistic desire would consist in requesting $N(||\varphi||) = 1$ instead of $\Pi(||\varphi||) = 1$, that is realistic desire would concern only propositions φ such that the agent is certain that φ is true. However one may question the fact that realistically desiring φ to be true may presuppose no risk at all for φ being false, namely the complete certainty that φ is true. This corresponds better to the idea of happiness. On the contrary, the preceding the notion of realistic desire defined by $\Delta(||\varphi||) = 1$ and $\Pi(||\varphi||) = 1$ corresponds to the notion of hope. Indeed, according to psychological theories of emotion (e.g., [24]), while happiness is triggered by *prospective consequences* (or *prospects*), hope is triggered by *actual consequences*. [5]

In the case of graded mental states, (π, δ), one may take a restritive point of view on possible states of affairs, evaluating desired statements overs pairs (π, δ) such that $E = \{w : \pi(w) = 1\} \cap \{w : \delta(w) > 0\} \neq \emptyset$. It comes down to working with pairs (E, δ) where only desire is graded. Then desires are expressed under the assumption that they can be achieved in at least one normal situation.

Alternatively one may compute the degree of realistic desire in the mental state (π, δ) as $\rho(||\varphi||) = \min(\Delta(||\varphi||), \Pi(||\varphi||))$. It presupposes that degrees of plausibility and degrees of desire are commensurate.

Note that the above proposal differs from the one that would restrict desired states to possible ones, that is, replacing δ by $\delta_\pi = \min(\delta, \pi)$ since $\rho(A) \geq \Delta_\pi(A) = \min_{w \in A} \min(\delta(w), \pi(w))$. For instance, if $E \cap D \neq \emptyset$, $A \cap E \neq \emptyset$, $\overline{E} \cap A \neq \emptyset$, and $A \subseteq D$, then $\rho(A) = 1$ but $\Delta_\pi(A) = 0$ since then $\delta_\pi(w) = 0$ for some $w \in A$.

4   Logics of Beliefs and Desires

In this section we introduce some variants of a modal logic of beliefs and desires, called here BDL, based on the ideas presented in the previous sections. Specifically, the logics presented here support reasoning about the notion of belief, as traditionally studied in the area of modal logic of belief (*alias* doxastic logic) [23,14,17], in combination with the notion of (Δ-based) desire discussed in Section 3.1 and the notion of (Δ-based) realistic desire discussed in Section 3.4. We first consider a simpler logic, in line with the previous sections, that, like MEL [1], does not support the nesting of modalities and allows us to reason about purely (non-graded) notions of belief and desire. The semantics will then be defined in terms of mental states (pairs of sets or distributions). Then, we present a simple generalization of this logic that allows us to formalize notions of graded belief and graded desire. Finally, we consider a full-fledged modal logic of graded beliefs and graded desires with multiple agents that supports the nesting of modalities. The nesting of modalities is crucial in order to represent an agent i's beliefs about the beliefs (or the desires) of a different agent j.

---

[5] Like [15], we here interpret the term 'prospect' as synonymous of 'uncertain consequence' (in contrast with 'actual consequence' as synonymous of 'certain consequence').

4.1   Minimal Modal Logic of Beliefs and Desires MBDL

Let us introduce a propositional language PL based on a countable set $Prop$ of atomic propositions (with typical members denoted $p, q, \ldots$), and defined by the following grammar: $\varphi ::= p \mid \neg\varphi \mid \varphi \wedge \psi$, where p ranges over a given countable set of atomic propositions $Prop = \{p, q, \ldots\}$, some of which can be decision variables. The other Boolean constructions $\top, \bot, \vee, \rightarrow$ and $\leftrightarrow$ are defined from p, $\neg$ and $\wedge$ in the standard way. A propositional valuation is defined in the standard way as a subset of atomic propositions considered as true, the other ones being false. Propositional valuations, also called *worlds* or *states*, are denoted by symbols w. The set W is identified with the set 2^{Prop} of all propositional valuations. Let $||p|| = \{w : p \in w\}$ be the set of models of p.

The extension of propositional formulas is defined in the standard way as follows:
$$||\neg\varphi|| = W \setminus ||\varphi||; \; ||\varphi \wedge \psi|| = ||\varphi|| \cap ||\psi||$$

We first consider the Boolean case and the most elementary language that may capture the previously introduced notions. The language $\mathcal{L}_{N,\Delta}$ of the logic MBDL is defined as follows: $\Phi ::= N\varphi \mid \Delta\varphi \mid \neg\Phi \mid \Phi \wedge \Psi$, where formulas φ range over PL. In other words, $N\varphi$ and $\Delta\varphi$ are atomic propositions, respectively referring to the statements "the agent believes that φ" and "the agent desires that φ". The two modal operators N and Δ have the following intuitive readings:

- $N\varphi$ means: the agent believes that φ is true (i.e., φ is true in all worlds that the agent envisages as possible),
- $\Delta\varphi$ means: the agent considers φ desirable in all worlds where φ is true.

The dual operators Π and ∇ are defined in the usual way as follows:
$$\Pi\varphi \overset{\text{def}}{=} \neg N\neg\varphi; \quad \nabla\varphi \overset{\text{def}}{=} \neg\Delta\neg\varphi$$

The set of axioms of MBDL is given in Figure 2 and provides a proof system for this logic MBDL. The first three modal axioms are those of KD, more specifically its subjective fragment where modalities are not nested. They account for Boolean necessity measures. The three following ones for the desire modality are the translation of the former when replacing necessity by guaranteed possibility, using the identity $N(A) = \min_{w \notin A} 1 - \pi(w) = \Delta(\overline{A})$ if $\delta = 1 - \pi$.

All tautologies of propositional calculus	(PC)
$(N\varphi \wedge N(\varphi \rightarrow \psi)) \rightarrow N\psi$	($\mathbf{K}_N$)
$\neg(N\varphi \wedge N\neg\varphi)$	($\mathbf{D}_N$)
$N\top$	($\mathbf{N}_N$)
$(\Delta\varphi \wedge \Delta(\neg\varphi \wedge \psi)) \rightarrow \Delta\psi$	($\mathbf{K}_\Delta$)
$\neg(\Delta\varphi \wedge \Delta\neg\varphi)$	($\mathbf{D}_\Delta$)
$\Delta\bot$	($\mathbf{N}_\Delta$)
$\dfrac{\varphi, \varphi \rightarrow \psi}{\psi}$	(MP)

Fig. 1. Sound and complete axiomatization of MBDL

The truth of a MBDL formula is evaluated w. r. t. a valuation w and a mental state $M = (E, D) \in \mathcal{M}$, by means of the following rules:

$$M \models N\varphi \iff \forall w \in E, w \in ||\varphi||; \quad M \models \Delta\varphi \iff \forall w \models \varphi, w \in D$$
$$M \models \neg\Phi \iff M \not\models \Phi; \quad M \models \Phi \wedge \Psi \iff M \models \Phi \text{ AND } M \models \Psi$$

We say that a formula Φ of the language $\mathcal{L}_{\mathsf{BDL}}(Prop)$ is valid, denoted by $\models_{\mathsf{MBDL}} \Phi$, if and only if for every mental state M in $\mathcal{M}$ $M \models \Phi$. We say that Φ is satisfiable if and only if $\neg\Phi$ is not valid. It can be checked that

- $M \models N\varphi$ if and only if $N(||\varphi||) = 1$ with respect to E.
- $M \models \Delta\varphi$ if and only if $\Delta(||\varphi||) = 1$ with respect to D.

In the logic MBDL we can also formally express the concept of realistic desire:

$$R\Delta\varphi \stackrel{\mathrm{def}}{=} \Delta\varphi \wedge \neg N\neg\varphi.$$

We can now prove the completeness theorem for this logic:

Theorem 1. *The axioms and the rules of inference given in Figure 1 provides a sound and complete axiomatization for the logic* MBDL.

Proof (Sketch). Soundness is easy to obtain. As to completeness, note that axioms K_N, D_N, and N_N imply the equivalence between $N(\varphi \wedge \psi)$ and $N\varphi \wedge N\psi$. Likewise, axioms K_Δ, D_Δ, and N_Δ imply the equivalence between $\Delta(\varphi \vee \psi)$ and $\Delta\varphi \wedge \Delta\psi$. Besides, a propositional valuation v of the language of MBDL assigns 0 or 1 to each $N\varphi$ and $\Delta\varphi$. Define two set functions ν and μ over W by letting $\nu(A) = 1$ if and only if $v(N\varphi) = 1$ if $A = ||\varphi||$ and 0 otherwise; and likewise $\mu(A) = 1$ if and only if $v(\Delta\varphi) = 1$ if $A = ||\varphi||$ and 0 otherwise. Axioms of propositional logic ensures these definitions are sound (truth assignments to $N\varphi$ and $\Delta\varphi$ do not change if φ is replaced by a logically equivalent proposition). Moreover the 6 first modal axioms imply that ν is a Boolean necessity measure and μ a Boolean guaranteed possibility measure. It means that there exists a mental state $M = (E_v, D_v)$ such that $v(N\varphi) = 1$ if and only if $E_v \subseteq ||\varphi||$ and $v(\Delta\varphi) = 1$ if and only if $||\varphi|| \subseteq D_v$. Using the completeness of propositional logic, we thus prove that, for any subset of formulas B in the logic MBDL:

$$B \vdash_{\mathsf{MBDL}} \Phi \iff B \cup \{K_N, D_N, N_N, K_\Delta, D_\Delta, N_\Delta\} \vdash_{PL} \Phi$$
$$\iff (\forall v, v \models_{PL} B \cup \{K_N, D_N, N_N, K_\Delta, D_\Delta, N_\Delta\} \Rightarrow v \models_{PL} \Phi$$
$$\iff \forall(E, D) \in \mathcal{M}, (E, D) \models_{\mathsf{MBDL}} B \Rightarrow (E, D) \models_{\mathsf{MBDL}} \Phi \iff B \models_{\mathsf{MBDL}} \Phi$$

An interesting aspect of the modal logic MBDL is that one can make syntactic proofs of some properties of the notions of (Δ-based) desire. For instance, we can give the following syntactic proof of the inference rule for desire given in Proposition 4 when $\alpha = 1$, namely $\{\Delta(\varphi \wedge \psi), \Delta(\neg\varphi \wedge \chi)\} \vdash_{\mathsf{MBDL}} \Delta(\psi \wedge \chi)$:

1. Applying K_Δ when $\psi \models \varphi$ yields theorem T1:$\vdash_{\mathsf{MBDL}} \Delta\varphi \rightarrow \Delta\psi$ if $\psi \models \varphi$.
2. K_Δ can be written as T2: if $\varphi \wedge \psi$ is a contradiction, $\vdash_{\mathsf{MBDL}} \Delta\varphi \wedge \Delta\psi \rightarrow \Delta(\varphi \vee \psi)$.
3. Applying T2: $\{\Delta(\varphi \wedge \psi), \Delta(\neg\varphi \wedge \chi)\} \vdash_{\mathsf{MBDL}} \Delta((\varphi \wedge \psi) \vee (\neg\varphi \wedge \chi))$
4. By T1, $\{\Delta(\varphi \wedge \psi), \Delta(\neg\varphi \wedge \chi)\} \vdash_{\mathsf{MBDL}} \Delta(\psi \wedge \chi)$ since $\psi \wedge \chi \models (\varphi \wedge \psi) \vee (\neg\varphi \wedge \chi)$

4.2   Outline of a Minimal Modal Logic of Graded Beliefs and Desires MGBDL

Assume a finite chain $L \subseteq [0, 1]$ containing the values 0 and 1 such that for every $\alpha \in L$ we have $1 - \alpha \in L$. For every $\alpha \in L$ such that $\alpha > 0$, let $p(\alpha)$ denote the number $\beta \in L$ such that $\beta < \alpha$ and there is no $\gamma \in L$ such that $\beta < \gamma < \alpha$. β is called the predecessor of α in L. Furthermore, let $p(0) = 0$. For every $\alpha \in L$ such that $\alpha < 1$, let $\sigma(\alpha)$ denote the number $\beta \in L$ such that $\alpha < \beta$ and there is no $\gamma \in L$ such that $\alpha < \gamma < \beta$. β is called the successor of α in L. Furthermore, let $\sigma(1) = 1$.

The language $\mathcal{L}^L_{N,\Delta}$ of the graded logic MGBDL is defined as follows:

$$\Phi ::= N^{\geq \alpha}\varphi \mid \Delta^{\geq \beta}\varphi \mid \neg\Phi \mid \Phi \wedge \Psi,$$

where formulas φ range over PL and $\alpha > 0, \beta > 0 \in L$. In other words, $N^{\geq \alpha}\varphi$ and $\Delta^{\geq \beta}\varphi$ are atomic propositions, respectively encoding the statements:

- $N^{\geq \alpha}\varphi$ means: the agent believes that φ is true to at least level α (i.e., φ is true in all worlds that the agent envisages as possible at level at least $\sigma(1 - \alpha)$),
- $\Delta^{\geq \beta}\varphi$: for the agent φ is desirable at least at level β in all worlds where φ is true.

The set of axioms of MGBDL are those of PL, the six first modal ones of Figure 1 for each $N^{\geq \alpha}$ and $\Delta^{\geq \beta}$. Finally, we must add weakening axioms for the two graded modalities:

$$[W_\Delta] : \Delta^{\geq \beta}\varphi \to \Delta^{\geq p(\beta)}\varphi \quad [W_N] : N^{\geq \alpha}\varphi \to N^{\geq p(\alpha)}\varphi.$$

The semantics is defined by means of graded mental states (π, δ) as defined in Section 3. A completeness theorem can be proved as for the Boolean case. Again the idea is to interpret any propositional valuation v of the language $\mathcal{L}^L_{N,\Delta}$ as a pair of set functions (g^v_N, g^v_Δ) on W stemming from a pair (π, δ). Namely we can let $g^v_N(\|\varphi\|) = \max\{\alpha : v(N^{\geq \alpha}\varphi) = 1\}$, and $g^v_\Delta(\|\varphi\|) = \max\{\beta : v(\Delta^{\geq \beta}\varphi) = 1\}$, which is meaningful due to the weakening axioms, and prove that the other axioms ensure that g^v_N is a necessity measure, and g^v_Δ is a guaranteed possibility measure.

4.3   Multi-agent Modal Logic of Graded Beliefs and Desires GBDL^n

Let $Agt = \{1, \ldots, n\}$ be a finite set of agents (or individuals). The language $\mathcal{L}^L_{N_i,\Delta_i}$ of the logic GBDL^n consists of a set of formulae and is defined as follows:

$$\varphi ::= p \mid \neg\varphi \mid \varphi \wedge \psi \mid N^{\geq \alpha}_i\varphi \mid \Delta^{\geq \beta}_i\varphi$$

where p ranges over the set of atomic propositions $Prop$, $\alpha, \beta \in L \setminus \{0\}$ and i ranges over the set of agents Agt. The two modal operators $N^{\geq \alpha}_i$ and $\Delta^{\geq \beta}_i$ have the following intuitive readings:

- $N^{\geq \alpha}_i\varphi$ means: agent i believes that φ is true with strength at least α (i.e., φ is true in all worlds that agent i considers possible at level at least $\sigma(1 - \alpha)$),
- $\Delta^{\geq \beta}_i\varphi$ means: agent i desires φ with strength at least β (i.e., all worlds in which φ is true are desirable for agent i at level at least β).

The interesting aspect of the logic GBDL^n is that it allows to represent what a given agent i believes about another j's beliefs and desires. For instance, the formula $N^{\geq \alpha_1}_1 N^{\geq \alpha_2}_2 \varphi$ expresses that agent 1 believes with strength at least α_1 that agent 2 believes φ with strength at least α_2, whereas the formula $N_1^{\geq \alpha_1} \Delta_2^{\geq \beta_2} \varphi$ expresses that agent 1 believes with strength at least α_1 that agent 2 desires φ with strength at least β_2.

The semantics of the logic $GBDL^n$ is defined in terms of multi-agent mental states of the form $M = (S, \{\pi_{i,s}\}_{i \in Agt, s \in S}, \{\delta_{i,s}\}_{i \in Agt, s \in S}, V)$ where:

- S is a set of states, including states of the agents (possibly more general than W);
- for all $s \in S$ and for all $i \in Agt$, $(\pi_{i,s}, \delta_{i,s})$ is a graded mental state over the set S as the one defined in Section 3;
- $V : S \longrightarrow 2^{Prop}$ is a valuation function for atomic propositions: $p \in V(s)$ means that proposition p is true at world $w = V(s)$.

Specifically, $\pi_{i,s}(s')$ captures how much, in state s, agent i thinks that state s' is (epistemically) possible, while $\delta_{i,s}(s')$ captures how much, in state s, agent i thinks that state s is desirable. Note that parameterizing possibility distributions $\pi_{i,s}$ and $\delta_{i,s}$ with states is one way to model an agent i's uncertainty about the beliefs and the desires of another agent j.

The truth of a $GBDL^n$ formula is evaluated with respect to a given state s in a multi-agent mental state $M = (S, \{\pi_{i,s}\}_{i \in Agt, s \in S}, \{\delta_{i,s}\}_{i \in Agt, s \in S}, V)$ by means of the following rules:

$$M, s \models p \iff p \in V(s)$$
$$M, s \models \neg \varphi \iff M, s \not\models \varphi$$
$$M, s \models \varphi \wedge \psi \iff M, s \models \varphi \text{ AND } M, s \models \psi$$
$$M, s \models N_i^{\geq \alpha} \varphi \iff \forall s' \in S : \text{IF } \pi_{i,s}(s') \geq \sigma(1 - \alpha) \text{ THEN } M, s \models \varphi$$
$$M, s \models \Delta_i^{\geq \beta} \varphi \iff \forall s' \in S : \text{IF } M, s \models \varphi \text{ THEN } \delta_{i,s}(s') \geq \beta$$

We say that a formula φ of the logic $GBDL^n$ is valid, denoted by $\models_{GBDL^n} \varphi$, if and only if for every multi-agent mental state $M = (S, \{\pi_{i,s}\}_{i \in Agt, s \in S}, \{\delta_{i,s}\}_{i \in Agt, s \in S}, V)$ and for every state s in S, $M, s \models \varphi$. We say that φ is $GBDL^n$ satisfiable if and only if $\neg \varphi$ is not $GBDL^n$ valid. For instance, $\neg N_1^{\geq \sigma(0)} \Delta_2^{\geq \sigma(0)} p \wedge \neg N_1^{\geq \sigma(0)} \neg \Delta_2^{\geq \sigma(0)} p$ is a satisfiable formula in the logic $GBDL^n$. It means that agent 1 is uncertain whether agent 2 desires p or not.

In the logic $GBDL^n$ we can also formally express the concept of realistic desire for a given agent i. For all all $i \in Agt$ we define:

$$R\Delta_i^{\geq \beta} \varphi \overset{\text{def}}{=} \Delta_i^{\geq \beta} \varphi \wedge \neg N_i^{\geq 1} \neg \varphi$$

where $R\Delta_i^{\geq \beta} \varphi$ has to be read "agent i realistically desires φ with strength at least β". The above realistic desire operator $R\Delta_i^{\geq \beta}$ exactly corresponds to the notion of realistic desire discussed in Section 3.4. The general idea is that agent i realistically desires φ with strength at least β (i.e., $R\Delta_i^{\geq \beta} \varphi$) if and only if agent i desires φ with strength at least β and is not completely certain that φ is false.

We can prove that the list of principles given in Figure 2 provides a proof system for the logic $GBDL^n$. In the axiomatization we use the following abbreviation for all $\gamma \in L \setminus \{1\}$:

$$\square_i^{\leq \gamma} \varphi \overset{\text{def}}{=} \Delta_i^{\geq \sigma(\gamma)} \neg \varphi$$

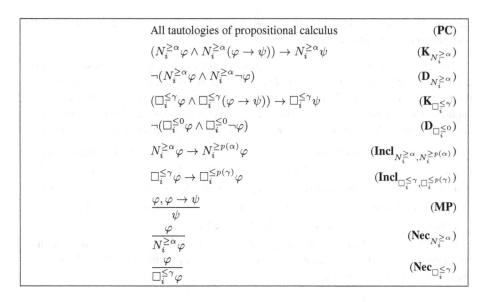

All tautologies of propositional calculus	**(PC)**
$(N_i^{\geq\alpha}\varphi \wedge N_i^{\geq\alpha}(\varphi \to \psi)) \to N_i^{\geq\alpha}\psi$	$(\mathbf{K}_{N_i^{\geq\alpha}})$
$\neg(N_i^{\geq\alpha}\varphi \wedge N_i^{\geq\alpha}\neg\varphi)$	$(\mathbf{D}_{N_i^{\geq\alpha}})$
$(\Box_i^{\leq\gamma}\varphi \wedge \Box_i^{\leq\gamma}(\varphi \to \psi)) \to \Box_i^{\leq\gamma}\psi$	$(\mathbf{K}_{\Box_i^{\leq\gamma}})$
$\neg(\Box_i^{\leq 0}\varphi \wedge \Box_i^{\leq 0}\neg\varphi)$	$(\mathbf{D}_{\Box_i^{\leq 0}})$
$N_i^{\geq\alpha}\varphi \to N_i^{\geq p(\alpha)}\varphi$	$(\mathbf{Incl}_{N_i^{\geq\alpha},N_i^{\geq p(\alpha)}})$
$\Box_i^{\leq\gamma}\varphi \to \Box_i^{\leq p(\gamma)}\varphi$	$(\mathbf{Incl}_{\Box_i^{\leq\gamma},\Box_i^{\leq p(\gamma)}})$
$\dfrac{\varphi,\varphi \to \psi}{\psi}$	**(MP)**
$\dfrac{\varphi}{N_i^{\geq\alpha}\varphi}$	$(\mathbf{Nec}_{N_i^{\geq\alpha}})$
$\dfrac{\varphi}{\Box_i^{\leq\gamma}\varphi}$	$(\mathbf{Nec}_{\Box_i^{\leq\gamma}})$

Fig. 2. Sound and complete axiomatization of GBDL^n

Theorem 2. *The axioms and the rules of inference given in Figure 2 provides a sound and complete axiomatization for the logic* BDL.

Proof (Sketch). It is a routine task to verify that the axioms given in Figure 2 are sound and that the rules of inference preserve validity. The proof of completeness has 2 steps. Step 1 consists in proving that the semantics of the logic GBDL^n given above is equivalent to an alternative semantics in terms of Kripke models with accessibility relations. Specifically, let us define the notion of Kripke GBDL^n model as a tuple $M = (S, \{T_{i,\geq\alpha}\}_{i\in Agt,\alpha\in L\setminus\{0\}}, \{R_{i,\leq\gamma}\}_{i\in Agt,\gamma\in L\setminus\{1\}}, V)$ where S and V are as defined above, and every $T_{i,\geq\alpha}$ and every $R_{i,\leq\gamma}$ are binary relations on S satisfying the constraints: (C1) $T_{i,\geq 1}$ is serial; (C2) $R_{i,\leq 0}$ is serial; (C3) forall $\alpha \in L\setminus\{0\}$, $T_{i,\geq\sigma(\alpha)} \subseteq T_{i,\geq\alpha}$; (C4) for all $\gamma \in L\setminus\{1\}$, $R_{i,\leq p(\gamma)} \subseteq R_{i,\leq\gamma}$.

In this alternative Kripke semantics for GBDL^n, the truth of a formula is evaluated w. r. t. a state s in a Kripke GBDL^n model M by means of the following rules:

$$M, s \models p \Longleftrightarrow p \in V(s) \;;\; M, s \models \neg\varphi \Longleftrightarrow M, s \not\models \varphi$$

$$M, s \models \varphi \wedge \psi \Longleftrightarrow M, s \models \varphi \text{ AND } M, s \models \psi$$

$$M, s \models N_i^{\geq\alpha}\varphi \Longleftrightarrow \forall s' \in T_{i,\geq\sigma(1-\alpha)}(s) : M, s \models \varphi$$

$$M, s \models \Delta_i^{\geq\beta}\varphi \Longleftrightarrow \forall s' \in R_{i,\leq p(\beta)}(s) : M, s \models \neg\varphi$$

where $T_{i,\geq\sigma(1-\alpha)}(s) = \{s' \in S \mid (s,s') \in T_{i,\geq\sigma(1-\alpha)}\}$ and $R_{i,\leq p(\beta)}(s) = \{s \in S \mid (s,s') \in R_{i,\leq p(\beta)}\}$. We say that a formula φ of the logic GBDL^n is valid with respect to the class of Kripke GBDL^n models if and only if for every Kripke GBDL^n model M and for every state s in S we have $M, s \models \varphi$.

Lemma 1. *For every formula φ of the logic GBDL^n, $\models_{\mathsf{GBDL}^n} \varphi$ if and only if φ is valid with respect to the class of Kripke GBDL^n models.*

The 2nd step of the proof consists in showing that the list of principles given in Figure 2 completely axiomatizes the set of validities of the logic GBDL^n whose language is interpreted over Kripke GBDL^n models. It is a routine task to check that the axioms in Figure 2 correspond one-to-one to their semantic counterparts on GBDL^n Kripke models. In particular, Axioms $\mathbf{K}_{N_i^{\geq\alpha}}$ and $\mathbf{K}_{\square_i^{\leq\gamma}}$ together with the rules of inference $\mathbf{Nec}_{N_i^{\geq\alpha}}$ and $\mathbf{Nec}_{\square_i^{\leq\gamma}}$ correspond to the fact that $N_i^{\geq\alpha}$ and $\square_i^{\leq\gamma}\varphi$ are normal modal operators interpreted by means of accessibility relations. Axiom $\mathbf{D}_{N_i^{\geq\alpha}}$ corresponds to the fact that the relation $T_{i,\geq 1}$ is serial (Constraint C1), while Axiom $\mathbf{D}_{\square_i^{\leq 0}}$ corresponds to the fact that the relation $R_{i,\leq 0}$ is serial (Constraint C2). Moreover, Axioms $\mathbf{Incl}_{N_i^{\geq\alpha},N_i^{\geq p(\alpha)}}$ and $\mathbf{Incl}_{\square_i^{\leq\gamma},\square_i^{\leq p(\gamma)}}$ correspond respectively to the Constraints C3 and C4.

It is routine, too, to check that all principles given in Figure 2 are in the so-called Sahlqvist class [25]. This means that they are complete with respect to the defined model classes, cf. [3, Th. 2.42]. □

5   Conclusive Remarks: Towards Emotions

In the previous sections, we have shown that possibility theory offers a unified logical framework in which both epistemic attitudes such as beliefs and motivational attitudes such as desires can be modeled. As a perspective along this line, we may study how the components of the approach, the epistemic one and the motivational one, can be combined in order to model basic emotion types such as hope and fear. Similar ideas on the logic of emotion intensity have been recently presented in [6] without making a connection with possibility theory.

Besides, we have described two extreme approaches to the problem: a minimal single agent logic of desire and belief and a maximal multi-agent one. The first one is completely faithful to the framework of possibility theory described in the first two sections of this paper. Its weighted version is an extension of possibilistic logic, but it has arguably a limited expressive power. On the other hand, the multi-agent logic GBDL^n is a graded extension of the full-fledged multimodal logic KD^n that is very expressive, but is arguably overexpressive as it contains formulas that can be make hardly intuitive sense, and its semantics is much richer than the framework of possibility theory. So, there is a need for more research on the bridge between modal and possibilistic logics. The logic GBDL^n:

- allows for objective formulas while MBDL does not. What is their role and can we dispense with them?
- allows for introspective formulas of the form $N_1^{\geq\alpha_1} N_1^{\geq\alpha_2}\varphi$ that are not part of the setting of possibility theory. How to make sense of them?
- presupposes that the epistemic state of an agent depends on the (objective) state this agent is in, which is not part of the formal framework described in the first sections. It enables standard techniques in modal logic to be applied, but it is not always easy to interpret.

More work is needed to come up with an epistemic logic framework which is at the same time expressive enough for our purpose, and where both semantic and syntactic aspects remain under control.

References

1. Banerjee, M., Dubois, D.: A simple modal logic for reasoning about revealed beliefs. In: Sossai, C., Chemello, G. (eds.) ECSQARU 2009. LNCS, vol. 5590, pp. 805–816. Springer, Heidelberg (2009)
2. Benferhat, S., Dubois, D., Kaci, S., Prade, H.: Bipolar possibility theory in preference modeling: Representation, fusion and optimal solutions. Information Fusion 7, 135–150 (2006)
3. Blackburn, P., de Rijke, M., Venema, Y.: Modal Logic. Cambridge University Press (2001)
4. Boutilier, C.: Towards a logic for qualitative decision theory. In: Principles of Knowledge Representation and Reasoning: Proc. of the 5th Int. Conf. (KR 1994), pp. 75–86. AAAI Press (1994)
5. Casali, A., Godo, L., Sierra, C.: A graded BDI agent model to represent and reason about preferences. Artif. Intell. 175, 1468–1478 (2011)
6. Dastani, M., Lorini, E.: A logic of emotions: from appraisal to coping. In: Proc. of AAMAS 2012, pp. 1133–1140. ACM Press (2012)
7. Dubois, D., Hajek, P., Prade, H.: Knowledge-driven versus data-driven logics. J. of Logic, Language, and Information 9, 65–89 (2000)
8. Dubois, D., Prade, H.: Resolution principles in possibilistic logic. Int. J. Appr. Reas. 4, 1–21 (1990)
9. Dubois, D., Prade, H.: Epistemic entrenchment and possibilistic logic. Artificial Intelligence 50, 223–239 (1991)
10. Dubois, D., Prade, H.: Possibility theory: qualitative and quantitative aspects. In: Gabbay, D., Smets, P. (eds.) Quantified Representation of Uncertainty and Imprecision. Handbook of Defeasible Reasoning and Uncertainty Management Systems, vol. 1, pp. 169–226. Kluwer (1998)
11. Dubois, D., Prade, H.: From Blanchés hexagonal organization of concepts to formal concept analysis and possibility theory. Logica Universalis 6(1), 149–169 (2012)
12. Dubois, D., Prade, H.: Possibilistic logic: a retrospective and prospective view. Fuzzy Sets and Systems 144, 3–23 (2004)
13. Fariñas del Cerro, L.: Resolution modal logic. Logique et Analyse 110-111, 153–172 (1985)
14. Fagin, R., Halpern, J.Y., Moses, Y., Vardi, M.Y.: Reasoning about Knowledge. MIT Press (1995)
15. Gratch, J., Marsella, S.: A domain independent framework for modeling emotion. Cognitive Systems Research 5(4), 269–306 (2004)
16. Goldszmidt, M., Pearl, J.: Qualitative probability for default reasoning, belief revision and causal modeling. Artificial Intelligence 84, 52–112 (1996)
17. Hintikka, J.: Knowledge and Belief: An Introduction to the Logic of the Two Notions. Cornell University Press (1962)
18. Hume, D.: A Treatise of Human Nature. In: Selby-Bigge, L.A., Nidditch, P.H. (eds.). Clarendon Press, Oxford (1978)
19. Lang, J., van der Torre, L., Weydert, E.: Hidden Uncertainty in the Logical Representation of Desires. In: Proc. of Int. Joint Conf. on Artificial Intelligence, pp. 685–690. Morgan Kaufmann (2003)
20. Lang, J., van der Torre, L., Weydert, E.: Utilitarian Desires. J. of Autonomous Agents and Multi-Agent Systems 5, 329–363 (2002)

21. Lorini, E.: A dynamic logic of knowledge, graded beliefs and graded goals and its application to emotion modelling. In: van Ditmarsch, H., Lang, J., Ju, S. (eds.) LORI 2011. LNCS, vol. 6953, pp. 165–178. Springer, Heidelberg (2011)

22. Lorini, E., Prade, H.: Strong possibility and weak necessity as a basis for a logic of desires. In: Godo, L., Prade, H. (eds.) Working Papers of the ECAI 2012 Workshop on Weighted Logics for Artificial Intelligence (WL4AI 2012), Montpellier, pp. 99–103 (August 28, 2012)

23. Meyer, J.J., van der Hoek, W.: Epistemic Logic for AI and Computer Science. Cambridge University Press (1995)

24. Ortony, A., Clore, G., Collins, A.: The Cognitive Structure of Emotions. Cambr. Univ. Pr. (1988)

25. Sahlqvist, H.: Completeness and correspondence in the first and second order semantics for modal logic. In: Proc. of 3rd Scandinavian Logic Symp., pp. 110–143. North-Holland (1975)

26. Shackle, G.L.S.: Decision, Order, and Time in Human Affairs. Cambridge Univ. Press (1961)

27. Spohn, W.: Ordinal conditional functions: a dynamic theory of epistemic states. In: Causation in Decision, Belief Change and Statistics, pp. 105–134. Kluwer (1988)

28. Zadeh, L.A.: Fuzzy sets as a basis for a theory of possibility. Fuzzy Sets & Syst. 1, 3–28 (1978)

A New Class of Lineage Expressions over Probabilistic Databases Computable in P-Time

Batya Kenig, Avigdor Gal, and Ofer Strichman

Technion, Israel Instutute of Technology, Haifa, Israel

Abstract. We study the problem of query evaluation over tuple-independent probabilistic databases. We define a new characterization of lineage expressions called *disjoint branch acyclic*, and show this class to be computed in P-time. Specifically, this work extends the class of lineage expressions for which evaluation can be performed in PTIME. We achieve this extension with a novel usage of junction trees to compute the probability of these lineage expressions.

1 Introduction

Applications in many areas such as data cleaning, data integration and event monitoring produce large volumes of uncertain data. Probabilistic databases in which the tuples' presence is uncertain, and known only with some probability, enable modeling and processing such uncertain data.

Answering queries over probabilistic databases has drawn much attention in the database community in recent years. A model of tuple-independent (or tuple-level semantics) probabilistic databases was introduced by Cavallo and Pittarelli [3] and was extensively discussed in the literature, *e.g.*, [9,6]. According to this model, each tuple t is annotated by an existence probability $p_t > 0$, meaning it appears in a possible world with probability p_t, independently of other tuples. This defines a probability distribution over all possible database instances.

Query evaluation over tuple-independent probabilistic databases is #P-hard in general, even for simple conjunctive queries without self-joins [6]. Dalvi and Suciu have introduced a dichotomy classification of queries over tuple-independent probabilistic databases, where any query with a *safe plan* can be computed *extensionally* by extending the query operators to enable an efficient computation of the result's probability [6,5]. The extensional approach is very efficient, but may be applied to a limited set of queries [13,6].

Even for queries without a safe plan there are database instances for which probabilities could be computed in PTIME. An intensional approach to evaluate queries over tuple-independent probabilistic databases considers both the query and the database instance. The query result is first computed and represented as a Boolean formula, termed a *lineage expression* [2], defined over Boolean variables corresponding to tuples in the database. The lineage describes how the answer was derived from the tuples in the database (see Table 1).

Various inference algorithms can be used to compute the result tuple probabilities, either exactly [16] or approximately [17]. Roy et al. [18] and Sen et al. [19]

W. Liu, V.S. Subrahmanian, and J. Wijsen (Eds.): SUM 2013, LNAI 8078, pp. 219–232, 2013.

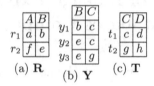

Fig. 1. Probabilistic Database with variables

showed a polynomial time algorithm for recognizing lineage expressions that can be transformed to a read-once form, and computing their probability. Their algorithm is applicable for lineages resulting from conjunctive queries without self joins.

Consider the tuple-independent probabilistic database of Figure 1 and the Boolean conjunctive query

$$Q1():\text{-}R(x,y),Y(y,z) \tag{1}$$

$Q1$ is a conjunctive query without self-joins that has a safe plan [6]. Olteanu and Huang [16] showed that the lineage resulting from conjunctive queries without self-joins, that have a safe plan, always have a read-once equivalent. Indeed, the lineage expression of $Q1$ over the database in Fig. 1 is $r_1y_1 + r_2y_2 + r_2y_3$, which has an equivalent read-once form, $r_1y_1 + r_2(y_2 + y_3)$.

$Q2$ is an example of a query that does not have a safe plan:

$$Q2():\text{-}R(x,y),Y(y,z),T(z,w) \tag{2}$$

The lineage expression of $Q2$ over the database is

$$r_1y_1t_1 + r_2y_2t_1 + r_2y_3t_2 \tag{3}$$

It was shown [18,12] that Expression 3 does not have an equivalent read-once form.

In this work we introduce a new class of lineage expressions called *disjoint branch acyclic lineage expressions*. Such lineage expressions are defined using restrictions on their respective hypergraph. Going back to Example 1, the lineage expression in Eq. 3 is disjoint branch acyclic, possessing a special structure that can be exploited for efficient computation.

We characterize disjoint branch acyclic lineage expressions and present, as part of the proof of the class computation time, an algorithm to compute the probability of this form in time that is polynomial in the size of the formula.

The rest of the paper is organized as follows: Section 2 introduces background on a specific class of chordal graphs, probability computation using probabilistic graphical models, and hypergraph acyclicity. Lineage acyclicity is presented in Section 3. Section 4 presents the main theorem of this work, proving it by showing an algorithm for probability computation of disjoint branch acyclic lineage expressions. We conclude in Section 5.

2   Preliminaries

At the heart of the proposed method for computing the probability of Boolean lineage expressions lies a graph with a specific structure termed *rooted directed path graph*. This class of graphs and its PTIME recognition algorithm were first introduced by Gavril [11]. This section discusses this class of graphs and other notions significant to our proposed approach. We present a class of chordal graphs, namely *rooted directed path graphs* (Section 2.1) and discuss probability computation using probabilistic graph models (Section 2.2). We conclude with the introduction of hypergraph acyclicity (Section 2.3).

A clique C of a graph $G(V, E)$ is a subset of V where every pair of nodes is adjacent. We denote by K_G the set of maximal cliques in G. For a vertex $v \in V$ we denote by K_v the set of maximal cliques in K_G that contain v. We use $T(V, E)$ to denote a tree. A subtree is a connected subgraph of a tree. In particular, a path in a tree can be viewed as a subtree. Whenever a subtree is induced from a subset of nodes $V' \subseteq V$ of a tree $T(V, E)$, we do not explicitly state its set of edges, but rather denote it using $T(V')$.

2.1   Classes of Chordal Graphs

A *chord* is an edge connecting two non-consecutive nodes in a cycle or path. G is *chordal* or *triangulated* if it does not contain any chordless cycles. Discovering whether G is chordal can be performed in time $O(|V| + |E|)$ [20].

A $P4$ denotes a chordless path with four vertices and three edges. A graph is considered to be *P4-free* if it does not contain a $P4$.

An *intersection graph* of a finite family of non-empty sets is obtained by representing each set by a vertex, and connecting two vertices if their corresponding sets intersect. Gavril [10] characterizes the connection between chordal graphs and intersection graphs, as follows.

Theorem 1 ([10]). *Let $G(V, E)$ be an undirected graph. The following statements are equivalent:*

1. *G is chordal.*
2. *There exists a tree $T(K_G)$ such that for every $v \in V$ the subgraph induced by K_v is a subtree $T(K_v)$.*
3. *G is the intersection graph of a family of subtrees of some tree T'.*

$T(K_G)$ is called a *junction tree* possessing the following *running intersection property*: for every pair of cliques $C_1, C_2 \in K_v$ every clique on the path from C_1 to C_2 in $T(K_G)$ belongs to K_v.

Let T' be a rooted directed tree, and consider a group of directed paths in T'. Let G be the intersection graph of directed paths in T'. Then G is a *Rooted Directed Path Graph*, and T' is called the *host tree* of G.

The following property (Theorem 2 [11]) defines a characteristic tree, associated with a rooted directed path graph. This tree is of prime concern in this work.

Theorem 2 ([11]). *A graph $G(V, E)$ is a rooted directed path graph (rdpg) iff there exists a rooted directed tree T_r whose vertex set is K_G, so that for every vertex $v \in V$, $T_r(K_v)$ is a directed path of T_r.*

Constructing the characteristic tree of a rooted directed path graph $G(V)$, if one exists, takes $O(|V|^4)$ [11].[1] The characteristic tree T_r of an rdpg $G(V, E)$ is, in fact, a special form of a junction tree (Definition 1), where every vertex $v \in V$ appears in exactly one branch of T_r. Such a junction tree is known as a *disjoint branch junction tree* (dbjt) [8].

Definition 1. *Let T_r be a junction tree with root r and children $r_1, r_2, ..., r_l$ roots of subtrees $T_{r_1}, ..., T_{r_l}$, respectively. A junction tree T_r is a dbjt if:*

1. *T_r contains a single node, r, i.e., $|T_r| = 1$, or*
2. *The following two conditions jointly hold: (a) $\forall r_i \neq r_j, C_{r_i} \cap C_{r_j} = \emptyset$; and (b) $\forall T_{r_i} \in \{T_{r_1}, T_{r_2}, ..., T_{r_l}\}$, T_{r_i} recursively complies with the conditions 1 and 2.*

An example of a rooted directed path graph and its corresponding characteristic tree (or dbjt) are presented in Figures 2b and 2c, respectively. Definition 1 characterizes the dbjt properties that enable the efficient computation we show in this work.

2.2 Probability Computation Using Probabilistic Graph Models

Probabilistic Graphical Models (PGMs) refer to a set of approaches for representing and reasoning about large joint probability distributions [15]. A PGM is a graph in which nodes represent random variables and edges represent direct dependencies between them. An example of a directed PGM is given in Figure 2a, representing the lineage expression of Eq. 3.

Inference in PGMs is the task of answering queries over the probability distribution described by the graph and is, in general, #P-complete [15]. One of the well-known inference algorithms is the junction tree algorithm [15]. The algorithm is designed for undirected PGMs in which for every maximal clique C in the PGM, there exists a factor F_C that is a function from the set of assignments of C to the set of non-negative reals. The algorithm consists of two parts, *compilation* and *message passing*. The compilation part includes three steps, namely *moralization, triangulation* and *construction*, as follows. *Moralization*, in the case of a directed PGM, involves connecting all parents of a given node and dropping the direction of edges (*e.g.*, Figure 2b). *Triangulation* adds extra edges to create a chordal graph. The example graph obtained after moralization in Figure 2b is already chordal and therefore no edges need to be added. We note that this example is also a rooted directed path graph (see Section 2.1). *Construction* forms

---

[1] To date, this is the most efficient published recognition algorithm for rooted directed path graphs [4]. There is also an unpublished linear time algorithm [7] for this class of graphs.

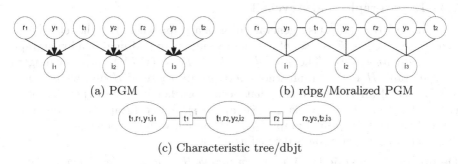

(a) PGM   (b) rdpg/Moralized PGM

(c) Characteristic tree/dbjt

Fig. 2. PGM, moralization and junction tree

a junction tree over the maximal cliques in the resulting graph $G(V)$. Figure 2c
shows the junction tree of the graph in Figure 2b. Note that since the graph is
a rooted directed path graph, its junction tree is in fact a dbjt.

The message-passing part has two steps. First, for each edge of the tree
$(C_1, C_2) \in T$ a factor, defined over the variables in the intersection $S = C_1 \cap C_2$, is
defined. The factor entries are initialized to 1. Then, neighboring nodes C_1, C_2 exchange messages through the factor defined on their intersection F_S, $S = C_1 \cap C_2$.
The message from C_1 to C_2 is:

$$\mu_{C_1, C_2}(S) = \sum_{x \in C_1 \backslash S} F_1(C_1)$$

and the message from C_2 to C_1 is:

$$\mu_{C_2, C_1}(S) = \frac{\sum_{x \in C_2 \backslash S} F_2(C_2)}{\mu_{C_1, C_2}(S)}$$

After every pair of adjacent nodes in the junction tree have exchanged messages, each factor holds the marginal of the joint probability distribution of the
entire variable set. The message-passing protocol is such that every edge in the
tree is processed once in each direction. Therefore, the runtime of the message
passing algorithm is $O(N \cdot D^k)$ where N is the number of nodes in the tree, D is
the domain of the variables, and k is the size of the largest clique. $k-1$ is referred
to as the *width* of the associated PGM and clearly, the inference algorithm is
exponential in the PGM's width so bounded width implies tractability in graphical models. PGMs may have several different triangulations, affecting the size
of the largest clique in the graph. The smallest width that can be obtained for a
PGM is its *treewidth*, where the treewidth of a chordal graph is simply its width.
Finding an optimal triangulation is known to be NP-complete. However, in this
work we introduce a method to compute the probability of a family of lineage
expressions in time that is polynomial in their treewidth.

2.3   Hypergraphs and Acyclicity

A *hypergraph* $H = (V, E)$ is a generalization of a graph where V is the set of nodes and the set of edges E is a set of non-empty subsets of V. Edges in a hypergraph are termed *hyperedges*. The *primal graph* $G(H) = (V, E^G)$ corresponding to a hypergraph H is the graph whose vertices are those of H and whose edges are the set of all pairs of nodes that occur together in some hyperedge of H ($E^G = \{(u, v) : \{u, v\} \subseteq V, \exists e \in E, \{u, v\} \subseteq e\}$). A hypergraph H is *conformal* if every clique in its primal graph $G(H)$ is contained in a hyperedge of H. Acyclicity in a hypergraph is defined as follows.

Definition 2 (acyclicity [1]). *A hypergraph H is acyclic (or α-acyclic) if H is conformal and its primal graph $G(H)$ is chordal.*

Beeri et. al [1] showed that a hypergraph is acyclic iff it has a junction tree. Duris [8] also showed that for a restricted form of acyclic hypergraphs (called γ-acyclic) there exists a dbjt rooted at every node. An algorithm that constructs a dbjt in time $O(|V|^2)$ for γ-acyclic hypergraphs was also introduced there.

3   Disjoint Branch Acyclic Lineage (DBAL) Expressions

We now introduce a class of Boolean lineage expressions, connecting it to rooted directed path graphs. Let $f(V)$ denote a lineage expression of a set of literals V, resulting from a query q, as derived by the query engine. Lineage expressions of conjunctive queries are monotone formulas, where all literals are positive and only conjunctions and disjunctions are used. An *implicant* $p \subseteq V$ of f is a set of literals such that whenever they are true, f is true as well. An implicant of f is called a *prime implicant* if it cannot be reduced. We denote by f_{IDNF} f's DNF form containing only prime implicants. f_{IDNF} can be modeled as a hypergraph $H_f(V, E)$, where each literal corresponds to a node in the graph and each prime implicant corresponds to a hyperedge.

f_{IDNF} is not always available, and expanding f to its DNF form may result in an exponentially larger formula. Therefore, we now propose the construction of an alternative graph $G(f)$, built over $f(V)$. The set of literals V is the set of nodes of $G(f)$ and two nodes are connected iff they belong to a common prime implicant. $G(f)$ is exactly H_f's primal graph, *i.e.*, $G(H_f) = G(f)$. For a restricted set of queries, $G(f)$ can be built directly from f by using a method proposed by Roy et al. [18]. We say that f is *conformal* if every maximal clique in $G(f)$ is contained in a prime implicant of f.

Definition 1 (Lineage expression acyclicity). *A lineage expression f is acyclic if $G(f)$ is chordal and f is conformal.*

Definition 2 (Disjoint Branch Acyclic Lineage Expressions). *A lineage expression f is a Disjoint Branch Acyclic Lineage Expression or DBAL if f is acyclic and $G(f)$ is a rooted directed path graph.*

Following the discussion in Section 2.1, DBAL expressions have a dbjt. The lineage expression in Eq. 3 (Section 1) is an example of a DBAL. Its corresponding dbjt is presented in Figure 2c.

4   DBAL Expression Probability Computation

In this section we prove that the probability of disjoint branch acyclic lineage (DBAL) expressions over tuple independent probabilistic databases can be computed in PTIME.

Theorem 1. *Let $f(V)$ be a DBAL expression. The probability $Pr(f = 1)$ can be computed in time $O(nk^2)$ where $n = |V|$ and k is the size of the largest clique in $G(f)$.*

At the heart of the proof is an algorithm for computing the probability of DBAL expressions in time that is quadratic in the size of the treewidth. This solution is unique since, to the best of our knowledge, it is the first time an algorithm that runs in time polynomial (quadratic) of the treewidth, as opposed to exponential, is introduced in the context of tuple independent probabilistic databases.

Section 4.1 introduces an example that will be used to demonstrate the algorithm and Section 4.2 discusses factor representation in our setting. Finally, Section 4.3 details the algorithm and presents lemmas 1 and 2 that argue for the correctness of the algorithm and its complexity, respectively, which together proves Theorem 1 above.

4.1   Illustrating Example

We first motivate and explain the algorithm approach using a simple example.

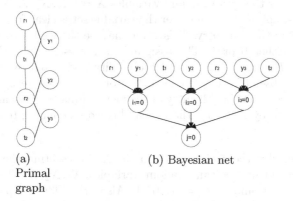

(a)
Primal
graph

(b) Bayesian net

Fig. 3. Illustration for Example 1

Example 1. Consider query $Q2() : -R(x, y), Y(y, z), T(z, w)$, presented earlier over the instance in Table 1. The lineage of the query is $j = r_1y_1t_1 + r_2y_2t_1 + r_2y_3t_2$. The primal graph corresponding to the query is given in Figure 3a. It is easy to see that this lineage expression is not read-once since it has a P4: (r_1, t_1, r_2, t_2). Let us denote by $i_1 = r_1y_1t_1$, $i_2 = r_2y_2t_1$, and $i_3 = r_2y_3t_2$. We are ultimately interested in calculating the probability: $Pr(j = 1) = 1 - Pr(j = 0)$. If $j = 0$ then we know that $i_1 = i_2 = i_3 = 0$. These values can be seen as introduction of evidence in a Bayesian network, illustrated in Figure 3b. After moralization (see Section 2.2), the network is chordal and conformal and therefore has a junction tree depicted in Figure 4. In the Junction-Tree algorithm, any node may be selected as root. For this example let us select node $\{r_2, y_3, t_2\}$ as the root.   □

In the classic junction tree algorithm, where factors are represented in tabular form, each entry in the factor table represents a single assignment, leading to a representation that is exponential in the number of variables in the table. We present linear sized factors, where each entry represents multiple assignments. See, for example, Figure 4, where asterisks represent wildcard assignments. Here, the number of entries in each factor is exactly the node's cardinality. The message passing is illustrated in Figure 4, and will be demonstrated in detail in Section 4.3. After the completion of the algorithm, the root node contains the marginal probability of its entries (see Section 2.2). Therefore, all the entries of the root node's factor are added in order to obtain the required probability.

4.2 Factor Representation and Projection

Hereinafter we shall use a tabular notation to represent a factor, where columns represent random variables, and rows correspond to a set of mutual exclusive value assignments. The notation introduced below is illustrated in Example 2. Given a factor over the set of random variables $X = (X_1, X_2, ..., X_n)$ we denote by $F_X[j, k]$ (or simply $F[j, k]$ whenever the variable set is clear from the context) the value of X_k in the jth entry. X_k's value may be the wildcard, '*', indicating that it can be either 0 or 1. The assignments represented by the jth entry are denoted by $F[j]$ and their overall probability is denoted by $Pr(F[j])$. Let $X' \subset X$ be a subset of the variables of factor F, we denote by $F[j, X']$ the values of variables X' in the jth entry of the factor. Finally, given an assignment $X = x$, we denote by $Pr(F[x])$ the probability corresponding to this entry in factor F.

Example 2. Consider the factor F in Table 2, representing the joint distribution of independent boolean random variables X_1, X_2, X_3. Using our notation, $F[2, 3] = *$ and $F[2] = [1, 0, *]$. Also, $Pr(F[3]) = p_{X_1} \cdot p_{\overline{X_3}}$ and $F[3, \{X_1, X_3\}] = [1, 0]$. Finally, $Pr(F[\{X_1 = 1, X_2 = 0, X_3 = *\}]) = p_{X_1}$.   □

Each maximal clique in the primal graph of a DBAL expression corresponds to exactly one prime implicant of the lineage's IDNF form. As a result, each node in the corresponding junction tree contains a factor that represents a single DNF prime implicant of the lineage. We will refer to these as *DNF factors*. For each variable X in the expression we define a *base factor*, F_X^b. Base factors contain exactly two entries, with values $0, 1$ and their appropriate probabilities $p_{\overline{X}}, p_X$, respectively. Each base factor is assigned to exactly one node in the junction tree.

Consider some DNF prime implicant d, containing k literals, $d = X_1 \cdot X_2 \cdot \ldots \cdot X_k$. The probability of $d = 0$ is computed as follows:

$$Pr(d = 0) = Pr(X_1 = 0) + Pr(X_1 = 1, X_2 = 0) + \ldots + Pr(X_1 = 1, \ldots, X_{k-1} = 1, X_k = 0). \tag{4}$$

The k summands in Eq. 4 create a mutually exclusive and exhaustive set of configurations.

For illustration, consider Table 1 over $X_1, \ldots, X_k$. The "Pr" values are initialized to 1.

The asterisks in the table represent wildcard assignments, as follows:

Table 1. Factor Table

X_1	X_2	...	...	X_k	Pr
0	*	*	*	*	1
1	0	*	*	*	1
1	1	0	*	*	1
1	...	...	...	*	1
1	1	1	...	0	1

$$Pr(X_1 = 1, \ldots, X_{i-1} = 1, X_i = 0, X_{i+1} = *, \ldots, X_k = *) =$$

$$\sum_{x_{i+1}, \ldots, x_k \in \{0,1\}} Pr(X_1 = 1, \ldots X_{i-1} = 1, X_i = 0, X_{i+1} = x_{i+1}, \ldots, X_k = x_k) =$$

$$\sum_{x_{i+1}, \ldots, x_k \in \{0,1\}} Pr(X_{i+1} = x_{i+1}, \ldots, X_k = x_k | X_1 = 1, \ldots, X_{i-1} = 1, X_i = 0) \cdot Pr(X_1 = 1, \ldots, X_{i-1} = 1, X_i = 0)$$

$$= Pr(X_1 = 1) \cdot \ldots \cdot Pr(X_{i-1} = 1) \cdot Pr(X_i = 0) \cdot \sum_{x_{i+1}, \ldots, x_k \in \{0,1\}} Pr(X_{i+1} = x_{i+1}, \ldots, X_k = x_k) \tag{5}$$

using the tuple independence assumption in the transition from the third to the fourth line of the equation. Informally, once we know that $X_j = 0$, then the implicant's value is false regardless of the values of its other literals.

At the beginning of the algorithm, the values in the "Pr" column of the factors depend on the assignment of the base factors to the nodes in the tree. For example, a factor over variables X_1, X_2, X_3 at the beginning of the algorithm is given in Table 2. For the sake of illustration, we assume that the base factor $F_{X_2}^b$ is assigned to a different node (DNF factor).

The proposed algorithm is actually a series of projections (defined below) over the linear-sized factors of the junction tree. In the general message passing algorithm [15], in which each entry in the factor represents a single configuration of the variables (and therefore the size of the factor is exponential in the number of variables), the probabilities of the entries with common values in the projected variables are simply added. This is

Table 2. Factor Table with partial base factors

X_1	X_2	X_3	Pr
0	*	*	$p_{\overline{X_1}}$
1	0	*	p_{X_1}
1	1	0	$p_{X_1} \cdot p_{\overline{X_3}}$

not the case for the linear sized factors used in our setting. Definition 1 formalizes this notion of projection in our setting, and Example 3 demonstrates it.

Definition 1 (factor projection). *Let $F_{X \cup X'}$ be a factor over variables $X \cup X'$ where $X = \{X_1, \ldots, X_m\}$ and $X' = \{X_1', \ldots, X_l'\}$. The projection of F over* the variables in X, denoted $F_X = \prod_X F_{X \cup X'}$, is a new factor containing only variables X. The probability column in F_X is computed as follows:

$$Pr(F_X[j]) = \sum_{i \in [1,|X \cup X'|]:F_{X \cup X'}[i,X]=F_X[j]} Pr(F_{X \cup X'}[i])$$

The projection $\prod_X F_{X \cup X'}$ may be applied to $F_{X \cup X'}$ under the following conditions:

1. The variables X', projected out of the factor $F_{X \cup X'}$, appear after (referring to column order) the variables X.
2. The base factors corresponding to the variables X' are included in factor $F_{X \cup X'}$ before the projection operation can be applied.

Example 3. Consider Table 2 and the factor F_{X_1,X_2,X_3} over the variable set $\{X_1, X_2, X_3\}$. We start by projecting out the variable X_3. Condition 1 of Definition 1 is satisfied. As for Condition 2, X_3's probability, p_{X_3}, is already available in the factor, and therefore

$$F_{X_1,X_2} = \prod_{X_1,X_2} F_{X_1,X_2,X_3} =$$

X_1	X_2	Pr
0	*	$p_{\overline{X_1}}$
1	0	p_{X_1}
1	1	$p_{X_1} \cdot p_{\overline{X_3}}$

Projecting out X_2 from F_{X_1,X_2} requires multiplying in X_2's base factor to satisfy Condition 2. Therefore,

$$F_{X_1} = \prod_{X_1} F_{X_1,X_2} =$$

X_1	Pr
0	$p_{\overline{X_1}}$
1	$p_{X_1}(p_{\overline{X_2}} + p_{X_2} \cdot p_{\overline{X_3}})$

□

4.3 Algorithm Description

Let C_i denote the set of variables in node i of the junction tree, $|C_i|$ its cardinality, and F_i its factor. In this section we use the factor notation defined in Section 4.2. B_r denotes the set of variables in node r, for which base factors have been assigned, i.e., $B_r = \{X : X \in C_r, F_X^b \text{ is assigned to } r\}$. We denote by $children(i)$ and $p(i)$ the children and parent of node i in the junction tree, respectively. A message between node i and node j, $\mu_{i,j}(C_i \cap C_j)$ is a factor over the intersection of the two nodes. The number of entries in $\mu_{i,j}(C_i \cap C_j)$ is $|C_i \cap C_j| + 1$ (including the entry containing all ones).

The algorithm uses a partial order $\preceq$ over the variables in the junction tree T. We denote by $vars(\preceq)$ the set of variables over which $\preceq$ is defined.

The pseudocode of the algorithm over linear sized factors is given in algorithms 1 and 2. After the initial call to Algorithm 2 (Line 1 of Algorithm 1), the

---

Algorithm 1. Message Passing: Initial Call

Input: dbjt (see Definition 1) $T_{r'}$ with root r' corresponding to a lineage expression f.
Output: $Pr(f = 0)$
1: Call Algorithm 2 with parameters: $T_{r'}$ and $\preccurlyeq \leftarrow \emptyset$.
2: **Return** $\sum_{j=1}^{|C_{r'}|} Pr(F_{r'}[j])$.

---

algorithm performs a series of recursive calls to update the probabilities in the node factors of the junction tree.

Each message from a node i to its parent, $p(i)$, is a projection on factor F_i over the variables $C_i \cap C_{p(i)}$. According to the definition of projection (Definition 1), variables $C_i \cap C_{p(i)}$ should appear before $C_i \setminus C_{p(i)}$ in the factor table representation. Therefore, lines 1-5 of Algorithm 2 define an order over the variables in the root node r that was given as a parameter (C_r), such that projection over variables $C_r \cap C_{p(r)}$ is made possible. In Example 1, Figure 4, the root node contains ordered variables $\{r_2, y_3, t_2\}$. In the factor for the child node with variables $\{t_1, r_2, y_2\}$, r_2 appears before t_1 and y_2 because the message between this node and its parent is over variable r_2. Likewise, in the factor with variables $\{t_1, r_1, y_1\}$, t_1 appears before r_1 and y_1. The order is updated in line 5.

Lines 6-10 initialize a factor for node r based on the ordering $\preccurlyeq$ that was updated in lines 1-5, and according to the base factors assigned to this node. In Example 1 (Figure 4), the base factors for variables y_3 and t_2 are assigned to the root node, while the base factor for r_2 is assigned to the middle node (with variables $\{t_1, r_2, y_2\}$).

Lines 11-21 initiate a recursive call on the children of r. In Line 13, the messages from all children of r are collected. Each one of the messages, $\mu_{i,r}(C_{r_i} \cap C_r)$, received by the node in line 13 contains $|C_{r_i} \cap C_r| + 1$ entries, which form an exhaustive and mutual exclusive set of configurations. For example, consider the message $\mu_{1,2}(t_1)$ from node $\{t_1, r_1, y_1\}$ to node $\{t_1, r_2, y_2\}$ (Figure 4).

As in the case of projection, in order to add the probabilities of the entries in the messages, the appropriate base factors need to be multiplied in before the addition can take place. For example, in Figure 4, the base factor for t_1, $F_{t_1}^b$, is not part of the factor of node $\{t_1, r_1, y_1\}$, and therefore not part of the original message, $\mu_{1,2}(t_1)$, (containing only entries where $t_1 = 0$ and $t_1 = 1$). However, in order to augment the factor with the entry $t_1 = *$, the values in the probability column of the entries corresponding to $t_1 = 0$ and $t_1 = 1$ need to be added. In order for the resulting probability to be correct, the probabilities of the entries corresponding to $t_1 = 0$ and $t_1 = 1$ are multiplied by $p_{\overline{t_1}}$ and p_{t_1} respectively. The entry where $t_1 = *$ in Figure 4 was appended to the message because it will be used by node $\{t_1, r_2, y_2\}$. Such entries are calculated in lines 14-21 of the algorithm. There are exactly $|C_{r_i} \cap C_r|$ such entries added to the message which correspond to partial sums over the entries in the original message $\mu_{i,r}(C_{r_i} \cap C_r)$.

Algorithm 2. Message Passing: Main Procedure

Input: dbjt T_r with root r and a partial order $\preceq$.
Output: Factor F_r with correct probabilities
1: **if** $r \neq r'$ **then**
2:   Define an order $\preceq_r$ over C_r s.t.: 1. $C_r \cap C_{p(r)}$ appear before $C_r \backslash C_{p(r)}$ 2. The order of
    variables $C_r \cap vars(\preceq)$ complies with $\preceq$.
3: **else**
4:   define an arbitrary order over variables C_r
5: Update $\preceq$ according to the steps above.
    {*Initialize node's factor based on $\preceq$*}
6: Define a linear-sized factor F_r based on $\preceq$.
7: **for** $j \leftarrow 1$ to $|C_r|$ **do**
8:   $Pr(F_r[j]) \leftarrow 1.0$ {initialize factor entries}
9:   **for** $X \in B_r$ **do**
10:     $Pr(F_r[j]) \leftarrow Pr(F_r[j]) \cdot Pr(X = F_r[j, \{X\}])$
    {*apply projection on subtrees*}
11: **for all** $r_i \in children(r)$ **do**
12:   Recursively call the algorithm on subtree T_{r_i} with root r_i and (updated) ordering $\preceq$.
13:   $\mu_{i,r}(C_r \cap C_{r_i}) \leftarrow \prod_{C_r \cap C_{r_i}}(F_{r_i})$ [project on the children's factor to get the message]
14:   **for** $j \leftarrow 1$ to $|C_r \cap C_{r_i}| + 1$ **do**
15:     $M_{i,r}[j] \leftarrow 1.0$ [iterate over entries in the message]
16:     **for** $X \in ((C_r \cap C_{r_i}) \backslash B_{r_i})$ **do**
17:       $M_{i,r}[j] \leftarrow M_{i,r}[j] \cdot Pr(X = \mu_{i,r}[j, \{X\}])$ $[Pr(X = *) = 1.0]$
18:     $prob \leftarrow Pr(\mu_{i,r}[|C_{r_i} \cap C_r| + 1) \cdot M_{i,r}[|C_{r_i} \cap C_r| + 1]$ [initialize $prob$ according to entry
       $[1,1,...,1]]$
19:     **for** $k \leftarrow |C_{r_i} \cap C_r|$ to 1 **do**
20:       $prob \leftarrow prob + Pr(\mu_{i,r}[k]) \cdot M_{i,r}[k]$ [update $prob$]
21:       $Pr(\mu_{i,r}[X_1 = 1, ..., X_{k-1} = 1, X_k = *, ..., X_{|C_{r_i} \cap C_r|} = *]) \leftarrow prob$
    {*update factor using children's projected factors*}
22: **for** $j \leftarrow 1$ to $|C_r|$ **do**
23:   **for all** $r_i \in children(r)$ **do**
24:     $Pr(F_r[j]) \leftarrow Pr(F_r[j]) \cdot Pr(\mu_{i,r}[F_r[j, C_{r_i} \cap C_r]])$

Finally, lines 22-24 update F_r according to the messages received from its children. The correctness of the algorithm for disjoint branch junction trees is given in Lemma 1. The proof is omitted due to space considerations. It is available in the full version of this paper [14].

Lemma 1. *Let $T_{r'}$ be a dbjt with root r', corresponding to lineage expression f. After running Algorithm 1 on $T_{r'}$, $Pr(F_{r'}[j])$, $j \in [1, |C_{r'}|]$ contains the marginal probability corresponding to the configurations represented by the jth entry of this factor.*

Lemma 2. *The complexity of algorithms 1 and 2 on a disjoint branch junction tree of size n is $O(n \cdot k_{MAX}^2)$ where $k_{MAX} = MAX_{i=1..n}|C_i|$.*

Proof. The loop in lines 6-10 is performed in $O(|C_r|^2)$ since $|B_r| \leq |C_r|$. Similarly, the loop in lines 14-17 is performed in $O((|C_r \cap C_{r_i}| + 1)^2)$, but since subsumption cannot occur in the junction tree, $|C_r| > |C_r \cap C_{r_i}|$, we arrive again at runtime of $O(|C_r|^2)$. The loop in lines 19-21 takes time $O(|C_r \cap C_{r_i}|)$.

A node r in the tree receives messages from all of its neighbors, except its parent in the algorithm. Since the children create a partition of a subset of the variables in the node, then the number of children can be at most $|C_r|$. The number of entries for which the probability is updated is exactly $|C_r|$, therefore the total runtime is $\sum_{i=1}^{n} O(|C_r|^2) = O(n \cdot (k_{MAX}^2))$.

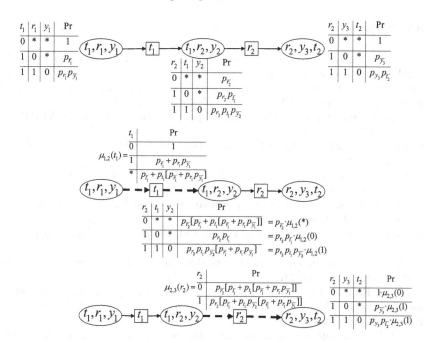

Fig. 4. Message Passing using Alg. 2 over $f = t_1 r_1 y_1 + t_1 r_2 y_2 + r_2 y_3 t_2$

Algorithms 1 and 2 along with Lemmas 1 and 2 complete the proof for the main theorem of this section, Theorem 1. Overall, we have shown that DBAL expressions, having a dbjt, can be evaluated in polynomial time.

5 Conclusions

We have presented *disjoint branch acyclic lineage expressions*, a new class of lineage expressions of queries over tuple independent probabilistic databases, and shown that probability computation over this class can be done in low polynomial data complexity.

As part of future research we plan to investigate queries and database instances that induce junction trees with structural properties that enable efficient probability calculation. Furthermore, we plan to explore how such queries relate to existing characterizations of tractability [13]. Since correlations between tuples can naturally arise in many applications, we intend to investigate how to extend the proposed approach to models without the tuple-independence assumption.

Acknowledgments. The work was carried out in and partially supported by the Technion–Microsoft Electronic Commerce research center.

References

1. Beeri, C., Fagin, R., Maier, D., Yannakakis, M.: On the desirability of acyclic database schemes. J. ACM 30, 479–513 (1983)
2. Benjelloun, O., Sarma, A., Halevy, A., Theobald, M., Widom, J.: Databases with uncertainty and lineage. VLDB Journal 17(2), 243–264 (2008)
3. Cavallo, R., Pittarelli, M.: The theory of probabilistic databases. In: VLDB, pp. 71–81 (1987)
4. Chaplick, S.: Path Graphs and PR-trees. PhD thesis, Charles University, Prague (January 2012)
5. Dalvi, N., Schnaitter, K., Suciu, D.: Computing query probability with incidence algebras. In: PODS, pp. 203–214. ACM (2010)
6. Dalvi, N., Suciu, D.: Efficient query evaluation on probabilistic databases. VLDB Journal 16, 523–544 (2007)
7. Dietz, P.F.: Intersection Graph Algorithms. PhD thesis, Cornell University (August 1984)
8. Duris, D.: Some characterizations of γ and β-acyclicity of hypergraphs. Inf. Process. Lett. 112(16), 617–620 (2012)
9. Fuhr, N., Rölleke, T.: A probabilistic relational algebra for the integration of information retrieval and database systems. ACM Transactions on Information Systems 15, 32–66 (1994)
10. Gavril, F.: The intersection graphs of subtrees in trees are exactly the chordal graphs. Journal of Combinatorial Theory, Series B 16(1), 47–56 (1974)
11. Gavril, F.: A recignition algorithm for the intersection graphs of directed paths in directed trees. Discrete Mathematics 13, 237–249 (1975)
12. Golumbic, M., Mintz, A., Rotics, U.: Factoring and recognition of read-once functions using cographs and normality. In: DAC (June 2001)
13. Jha, A.K., Suciu, D.: Knowledge compilation meets database theory: compiling queries to decision diagrams. In: ICDT, pp. 162–173 (2011)
14. Kenig, B., Gal, A., Strichman, O.: A new class of lineage expressions over probabilistic databases computable in p-time. Technical Report IE/IS-2013-01, Technion – Israel Institute of Technology (January 2013),
    http://ie.technion.ac.il/tech_reports/1365582056_TechReportSUM2013.pdf
15. Koller, D., Friedman, N.: Probabilistic Graphical Models: Principles and Techniques. The MIT Press (August 2009)
16. Olteanu, D., Huang, J.: Using OBDDs for efficient query evaluation on probabilistic databases. In: Greco, S., Lukasiewicz, T. (eds.) SUM 2008. LNCS (LNAI), vol. 5291, pp. 326–340. Springer, Heidelberg (2008)
17. Olteanu, D., Huang, J., Koch, C.: Approximate confidence computation in probabilistic databases. In: ICDE, pp. 145–156 (2010)
18. Roy, S., Perduca, V., Tannen, V.: Faster query answering in probabilistic databases using read-once functions. In: ICDT, pp. 232–243 (2011)
19. Sen, P., Deshpande, A., Getoor, L.: Read-once functions and query evaluation in probabilistic databases. PVLDB 3(1), 1068–1079 (2010)
20. Tarjan, R.E., Yannakakis, M.: Addendum: Simple linear-time algorithms to test chordality of graphs, test acyclicity of hypergraphs, and selectively reduce acyclic hypergraphs. SIAM J. Comput. 14(1), 254–255 (1985)

The Semantics of Aggregate Queries in Data Exchange Revisited[*]

Phokion G. Kolaitis[1] and Francesca Spezzano[2]

[1] University of California Santa Cruz & IBM Research - Almaden, USA
[2] DIMES, Università della Calabria, Italy
kolaitis@cs.ucsc.edu, fspezzano@dimes.unical.it

Abstract. Defining "good" semantics for non-monotonic queries and for aggregate queries in the context of data exchange has turned out to be a challenging problem for a number of reasons, including the dependence of the semantics of the concrete syntactic representation of the schema mapping at hand. In this paper, we revisit the semantics of aggregate queries in data exchange by introducing the *aggregate most-certain answers*, a new semantics that is invariant under logical equivalence. Informally, the aggregate most-certain answers are obtained by taking the intersection of the aggregate certain answers over all schema mappings that are logically equivalent to the given schema mapping. Our main technical result is that for schema mappings specified by source-to-target tuple-generating dependencies only (no target constraints), the aggregate most-certain answers w.r.t. a schema mapping coincide with the aggregate certain answers w.r.t. the schema mapping in normal form associated with the given schema mapping. This result provides an intrinsic justification for using schema mappings in normal form and, at the same time, implies that the aggregate most-certain answers are computable in polynomial time. We also consider the semantics of aggregate queries w.r.t. schema mappings whose specification includes target constraints, and discuss some of the delicate issues involved in defining rigorous semantics for such schema mappings.

1 Introduction

Data exchange is the problem of restructuring and translating data structured under one schema, called the *source schema*, into data structured under a different schema, called *target schema* [5]. Data exchange is formalized using schema mappings, i.e., quadruples of the form $\mathcal{M} = (\mathbf{S}, \mathbf{T}, \Sigma_{\mathsf{st}}, \Sigma_{\mathsf{t}})$, where $\mathbf{S}$ is the source schema, $\mathbf{T}$ is the target schema, Σ_{st} is a set of constraints describing the relationship between source and target, and Σ_{t} is a set of target constraints. The constraints between source and target are typically specified using *source-to-target tuple generating dependencies* (s-t tgds). Furthermore, the target constraints are typically specified using *target tuple-generating dependencies* (target tgds) and *target equality generating dependencies* (target egds).

[*] This research was partially supported by NSF Grants IIS-0905276 and IIS-1217869, and the project *Cardiotech* funded by the Italian Research Ministry.

W. Liu, V.S. Subrahmanian, and J. Wijsen (Eds.): SUM 2013, LNAI 8078, pp. 233–246, 2013.
© Springer-Verlag Berlin Heidelberg 2013

Two different algorithmic problems have been extensively studied in the context of data exchange: given a source instance, materialize a "good solution" for the given instance and answer queries posed over the target schema. If $\mathcal{M}$ is a schema mapping as above and if I is a source instance, then a target instance J is called a *solution for I w.r.t.* $\mathcal{M}$ if $\langle I, J \rangle \models \Sigma_{\text{st}} \cup \Sigma_{\text{t}}$. In general, a source instance may have more than one solutions, so this raises the questions: which solution should be materialized and what are the semantics of target queries? *Universal* solutions, first introduced in [5], have turned out to be the preferred solutions to materialize in the context of data exchange. Moreover, under the condition of *weak acyclicity* [5] or some other mild structural condition on the target tgds, a *canonical* universal solution for a given source instance can be efficiently constructed using the chase procedure. As regards to the second problem, the notion of the *certain answers* has been adopted as semantics of answering queries over the target schema. More precisely, assume that $\mathcal{M}$ is a schema mapping and that for every source instance I, we have a set $\mathcal{W}(\mathcal{M}, I)$ (or, simply, $\mathcal{W}(I)$) of "possible worlds" for I w.r.t. $\mathcal{M}$, that is, a set of solutions of interest for I w.r.t. $\mathcal{M}$. If Q is a query over the target schema, then the *certain answers* of Q on I is the set $certain(Q, I, \mathcal{W}(I)) = \bigcap \{Q(J) | J \in \mathcal{W}(I)\}$.

Several different types of solutions have been considered and studied in depth as possible worlds in the semantics of the certain answers; these include the set of all solutions [5], the set of all universal solutions [4], the set of *CWA*-solutions [9,10], and the set of all *GCWA**-solutions [8]. For (unions of) conjunctive queries, all these different types of possible worlds lead to the same certain answers, but they often lead to different certain answers on non-monotonic queries, that is, queries that involve universal quantification and/or some form of negation. Moreover, when the set of all solutions is used as the set of possible worlds, then the certain answers of non-monotonic queries can be counterintuitive. In fact, this was one of the main motivations for the introduction and study of the *CWA*-solutions and their variants.

The preceding discussion concerns relational queries without aggregate operators. The study of aggregate queries in data exchange was initiated in [1], where the notion of *range semantics* [2] of aggregate queries on inconsistent databases was suitably adapted in the context of data exchange to define the notion of *aggregate certain answers* with respect to a schema mapping $\mathcal{M}$ with no target constraints and a collection $\{\mathcal{W}(\mathcal{M}, I) : I \text{ a source instance}\}$ of possible worlds. In [1], it was pointed out that if the set of all solutions or the set of all *CWA*-solutions is taken as the set of possible worlds, then the resulting aggregate certain answers are rather trivial. For this reason, a different set of possible worlds was proposed as an alternative, namely, the set of all solutions that are endomorphic images of the canonical universal solution obtained using the *oblivious* chase (also known as the *naive* chase). In [1] it was shown that this approach gives rise to non-trivial and meaningful aggregate certain answers. Moreover, the aggregate certain answers of the operators min, max, count, sum, and avg were shown to be computable in polynomial time.

Even though the semantics of aggregate queries based on the endomorphic images of the canonical universal solution have many desirable features, they suffer from the following shortcoming: they are *not* invariant under logical equivalence, which means that they depend on the concrete syntactic representation of the s-t tgds defining the schema mapping at hand[1].

Example 1. Consider the logically equivalent schema mappings $\mathcal{M}_1 = (\{P(A)\}, \{R(B,C)\}, \{\sigma_1\}, \emptyset)$ and $\mathcal{M}_n = (\{P(A)\}, \{R(B,C)\}, \{\sigma_n\}, \emptyset)$, $n \geq 2$, where

$$\sigma_1 : \forall x (P(x) \to \exists y\ R(x,y))\ and$$
$$\sigma_n : \forall x (P(x) \to \exists y_1 \ldots y_n\ R(x,y_1), \ldots, R(x,y_n))$$

and the database instance $D = \{P(1)\}$. We have that, for example, the aggregate certain answers for $\texttt{count}(R.B)$ is $[1,1]$ on $\mathcal{M}_1$, while they are $[1,n]$ on $\mathcal{M}_n$. $\square$

In [7], it was shown that every schema mapping specified by s-t tgds only can be transformed via rewrite rules into a logically equivalent one that is also specified by s-t tgds only, is *minimal* with respect to four particular criteria, and is unique up to variable renaming. This transformation can thus be thought of as a *normal form* for a schema mapping specified by s-t tgds. In [7], this normal form was used to give unambiguous semantics for the aggregate certain answers in data exchange. Specifically, given a schema mapping $\mathcal{M}$ specified by s-t tgds, one first obtains a logically equivalent schema mapping $\mathcal{M}_{NF}$ in normal form and then uses the certain aggregate answers on $\mathcal{M}_{NF}$ with the endomorphic images of the canonical universal solution w.r.t. $\mathcal{M}_{NF}$ as possible worlds. This approach bypasses the dependence of the semantics of the concrete syntactic representation of the schema mapping at hand; however, it raises the question as to whether there is some other, and perhaps deeper, justification for the particular choice of the normal form. Furthermore, there is a broader question concerning the semantics of aggregate queries for schema mappings with target tgds, since, for, among other difficulties, no satisfactory normal form for such schema mappings has been found thus far.

In this paper, we revisit the semantics of aggregate queries in data exchange aiming to address the two preceding questions. To this effect, we introduce a new semantics for aggregate queries in data exchange, which we call the *aggregate most-certain answers*. Informally, the aggregate most-certain answers are obtained by taking the intersection of the aggregate certain answers over all schema mappings that are logically equivalent to the schema mapping at hand. Clearly, this semantics is invariant under logical equivalence and does not assume the existence of a normal form. Our main technical result is that for schema mappings specified by s-t tgds only (no target constraints) the aggregate most-certain answers w.r.t. a schema mapping $\mathcal{M}$ coincide with the aggregate certain answers w.r.t. the schema mapping $\mathcal{M}_{NF}$ in normal form associated

---

[1] Note that the certain answers of non-monotonic queries based on *CWA*-solutions suffer from the same shortcoming.

with $\mathcal{M}$. This result provides an intrinsic justification for using schema mappings in normal form and also implies that the aggregate most-certain answers are computable in polynomial time.

After this, we consider the semantics of aggregate queries w.r.t. schema mappings specified by s-t tgds and target tgds. For such schema mappings, it is well known that solutions are not closed under endomorphisms, but are closed under retractions (see, e.g., [6]). For this reason, we propose to use the retractions of the canonical universal solution as possible worlds, instead of using the endomorphic images of it (since some of these endomorphic images may not even be solutions). We discuss some of the delicate issues involved in defining rigorous semantics based on retractions for schema mappings specified by s-t tgds and tgds. Finally, we compare the semantics based on retractions against the semantics based on endomorphic images in the case of schema mappings specified by s-t tgds only. In particular, the semantics based on retractions coincide with those based on endomorphic images for queries involving `max`, `min`, and `count` (hence, they can be computed in polynomial time), but, in general, differ on queries involving `sum` and `avg`. However, we show that for queries involving `sum` and `avg`, the aggregate certain answers and the aggregate most-certain answers with retractions as possible worlds can be computed in polynomial time.

2   Preliminaries

Let $Const$ be an infinite set of *constants* and let $Null$ be an infinite set of *nulls* that is disjoint from $Const$. All values in source instances are assumed to be constants. In contrast, target instances have values from $Const \cup Null$.

Homomorphisms, Endomorphisms, and Retractions. Let K and J be two instances over a relational schema $\mathbf{R}$ with values in $Const \cup Null$. A *homomorphism* $h : K \to J$ is a mapping from the active domain $adom(K)$ to the active domain $adom(J)$ such that: (1) $h(c) = c$, for every $c \in Const$; and (2) for every fact $R_i(t)$ of K, we have that $R_i(h(t))$ is a fact of J, where if $t = (a_1, ..., a_s)$, then $h(t) = (h(a_1), ..., h(a_s))$. An *endomorphism* of J is a homomorphism from J to J. Clearly, the composition of two endomorphisms is also an endomorphism. A subinstance $J' \subseteq J$ is called a *retract* of J if there is a homomorphism $h : J \to J'$ such that for all $a \in adom(J')$, $h(a) = a$. Such a homomorphism is called a *retraction*. Note that the composition $h' \circ h$ of two retractions $h : J \to J'$ and $h' : J' \to J''$ is also a retraction. A target instance J is said to be a *core* if there is no proper subinstance $J' \subseteq J$ and homomorphism $h : J \to J'$.

Schema Mappings, Universal Solutions, Oblivious Chase. Let $\mathbf{S}$ be a source schema and $\mathbf{T}$ a target schema. A *source-to-target tuple-generating dependency* (s-t tgd) is an expression of the form $\forall \mathbf{x} \forall \mathbf{z}(\phi_{\mathbf{S}}(\mathbf{x}, \mathbf{z}) \to \exists \mathbf{y} \psi_{\mathbf{T}}(\mathbf{x}, \mathbf{y}))$, where $\phi_{\mathbf{S}}(\mathbf{x}, \mathbf{z})$ is a conjunction of atoms over $\mathbf{S}$, and $\psi_{\mathbf{T}}(\mathbf{x}, \mathbf{y}))$ is a conjunction of atoms over $\mathbf{T}$. A *target tuple-generating-dependency* (target tgd) is an expression of the form $\forall \mathbf{x} \forall \mathbf{z}(\phi_{\mathbf{T}}(\mathbf{x}, \mathbf{z}) \to \exists \mathbf{y} \, \psi_{\mathbf{T}}(\mathbf{x}, \mathbf{y}))$, where $\phi_{\mathbf{T}}(\mathbf{x}, \mathbf{z})$ and $\psi_{\mathbf{T}}(\mathbf{x}, \mathbf{y})$ are conjunctions of atoms over $\mathbf{T}$. Finally, a *target equality-generating dependency* (target egd) is an expression of the form $\forall \mathbf{x}(\phi_{\mathbf{T}}(\mathbf{x}) \to (x_i = x_j))$, where $\phi_{\mathbf{T}}(\mathbf{x})$ is

---

Rule 1 (Core of the conclusion)
$\tau : \phi(\mathbf{x}) \to (\exists \mathbf{y})\psi(\mathbf{x}, \mathbf{y}) \Rightarrow \tau' : \phi(\mathbf{x}) \to (\exists \mathbf{y})\psi(\mathbf{x}, \mathbf{y}\sigma)$, s.t. $\psi(\mathbf{x}, \mathbf{y}\sigma)$ is the core of $\psi(\mathbf{x}, \mathbf{y})$.

Rule 2 (Core of the antecedent)
$\tau : \phi(\mathbf{x}, \mathbf{z}) \to (\exists \mathbf{y})\psi(\mathbf{x}, \mathbf{y}) \Rightarrow \tau' : \phi(\mathbf{x}, \mathbf{z}\sigma) \to (\exists \mathbf{y})\psi(\mathbf{x}, \mathbf{y})$,
s.t. $\phi(\mathbf{x}, \mathbf{z}\sigma)$ is the core of $\phi(\mathbf{x}, \mathbf{z})$.

Rule 3 (Splitting)
$\tau : \phi(\mathbf{x}) \to (\exists \mathbf{y})\psi(\mathbf{x}, \mathbf{y}) \Rightarrow \{\tau_1, \ldots, \tau_n\}$, where $\{\psi_1(\mathbf{x}, \mathbf{y}_1), \ldots, \psi_n(\mathbf{x}, \mathbf{y}_n)\}$ are the
components of $\psi(\mathbf{x}, \mathbf{y})$, and $\tau_i : \phi(\mathbf{x}) \to (\exists \mathbf{y}_i)\psi_i(\mathbf{x}, \mathbf{y}_i)$, for $i \in \{1, \ldots, n\}$.

Rule 4 (Implication of a s-t tgd)
$\Sigma \Rightarrow \Sigma \setminus \{\tau\}$ if $\Sigma \setminus \{\tau\} \models \tau$.

Rule 5 (Implication of atoms in the conclusion)
$\Sigma \Rightarrow (\Sigma \setminus \{\tau\}) \cup \{\tau'\}$ if $\tau : \phi(\mathbf{x}) \to (\exists \mathbf{y})\psi(\mathbf{x}, \mathbf{y})$ and $\tau' : \phi(\mathbf{x}) \to (\exists \mathbf{y}')\psi'(\mathbf{x}, \mathbf{y}')$,
s.t. $At(\psi'(\mathbf{x}, \mathbf{y}')) \subset At(\psi(\mathbf{x}, \mathbf{y}))$ and $(\Sigma \setminus \{\tau\}) \cup \{\tau'\} \models \tau$.

Fig. 1. Rules for redundancy elimination from s-t TGDs

a conjunction of atoms over $\mathbf{T}$. In what follows, we will often omit the universal quantifiers in front of s-t tgds, target tgds, and target egds.

Let $\mathcal{M} = (\mathbf{S}, \mathbf{T}, \Sigma_{\mathrm{st}}, \Sigma_{\mathrm{t}})$ be a schema mapping, where Σ_{st} is a set of s-t tgds and Σ_{t} is a set of target tgds and target egds. If I is a source instance, then a *universal* solution [5] for I is a solution J such that for every solution J' of J, there is a homomorphism from J to J'. In general, a source instance may have no universal solutions, even if it has solutions. However, if the set of target tgds is *weakly acyclic* [5], then universal solutions exist if and only if solutions exist; moreover, if a solution exists, then a universal solution can be efficiently constructed using the *standard chase* procedure (see [5] for details). In particular, if the schema mapping has no target constraints (i.e., it is of the form $\mathcal{M} = (\mathbf{S}, \mathbf{T}, \Sigma_{\mathrm{st}}, \emptyset)$), then, given a source instance I, a *canonical universal* solution can be constructed using the *oblivious* (also known as the *naive*) chase [3,11]. The oblivious chase is the variant of the chase procedure in which, when a s-t tgd is triggered, then new nulls are introduced each time to witness the existential quantifiers in the right-hand side of that s-t tgd. For a given instance I, the oblivious chase produces a unique (up to renaming of nulls) canonical universal solution, which will be denoted as $CanSol(\mathcal{M}, I)$ or, simply, $CanSol(I)$ whenever $\mathcal{M}$ is understood from the context.

Logical Equivalence and Normal Forms. We will often identify a schema mapping $\mathcal{M} = (\mathbf{S}, \mathbf{T}, \Sigma_{\mathrm{st}}, \Sigma_{\mathrm{t}})$ with the set of dependencies $\Sigma = \Sigma_{\mathrm{st}} \cup \Sigma_{\mathrm{t}}$, without explicitly mentioning the schemas. Two schema mappings Σ and Σ' over $\langle \mathbf{S}, \mathbf{T} \rangle$ are *logically equivalent* (denoted as $\Sigma \equiv \Sigma'$) if for every source instance I and target instance J, we have that $\langle I, J \rangle \models \Sigma \Leftrightarrow \langle I, J \rangle \models \Sigma'$.

Let Σ_{st} be a set of s-t tgds. A normal form $\Sigma_{\mathrm{st}}^{NF}$ for Σ_{st} that is logically equivalent to Σ_{st} and unique up to variable renaming has been defined in [7].

Definition 1. [7] Let Σ_{st} be a set of s-t TGDs. The *s-t tgds normal form* $\Sigma_{\mathrm{st}}^{NF}$ associated with Σ_{st} is the set of s-t tgds obtained from Σ_{st} by the exhaustive application of the five rules in Figure 1. $\quad\square$

We now explain briefly some of the notions used in Figure 1. Detailed definitions can be found in [7]. A *substitution* σ is a function that sends variables to other domain elements (i.e., variables or constants). We write $\sigma = \{x_1 \leftarrow a_1, \ldots, x_n \leftarrow a_n\}$ if σ maps each x_i to a_i and is the identity outside $\{x_1, \ldots, x_n\}$. The application of a substitution is usually denoted in post-fix notation. Given a conjunctive query $\phi(\mathbf{x})$, we denote by $At(\phi(\mathbf{x}))$ the instance consisting of exactly the atoms of $\phi(\mathbf{x})$. Let $\phi(\mathbf{x}, \mathbf{y})$ be a conjunctive query with variables in $\mathbf{x} \cup \mathbf{y}$ and let $\mathcal{A}$ denote the instance consisting of the atoms $At(\phi(\mathbf{x}, \mathbf{y}))$, where the variables $\mathbf{x}$ are considered as constants and the variables $\mathbf{y}$ as labelled nulls. Let $\mathcal{A}'$ be the core of $\mathcal{A}$ and let $\sigma : \mathbf{y} \to Const \cup \mathbf{x} \cup \mathbf{y}$ be a substitution such that $At(\phi(\mathbf{x}, \mathbf{y}\sigma)) = \mathcal{A}'$. Then, the *core of* $\phi(\mathbf{x}, \mathbf{y})$ is defined as the conjunctive query $\phi(\mathbf{x}, \mathbf{y}\sigma)$.

In [7], it was shown that the normal form $\Sigma_{\mathrm{st}}^{NF}$ associated with a set Σ_{st} of s-t tgds is optimal in the sense that it is minimal with respect to the following criteria: number of tgds, antecedent and conclusion size, and number of existentially quantified variables in the conclusion.

Answering Aggregate Queries in Data Exchange. We now recall the semantics for answering aggregate queries in data exchange proposed in [1].

Definition 2. Let $\mathcal{M}$ be a schema mapping and suppose that, for every source instance I, we have a set $\mathcal{W}(I)$ of solutions of I w.r.t $\mathcal{M}$. Let Q be a query of the form SELECT $\mathbf{f}(R.A)$ FROM R, where R is a first-order query over the target schema T, A is an attribute of R, and $\mathbf{f} \in \{\min, \max, \mathrm{count}, \mathrm{count}(*), \mathrm{sum}, \mathrm{avg}\}$[2].
(1) A *possible answer* of Q w.r.t. I and $\mathcal{W}(I)$ is a value r s.t. there exists an instance $J \in \mathcal{W}(I)$ for which $Q(J) = r$.
(2) $poss(Q, I, \mathcal{W}(I))$ denotes the set of all possible answers of Q w.r.t. I and $\mathcal{W}(I)$.
(3) For the aggregate query Q, the *aggregate certain answers of Q with respect to I and $\mathcal{W}(I)$*, denoted by $agg\text{-}certain(Q, I, \mathcal{W}(I))$, is the interval
$$[\mathrm{glb}(poss(Q, I, \mathcal{W}(I))), \mathrm{lub}(poss(Q, I, \mathcal{W}(I)))],$$
where glb and lub stand for greatest lower bound and least upper bound. □

Let $\mathcal{M} = (\mathbf{S}, \mathbf{T}, \Sigma_{\mathbf{st}}, \emptyset)$ be a schema mapping with no target constraints. If I is a source instance, then $Endom(I, \mathcal{M})$ (or simply $Endom(I)$) stands for the set of all endomorphic images of $CanSol(\mathcal{M}, I)$. As mentioned earlier, it was shown in [1] that the sets $Endom(I, \mathcal{M})$, I source instance, form a collection of possible worlds that give rise to meaningful and non-trivial semantics of aggregate queries. Moreover, the aggregate certain answers are computable in PTIME.

Theorem 1. *[1] Let $\mathcal{M} = (\mathbf{S}, \mathbf{T}, \Sigma_{\mathbf{st}}, \emptyset)$ be a schema mapping with no target constraints, let R be a CQ over $\mathbf{T}$ or a relational symbol in $\mathbf{T}$, let A be an attribute in R and $\mathbf{f}$ the aggregate operator $min(R.A)$, $max(R.A)$, $count(R.A)$, $count(*)$, $sum(R.A)$, or $avg(R.A)$. Then, $agg\text{-}certain(\mathbf{f}(R.A), I, Endom(I))$ is in PTIME in each of the following cases:*

---

[2] For all aggregate operators but count(*), tuples with a null value in attribute $R.A$ are ignored in the computation.

1. R is a CQ and $\mathtt{f} \in \{min(R.A), max(R.A), \mathtt{count}(R.A), \mathtt{count}(*)\}$.
2. R is a CQ and $\mathtt{f} = sum(R.A)$ and A is an attribute with non negative values only.
3. R is a target relation symbol and $\mathtt{f} = avg\,(R.A)$. □

More specifically, $CanSol(I)$ and its core are sufficient for computing the certain answers to min, max, count, count(*), as well as the certain answers to the special case of sum in which the sum is over an attribute taking non-negative values only; in what follows, we will refer to this special case as special-sum. The case of the avg requires a more sophisticated PTIME algorithm based on the concepts of blocks of nulls and *local endomorphism*. For sum over an attribute taking arbitrary values, a simpler version of the algorithm for avg is used.

3 Aggregate Most-Certain Answers

In this section, we introduce a semantics for aggregate queries in data exchange that is independent from the concrete syntactic representation of the tgds, and, consequently, it is preserved between logically equivalent schema mappings.

Let $\mathbf{S}$ be a source schema and $\mathbf{T}$ a target schema. We will consider schema mappings $\mathcal{M} = (\mathbf{S}, \mathbf{T}, \Sigma_{\mathtt{st}}, \Sigma_{\mathtt{t}})$ and, as before, we will identify each such schema mapping with the set $\Sigma = \Sigma_{\mathtt{st}} \cup \Sigma_{\mathtt{t}}$ of the constraints that define $\mathcal{M}$. We also assume that, for every source instance I and every schema mapping Σ, we have a set $\mathcal{W}(I, \Sigma)$ of solutions for I w.r.t. Σ.

Definition 3. Let Σ be a schema mapping and let Q be a query of the form SELECT $\mathtt{f}(R.A)$ FROM R, where R is a first-order query over the target schema $\mathbf{T}$, A is an attribute of R, and $\mathtt{f} \in \{min, max, \mathtt{count}, \mathtt{count}(*), sum, avg\}$.

If I is a source instance, then the *aggregate most-certain answers* of the query Q on I w.r.t. $\mathcal{W}(\Sigma, I)$ is the set

$$agg\text{-}most\text{-}certain(Q, I, \mathcal{W}(\Sigma, I)) = \bigcap_{\Sigma' \equiv \Sigma} agg\text{-}certain(Q, I, \mathcal{W}(\Sigma', I)) \quad □$$

For every schema mapping $\Sigma' \equiv \Sigma$, we have that the aggregate certain answer $agg\text{-}certain(Q, I, \mathcal{W}(\Sigma', I))$ is the interval of values $[l_{\Sigma'}, u_{\Sigma'}]$, where $l_{\Sigma'} = \mathtt{glb}(poss(Q, I, \mathcal{W}(\Sigma', I)))$ and $u_{\Sigma'} = \mathtt{lub}(poss(Q, I, \mathcal{W}(\Sigma', I)))$. It follows that an equivalent definition of the aggregate most-certain answer is

$$agg\text{-}most\text{-}certain(Q, I, \mathcal{W}(\Sigma, I)) = \bigcap_{\Sigma' \equiv \Sigma}[l_{\Sigma'}, u_{\Sigma'}] = [L, U],$$

where $L = \mathtt{glb}(\{l_{\Sigma'} | \Sigma' \equiv \Sigma\})$ and $U = \mathtt{lub}(\{u_{\Sigma'} | \Sigma' \equiv \Sigma\})$.

Computing the Most-Certain Answers for s-t Tgds. On the face of it, the definition of the aggregate most-certain answers is not effective, since it entails an intersection of infinitely many sets. In what follows, we will show that if Σ is a schema mapping specified by s-t tgds only, then the aggregate most-certain answers w.r.t Σ coincide with the aggregate certain answers w.r.t. the normal form Σ^{NF} associated with Σ. To prove this result, we will analyze separately each normalization rule in Figure 1. We state a series of relevant lemmas, whose proofs will be given in the full version.

Lemma 1. *Let Σ be a set of s-t tgds and Σ' the set of s-t tgds obtained by the application of Rule 2 (core of the antecedent) to Σ. If I is a source instance, then $CanSol(I, \Sigma')$ is a retraction of $CanSol(I, \Sigma)$ (up to renaming of nulls).*   □

Fact 2. *Let Σ be a set of s-t tgds and Σ' the set of s-t tgds obtained by the application of Rule 3 (splitting) to Σ. if I is a source instance, then $CanSol(I, \Sigma')$ is isomorphic to $CanSol(I, \Sigma)$.*   □

Lemma 2. *Let Σ be a set of s-t tgds and Σ' the set of s-t tgds obtained by the application of Rule 5 (implication of atoms in the conclusion) to Σ. If I is a source instance, then $CanSol(I, \Sigma')$ is a retraction of $CanSol(I, \Sigma)$ (up to renaming of nulls).*   □

The next result follows from the fact that Rule 1 is a particular case of Rule 5, since it corresponds to the application of Rule 5 to the singleton $\Sigma = \{\tau\}$.

Corollary 1. *Let Σ be a set of s-t tgds and Σ' the set of s-t tgds obtained by the application of Rule 1 (core of the conclusion) to Σ. If I is a source instance, then $CanSol(I, \Sigma')$ is a retraction of $CanSol(I, \Sigma)$ (up to renaming of nulls).* □

Lemma 3. *Let Σ be a set of s-t tgds and Σ' the set of s-t tgds obtained by the application of Rule 4 (implication of a tgd) to Σ. If I is a source instance, then $CanSol(I, \Sigma')$ is a retraction of $CanSol(I, \Sigma)$ (up to renaming of nulls).*   □

Theorem 3. *Let Σ be a set of s-t tgds, Σ^{NF} its associated normal form, I a source instance and $J_{NF} = CanSol(\Sigma^{NF}, I)$. Consider a set of s-t tgds $\Sigma' \equiv \Sigma$ and $J' = CanSol(\Sigma', I)$. Then J_{NF} is a retraction, hence also an endomorphic image, of J' (up to renaming of nulls).*

Proof. If Σ' is not in normal form, then we can normalize it by applying one or more normalization steps: $\Sigma' \to \Sigma_1 \to \ldots \to \Sigma_k = \Sigma_{NF}$. Let I an instance and $J_i = CanSol(I, \Sigma_i)$ for $i = 1, \ldots, k$. Then, by Lemmas 1, 2 and 3, and Corollary 1, J_1 is a retraction of J' and J_{i+1} is a retraction of J_i for $i = 1, \ldots, k-1$ (all up to renaming of nulls). The result now follows from the fact that the composition of retractions is a retraction.   □

Corollary 2. *Let Σ be a set of s-t tgds and let Σ^{NF} be the normal form of Σ. If I is a source instance, then $Endom(I, \Sigma^{NF}) \subseteq Endom(I, \Sigma)$.*

Proof. This follows from Theorem 3 and the fact that the composition of endomorphisms is an endomorphism.   □

The main theorem of this section now follows easily.

Theorem 4. *Let Σ be a set of s-t tgds, let Σ^{NF} be the associated normal form of Σ, and let Q be a query of the form SELECT $f(R.A)$ FROM R, where R is a first-order query over the target schema $\mathbf{T}$, A is an attribute of R, and $f \in \{min, max, count, count(*), sum, avg\}$. If I is a source instance, then*

$$agg\text{-}most\text{-}certain(Q, I, Endom(I, \Sigma)) = agg\text{-}certain(Q, I, Endom(I, \Sigma^{NF})).$$ □

As mentioned in the Introduction, the preceding Theorem 4 provides an intrinsic justification for working with schema mappings in normal form. Moreover, combined with Theorem 1, it implies that the aggregate most-certain answers are computable in polynomial time.

4   Semantics of Aggregate Queries Based on Retractions

In this section, we consider schema mappings $\mathcal{M} = (\mathbf{S}, \mathbf{T}, \Sigma_{st}, \Sigma_t)$, where Σ_t is a set of target tgds, and explore the delicate issues involved in defining meaningful semantics for aggregate queries over the target schema. The first issue is that, unlike the case of schema mappings with no target dependencies, the semantics based on the endomorphic images of the canonical universal solutions are no longer suitable. The reason is that solutions need not be closed under endomorphisms. In particular, an endomorphic image of the canonical universal solution need not be a solution (see [6] for an example). On the other hand, in the presence of target tgds, solutions are closed under retractions, which form an important special case of endomorphisms. Thus, given a source instance I, it is natural to consider the set of all retracts of the canonical universal solution for I as possible worlds, and then base the semantics of aggregate queries over the target schema on this set.

The second and arguably more delicate issue is whether, given a source instance I, there is a unique (up to renaming of nulls) canonical universal solution. To begin with, if Σ_t is an arbitrary set of target tgds, then the standard chase procedure may not terminate on a given source instance; for example, it is easy to see that this is the case with the target tgd $\forall x \forall y (T(x, y) \rightarrow \exists z T(z, x))$. As mentioned earlier, structural conditions on target tgds (such as weak acyclicity) that guarantee the termination of the standard chase have been considered. However, if the target tgds form a weakly acyclic set, then different runs of the standard chase may produce universal solutions that are not unique (up to renaming of nulls); in fact, this is true even when Σ_t is empty. For this reason, we need to consider the extension of the oblivious chase procedure for schema mappings $\mathcal{M} = (\mathbf{S}, \mathbf{T}, \Sigma_{st}, \Sigma_t)$ specified by s-t tgds and target tgds. In [11], it was shown that if $\mathcal{M}$ is such a schema mapping and if the oblivious chase terminates on some instance I, then the canonical universal solution produced by the oblivious chase is unique (up to renaming of nulls). This raises the question: what are broad sufficient conditions for the termination of the oblivious chase? As pointed out in [11], unlike the standard chase, the oblivious chase need not terminate if the set of target tgds form a weakly acyclic set. To see this, consider the singleton set Σ_t consisting of the target tgd $\forall x \forall y (T(x, y) \rightarrow \exists z T(x, z))$. This set is weakly acyclic, yet the oblivious chase does not terminate on an instance containing the fact $T(1, 2)$, since the oblivious chase produces the infinite sequence of facts $T(1, N_k)$, $k \geq 1$, where each N_k is a null. Nonetheless, it turns out that the oblivious chase always terminates if target tgds form a *richly acyclic* set, a strengthening of weak acyclicity considered in [6,9].

Definition 4. A *position* is a pair (R, i) (which we write as R^i), where R is a relation symbol of arity r and $1 \leq i \leq r$. We say that x *occurs* in R^i in a tgd ϕ if there is an atom of the form $R(\ldots, x, \ldots)$ in ϕ such that x appears in the ith position. The *dependency graph* of a set Σ of target tgds is the directed graph whose vertices are the positions of the relation symbols in Σ and, for every target tgd σ of the form $\phi(\mathbf{x}, \mathbf{z}) \rightarrow \exists \mathbf{y} \, \psi(\mathbf{x}, \mathbf{y})$, there is

(1) an edge between R^i and S^j, whenever some $x \in \mathbf{x}$ occurs in R^i in ϕ and in S^j in ϕ, and

(2) an edge between R^i and S^j, whenever some $x \in \mathbf{x}$ appears in R^i in ϕ and some $y \in \mathbf{y}$ occurs in S^j in ψ. Furthermore, these latter edges are labeled with $\exists$, and are called *existential edges*.

We say that a set Σ of target tgds is *richly acyclic* if its dependency graph has no cycle going through an existential edge. $\qquad \square$

Note that this definition differs from the definition of a weakly acyclic set of target tgds, as weak acyclicity stipulates in condition (2) that x must occur in ψ. Thus, every richly acyclic set is also weakly acyclic. Some examples of richly acyclic sets of tgds are sets of full tgds and acyclic sets of tgds.

Theorem 5. *[11] Let $\mathcal{M} = (\mathbf{S}, \mathbf{T}, \Sigma_{\mathtt{st}}, \Sigma_{\mathtt{t}})$ be a schema mapping such that $\Sigma_{\mathtt{t}}$ is a richly acyclic set of target tgds. For every source instance I, the oblivious chase terminates in polynomial time in the size of I, and produces a unique (up to renaming of nulls) canonical universal solution $CanSol(\mathcal{M}, I)$ for I.* $\qquad \square$

Before proceeding further, we note that the preceding Theorem 5 does not extend to schema mappings whose specification also includes target egds.

Example 2. [12] Let $\mathcal{M}$ be the schema mapping specified by the copy s-t tgd $S(x) \to P(x)$, the target tgds σ_1, σ_2, and the target egd σ_3, where

$$\sigma_1 : \ P(x) \to \exists y R(x, y); \quad \sigma_2 : \ R(x, y) \to \exists z T(y, z); \quad \sigma_3 : R(x, y) \to x = y.$$

Consider the source instance $I = \{S(a)\}$. By applying the dependencies in the order $\sigma_1, \sigma_2, \sigma_3, \sigma_1$, the instance produced by the oblivious chase algorithm is $J_1 = \{P(a), R(a, a), T(a, \eta_1), T(a, \eta_2)\}$. If, however, we apply σ_1, σ_3 first, and then σ_2, we obtain the instance $J_2 = \{P(a), R(a, a), T(a, \eta_3)\}$, which is not isomorphic to J_1. $\qquad \square$

Let $\mathcal{M} = (\mathbf{S}, \mathbf{T}, \Sigma_{\mathtt{st}}, \Sigma_{\mathtt{t}})$ be a schema mapping in which $\Sigma_{\mathtt{t}}$ is a finite richly acyclic set of target tgds. If I is a source instance, then $Retract(I, \mathcal{M})$ denotes the set of all retractions of the canonical universal solution $CanSol(I)$ for I. If $\mathcal{M}$ is understood from the context, we simply write $Retract(I)$ instead of $Retract(I, \mathcal{M})$. By using the sets $Retract(I)$ as sets $\mathcal{W}(I)$ of possible worlds in Definition 2 and in Definition 3, we obtain the notions of the aggregate certain answers $agg\text{-}certain(Q, I, Retract(I))$ and of the aggregate most-certain answers $agg\text{-}most\text{-}certain(Q, I, Retract(I))$, for an aggregate query Q.

Theorem 5 implies that the notions of $agg\text{-}certain(Q, I, Retract(I))$ and $agg\text{-}most\text{-}certain(Q, I, Retract(I))$ are well defined. Moreover, it is clear that the aggregate most-certain answers $agg\text{-}most\text{-}certain(Q, I, Retract(I))$ are independent of the syntactic representation of the schema mapping at hand. The following results follows easily from the preceding discussion in this section, the fact that both $CanSol(I)$ and its core belongs to $Retract(I)$, and the fact that the core of $CanSol(I)$ can be computed in polynomial time for a weakly acyclic set of tgds [6] (hence also for a richly acyclic one).

Proposition 1. *Let $\mathcal{M} = (\mathbf{S}, \mathbf{T}, \Sigma_{st}, \Sigma_t)$ be a schema mapping where Σ_t is a richly acyclic set of target tgds, let Q be a conjunctive query over $\mathbf{T}$, and let $\mathtt{f}$ be one of the aggregate operator $\mathtt{min}(A)$, $\mathtt{max}(A)$, $\mathtt{count}(A)$, $\mathtt{count}(*)(A)$ or $\mathtt{special\text{-}sum}(A)$, where A is an attribute of Q. Then $agg\text{-}certain(\mathtt{f}(Q), I, Retract(I))$ can be computed in PTIME.* □

It is an open problem to determine whether $agg\text{-}certain(\mathtt{avg}(Q), I, Retract(I))$ is in PTIME. The same question is also open for the aggregate most-certain answers $agg\text{-}most\text{-}certain(\mathtt{f}(Q), I, Retract(I))$. Here, we note that the existence of a unique normal form for schema mappings specified by s-t tgds only does not extend (at least in a straightforward way) to schema mappings specified by s-t tgds and a richly acyclic set of target tgds. We refer the reader to the full paper where it is shown that the most natural modification of the normalization rules to handle target full tgds does not give rise to a unique normal form.

4.1 Endomorphism-Based Semantics vs. Retraction-Based Semantics

If $\mathcal{M} = (\mathbf{S}, \mathbf{T}, \Sigma_{st}, \emptyset)$ is a schema mapping specified by s-t tgds only, then aggregate queries can be given meaningful semantics using either the endomorphic images or the retracts of the canonical universal solution. Thus, it is natural to ask: how do the endomorphism-based semantics and the retraction-based semantics compare in this case?

We begin by pointing out that, as is the case with $Endom(I)$, both $CanSol(I)$ and its core are members of $Retract(I)$. Moreover, since $Retract(I) \subseteq Endom(I)$, we have that every member of $Retract(I)$ is a subinstance of $CanSol(I)$ and also a CWA-solution for I. Since both $CanSol(I)$ and its core are members of $Retract(I)$, it follows that the the endomorphism-based semantics and the retraction-based semantics give rise to the same certain answers for aggregate queries with $\mathtt{min}$, $\mathtt{max}$, $\mathtt{count}$, $\mathtt{count}(*)$, and $\mathtt{special\text{-}sum}$ (as explained earlier, $\mathtt{special\text{-}sum}$ is the special case of $\mathtt{sum}$ in which the sum is over an attribute taking non-negative values only). By Theorem 4, this also holds for the most-certain answers. Thus, we have the following result.

Proposition 2. *Let $\mathcal{M} = (\mathbf{S}, \mathbf{T}, \Sigma_{st}, \emptyset)$ be a schema mapping specified by s-t tgds only, let Q be a conjunctive query over $\mathbf{T}$, and let $\mathtt{f}$ be one of the aggregate operator $\mathtt{min}(A)$, $\mathtt{max}(A)$, $\mathtt{count}(A)$, $\mathtt{count}(*)(A)$ or $\mathtt{special\text{-}sum}$, where A is an attribute of Q. Then*
$$agg\text{-}certain(\mathtt{f}(Q), I, Retract(I)) = agg\text{-}certain(\mathtt{f}(Q), I, Endom(I)), \text{ and}$$
$$agg\text{-}most\text{-}certain(\mathtt{f}(Q), I, Retract(I)) = agg\text{-}most\text{-}certain(\mathtt{f}(Q), I, Endom(I)).$$
□

In contrast, for aggregate queries with average and (general) sum, the endomorphisms and the retractions may give rise to different certain answers.

Proposition 3. *Let $\mathtt{f}$ be the $\mathtt{avg}$ or the $\mathtt{sum}$ aggregate operator. If $\mathcal{M} = (\mathbf{S}, \mathbf{T}, \Sigma_{st}, \emptyset)$ is a schema mapping specified by s-t tgds only, then*

1. $agg\text{-}certain(\mathtt{f}(Q), I, Retract(I)) \subseteq agg\text{-}certain(\mathtt{f}(Q), I, Endom(I))$;

2. $agg\text{-}most\text{-}certain(\mathtt{f}(Q), I, Retract(I)) \subseteq agg\text{-}most\text{-}certain(\mathtt{f}(Q), I, Endom(I))$.

Moreover, these containments may be proper.

Proof. The containments follow from the fact that $Retract(I) \subseteq Endom(I)$, for every source instance I. To show that the containment for the certain answers of the average operator can be proper, consider the schema mapping $\mathcal{M}$ specified by the s-t tgds $R(x,y) \to T(x,y), T(y,y); \quad P(x,y) \to T(x,w_1), T(w_1,w_2), T(y,w1)$ and the source instance $I = \{R(1,2), P(1,2)\}$. We have that

$$J = CanSol(I) = \{R(1,2), P(1,2), T(1,2), T(2,2), T(1,\eta_1), T(\eta_1,\eta_2), T(2,\eta_1)\},$$

$Endom(J) = \{J, J_1, J_{core}\}$, and $Retract(J) = \{J, J_{core}\}$, where

$$J_1 = \{R(1,2), P(1,2), T(1,2), T(2,2), T(2,\eta_1)\} \text{ and}$$

$$J_{core} = \{R(1,2), P(1,2), T(1,2), T(2,2)\}.$$

It easy to check that $agg\text{-}certain(\mathtt{avg}(T.A), I, Endom(I)) = [\frac{3}{2}, \frac{5}{3}]$, while $agg\text{-}certain(\mathtt{avg}(T.A), I, Retract(I)) = [\frac{3}{2}, \frac{3}{2}]$. $\qquad\square$

Computing the Average under Retraction-Based Semantics. As mentioned earlier, in the case of the endomorphism-based semantics, a polynomial-time algorithm for computing $\mathtt{avg}(R.A)$, where R is a target relation and A an its attribute, was given in [1]. Here, we show that this algorithm can be adapted to the case of the retraction-based semantics.

At this point, let us recall the concept of block homomorphism. The *Gaifman graph of the nulls of K* is an undirected graph in which the nodes are all the nulls of K, and there exists an edge between two nulls if there exists some tuple in K in which both nulls occur. A *block* of nulls is the set of nulls in a connected component of the Gaifman graph of the nulls. A *block homomorphism* for a block B is a homomorphism from $K[B]$ to K, where $K[B]$ denotes the subinstance of K induced by the nulls of B and the constants of K.

The algorithm for the average given in [1] is based on the following two facts: (a) the union of block homomorphisms, one for each block of the canonical universal solution J, is an endomorphic image of J; (b) given an endomorphism h for J, there exists a set of block homomorphisms such that h is equal to their union (Proposition 5.2 in [1]). Then, the research space for the optimal endomorphic images is given by those block homomorphisms that allow to define all the endomorphic images of J.

In order to work with retracts instead of endomorphic images, we have to restrict the type of block homomorphism that we have to consider, so that their combinations give only retractions.

Definition 5. A *block retraction* is a block homomorphisms $h_r : K[B] \to K$ such that $h_r(x) = x$, for all $x \in h_r(K[B]) \cap K[B]$. $\qquad\square$

The previous definition stipulates that if a null of B also appears in $h_r(K[B])$, then it must be mapped to itself.

Example 3. Consider the schema mapping $\mathcal{M}$ in the proof of Proposition 3. We have only one block containing η_1 and η_2. The target instance J_1 is generated via the endomorphism $h_1 = \{\eta_1/2, \eta_2/\eta_1\}$. Observe that h_1 is not a block retraction, so J_1 will not be taken into account as a possible world. □

Proposition 4. *Let K be a target instance with b many blocks.*

1. *If h is a retraction of K, then h is a union of block retractions.*
2. *If h is a union of block retractions, then $h(K)$ is isomorphic to a retraction of K.*

Proof. For the first part, for each block B of K, take the restriction h_B of h to B. Since h is a retraction, we have that $h_B(x) = x$, for all $x \in h_B(K[B]) \cap K[B]$. The union of these block retractions is equal to h.

For the second part, since a union of block homomorphisms is an endomorphism and a block retraction is also a block homomorphism, we have that $h(K)$ is an endomorphism of K. If $h(K)$ is by itself a retraction of K, then the statement holds. Suppose that $h(K)$ is not a retraction of K. This means that there exists at least one null value η_i in K such that $h(\eta_i) = \eta_j$ with $\eta_i \neq \eta_j$ and $h(\eta_j) \neq \eta_j$ (otherwise $h(K)$ is a retraction). Consider now the homomorphism h', where $h'(\eta_i) = \eta_i$, for all η_i as above in h, and $h'(\eta_i) = h(\eta_i)$, otherwise. Clearly, $h'(K)$ is a retraction for K. Moreover, by means of h', we are substituting $h(\eta_i)$ with a null value that is different from $h(\eta_j)$. In fact, we have that if $h(n_j) \neq n_i$, then $h'(n_j) \neq h'(n_i)$ and if $h(n_j) = n_i$, then $h(n_j) = n_i = h'(n_i)$, and vice versa $(h'(n_i) = n_i = h(n_j))$.

It is now easy to see that $h(K)$ and $h'(K)$ are isomorphic, as it is sufficient to rename $h(\eta_i)$ as $h'(\eta_i)$, and vice versa. □

Example 4. Consider the instance $I = \{N(a), P(a)\}$ and the schema mapping

$$\Sigma_{\mathsf{st}} = \{N(x) \to R(x, y, z), \ P(x) \to R(x, y, z)\}$$

Let $K = CanSol(I, \Sigma_{st}) = \{R(a, \eta_1, \eta_2), R(a, \eta_3, \eta_4)\}$. We have two blocks in K, namely, $B_1 = \{\eta_1, \eta_2\}$ and $B_2 = \{\eta_3, \eta_4\}$. Then $h_{B_1} = \{\eta_1/\eta_3, \eta_2/\eta_4\}$ and $h_{B_2} = \{\eta_3/\eta_1, \eta_4/\eta_2\}$ are two block retractions, their union is not a retraction for K, but it is isomorphic to a retraction, namely to K itself. □

Theorem 6. *Let $\mathcal{M} = (\mathbf{S}, \mathbf{T}, \Sigma_{\mathsf{st}}, \emptyset)$ be a schema mapping specified by s-t tgds only, let R be a target relation symbol, and let A be an attribute of R. Then agg-certain($\mathsf{avg}(R.A), I, Retract(I)$) and agg-most-certain($\mathsf{avg}(R.A), I, Retract(I)$) are computable in PTIME.*

Proof. Let $\mathcal{M} = (\mathbf{S}, \mathbf{T}, \Sigma_{\mathsf{st}}, \emptyset)$ be a schema mapping specified by s-t tgds only. The answer to the query $\mathsf{avg}(R.A)$, where R is a target relation symbol, can be computed in PTIME by using the same algorithm for the average given by [1], but by considering block retractions instead of block homomorphisms and using the fact that the algorithm for the average in [1] runs in polynomial time. □

5 Concluding Remarks

In this paper, we revisited the semantics for aggregate queries in the context of data exchange by introducing and investigating the notion of the aggregate most-certain answers. For schema mappings specified by s-t tgds only, we obtained a fairly complete picture for these semantics by relating them to the most-certain answers for aggregate queries with respect to schema mappings in normal form. We also discussed some of the challenges and subtleties involved in extending the semantics of aggregate queries to schema mappings whose specification also includes target constraints. Several problems remain open for such schema mappings. One of them is whether or not the aggregate most-certain answers under retraction-based semantics are computable in polynomial time for schema mappings specified by s-t tgds and a richly acyclic set of target tgds. Another problem, which will require both conceptual and technical advances, is the study of the semantics of aggregate queries for schema mappings whose target constraints also include key constraints or, more generally, target equality-generating dependencies.

References

1. Afrati, F.N., Kolaitis, P.G.: Answering aggregate queries in data exchange. In: PODS, pp. 129–138 (2008)
2. Arenas, M., Bertossi, L.E., Chomicki, J., He, X., Raghavan, V., Spinrad, J.: Scalar aggregation in inconsistent databases. Theor. Comput. Sci. 296(3), 405–434 (2003)
3. Calì, A., Gottlob, G., Kifer, M.: Taming the infinite chase: Query answering under expressive relational constraints. In: Description Logics (2008)
4. Fagin, R., Kolaitis, P.G., Popa, L.: Data Exchange: Getting to the Core. ACM Transactions on Database Systems (TODS) 30(1), 174–210 (2005); A preliminary version of this paper appeared in the 2003 PODS conference
5. Fagin, R., Kolaitis, P.G., Miller, R.J., Popa, L.: Data exchange: semantics and query answering. Theor. Comput. Sci. 336(1), 89–124 (2005)
6. Gottlob, G., Nash, A.: Efficient core computation in data exchange. J. ACM 55(2) (2008)
7. Gottlob, G., Pichler, R., Savenkov, V.: Normalization and optimization of schema mappings. VLDB J. 20(2), 277–302 (2011)
8. Hernich, A.: Answering non-monotonic queries in relational data exchange. In: ICDT, pp. 143–154 (2010)
9. Hernich, A., Schweikardt, N.: CWA-solutions for data exchange settings with target dependencies. In: PODS, pp. 113–122 (2007)
10. Libkin, L.: Data exchange and incomplete information. In: PODS, pp. 60–69 (2006)
11. Onet, A.: The chase procedure and its applications. PhD thesis, Concordia University (2012)
12. Onet, A.: The chase procedure and its applications to data exchange. In: Kolaitis, P.G., Lenzerini, M., Schweikardt, N. (eds.) Data Exchange, Integration, and Streams. Dagstuhl Follow-Ups, Schloss Dagstuhl - Leibniz-Zentrum für Informatik, Germany (to appear, 2013)

PossDB: An Uncertainty Database Management System

Gösta Grahne, Adrian Onet, and Nihat Tartal

Concordia University, Montreal, QC, H3G 1M8, Canada
{grahne,a_onet,m_tartal}@cs.concordia.ca

1 Introduction

Management of uncertain and imprecise data has long been recognized as an important direction of research in data bases. With the tremendous growth of information stored and shared over the Internet, and the introduction of new technologies able to capture and transmit information, it has become increasingly important for Data Base Management Systems to be able to handle uncertain and probabilistic data. As a consequence, there has lately been significant efforts by the database research community to develop new systems able to deal with uncertainty, either by annotating values with probabilistic measures or defining new structures capable of capturing missing information (e.g. Trio [3] and MayBMS [2]).

Uncertainty management is an important topic also in data exchange and information integration. In these scenarios the data stored in one database has to be restructured to fit the schema of a different database. The restructuring forces the introduction of "null" values in the translated data, since the second schema can contain columns not present in the first. In the currently commercially available relational DBMS's the missing or unknown information is stored with a placeholder value here denoted null. It is well known that this representation has drawbacks when it comes to query answering, and that a logically coherent treatment of the null is still lacking from most DBMS's.

To illustrate the above mentioned drawbacks, consider a merger of companies "Acme" and "Ajax." Both companies keep an employee database. Let $Emp1(Name,Mstat,Dept)$ and $Emp2(Name,Gender, Mstat)$, where $Mstat$ stands for marital status, be the schemas used by Acme and Ajax, respectively. The merged company decides to use the schema $Emp(Name,Gender,Mstat,Dept)$, and it is known that all the employees from Ajax will work under the same department, which will either be 'IT' or 'PR'. Consider now the initial data from both companies:

Emp1		
Name	Mstat	Dept
Alice	married	IT
Bob	married	HR
Cecilia	married	HR

Emp2		
Name	Gender	Mstat
David	M	married
Ella	F	single

W. Liu, V.S. Subrahmanian, and J. Wijsen (Eds.): SUM 2013, LNAI 8078, pp. 247–254, 2013.

The merged company database instance would be represented as the following database in a standard relational DBMS:

Emp			
Name	Gender	Mstat	Dept
Alice	null	married	IT
Bob	null	married	HR
Cecilia	null	married	HR
David	M	married	null
Ella	F	single	null

With this incomplete database consider now the following two simple queries:

Q_1: SELECT Name FROM Emp WHERE
   (Gender = 'M' AND Mstat = 'married') OR Gender = 'F'

Q_2: SELECT E.Name, F.Name FROM Emp E, Emp F
       WHERE E.Dept=F.Dept AND E.Name != F.Name

Having in mind $Emp1$ and $Emp2$, one would expect the first query to return all employee names and the second query to return the set of tuples $\{(Bob,\ Cecilia),$ $(David,\ Ella)\}$. Unfortunately by the default way null values are treated in standard systems the tuples returned by the first query would return the set $\{(David,\ Ella)\}$ and the second query would return the set $\{(Bob,\ Cecilia)\}$.

2   PossDB and Conditional Tables

In this paper we introduce a new database management system called PossDB (Possibility Data Base) able to fully support incomplete information. The purpose of the PossDB system is to demonstrate that scalable processing of semantically meaningful null values is indeed possible, and can be built on top of a standard DBMS (PostgreSQL in our case).

Irrespectively of how an incomplete database instance $\mathcal{I}$ is represented, conceptually it is a (finite or infinite) *set* of possible complete database instances I (i.e. databases without null values), denoted $Poss(\mathcal{I})$. Each $I \in Poss(\mathcal{I})$ is called a *possible world* of $\mathcal{I}$. A query Q over a complete instance I gives a complete instance $Q(I)$ as answer. For incomplete databases there are three semantics for query answers:

1. *The exact answer.* The answer is (conceptually) a set of complete instances, each obtained by querying a possible world of $\mathcal{I}$, i.e. $\{Q(I) : I \in Poss(\mathcal{I})\}$. The answer should be represented in the same way as the input database, e.g. as a relation with meaningful nulls.

2. *The certain answer.* This answer is a complete database containing only the (complete) tuples that appear in in the query answer in *all* possible worlds. In other words, $Cert(Q(\mathcal{I})) = \bigcap_{I \in Poss(\mathcal{I})} Q(I)$.

3. *The possible answer.* $Poss(Q(\mathcal{I})) = \bigcup_{I \in Poss(\mathcal{I})} Q(I)$.

The PossDB system is based on conditional tables (c-tables) [6] which generalize relations in three ways. First, in the entries in the columns, *variables*, representing unknown values are allowed in addition to the usual *constants*. The same variable may occur in several entries, and it represents the *same* unknown value wherever it occurs. A c-table T represents a set of complete instances, each obtained by substituting each variable with a constant, that is, applying a valuation v to the table, where v is a mapping from the variables to constants. Each valuation v then gives rise to a possible world $v(T)$. The second generalization is that each tuple t is associated with a *local condition* $\varphi(t)$, which is a Boolean formula over equalities between constants and variables, or variables and variables. The final generalization introduces a *global condition* $\Phi(T)$, which has the same form as the local conditions. In obtaining complete instances from a table T, we consider only those valuations v, for which $v(\Phi(T))$ evaluates to *True*, and include in $v(T)$ only tuples $v(t)$, where $v(\varphi(t))$ evaluates to *True*.

In our previous example the merged incomplete database would be represented as the following c-table:

Emp				
Name	Gender	Mstat	Dept	$\varphi(t)$
Alice	x_1	married	IT	*True*
Bob	x_2	married	HR	*True*
Cecilia	x_3	married	HR	*True*
David	M	married	x_4	*True*
Ella	F	single	x_4	*True*

The global condition $\Phi(\text{Emp})$ is $(x_i = \text{'M'}) \vee (x_i = \text{'F'})$, for $i = 1, 2, 3$, and $(x_4 = \text{'IT'}) \vee (x_4 = \text{'PR'})$. Under this interpretation PossDB will return the expected results for both queries. Note that in this example the exact, possible, and certain answers are the same.

The c-tables support the full relational algebra [6], and are capable of returning the possible, the certain and the exact answers. A (complete) tuple t is in the possible answer to a query Q, if $t \in Q(v(T))$ for *some* valuation v, and t is in the certain answer if $t \in Q(v(T))$ for *all* valuations v. The exact answer of a query Q on a c-table T is a c-table $Q(T)$ such that $v(Q(T)) = Q(v(T))$, for all valuations v.

C-tables are the oldest and most fundamental instance of *semiring-labeled* databases [5]. By choosing the appropriate semiring, labeled databases can model a variety of phenomena in addition to incomplete information. Examples are probabilistic databases, various forms of database provenance, databases with bag semantics, etc. It is our view that the experiences obtained from the PossDB project will also be applicable to other semiring based databases.

To the best of our knowledge, PossDB is the first implemented system based on c-tables. In the future we plan to extend our system to support the Conditional Chase [4], a functionality which is highly relevant in data exchange and information integration.

In order to gauge the scalability of our system, we have run some experiments comparing the performance of PossDB with MayBMS [2]. The MayBMS system uses a representation mechanism called *World Set Decompositions*, which is fundamentally different from c-tables. For details we refer to [2]. Similarly to PossDB, the MayBMS system is build on top of PostgreSQL, an open source relational database management system. In the case where there is no incomplete information, both PossDB and MayBMS work exactly like classical DBMS's. However, at this point we have restricted, similarly to MayBMS, our data to be encoded as positive integers. In the future we will extend the allowed data types to include all standard base data types.

In this current stage PossDB allows the following operations: Creation of c-tables, Querying c-tables, Inserting into c-tables, Materializing c-tables representing the exact answers to queries, and Testing for tuple possibility and certainty in c-tables. All these operations are expressed using an extension of the ANSI SQL language, called C-SQL (Conditional SQL).

3 Features of PossDB

The PossDB system has system specific operations and functions related to c-tables. To illustrate these operations, let us continue with the example from the previous section. The global condition in our example is $\Phi(Emp) =_{\text{def}} \{(x_i = \text{'M'} \lor x_i = \text{'F'}) : i = 1, 2, 3\} \cup \{x_4 = \text{'IT'} \lor x_4 = \text{'PR'}\}$. This set corresponds to a CNF formula, where each disjunct contains all possible values for a given variable. It is stored in a hash structure such that for each variable the hash function will return all possible values for that variable. This representation speeds up the processing when checking for contradictory and tautological local conditions.

Next, we present the operations of the PossDB system. Note that none of these operations affect the global condition.

Relational Selection. The select statement generalizes the standard SQL select statement. The generalized select statement will work on c-tables rather than relations with **null**'s. Beside returning the exact answer, the select statement also optimizes the c-table by removing tuples t, where $\varphi(t) \land \Phi(T)$ is a contradiction, and replacing with *True* local conditions of tuples t, where $\Phi(T)$ logically implies $\varphi(t)$.

Consider e.g. the query that returns all employee from the 'IT' department:

	Name	Gender	Mstat	Dept	$\varphi(t)$
SELECT * FROM Emp	Alice	x_1	married	IT	*True*
WHERE Dept = 'IT'	David	M	married	x_4	$x_4 = \text{'IT'}$
	Ella	F	single	x_4	$x_4 = \text{'IT'}$

Note that the query returns a representation of the exact answer, that is a c-table that represents the set of all possible answer instances. This pertains to all query operations in the PossDB system.

Relational Projection. The operation is implemented, as expected, as an extension of the SQL **SELECT** statement.

Relational Join. The join and cross product operations work similarly with their standard SQL counterparts, with the extension that the joined or concatenated tuple will have a local condition that is the conjunction of the local conditions of the two component tuples.

Query Answers. We return the exact answer as a c-table. This is comparable with MayBMS that returns all the tuples that can occur in the query answer on some complete instance corresponding to the input database. This has the drawback that the answer may contain two mutually exclusive tuples. On the other hand PossDB returns a c-table representing the exact answer. In some cases this c-table might have convoluted local conditions, and it might be difficult for the user to understand the structure. In order to overcome this, we have included two special functions IS POSSIBLE and IS CERTAIN described next. Both functions work in polynomial time.

Special Functions. We have two new functions unique to PossDB. These functions are used to query for certainty and possibility of a tuple in a c-table or in the result of a query.

    IS POSSIBLE(Tuple) IN C-Table

The IS POSSIBLE is a Boolean function that takes a tuple Tuple and decides if the tuple is possible in the c-table. Intuitively a tuple is possible in a given c-table if there exists a valuation for the c-table that contains that tuple. The Tuple has to be specified as a list of (Name, Value) pairs. As an example consider the following function call:

    IS POSSIBLE(Name, 'Bob', Gender, 'M', Dept, 'HR') IN Emp

With the data from the previous example the IS POSSIBLE function returns *True*, because the given tuple is possible in the system. However, it is not certain because it depends on the condition ($x_2 = $'M').

    IS CERTAIN(Tuple) IN C-Table

Similarly to IS POSSIBLE the IS CERTAIN function takes as parameter a tuple, and a c-table and returns *True* if the tuple is certain in the given c-table. Certain means that the tuple appears under all possible interpretations of the nulls. The following is an example of the usage of the IS CERTAIN function

    IS CERTAIN (Name, 'Bob', Dept, 'HR') IN Emp

This function returns *True*, because the given tuple appears under any interpretation for the nulls. Note also that the function would return *False* if we also included the Gender column in the query.

4 Implementation

Without loss of generality, the information in our conditional tables are encoded as integers. Positive integers denote constants and negative integers denote nulls. Consequently, without variables, the PossDB system works as a regular RDBMS, and the performance in this case will be the same as that of PostgreSQL. With this encoding we need to be sure that the queries are properly evaluated. Thus, each equality condition of the form $A = c$ part of a C-SQL query, where A is a column name and c a constant, is rewritten as $(A = c \lor A < 0)$ in SQL. This is necessary in order to check that the column A is either constant c or that it represents a null value, here encoded as negative integers. In order to check for satisfiability of a local conditions and its conjunction with the global condition, the local conditions are converted into DNF (Disjunctive Normal Form). To make a faster satisfiability test we store the global condition as hash based representation of its CNF (Conjunctive Normal Form).

After the Satisfiability and Tautology checks, the system decides which tuples to show in the result of the query by adding some annotations in the local condition column. Our application has a GUI capable of generating the query result in a human readable interface. From a technical perspective PossDB is a two-layer system, the application layer built in Java and as a database layer it uses PostgreSQL database engine. When the user types a C-SQL query, the system interprets it and execute it by a series of SQL statements against the database and a series of Java calls needed to make sure that the c-tables are correctly manipulated and displayed to the user.

System Architecture. PossDB system is built on top of PostgreSQL. On the middle tier Java® and ANTLR [7] are being used. ANTLR is used to parse the C-SQL queries and database conditions, while Java is used to implement the C-SQL processing part, displaying the results, evaluating conditions, and connecting to the PostgreSQL database server.

This Java application is working with input and output streams, hence it can be easily ported to the any kind of application server or simply used through a console. The connection between the Java middle tier and PostgreSQL database server is done through JDBC.

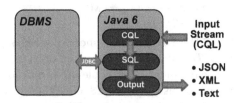

Fig. 1. System Workflow

5   Experimental Results

We compared PossDB with MayBMS, as MayBMS also returns the exact answer to queries, and the scalability of MayBMS has been proven [2]. Furthermore, both PossDB and MayBMS are built on top of PostgeSQL.

Our experiments are based on the queries and data which were used for the MayBMS experimental evaluation [2]. Those experiments used a large census database encrypted as integers [8]. Noise was introduced by replacing some values with variables that could take between 2 and 8 possible values. A noise ratio of $n\%$ meant that $n\%$ of the values were perturbed in this fashion. In our experiments we used the same data and noise generator as MayBMS. The MayBMS system and the noise generator were obtained from [1].

We tested both systems with up to 10 million tuples. The charts below contain the result of the test using queries Q_1 and Q_2 from the experiments in [2]. The results show that PossDB clearly outperforms MayBMS. This is expected because MayBMS needs to perform joins in order to check valuations against the constant values, whereas for PossDB there is no such need.

System Configuration. We conducted all our experiments on Intel®Core™i5-760 processor machine with 8 GB RAM, running Windows 7 Enterprise and PostgreSQL 9.0.

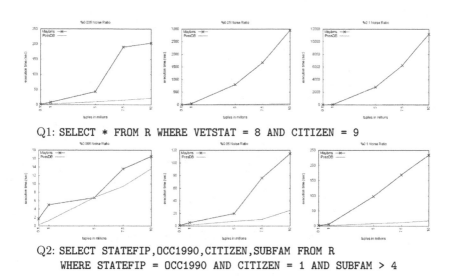

Q1: SELECT * FROM R WHERE VETSTAT = 8 AND CITIZEN = 9

Q2: SELECT STATEFIP,OCC1990,CITIZEN,SUBFAM FROM R
WHERE STATEFIP = OCC1990 AND CITIZEN = 1 AND SUBFAM > 4

For a demonstration of the system and extended version of this paper you can visit http://triptych.encs.concordia.ca/PossDB. In future work we will consider integrity constraints, extend the language to full SQL (e.g. queries involving negation and aggregation) and integrate an optimized SAT-solver.

References

1. Maybms system and the noise generator, http://pdbench.sourceforge.net/
2. Antova, L., Koch, C., Olteanu, D.: 10^10^6 worlds and beyond: efficient representation and processing of incomplete information. The VLDB Journal 18(5), 1021–1040 (2009)
3. Benjelloun, O., Das Sarma, A., Halevy, A., Theobald, M., Widom, J.: Databases with uncertainty and lineage. The VLDB Journal 17(2), 243–264 (2008)
4. Grahne, G., Onet, A.: Closed world chasing. In: Proceedings of the 4th International Workshop on Logic in Databases, LID 2011, pp. 7–14. ACM, New York (2011)

5. Green, T.J., Karvounarakis, G., Tannen, V.: Provenance semirings. In: Proceedings of the Twenty-Sixth ACM SIGMOD-SIGACT-SIGART Symposium on Principles of Database Systems, PODS 2007, pp. 31–40. ACM, New York (2007)
6. Imielinski, T., Lipski, W.: Incomplete information in relational databases. J. ACM 31(4), 761–791 (1984)
7. Parr, T.J., Parr, T.J., Quong, R.W.: Antlr: A predicated-ll(k) parser generator (1995)
8. Ruggles, S.: Integrated public use microdata series: Version 3.0 (2004)

Aggregate Count Queries in Probabilistic Spatio-temporal Databases

John Grant[1], Cristian Molinaro[2], and Francesco Parisi[2]

[1] Towson University and University of Maryland at College Park, USA
jgrant@towson.edu
[2] DIMES Department, Università della Calabria, Italy
{cmolinaro,fparisi}@dimes.unical.it

Abstract. The SPOT database concept was defined several years ago to provide a declarative framework for probabilistic spatio-temporal databases where even the probabilities are uncertain. Earlier work on SPOT focused on the efficient processing of selection queries and updates. In this paper, we deal with aggregate count queries. First, we propose three alternative semantics for the meaning of such a query. Then, we provide polynomial time algorithms for answering count queries under the various semantics and discuss complexity issues.

1 Introduction

Recent years have seen a great deal of interest in tracking moving objects. For this reason, researchers have investigated in detail the representation and processing of spatio-temporal databases (see, for instance, [29,34,20,3,30,18]). However, in many cases the location of objects is uncertain: such cases can be handled by using probabilities [35,10,7,6]. Sometimes the probabilities themselves are uncertain. The SPOT (Spatial PrObabilistic Temporal) database concept was introduced in [26] to provide a declarative framework for the representation and processing of probabilistic spatio-temporal databases with uncertain probabilities. Previous work included a formal syntax and semantics as well as checking for consistency: an object cannot be in two places at the same time. Additional research focused on the efficient processing of selection queries and database updates [27,23,16].

In this paper, we study a different kind of query: the aggregate count query, one that has not been considered previously in the SPOT framework. A count query asks how many objects are in a certain region at a given time. Answering this kind of query is useful in several applications involving probabilistic spatio-temporal data. As an example, a military agency might be interested in counting the number of enemy vehicles that may be in a region at a given time point in order to adequately arrange its defense line. As a second example, a cell phone provider might be interested in knowing load on its cell towers by determining the number of cell phones that will be in the range of some towers at a given time. As a third example, a transportation company might be interested in predicting the number of vehicles that will be on a given road at a given time in order avoid congestion.

We provide three alternative semantics for interpreting count queries: the *expected value semantics*, the *extreme values semantics*, and the *ranking semantics*. For a query

W. Liu, V.S. Subrahmanian, and J. Wijsen (Eds.): SUM 2013, LNAI 8078, pp. 255–268, 2013.

region r and a time t, the first semantics looks at the minimum and maximum expected values of the number of objects that can be inside region r at time t; the second semantics returns the lowest and the highest numbers of objects that can be inside region r at time t; the last semantics returns a confidence interval for each number of objects that may be inside r at time t. A user can choose the most suitable semantics for the application of interest. We propose polynomial time algorithms for evaluating count queries under these three semantics, discuss relationships among them and complexity issues.

2   SPOT Databases

This section reviews the syntax and semantics of SPOT databases given in [26].

2.1   Syntax

We assume the existence of a set ID of objects ids, a set T of time points ranging over the integers, and a finite set $Space$ of points. We assume that $Space$ is a grid of size $N \times N$ where we only consider integer coordinates (the framework is easily extensible to higher dimensions). We assume that an object can be in only one location at a time, but that a single location may contain more than one object. The initial SPOT definition used only rectangular regions; we allow a region to be any non-empty set of points.

Definition 1 (SPOT atom/database). *A* SPOT *atom is a tuple* $(id, r, t, [\ell, u])$, *where* $id \in ID$ *is an object id,* $r \subseteq Space$ *is a region in the space,* $t \in T$ *is a time point, and* $[\ell, u] \subseteq [0, 1]$ *is a probability interval. A* SPOT *database is a finite set of* SPOT *atoms.*

Intuitively, the SPOT atom $(id, r, t, [\ell, u])$ says that object id is/was/will be inside region r at time t with probability in the interval $[\ell, u]$. Hence, SPOT atoms can represent information about the past and the present, but also information about the future, such as that deriving from methods for predicting the destination of moving objects [22,19,32], or from querying predictive databases [4,12,14,2,24,25].

Example 1. Consider a lab where data coming from biometric sensors are collected and analyzed. Biometric data such as faces, voices, and fingerprints recognized by sensors are matched against given profiles (such as those of people having access to the lab) and tuples like those in Fig. 1(a) are obtained. Every tuple consists of the profile id resulting from the matching phase, the area of the lab where the sensor recognizing the profile is operating, the time point at which the profile has been recognized, and the lower and upper probability bounds of the recognizing process getting the tuple. For instance, the tuple in the first row of the table in Fig. 1(a), representing the SPOT atom $(id_1, d, 1, [0.9, 1])$, says that the profile having id id_1 was in region d at time 1 with probability in the interval $[0.9, 1]$. In Fig. 1(b), the plan of the lab and the areas covered by sensors are shown. In area d a fingerprint sensor is located, whose high accuracy entails a narrow probability interval with upper bound equal to 1. After fingerprint authentication, id_1 was recognized at time 3 in areas b and c with probability in $[0.6, 1]$ and $[0.7, 0.8]$, respectively. □

Given a SPOT database $\mathscr{S}$, an object id, and a time t, we use $\mathscr{S}^{id,t}$ to refer to the set $\mathscr{S}^{id,t} = \{(id', r', t', [\ell', u']) \in \mathscr{S} \mid id' = id \wedge t' = t\}$. Moreover, we will use $ID(\mathscr{S})$ to denote the set of ids that appear in $\mathscr{S}$, that is $ID(\mathscr{S}) = \{id \mid (id, r, t, [\ell, u]) \in \mathscr{S}\}$.

Id	Area	Time	Lower Probability	Upper Probability
id_1	d	1	0.9	1
id_1	b	3	0.6	1
id_1	c	3	0.7	0.8
id_2	b	1	0.5	0.9
id_2	e	2	0.2	0.5
id_3	e	1	0.6	0.9

(a)

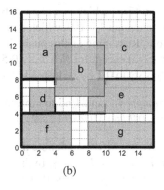

(b)

Fig. 1. (a) SPOT database $\mathscr{S}_{lab}$; (b) Areas of the lab

2.2 Semantics

The meaning of a SPOT database is given by the set of interpretations that satisfy it.

Definition 2 (SPOT interpretation). *A* SPOT *interpretation is a function* $I : ID \times Space \times T \to [0,1]$ *such that for each* $id \in ID$ *and* $t \in T$, $\sum_{p \in Space} I(id,p,t) = 1$.

For a given interpretation I, we sometimes abuse notation and write $I^{id,t}(p) = I(id,p,t)$. In this case, $I^{id,t}$ is a probability distribution function (PDF). The set of all interpretations for a SPOT database $\mathscr{S}$ will be denoted as $\mathbf{I}(\mathscr{S})$.

Example 2. Interpretation I_1 for the SPOT database $\mathscr{S}_{lab}$ of Example 1 is as follows.

$I_1(id_1,(3,6),1) = 0.4$ $I_1(id_1,(2,5),1) = 0.2$ $I_1(id_1,(3,5),1) = 0.3$
$I_1(id_1,(7,7),1) = 0.1$ $I_1(id_1,(7,5),2) = 0.5$ $I_1(id_1,(4,2),2) = 0.5$
$I_1(id_1,(10,10),3) = 0.7$ $I_1(id_1,(7,5),3) = 0.3$ $I_1(id_2,(5,7),1) = 0.1$
$I_1(id_2,(12,12),1) = 0.9$ $I_1(id_2,(9,7),2) = 0.3$ $I_1(id_2,(12,13),2) = 0.7$
$I_1(id_2,(8,7),3) = 0.9$ $I_1(id_2,(11,15),3) = 0.1$ $I_1(id_3,(10,5),1) = 0.8$
$I_1(id_3,(5,6),1) = 0.2$ $I_1(id_3,(5,5),2) = 0.5$ $I_1(id_3,(6,5),2) = 0.5$
$I_1(id_3,(5,3),3) = 0.6$ $I_1(id_3,(5,6),3) = 0.4$

Moreover, $I_1(id,p,t) = 0$ for all triplets (id,p,t) not mentioned above.   □

Given an interpretation I and region r, the probability that object id is in r at time t *according to* I is $\sum_{p \in r} I(id,p,t)$. We now define satisfaction and SPOT models.

Definition 3 (Satisfaction and SPOT model). *Let* $A = (id,r,t,[\ell,u])$ *be a* SPOT *atom and let* I *be a* SPOT *interpretation. We say that* I *satisfies* A *(denoted* $I \models A$*) iff* $\sum_{p \in r} I(id,p,t) \in [\ell,u]$. I *satisfies a* SPOT *database* $\mathscr{S}$ *(denoted* $I \models \mathscr{S}$*) iff* $\forall A \in \mathscr{S}$, $I \models A$. *If* I *satisfies a* SPOT *atom* A *(resp.* SPOT *database* $\mathscr{S}$*), we say that* I *is a model for* A *(resp.* $\mathscr{S}$*).*

Example 3. In our running example, interpretation I_1 is a model for the SPOT atom $(id_1,d,1,[0.9,1])$ as, for id id_1 and time point 1, I_1 assigns probability 0.4 to point $(3,6)$, 0.2 to point $(2,5)$, and 0.3 to point $(3,5)$ (which are points in area d), and probability

0.1 to $(7,7)$ which is a point outside area d. Hence, the probability that id_1 is in area d at time point 1 is 0.9, which is in the interval $[0.9,1]$ specified by the considered SPOT atom. Reasoning analogously, it is easy to see that I_1 is a model for all of the atoms in Fig. 1(a) except for $(id_2,b,1,[0.5,0.9])$ as the probability to be in area b at time 1 for id id_2 is set to 0.1 by I_1, instead of a value in $[0.5,0.9]$. Hence I_1 is not a model for $\mathscr{S}_{lab}$. $\square$

Example 4. Let M be the interpretation which is equal to I_1 except that $M(id_2,(5,7),1) = 0.7$ and $M(id_2,(12,12),1) = 0.3$. It is easy to check that M is a model for $\mathscr{S}_{lab}$. $\square$

We use $\mathbf{M}(\mathscr{S})$ to denote the set of models for a SPOT database $\mathscr{S}$, that is, $\mathbf{M}(\mathscr{S}) = \{I \mid I \in \mathbf{I}(\mathscr{S}) \land I \models \mathscr{S}\}$. In the following we will use the symbol M to refer to interpretations that are models, that is, elements in $\mathbf{M}(\mathscr{S})$.

Definition 4 (Consistency). *A* SPOT *database $\mathscr{S}$ is consistent iff $\mathbf{M}(\mathscr{S}) \neq \emptyset$.*

Example 5. Model M of Example 4 proves that $\mathscr{S}_{lab}$ is consistent. $\square$

As shown in [26], the consistency of a SPOT database can be checked by means of a linear programming algorithm whose complexity is $O(|ID(\mathscr{S})| \cdot |T| \cdot (|Space| \cdot |\mathscr{S}|)^3)$.
Throughout the paper we assume that SPOT *databases are consistent.*

3   Count Queries in SPOT

In this section, we define the syntax of count queries over SPOT databases and propose three alternative semantics. Intuitively, a count query asks for the number of objects inside a specified region at a given time point.

Definition 5 (Count query). *A* count query *is an expression of the form $Count(r,t)$ where r is a region of Space and t is a time point in T.*

The count query $Count(r,t)$ asks: "How many objects are inside region r at time t?". We propose three different semantics for interpreting this kind of query, namely the *expected value semantics*, the *extreme values semantics*, and the *ranking semantics*, which are introduced in the following three subsections. The first semantics looks at the expected value of the random variable representing the number of objects that are inside the query region r at time t. Since such an expected value varies from model to model and a SPOT database can have multiple models, we take the lowest and highest values across all models. The second semantics returns two integers z and Z, which are, respectively, the lowest and the highest numbers of objects that can be inside region r at time t (according to the different models of the given SPOT database). The last semantics returns a confidence interval for each number of objects that may be inside the query region r at time t.

In the rest of the paper, we assume that a SPOT database $\mathscr{S}$ and a count query $Q = Count(r,t)$ are given.

3.1   Expected Value Semantics

For the count query $Count(r,t)$, we first define the expected number of objects in region r at time t w.r.t. a model M; then, as there may be many models for a SPOT database, we define the expected value semantics as the tightest interval $[c,C]$ s.t. for any $M \in \mathbf{M}(\mathscr{S})$ the expected number of objects in r at time t w.r.t. M is in $[c,C]$.

Let M be a model for $\mathscr{S}$, and let X_M be a random variable representing the number of objects in region r at time t according to M. The expected answer to Q w.r.t. M is the expected number of objects in r at time t w.r.t. M, that is: $Q^{exp}(M) = \mathbb{E}[X_M] = \sum_{k=0}^{|ID(\mathscr{S})|} k \cdot \Pr(X_M = k)$, where $\Pr(X_M = k)$ denotes the probability that there are exactly k objects in r at time t according to M.

Example 6. Continuing our running example, one may be interested in knowing the expected number of people who are at time 1 in the region $r = \{(x,y) \in Space \mid (0 \le x \le 6) \wedge (4 \le y \le 8)\}$ (this region includes the whole area d, a portion of area b, and some other points). This can be expressed by the count query $Q = Count(r,1)$. The expected answer to Q w.r.t. model M of Example 4 is $Q^{exp}(M) = 0.9 + 0.7 + 0.2 = 1.8$.   □

In general, the expected number of objects in a given region at a given time point may vary in different models of $\mathbf{M}(\mathscr{S})$. The expected value answer is the tightest interval that includes the expected numbers w.r.t. every model for $\mathscr{S}$.

Definition 6 (Expected value answer). *The* expected value *answer to Q w.r.t. $\mathscr{S}$ is $Q^{exp}(\mathscr{S}) = [c,C]$, where:*

$$c = \min_{M \in \mathbf{M}(\mathscr{S})} Q^{exp}(M) \quad and \quad C = \max_{M \in \mathbf{M}(\mathscr{S})} Q^{exp}(M).$$

Example 7. The expected value answer to the query $Q = Count(r,1)$ of Example 6 is as follows. A model M_c for $\mathscr{S}_{lab}$ such that $Q^{exp}(M_c)$ is less than or equal to the expected answer w.r.t. any other model for $\mathscr{S}_{lab}$ is the following:

- $M_c(id_1,(3,6),1) = 0.9$, $M_c(id_1,(8,2),1) = 0.1$, $M_c(id_2,(8,10),1) = 0.9$, $M_c(id_2,(12,1),1) = 0.1$, $M_c(id_3,(10,5),1) = 0.9$, and $M_c(id_3,(12,1),1) = 0.1$;
- $M_c(id_i,p,1) = 0$ for all other triplets not mentioned above, and $M_c(id_i,p,t) = M(id_i,p,t)$ for all triplets (id_i,p,t) with $t \in [2,3]$, where M is the model of Example 4.

Reasoning as in Example 6, we obtain that $Q^{exp}(M_c) = 0.9$ (note that, according to M_c, both id_2 and id_3 are not in r).

A model M_C for $\mathscr{S}_{lab}$ such that $Q^{exp}(M_C)$ is greater than or equal to the expected answer w.r.t. any other model for $\mathscr{S}_{lab}$ is the following:

- $M_C(id_1,(3,6),1) = 1$, $M_C(id_2,(5,7),1) = 0.9$, $M_C(id_2,(5,5),1) = 0.1$, $M_C(id_3,(10,5),1) = 0.6$, and $M_C(id_3,(5,5),1) = 0.4$;
- $M_C(id_i,p,1) = 0$ for all other triplets not mentioned above, and $M_C(id_i,p,t) = M(id_i,p,t)$ for all triplets (id_i,p,t) with $t \in [2,3]$, where M is the model of Example 4.

In this case, we get $Q^{exp}(M_C) = 2.4$ and thus $Q^{exp}(\mathscr{S}_{lab}) = [0.9, 2.4]$.   □

By Definition 6, if $[c,C]$ is the expected value answer to a count query $Q = Count(r,t)$ w.r.t. a SPOT database $\mathscr{S}$, then the expected number of objects in r at time t w.r.t. any model $M \in \mathbf{M}(\mathscr{S})$ belongs to the interval $[c,C]$. The following proposition states that, for each value v in the interval $[c,C]$, there exists a model M such that the expected number of objects in r at time t w.r.t. M is v.

Proposition 1. *If $Q^{exp}(\mathscr{S}) = [c,C]$, then $\forall v \in [c,C]$, $\exists M \in \mathbf{M}(\mathscr{S})$ s.t. $Q^{exp}(M) = v$.*

Hence, the interval $[c,C]$ makes good sense as query answer because for every value in the interval there is a model whose expected number of objects is that value.

3.2 Extreme Values Semantics

Given a model M for a SPOT database, an object id, a region r, and a time point t, in the following, with a slight abuse of notation we use $M(id,r,t)$ to denote $\sum_{p \in r} M(id,p,t)$, i.e. the probability that id is in r at time t w.r.t. M.

Definition 7 (Extreme values answer). *The* extreme values *answer to Q w.r.t. $\mathscr{S}$ is $Q^{extreme}(\mathscr{S}) = [z,Z]$, where:*

$$z = \min_{M \in \mathbf{M}(\mathscr{S})} |\{id \text{ such that: } id \in ID(\mathscr{S}) \wedge M(id,r,t) = 1\}|, \text{ and}$$
$$Z = \max_{M \in \mathbf{M}(\mathscr{S})} |\{id \text{ such that: } id \in ID(\mathscr{S}) \wedge M(id,r,t) \neq 0\}|.$$

Note that, unlike the expected value semantics, the extreme values answer gives a pair of integers.

Example 8. Continuing our running example, one may be interested in knowing the min and max number of people who are at time 1 in region r of Example 6. We can get this from the extreme values answer $Q^{extreme}(\mathscr{S}_{lab})$ to the count query $Q = Count(r,1)$.

A model for $\mathscr{S}_{lab}$ such that the number of ids that are in r with probability 1 at time 1 is minimum w.r.t. any other model for $\mathscr{S}_{lab}$ is model M_c of Example 7. In fact, according this model no id is in r at time 1 with probability 1.

A model for $\mathscr{S}_{lab}$ such that the number of ids that are in r with non-zero probability at time 1 is maximum w.r.t. any other model for $\mathscr{S}_{lab}$ is model M_C of Example 7. In fact, according to M_C, ids id_1, id_2, and id_3 are in r at time 1. Hence, $Q^{extreme}(\mathscr{S}) = [0,3]$. $\square$

The relationship between the expected and extreme answers is the following.

Proposition 2. *If $Q^{exp}(\mathscr{S}) = [c,C]$ and $Q^{extreme}(\mathscr{S}) = [z,Z]$, then $z \leq c \leq C \leq Z$.*

3.3 Ranking Semantics

The answers provided by the expected value semantics and the extreme values semantics are numeric intervals. In this section, we propose an alternative semantics for a count query $Count(r,t)$ which gives a set of pairs of the form $\langle i, [\ell_i, u_i] \rangle$ (with $0 \leq i \leq |ID(\mathscr{S})|$) where i is the number of objects that may be in the given region r at time point t, and $[\ell_i, u_i]$ is the corresponding probability interval for each number.

We follow an approach similar to the one adopted in the definition of the expected value semantics, that is, for each $0 \leq i \leq |ID(\mathcal{S})|$ we first define the probability of having exactly i objects in a certain region at a certain time point w.r.t. a model M. Then, we consider the tightest interval $[\ell_i, u_i]$ that includes such a value for every model of a SPOT database. *For this case only, we assume independence of events involving the locations of different objects.* This assumption is often adopted in probabilistic databases [33], and is reasonable in many applications (e.g., when the movement of unrelated objects is tracked). Thus, applying the independence assumption, given a model M, the probability that two objects id_1 and id_2 are in locations p_1 and p_2, respectively, at time t is given by $M(id_1, p_1, t) \cdot M(id_2, p_2, t)$ (i.e., the product of the probability that id_1 is on p_1 at time t and the probability that id_2 is on p_2 at time t according M).

Below we define the probability that exactly i objects are in a region r at a time point t according to a given model M.

Definition 8 (Probability of exactly i objects with respect to a model). *Let M be a model for $\mathcal{S}$. For $0 \leq i \leq |ID(\mathcal{S})|$, the probability of having exactly i objects in r at time t w.r.t. M is as follows:*
$$Prob_M(r,t,i) = \Sigma_{S \subseteq ID(\mathcal{S}) \wedge |S|=i} \left(\Pi_{id \in S} M(id,r,t) \cdot \Pi_{id \in ID(\mathcal{S}) \backslash S} (1 - M(id,r,t)) \right).$$

Example 9. Consider the count query $Q = Count(r,1)$ of Example 6 and the model M of Example 4. As observed in Example 6, according to M, ids id_1, id_2, and id_3 can be in region r at time 1 with probabilities $M(id_1, r, 1) = \alpha_1 = 0.9$, $M(id_2, r, 1) = \alpha_2 = 0.7$, and $M(id_3, r, 1) = \alpha_3 = 0.2$, respectively. The probability of having exactly 2 objects in r at time 1 w.r.t. M is $Prob_M(r,1,2) = \alpha_1 \alpha_2 (1 - \alpha_3) + \alpha_1 (1 - \alpha_2) \alpha_3 + (1 - \alpha_1) \alpha_2 \alpha_3 = 0.9 \cdot 0.7 \cdot 0.8 + 0.9 \cdot 0.3 \cdot 0.2 + 0.1 \cdot 0.7 \cdot 0.2 = 0.572$. □

Before defining the ranking semantics we need the following auxiliary definition.

Definition 9 (Probability range for exactly i objects). *For $0 \leq i \leq |ID(\mathcal{S})|$, define*
$$Prob_{\mathcal{S}}^{min}(r,t,i) = \min_{M \in \mathbf{M}(\mathcal{S})} Prob_M(r,t,i) \quad and \quad Prob_{\mathcal{S}}^{max}(r,t,i) = \max_{M \in \mathbf{M}(\mathcal{S})} Prob_M(r,t,i).$$

Example 10. Consider again the count query $Q = Count(r,1)$ of our running example. It is easy to see that the minimum probability $Prob_{\mathcal{S}_{lab}}^{min}(r,1,0)$ of having no objects in r at time 1 is equal to zero. In fact, the SPOT atom in the first row of the table in Fig. 1(a) makes it possible to have id_1 is in region r with probability 1 at time 1, and in such a case the probability of having no objects in r at time 1 is 0.

The value of $Prob_{\mathcal{S}_{lab}}^{max}(r,1,0)$ can be obtained by considering model M_c of Example 7 according to which only id_1 is in r at time 1 with probability equal to 0.9, and both id_2 and id_3 are outside r. In this case, $Prob_{M_c}(r,1,0) = 0.1$, which is the maximum value of $Prob_M(r,1,0)$ w.r.t. any model M for $\mathcal{S}_{lab}$. Thus, $Prob_{\mathcal{S}_{lab}}^{max}(r,1,0) = 0.1$. □

Definition 10 (Ranking answer). *The ranking answer to Q w.r.t. $\mathcal{S}$ is as follows:*
$$Q^{rank}(\mathcal{S}) = \{ \langle i, [\ell_i, u_i] \rangle \mid 0 \leq i \leq |ID(\mathcal{S})| \wedge \ell_i = Prob_{\mathcal{S}}^{min}(r,t,i) \wedge u_i = Prob_{\mathcal{S}}^{max}(r,t,i) \}$$

Example 11. From Example 10, we have that $\langle 0, [0, 0.1] \rangle$ belongs to $Q^{rank}(\mathcal{S}_{lab})$, meaning that the minimum/maximum probability of having no objects in r at time 1 ranges between 0 and 0.1. With a little effort, the reader can check that the minimum probability $Prob_{\mathscr{S}}^{min}(r,1,1)$ of having exactly one object in r at time 1 is also 0. In fact, the model M_C introduced in Example 7 is such that $Prob_{M_C}(r,1,1) = M_C(id_1,r,1) \cdot (1 - M_C(id_2,r,1)) \cdot (1 - M_C(id_3,r,1)) + (1 - M_C(id_1,r,1)) \cdot M_C(id_2,r,1) \cdot (1 - M_C(id_3,r,1)) + (1 - M_C(id_1,r,1)) \cdot (1 - M_C(id_2,r,1)) \cdot M_C(id_3,r,1) = 1 \cdot 0 \cdot 0.6 + 0 \cdot 1 \cdot 0.6 + 0 \cdot 0 \cdot 0.4 = 0$, which is the minimum probability value w.r.t. the models for $\mathscr{S}_{lab}$. The maximum probability $Prob_{\mathscr{S}}^{max}(r,1,1)$ of having exactly one object in r at time 1 is equal to 1 by considering any model M' placing id_1 at any point in *Area d* with probability equal to 1, and placing neither id_2 nor id_3 in region r. Hence, the pair $\langle 1, [0,1] \rangle$ is in $Q^{rank}(\mathscr{S}_{lab})$. $\square$

The following proposition states the relationship between the expected value (resp. extreme values) answer and the ranking answer to a count query for the class of *simple* SPOT *databases* introduced in [26], that is SPOT databases admitting a single model. Notice that if $[c,C]$ is the expected value answer to a count query w.r.t. a simple SPOT database, then $c = C$. Furthermore, if $\{\langle 0, [\ell_0, u_0] \rangle, \ldots, \langle n, [\ell_n, u_n] \rangle\}$ is the ranking answer to a count query w.r.t. a simple SPOT database, then $\ell_i = u_i$ for $0 \leq i \leq n$.

Proposition 3. *Let $\mathscr{S}$ be a simple* SPOT *database, and let* $Q^{rank}(\mathscr{S}) = \{\langle 0, [\ell_0, u_0] \rangle, \ldots, \langle n, [\ell_n, u_n] \rangle\}$, *where* $n = |ID(\mathscr{S})|$. *Then,*

- $Q^{exp}(\mathscr{S}) = \left[\sum_{i=0}^{|ID(\mathscr{S})|} i \cdot \ell_i, \ \sum_{i=0}^{|ID(\mathscr{S})|} i \cdot u_i \right]$
- $Q^{extreme}(\mathscr{S}) = [\min\{i \mid 0 \leq i \leq n \wedge \ell_i = 1\}, \ \max\{i \mid 0 \leq i \leq n \wedge \ell_i \neq 0\}]$

4   Computing Count Queries

In this section, we provide techniques for computing count queries under the three semantics presented in the previous section. It turns out that both the expected and extreme values semantics are polynomial-time computable for general SPOT databases, while the ranking semantics can be computed in polynomial time for the class of simple SPOT databases. Notice that the proposed semantics cannot be computed by directly applying their definitions as the number of models of a SPOT database can be infinite.

Our methods for evaluating count queries exploit the definition, introduced in [26], of a set of linear constraints $LC(\mathscr{S}, id, t)$ associated with a SPOT database $\mathscr{S}$, an object id, and a time point t. $LC(\mathscr{S}, id, t)$ is reported in the next definition, where the variable v_p denotes the probability that object id is at point $p \in Space$ at time t.

Definition 11 $(LC(\cdot))$. *For* SPOT *database* $\mathscr{S}$, $id \in ID$, *and* $t \in T$, $LC(\mathscr{S}, id, t)$ *contains exactly the linear constraints defined below:*

- $\forall (id, r, t, [\ell, u]) \in \mathscr{S}^{id,t}, \ (\ell \leq \sum_{p \in r} v_p \leq u) \in LC(\mathscr{S}, id, t)$;
- $(\sum_{p \in Space} v_p = 1) \in LC(\mathscr{S}, id, t)$;
- $\forall p \in Space \ \ (v_p \geq 0) \in LC(\mathscr{S}, id, t)$.

$LC(\mathscr{S}, id, t)$ was exploited in [26] to check the consistency of a SPOT database. In particular, it was shown that $\mathscr{S}$ is consistent iff $LC(\mathscr{S}, id, t)$ is feasible for all $\langle id, t \rangle$ pairs. We build on $LC(\mathscr{S}, id, t)$ to devise strategies for computing count queries.

4.1 Computing Expected Value Semantics

The following theorem provides a method to compute the answer to a count query under the expected value semantics.

Theorem 1. *If* $Q^{exp}(\mathscr{S}) = [c,C]$, *then*

$$c = \Sigma_{id \in ID(\mathscr{S})} \left(\textit{minimize } \Sigma_{p \in r} v_p \textit{ subject to } LC(\mathscr{S}, id, t) \right)$$
$$C = \Sigma_{id \in ID(\mathscr{S})} \left(\textit{maximize } \Sigma_{p \in r} v_p \textit{ subject to } LC(\mathscr{S}, id, t) \right)$$

From this theorem, a straightforward algorithm for computing count queries under the expected value semantics follows: sum up the values obtained by solving, for each $id \in ID(\mathscr{S})$, the linear programs reported in Theorem 1.

Example 12. The expected answer to the query $Q = Count(r,1)$ of Example 7 can be determined as follows. Consider the following linear programs:

- $LC(\mathscr{S}_{lab}, id_1, 1) = \{0.9 \leq \Sigma_{p \in \text{Area d}} v_p \leq 1; \Sigma_{p \in Space} v_p = 1; \forall p \in Space, v_p \geq 0\}$.
- $LC(\mathscr{S}_{lab}, id_2, 1) = \{0.5 \leq \Sigma_{p \in \text{Area b}} v_p \leq 0.9; \Sigma_{p \in Space} v_p = 1; \forall p \in Space, v_p \geq 0\}$.
- $LC(\mathscr{S}_{lab}, id_3, 1) = \{0.6 \leq \Sigma_{p \in \text{Area e}} v_p \leq 0.9; \Sigma_{p \in Space} v_p = 1; \forall p \in Space, v_p \geq 0\}$.

where $Space$ is the set of points in the grid of Fig. 1(b). By minimizing $(\Sigma_{p \in r} v_p)$ subject to the three LCs above, we obtain 0.9, 0, and 0, respectively. Hence, applying Theorem 1, the lower bound of $Q^{exp}(\mathscr{S}_{lab})$ is the sum of these values, that is, $c = 0.9$. Similarly, maximizing $(\Sigma_{p \in r} v_p)$ subject to the LCs results in 1, 1, and 0.4, respectively, and applying Theorem 1 we get that the upper bound of $Q^{exp}(\mathscr{S}_{lab})$ is $C = 1 + 1 + 0.4 = 2.4$. Indeed, as discussed in Example 7, $Q^{exp}(\mathscr{S}_{lab}) = [0.9, 2.4]$. □

The following corollary states that $Q^{exp}(\mathscr{S})$ can be computed in polynomial time w.r.t. the size of $\mathscr{S}$ and the number of points in $Space$.

Corollary 1. $Q^{exp}(\mathscr{S})$ *can be computed in time* $O(|ID(\mathscr{S})| \cdot (|Space| \cdot |\mathscr{S}|)^3)$.

The above corollary entails that computing count queries under the expected value semantics is not more expensive than checking the consistency of a SPOT database.

Another interesting consequence of Theorem 1 is the following. An optimized version of $LC(\cdot)$, where the number of variables is drastically reduced, was introduced in [27]. The experimental results of [27] show that solving the reduced-size $LC(\cdot)$ is much more efficient than solving the equivalent system of linear inequalities in Definition 11. In this paper, we can take advantage of the result of [27] to make more efficient the computation of the expected value semantics. In fact, even though we used $LC(\cdot)$ in Theorem 1 for the sake of simplicity, it is easy to see that nothing changes if we replace $LC(\cdot)$ with its equivalent optimized version of [27]. Thus, the efficiency improvements of using the optimized version of $LC(\cdot)$ immediately apply to the case of computing expected value semantics.

4.2 Computing Extreme Values Semantics

We now address the problem of computing the extreme values answer.

Theorem 2. *If $Q^{extreme}(\mathscr{S}) = [z,Z]$, then*

$z = |\{id \text{ such that } id \in ID(\mathscr{S}) \wedge (\textbf{minimize } \sum_{p \in r} v_p \textbf{ subject to } LC(\mathscr{S},id,t)) = 1\}|$
$Z = |\{id \text{ such that } id \in ID(\mathscr{S}) \wedge (\textbf{maximize } \sum_{p \in r} v_p \textbf{ subject to } LC(\mathscr{S},id,t)) \neq 0\}|$

Theorem 2 provides a way of computing the extreme values answer. Specifically, in order to compute z (resp. Z), it suffices to solve the min- (resp. max-) version of the linear program in Theorem 2 for each $id \in ID(\mathscr{S})$, and then count for how many object ids the optimization program gives a solution equal to one (resp., a positive solution).

Example 13. The extreme values answer to the query $Q = Count(r,1)$ of Example 8 can be determined as follows. Consider $LC(\mathscr{S}_{lab},id_1,1)$, $LC(\mathscr{S}_{lab},id_2,1)$, and $LC(\mathscr{S}_{lab},id_3,1)$ as reported in Example 12. Since minimizing the function $(\sum_{p \in r} v_p)$ subject to any of these three LCs we obtain all zero values, Theorem 2 entails that the lower bound of $Q^{extreme}(\mathscr{S}_{lab})$ is 0. Similarly, maximizing $(\sum_{p \in r} v_p)$ subject to these LCs results in three non-zero values (i.e., 1, 1, and 0.4, respectively). Hence, by applying Theorem 2, we obtain that the upper bound of $Q^{extreme}(\mathscr{S}_{lab})$ is 3.   □

As stated in the following corollary, the complexity of computing $Q^{extreme}(\mathscr{S})$ is polynomial w.r.t. the size of $\mathscr{S}$ and the number of points in *Space*.

Corollary 2. $Q^{extreme}(\mathscr{S})$ *can be computed in time* $O(|ID(\mathscr{S})| \cdot (|Space| \cdot |\mathscr{S}|)^3)$.

It is worth noting that, also in this case, the optimized version of $LC(\cdot)$ can be exploited to make more efficient the computation of the extreme values answer.

4.3   Computing Ranking Semantics

In the following proposition, given a SPOT database $\mathscr{S}$, an object identifier $id \in ID(\mathscr{S})$, and a time point t, we use v_p^{id} to denote variable v_p of $LC(\mathscr{S},id,t)$.

Proposition 4. *If $Q^{rank}(\mathscr{S}) = \{\langle 0, [\ell_0,u_0]\rangle,\ldots,\langle n,[\ell_n,u_n]\rangle\}$ where $n = |ID(\mathscr{S})|$, then for $0 \leq i \leq n$, the following holds*

$\ell_i = \textbf{minimize } \sum_{S \subseteq ID(\mathscr{S}) \wedge |S|=i} \left(\prod_{id \in S} \sum_{p \in r} v_p^{id} \cdot \prod_{id \in ID(\mathscr{S}) \backslash S}(1 - \sum_{p \in r} v_p^{id})\right)$
    $\textbf{subject to } LC(\mathscr{S},id_1,t) \cup \cdots \cup LC(\mathscr{S},id_n,t)$

$u_i = \textbf{maximize } \sum_{S \subseteq ID(\mathscr{S}) \wedge |S|=i} \left(\prod_{id \in S} \sum_{p \in r} v_p^{id} \cdot \prod_{id \in ID(\mathscr{S}) \backslash S}(1 - \sum_{p \in r} v_p^{id})\right)$
    $\textbf{subject to } LC(\mathscr{S},id_1,t) \cup \cdots \cup LC(\mathscr{S},id_n,t)$

The previous proposition provides a method for computing ranking answers for general SPOT databases. However, this method has exponential complexity w.r.t. the size of $\mathscr{S}$. Below we propose a dynamic programming algorithm to compute $Q^{rank}(\mathscr{S})$ w.r.t. simple SPOT databases, and show that the algorithm is polynomial time.

Consider a simple SPOT database $\mathscr{S}$, the unique model M for $\mathscr{S}$, and a count query $Q = Count(r,t)$. Assume an arbitrary but fixed ordering of the object identifiers in $ID(\mathscr{S})$—we will consider the lexicographic ordering $id_1,\ldots,id_n$, where $n = |ID(\mathscr{S})|$. For $1 \leq j \leq n$ and $0 \leq i \leq j$, we define $Prob_M(q,t,i,j)$ as follows:

$$Prob_M(r,t,0,j) = \prod_{k=1}^{j}(1 - M(id_k,r,t)) \qquad 1 \leq j \leq n$$
$$Prob_M(r,t,j,j) = \prod_{k=1}^{j} M(id_k,r,t) \qquad 1 \leq j \leq n$$
$$Prob_M(r,t,i,j) = M(id_j,r,t) \cdot Prob_M(r,t,i-1,j-1) +$$
$$(1 - M(id_j,r,t)) \cdot Prob_M(r,t,i,j-1) \qquad 2 \leq j \leq n,\ 1 \leq i \leq j-1$$

The following example shows how $Prob_M(q,t,i,j)$ is computed for a given model M.

Example 14. Consider the count query $Q = Count(r,1)$ of Example 6 and the model M of Example 4. As observed in Example 6, according to M, ids id_1, id_2, and id_3 can be in region r at time 1 with probabilities $M(id_1,r,1) = \alpha_1 = 0.9$, $M(id_2,r,1) = \alpha_2 = 0.7$, and $M(id_3,r,1) = \alpha_3 = 0.2$, respectively. Hence, we have that:

$Prob_M(r,1,0,1) = (1 - \alpha_1) = 0.1$; $Prob_M(r,1,0,2) = (1 - \alpha_1)(1 - \alpha_2) = 0.03$;

$Prob_M(r,1,0,3) = (1 - \alpha_1)(1 - \alpha_2)(1 - \alpha_3) = 0.024$; $Prob_M(r,1,1,1) = \alpha_1 = 0.9$;

$Prob_M(r,1,2,2) = \alpha_1 \alpha_2 = 0.63$; $Prob_M(r,1,3,3) = \alpha_1 \alpha_2 \alpha_3 = 0.126$;

$Prob_M(r,1,1,2) = \alpha_2 \cdot Prob_M(r,1,0,1) + (1 - \alpha_2) \cdot Prob_M(r,1,1,1) = 0.34$.

$Prob_M(r,1,1,3) = \alpha_3 \cdot Prob_M(r,1,0,2) + (1 - \alpha_3) \cdot Prob_M(r,1,1,2) = 0.278$.

$Prob_M(r,1,2,3) = \alpha_3 \cdot Prob_M(r,1,1,2) + (1 - \alpha_3) \cdot Prob_M(r,1,2,2) = 0.572$. □

Theorem 3. *For a simple* SPOT *database* $\mathscr{S}$, *if* $Q^{rank}(\mathscr{S}) = \{\langle 0, [\ell_0, u_0]\rangle, \ldots, \langle n, [\ell_n, u_n]\rangle\}$, *where* $n = |ID(\mathscr{S})|$, *then* $\ell_i = u_i = Prob_M(r,t,i,n)$ *for* $0 \leq i \leq n$.

Example 15. Assume we have a simple SPOT database $\mathscr{S}$ whose unique model is model M of Example 14. Theorem 3 entails that the ranking answer to $Q = Count(r,1)$ can be derived from $Prob_M(r,1,0,3) = 0.024$, $Prob_M(r,1,1,3) = 0.278$, $Prob_M(r,1,2,3) = 0.572$, and $Prob_M(r,1,3,3) = 0.126$. Hence, $Q^{rank}(\mathscr{S}) = \{\langle 0, [0.024, 0.024]\rangle, \langle 1, [0.278, 0.278]\rangle, \langle 2, [0.572, 0.572]\rangle, \langle 3, [0.126, 0.126]\rangle\}$. □

As stated in the following corollary, computing the ranking answer using the result of Theorem 3 results in a polynomial time algorithm.

Corollary 3. *Given a simple* SPOT *database* $\mathscr{S}$ *and a count query* $Q = Count(r,t)$, *the complexity of computing* $Q^{rank}(\mathscr{S})$ *is* $O(|ID(\mathscr{S})| \cdot (|Space| \cdot |\mathscr{S}|)^3)$.

5 Related Work

Aggregates in probabilistic databases were deeply investigated in [31], where all standard aggregate queries including finding the mean and standard deviation over an attribute for specified values of other attributes were considered. In the SPOT framework this does not apply directly as we do not actually deal with attributes whose domain is numeric (although we use integers for time values). Thus, in this paper we focused on counting the number of objects in a certain region at a certain time.

One aspect that distinguishes our work from previous work on aggregates in probabilistic databases is that we rely on the SPOT framework, the only *declarative* framework we are aware of that deals with spatio-temporal data with *uncertain probabilities*.

The standard reference for computing aggregates in spatio-temporal databases is Lopez et al. [21]. This paper studies in detail the most efficient techniques for evaluating aggregate queries on spatial, temporal, and spatio-temporal databases. Gomez et al. [15] address aggregate queries over GIS and moving object data where the non-spatial information is stored in a data warehouse. Both these papers deal with definite data; no probabilities are involved.

There has been recent research on probabilistic spatio-temporal databases [9,37,36,8,38]. Chung et al. [9] derive a PDF for the location of an object moving in a one-dimensional space. Their probabilistic range queries find objects that are in a specified region of space within a specified time interval, and with a probability that is at least a threshhold value. Zhang et al. [37] provide a framework that allows their model to be incorporated into existing DBMSs and work for all objects even if their location and velocity are uncertain. Yang et al. [36] work with moving objects in indoor space, and their query asks for all sets of k objects that have at least a threshold probability of containing the k nearest objects to a given object. Chen et al. [8] deal with a similar problem, and also deal with the query result quality by using both a false positive and a false negative rate. Finally, [38] deals primarily with objects moving along road networks, and introduce an indexing mechanism called UTH (Uncertain Trajectories Hierarchy) to efficiently process probabilistic range queries.

Our approach to define the semantics of count queries is somewhat similar to the approach in [5], also adopted in other works such as [1] and [13], to define the semantics of aggregate queries over inconsistent databases. While we derive some information (e.g., expected answer) from each model of a SPOT database and then combine it over all models, [5] evaluates an aggregate query over each repair and combines the resulting values by considering the tightest interval that includes all of them.

Past work on the SPOT framework investigated selection queries [27,23], that is, queries asking for the objects *id* and times *t* such that *id* is inside a given query region *r* at time *t* with a probability in the given interval $[\ell, u]$. Optimistic and cautious semantics were proposed for interpreting these queries, and efficient algorithms for computing optimistic and cautious answers were proposed in [27] and [23], respectively. A more general version of the SPOT framework adopted in this paper was presented in [28] and [16] with velocity constraints on moving objects as well as points in $\mathscr{S}$ that may not be reachable from all other points by all objects. Also, as SPOT databases provide information on moving objects, one aspect addressed in [28] and then further investigated in [16] is that of revising SPOT data so that information on these objects may be changed as objects move. A comprehensive survey of the results on the SPOT framework can be found in [17] where several open problems were identified. One of these problems is that of devising a full logic (including negation, disjunction and quantifiers) for managing SPOT data. This problem was recently addressed in [11].

6   Conclusion and Topics for Further Research

We started the study of computing aggregates in the SPOT framework for probabilistic spatio-temporal databases by focusing on count queries. We defined three semantics for interpreting queries that ask for the number of objects in a region at a specified time and developed polynomial time algorithms for them. However, for the ranking semantics we assume independence of events involving the location of different objects. Also, in this case, the polynomial time algorithm requires a SPOT database with a unique model.

There are additional topics to investigate in addition to generalizing our result for the ranking semantics. We have only considered count queries for a single time value, but such queries are meaningful for time intervals as well. A simple way of dealing with a count query specifying a time interval is to evaluate the count query for each time point in the interval and return the set of answers obtained in this way. More complex semantics would require combining such results. Additional types of count queries may be useful for other purposes. For example, we may want to count the number of locations visited by a moving object in a time interval. So in this case we count location points instead of objects. Alternately, we may wish to count the number of time points during which a location contained an object. The SPOT framework does not have a true numeric component for asking other types of aggregate queries. If we extend the SPOT framework by adding such a component (e.g., the number of items carried by a moving object), then computing other types of aggregates becomes meaningful.

References

1. Afrati, F.N., Kolaitis, P.G.: Answering aggregate queries in data exchange. In: Proc. PODS, pp. 129–138 (2008)
2. Agarwal, D., Chen, D., Lin, L.-J., Shanmugasundaram, J., Vee, E.: Forecasting high-dimensional data. In: Proc. SIGMOD, pp. 1003–1012 (2010)
3. Agarwal, P.K., Arge, L., Erickson, J.: Indexing moving points. Journal of Computer and System Sciences 66(1), 207–243 (2003)
4. Akdere, M., Cetintemel, U., Riondato, M., Upfal, E., Zdonik, S.: The case for predictive database systems: Opportunities and challenges. In: Proc. CIDR, pp. 167–174 (2011)
5. Arenas, M., Bertossi, L.E., Chomicki, J., He, X., Raghavan, V., Spinrad, J.: Scalar aggregation in inconsistent databases. Theor. Comput. Sci. 296(3), 405–434 (2003)
6. Benjelloun, O., Sarma, A.D., Halevy, A.Y., Widom, J.: Uldbs: Databases with uncertainty and lineage. In: Proc. VLDB, pp. 953–964 (2006)
7. Cao, H., Wolfson, O., Trajcevski, G.: Spatio-temporal data reduction with deterministic error bounds. VLDB Journal 15, 211–228 (2006)
8. Chen, Y.F., Qin, X.L., Liu, L.: Uncertain distance-based range queries over uncertain moving objects. J. Comput. Sci. Technol. 25(5), 982–998 (2010)
9. Chung, B.S.E., Lee, W.C., Chen, A.L.P.: Processing probabilistic spatio-temporal range queries over moving objects with uncertainty. In: Proc. EDBT, pp. 60–71 (2009)
10. Dai, X., Yiu, M.L., Mamoulis, N., Tao, Y., Vaitis, M.: Probabilistic spatial queries on existentially uncertain data. In: Medeiros, C.B., Egenhofer, M., Bertino, E. (eds.) SSTD 2005. LNCS, vol. 3633, pp. 400–417. Springer, Heidelberg (2005)
11. Doder, D., Grant, J., Ognjanović, Z.: Probabilistic logics for objects located in space and time. J. of Logic and Computation 23(3), 487–515 (2013)
12. Duan, S., Babu, S.: Processing forecasting queries. In: Proc. VLBD (2007)
13. Flesca, S., Furfaro, F., Parisi, F.: Range-consistent answers of aggregate queries under aggregate constraints. In: Deshpande, A., Hunter, A. (eds.) SUM 2010. LNCS, vol. 6379, pp. 163–176. Springer, Heidelberg (2010)
14. Ge, T., Zdonik, S.: A skip-list aproach for efficiently processing forecasting queries. In: Proc. VLDB (2008)
15. Gomez, L.I., Kuijpers, B., Vaisman, A.A.: Aggregate languages for moving object and places of interest. In: Proc. SAC, pp. 857–862 (2008)
16. Grant, J., Parisi, F., Parker, A., Subrahmanian, V.S.: An agm-style belief revision mechanism for probabilistic spatio-temporal logics. Artif. Intell. 174(1), 72–104 (2010)
17. Grant, J., Parisi, F., Subrahmanian, V.S.: Research in Probabilistic Spatiotemporal Databases: The SPOT Framework. In: Ma, Z., Yan, L. (eds.) Advances in Probabilistic Databases. STUDFUZZ, vol. 340, pp. 1–22. Springer, Heidelberg (2013)

18. Hadjieleftheriou, M., Kollios, G., Tsotras, V.J., Gunopulos, D.: Efficient indexing of spatiotemporal objects. In: Jensen, C.S., Jeffery, K., Pokorný, J., Šaltenis, S., Bertino, E., Böhm, K., Jarke, M. (eds.) EDBT 2002. LNCS, vol. 2287, pp. 251–268. Springer, Heidelberg (2002)
19. Hammel, T., Rogers, T.J., Yetso, B.: Fusing live sensor data into situational multimedia views. In: Proc. MIS, pp. 145–156 (2003)
20. Kollios, G., Gunopulos, D., Tsotras, V.J.: On indexing mobile objects. In: Proc. PODS, pp. 261–272 (1999)
21. Lopez, I.F.V., Snodgrass, R.T., Moon, B.: Spatiotemporal aggregate computation: A survey. IEEE TKDE 17(2), 271–286 (2005)
22. Mittu, R., Ross, R.: Building upon the coalitions agent experiment (coax) - integration of multimedia information in gccs-m using impact. In: Proc. MIS, pp. 35–44 (2003)
23. Parisi, F., Parker, A., Grant, J., Subrahmanian, V.S.: Scaling cautious selection in spatial probabilistic temporal databases. In: Jeansoulin, R., Papini, O., Prade, H., Schockaert, S. (eds.) Methods for Handling Imperfect Spatial Information. STUDFUZZ, vol. 256, pp. 307–340. Springer, Heidelberg (2010)
24. Parisi, F., Sliva, A., Subrahmanian, V.S.: Embedding forecast operators in databases. In: Benferhat, S., Grant, J. (eds.) SUM 2011. LNCS, vol. 6929, pp. 373–386. Springer, Heidelberg (2011)
25. Parisi, F., Sliva, A., Subrahmanian, V.S.: A temporal database forecasting algebra. Int. J. of Approximate Reasoning 54(7), 827–860 (2013)
26. Parker, A., Subrahmanian, V.S., Grant, J.: A logical formulation of probabilistic spatial databases. IEEE TKDE, 1541–1556 (2007)
27. Parker, A., Infantes, G., Grant, J., Subrahmanian, V.S.: Spot databases: Efficient consistency checking and optimistic selection in probabilistic spatial databases. IEEE TKDE 21(1), 92–107 (2009)
28. Parker, A., Infantes, G., Grant, J., Subrahmanian, V.S.: An AGM-based belief revision mechanism for probabilistic spatio-temporal logics. In: Proc. AAAI (2008)
29. Pelanis, M., Saltenis, S., Jensen, C.S.: Indexing the past, present, and anticipated future positions of moving objects. ACM Trans. Database Syst. 31(1), 255–298 (2006)
30. Pfoser, D., Jensen, C.S., Theodoridis, Y.: Novel approaches to the indexing of moving object trajectories. In: Proc. VLDB (2000)
31. Ross, R., Subrahmanian, V.S., Grant, J.: Aggregate operators in probabilistic databases. Journal of the ACM 52(1), 54–101 (2005)
32. Southey, F., Loh, W., Wilkinson, D.F.: Inferring complex agent motions from partial trajectory observations. In: Proc. IJCAI, pp. 2631–2637 (2007)
33. Suciu, D., Olteanu, D., Ré, C., Koch, C.: Probabilistic Databases. Synthesis Lectures on Data Management. Morgan & Claypool Publishers (2011)
34. Tao, Y., Papadias, D., Sun, J.: The TPR*-tree: an optimized spatio-temporal access method for predictive queries. In: Proc. VLDB, pp. 790–801 (2003)
35. Tao, Y., Cheng, R., Xiao, X., Ngai, W.K., Kao, B., Prabhakar, S.: Indexing multi-dimensional uncertain data with arbitrary probability density functions. In: Proc. VLDB, pp. 922–933 (2005)
36. Yang, B., Lu, H., Jensen, C.S.: Probabilistic threshold k nearest neighbor queries over moving objects in symbolic indoor space. In: Proc. EDBT, pp. 335–346 (2010)
37. Zhang, M., Chen, S., Jensen, C.S., Ooi, B.C., Zhang, Z.: Effectively indexing uncertain moving objects for predictive queries. PVLDB 2(1), 1198–1209 (2009)
38. Zheng, K., Trajcevski, G., Zhou, X., Scheuermann, P.: Probabilistic range queries for uncertain trajectories on road networks. In: Proc. EDBT, pp. 283–294 (2011)

Approximate Reasoning about Generalized Conditional Independence with Complete Random Variables

Sebastian Link

Department of Computer Science, The University of Auckland, New Zealand

Abstract. The implication problem of conditional statements about the independence of finitely many sets of random variables is studied in the presence of controlled uncertainty. Uncertainty refers to the possibility of missing data. As a control mechanism random variables can be declared complete, in which case data on these random variables cannot be missing. While the implication of conditional independence statements is not axiomatizable, a finite Horn axiomatization is established for the expressive class of saturated conditional independence statements under controlled uncertainty. Complete random variables allow us to balance the expressivity of sets of saturated statements with the efficiency of deciding their implication. This ability can soundly approximate reasoning in the absence of missing data. Delobel's class of full first-order hierarchical database decompositions are generalized to the presence of controlled uncertainty, and their implication problem shown to be equivalent to that of saturated conditional independence.

1 Introduction

Background. Conditional independence is an important concept to capture structural aspects of probability distributions, to deal with knowledge and uncertainty in AI, and for learning and reasoning in intelligent systems [6,30,5]. Application areas include computational biology, computer vision, databases, error-control coding, natural language processing, speech processing, and robotics [17,5]. Recently, independence logic has been introduced as an extension of classical first-order logic to capture independence statements [15,20]. A conditional independence (CI) statement $I(Y, Z \mid X)$ represents the independence of two sets of random variables relative to a third: given three mutually disjoint subsets X, Y, and Z of a set V of random variables, if we have knowledge about the state of X, then knowledge about the state of Y does not provide additional evidence for the state of Z and vice versa. Fundamental to the task of building a Bayesian network is the implication problem of CI statements, which is to decide for an arbitrary set V of random variables, and an arbitrary set $\Sigma \cup \{\varphi\}$ of CI statements over V, whether every probability model that satisfies every CI statement in Σ also satisfies φ. Indeed, if an important CI statement φ is not implied by Σ, then adding φ to Σ results in new opportunities to construct complex probability models with polynomially many parameters and to

W. Liu, V.S. Subrahmanian, and J. Wijsen (Eds.): SUM 2013, LNAI 8078, pp. 269–282, 2013.
© Springer-Verlag Berlin Heidelberg 2013

efficiently organize distributed probability computations [14]. The implication problem for CI statements is not axiomatizable by a finite set of Horn rules [33], and every axiom for CI statements is an axiom for graph separation, but not vice versa [13]. Recently, the implication problem of stable CI statements [27,35] has been shown to be finitely axiomatizable [29], and *coNP*-complete to decide [28]. An important subclass of CI statements are saturated conditional independence (SCI) statements. These are CI statements $I(Y, Z \mid X)$ over V that satisfy $XYZ = V$. Indeed, graph separation and SCI statements enjoy the same axioms [13], their implication problem is equivalent to that of a propositional fragment and to that of multivalued data dependencies [23], and decidable in almost linear time [12,14,36]. These results contribute to the success story of Bayesian networks in AI and machine learning [5,13].

Motivation. In the real world most data samples contain missing data. AI has long recognized the need to reveal missing data and to explain where they come from. Significant contributions towards that aim have been made in the literature, including [2,4,8,9,11,21,26,31,34]. However, for many missing data it is either impossible to reveal them at all, or the process is too expensive or inaccurate. Consequently, the concept of conditional independence should be sufficiently robust to deal with missing data. Consider, for example, the set V with binary random variables *Cancer*, *Cavity*, *Gender*, and *Smoking*, and the probability model where

Smoking	Gender	Cancer	Cavity	P
true	μ	true	true	0.5
true	μ	false	false	0.5

are two assignments of probability one half each. Here, μ denotes a marker saying that "no information" is available about the value of a random variable in an assignment. It is natural to ask which (saturated) CI statements this probability model satisfies, for example, $\sigma_1 = I(\{Cancer, Cavity\}, \{Gender\} \mid \{Smoking\})$, or $\sigma_2 = I(\{Cancer\}, \{Cavity\} \mid \{Smoking, Gender\})$, or $\varphi = I(\{Cancer\}, \{Cavity, Gender\} \mid \{Smoking\})$. Having given CI statements a suitable semantics in the presence of missing data, one may wonder about the associated implication problem. For example, is there any probability model that satisfies σ_1 and σ_2, but violates φ? In practice, it is also desirable to control the occurrence of missing data. For this purpose, we propose the possibility to declare random variables as complete. Assignments that carry missing data on complete random variables are excluded from data samples. Complete variables provide thus a mechanism to *soundly* approximate the classical concept of conditional independence. This is achieved by defining that conditional independence is only violated by some probability model, if the model features complete evidence of its violation. That is, assignments with missing data on some random variables in the condition are not taken into account when judging conditional independence. It is the goal of this paper to investigate the impact of complete random variables on the implication problem of generalized conditional independence between an arbitrary finite number of sets of random variables.

Our findings are the starting point of a foundation for reasoning about conditional independence in the presence of missing data, much like the findings of [6,14] for complete data.

Contributions. As a first contribution we assign a suitable semantics to generalized CI statements in the presence of complete random variables. While the associated implication problem is infeasible in general, our second contribution establishes an axiomatization for generalized SCI statements. Complete random variables serve as a control mechanism to soundly approximate classical reasoning about generalized conditional independence. Our completeness argument is based on special probability models in which two assignments are assigned probability one half. This insight has remarkable consequences. As a third contribution we establish an equivalence between instances of the implication problem and sliced versions of the same instance in the absence of missing data. Hence, classical tools, can even be used in the presence of missing data. As a fourth contribution, we establish a characterization of the implication problem in terms of that for Delobel's class of full first-order hierarchical decompositions [7], which we generalize to the presence of null values. Finally, we combine the third and fourth contributions to exploit an almost linear time algorithm, originally designed to decide the implication of multivalued dependencies over complete database relations [12], for deciding the implication of generalized SCI statements. It is shown that attempts to increase the expressivity of generalized SCI statements result in the failure of our equivalences, the intractability or even the infeasibility of the associated implication problem.

2   Conditional Independence and Uncertainty

We denote by V a finite set $\{v_1, \ldots, v_n\}$ of *random variables*. A *domain mapping* is a mapping that associates a set, $dom(v_i)$, with each random variable v_i. This set is called the *domain* of v_i and each of its elements is a *data value* of v_i. We assume that each domain $dom(v_i)$ contains the element μ, which we call the *marker*. Although we use μ like any other domain value, we prefer to think of μ as a marker, denoting that no information is currently available about the data value of v_i. The interpretation of this marker means that a data value does either not exist (known as a structural zero in statistics, and the null value inapplicable in databases), or a data value exists but is currently unknown (known as a sampling zero in statistics, and the null value *applicable* in databases). The disadvantage is a loss in knowledge when representing data values known to not exist or known to exist but currently unknown. One advantage is its simplicity. As another advantage one can represent missing values, even if it is unknown whether they do not exist or exist but are currently unknown. Further advantages will be revealed by the results established later. For $X = \{v_1, \ldots, v_k\} \subseteq V$ we say that $\mathbf{x}$ is an *assignment* of X, if $\mathbf{x} \in dom(v_1) \times \cdots \times dom(v_k)$. For an assignment $\mathbf{x}$ of X we write $\mathbf{x}(y)$ for the projection of $\mathbf{x}$ onto $Y \subseteq X$. We say that $\mathbf{x}$ is Y-*complete*, if $\mathbf{x}(v) \neq \mu$ for all $v \in Y$.

It is beneficial to gain control over the occurrences of markers. For this purpose we introduce *complete random variables*. If a random variable v is declared *complete*, then $\mu \notin dom(v)$. For a given V we define C to be the set of random variables of V that are complete. It is a goal of this article to investigate the properties of CI statements under controlled uncertainty, i.e., in the presence of complete random variables. These variables provide an effective means to not just control the degree of uncertainty, but also to soundly approximate reasoning about CI statements in the absence of missing data.

A *probability model* over $(V = \{v_1, \ldots, v_n\}, C)$ is a pair (dom, P) where dom is a domain mapping that maps each v_i to a finite domain $dom(v_i)$, and P : $dom(v_1) \times \cdots \times dom(v_n) \to [0, 1]$ is a probability distribution having the Cartesian product of these domains as its sample space. Note that $\mu \notin dom(v_i)$ whenever $v_i \in C$.

The expression $I(Y_1, \ldots, Y_k \mid X)$ where k is a non-negative integer, and $X, Y_1, \ldots, Y_k$ are mutually disjoint subsets of V, is called a *generalized conditional independence* (GCI) *statement* over (V, C). The set X is called the *condition*. If $XY_1 \cdots Y_k = V$, we call $I(Y_1, \ldots, Y_k \mid X)$ a *saturated* GCI (GSCI) statement. Let (dom, P) be a probability model over V. A generalized CI statement $I(Y_1, \ldots, Y_k \mid X)$ is said to *hold for* (dom, P) if for every complete assignment $\mathbf{x}$ of X, and for every assignment $\mathbf{y}_i$ of Y_i for $i = 1, \ldots, k$,

$$P(\mathbf{y_1}, \ldots, \mathbf{y_k}, \mathbf{x}) \cdot P(\mathbf{x})^{k-1} = P(\mathbf{y_1}, \mathbf{x}) \cdot \ldots \cdot P(\mathbf{y_k}, \mathbf{x}). \tag{1}$$

Equivalently, (dom, P) is said to *satisfy* $I(Y_1, \ldots, Y_k \mid X)$.

The satisfaction of $I(Y_1, \ldots, Y_k \mid X)$ requires (1) to hold for *complete* assignments $\mathbf{x}$ of X only. The reason is that the pairwise independence between the Y_i is conditional on X. That is, assignments that have *no information* about some random variable in X are not taken into account when judging the pairwise independence between the Y_i.

The probability model from the introduction satisfies σ_1 and σ_2, but not φ. In particular, the two assignments are not complete on *Gender*, i.e., σ_2 is satisfied.

The expressions $I(Y_1, \ldots, Y_k \mid X)$ are generalized in the sense that they cover CI statements as the special case where $k = 2$. We assume w.l.o.g. that the sets Y_i are non-empty. Indeed, for all $k > 1$ we have the property that a probability distribution satisfies $I(\emptyset, Y_2, \ldots, Y_k \mid X)$ if and only if the probability distribution satisfies $I(Y_2, \ldots, Y_k \mid X)$. If $k = 1$, then $I(Y \mid X)$ is always satisfied, and we identify $I(\emptyset \mid X)$ with $I(\cdot \mid X)$. One may now define an equivalence relation over the set of GCI statements. Indeed, two such GCI statements are equivalent whenever they are satisfied by the same probability distributions. Strictly speaking, we will apply inference rules to these equivalence classes. For the sake of simplicity, however, we assume that in GCI statements $I(Y_1, \ldots, Y_k \mid X)$ the sets Y_i are non-empty. As we have just argued this is just a suitable choice of a representative from the equivalence classes.

Let $\Sigma \cup \{\varphi\}$ be a set of GCI statements over (V, C). We say that Σ C-*implies* φ, denoted by $\Sigma \models_C \varphi$, if every probability model over (V, C) that satisfies Σ also satisfies φ. The *implication problem* for GCI statements under controlled uncertainty is defined as follows.

PROBLEM:	Implication problem
INPUT:	(S,C), Set $\Sigma \cup \{\varphi\}$ of GCI statements over (S,C)
OUTPUT:	Yes, if $\Sigma \models_C \varphi$; No, otherwise

For $\Sigma = \{\sigma_1, \sigma_2\}$, one can observe that Σ C-implies φ if and only if $Gender \in C$. Indeed, if $Gender \notin C$, then the two assignments from the introduction form a probability model over (V,C) that satisfies Σ, but violates φ.

For Σ we let $\Sigma_C^* = \{\varphi \mid \Sigma \models_C \varphi\}$ be the *semantic closure* of Σ, i.e., the set of all GCI statements C-implied by Σ. In order to determine the C-implied GCI statements we use a syntactic approach by applying inference rules. These inference rules have the form $\dfrac{premise}{conclusion}$ *condition* and inference rules without any
premises and any condition are called axioms. The premise consists of a finite set of GCI statements, and the conclusion is a singleton GCI statement. The condition of the rule is simple in the sense that it stipulates a simple syntactic restriction on the application of the rule. Instead of using this graphical representation, we could also state our rules in the form of Horn rules. An inference rule is called *sound*, if every probability model over (V,C) that satisfies every GCI statement in the premise of the rule also satisfies the GCI statement in the conclusion of the rule, given that the condition is satisfied. We let $\Sigma \vdash_{\mathfrak{R}} \varphi$ denote the *inference* of φ from Σ by the set $\mathfrak{R}$ of inference rules. That is, there is some sequence $\gamma = [\sigma_1, \ldots, \sigma_n]$ of GCI statements such that $\sigma_n = \varphi$ and every σ_i is an element of Σ or results from an application of an inference rule in $\mathfrak{R}$ to some elements in $\{\sigma_1, \ldots, \sigma_{i-1}\}$. For Σ, let $\Sigma_{\mathfrak{R}}^+ = \{\varphi \mid \Sigma \vdash_{\mathfrak{R}} \varphi\}$ be its *syntactic closure* under inferences by $\mathfrak{R}$. A set $\mathfrak{R}$ of inference rules is said to be *sound* (*complete*) for the implication of GCI statements under controlled uncertainty, if for every V, every $C \subseteq V$ and for every set Σ of GCI statements over (V,C) we have $\Sigma_{\mathfrak{R}}^+ \subseteq \Sigma_C^*$ ($\Sigma_C^* \subseteq \Sigma_{\mathfrak{R}}^+$). The (finite) set $\mathfrak{R}$ is said to be a (finite) *axiomatization* for the implication of GCI statements under controlled uncertainty if $\mathfrak{R}$ is both sound and complete.

The focus of the paper is the implication problem of GSCI statements under controlled uncertainty. There are good reasons for this limitation. Already for the special case where $k = 2$ and where all variables are complete, CI statements do not enjoy a finite ground axiomatization in the form of Horn rules [33]. While, in this special case, stable CI statements do enjoy a finite ground axiomatization, their implication problem is coNP-hard to decide [28]. In the idealized special case, Geiger and Pearl have established a finite axiomatization for the implication of SCI statements [13]. That is, if for any given V, C is assumed to be V, then the set $\mathfrak{C}$ of inference rules from Table 1 forms an axiomatization for the implication of SCI statements in form of a finite set of Horn rules. Note that the algebra rule ($\mathcal{A}$):

$$\frac{I(YY', ZZ' \mid X) \quad I(YZ, Y'Z' \mid X)}{I(YY'Z, Z' \mid X)}$$

can be derived from $\mathfrak{C}$, and is thus sound for the implication of SCIs under certainty.

Table 1. Axiomatization $\mathfrak{C}$ of SCI statements over V

$$\frac{}{I(V - X, \emptyset \mid X)} \qquad \frac{I(Y, Z \mid X)}{I(Z, Y \mid X)}$$

(saturated trivial independence, $\mathcal{T}'$)   (symmetry, $\mathcal{S}'$)

$$\frac{I(ZW, Y \mid X) \quad I(Z, W \mid XY)}{I(Z, YW \mid X)} \qquad \frac{I(Y, ZW \mid X)}{I(Y, Z \mid XW)}$$

(weak contraction, $\mathcal{C}$)   (weak union, $\mathcal{W}'$)

The following lemma shows that $\mathfrak{C}$ does not form a finite axiomatization for the implication of SCI statements under controlled uncertainty.

Lemma 1. *The weak contraction rule* $(\mathcal{C})$ *is not sound for the implication of SCI statements under controlled uncertainty.*

Proof. It suffices to find some probability model (dom, P) over (V, C) that satisfies $I(ZW, Y \mid X)$ and $I(Z, W \mid XY)$, but violates $I(Z, YW \mid X)$. Such a probability model has already been given in the introduction where $V = \{Cancer, Cavity, Gender, Smoking\}$, $C = \{Cancer, Cavity, Gender\}$, $Z = \{Cancer\}$, $W = \{Cavity\}$, $Y = \{Gender\}$, and $X = \{Smoking\}$.   □

The reason that the probability model from the introduction satisfies $\sigma_2 = I(\{Cancer\}, \{Cavity\} \mid \{Smoking, Gender\})$ is that the assignments carry μ on $Gender$. That is, there is no complete evidence for the violation of σ_2. This semantics of GCI statements achieves that the classical implication problem in the absence of missing data is soundly approximated by the implication problem under controlled uncertainty. The price to pay for permitting missing data is therefore a loss in completeness, e.g., while φ is implied by σ_1 and σ_2 classically, it is no longer implied when $Gender$ is not a complete random variable.

3   Axiomatization

We show that the set $\mathfrak{G}$ of rules in Table 2 is a finite axiomatization for the implication of GSCI statements under controlled uncertainty. We first argue that the rules in $\mathfrak{G}$ are sound. There are three key observations. The first is that a GSCI statement $I(Y_1, \ldots, Y_k \mid X)$ is satisfied by a probability model $\pi = (dom, P)$ over (V, C) if and only if for $i = 1, \ldots, k$, π satisfies the SCI statement $I(Y_i, Y_1 \cdots Y_k - Y_i \mid X)$. The second key observation is that, for $(\mathcal{T})$, $(\mathcal{P})$, and $(\mathcal{W})$, the conditions in the conclusion contain the conditions of all its premises. If there is a probability model that violates the conclusion, then there is an assignment which is complete on the condition and violates Equation (1). Hence, the same assignment is complete on the conditions of all premises. The soundness of the inference rules $(\mathcal{T}')$, $(\mathcal{S})$, and $(\mathcal{W}')$ [13] means that one of the premises is also violated. This shows the soundness of $(\mathcal{T})$, $(\mathcal{P})$, and

Table 2. GCSI-Axiomatization $\mathfrak{S} = \{(\mathcal{T}), (\mathcal{P}), (\mathcal{M}), (\mathcal{W}), (\mathcal{R})\}$ over (V, C)

$$\frac{}{I(V \mid \emptyset)}$$
(saturated trivial independence, $\mathcal{T}$)

$$\frac{I(Y_1, \ldots, Y_k \mid X)}{I(Y_{\pi(1)}, \ldots, Y_{\pi(k)} \mid X)}$$
(permutation, $\mathcal{P}$)

$$\frac{I(Y_1, \ldots, Y_{k-1}, Y, Z \mid X)}{I(Y_1, \ldots, Y_{k-1}, YZ \mid X)}$$
(merging, $\mathcal{M}$)

$$\frac{I(Y_1, \ldots, Y_{k-1}, ZW \mid X)}{I(Y_1, \ldots, Y_{k-1}, Z \mid XW)}$$
(weak union, $\mathcal{W}$)

$$\frac{I(Z_1 \cdots Z_k, YW_1 \cdots W_k \mid X) \quad I(Z_1 W_1, \ldots, Z_k W_k \mid XY)}{I(Z_1, \ldots, Z_k, YW_1 \cdots W_k \mid X)} \, Y \subseteq C$$
(restricted weak contraction, $\mathcal{R}$)

($\mathcal{W}$). The soundness of the merging rule ($\mathcal{M}$) is a consequence of the first key observation, and the soundness of the algebra rule ($\mathcal{A}$), already when all variables are complete. It remains to verify the soundness of ($\mathcal{R}$), to which the second key observation does not apply. If all variables are complete, the rule ($\mathcal{C}'$) is sound:

$$\frac{I(ZW, YUV \mid X) \quad I(ZU, VW \mid XY)}{I(Z, YUVW \mid X)}.$$

The rule ($\mathcal{R}'$), resulting from ($\mathcal{C}'$) by adding the condition $Y \subseteq C$, is sound for SCI statements under controlled uncertainty. Indeed, assignments which are complete on the condition X are also complete on XY as $Y \subseteq C$ holds. The soundness of ($\mathcal{R}$) follows from the first key observation and that of ($\mathcal{R}'$).

For some V, some $C \subseteq V$, and some set Σ of GSCI statements over (V, C), and some $X \subseteq V$ let $IDep_{\Sigma,C}(X) := \{Y \subseteq V - X \mid \Sigma \vdash_{\mathfrak{S}} I(Y, Z \mid X)\}$ denote the set of all $Y \subseteq V - X$ such that $I(Y, Z \mid X)$ can be inferred from Σ by $\mathfrak{S}$. The soundness of the algebra rule ($\mathcal{A}$) implies that

$$(IDep_{\Sigma,C}(X), \subseteq, \cup, \cap, (\cdot)^{\mathcal{C}}, \emptyset, V - X)$$

forms a finite Boolean algebra where $(\cdot)^{\mathcal{C}}$ maps a set Y to its complement $V - XY$. Recall that an element $a \in P$ of a poset $(P, \sqsubseteq, 0)$ with least element 0 is called an *atom* of $(P, \sqsubseteq, 0)$ precisely when $a \neq 0$ and every element $b \in P$ with $b \sqsubseteq a$ satisfies $b = 0$ or $b = a$ [16]. Further, $(P, \sqsubseteq, 0)$ is said to be *atomic* if for every element $b \in P - \{0\}$ there is an atom $a \in P$ with $a \sqsubseteq b$. In particular, every finite Boolean algebra is atomic [16]. Let $IDepB_{\Sigma,C}(X)$ denote the set of all atoms of $(IDep_{\Sigma,C}(X), \subseteq, \emptyset)$. We call $IDepB_{\Sigma,C}(X)$ the *independence basis* of X with respect to Σ and C. The importance of this notion is manifested in the following result.

Theorem 1. *Let Σ be a set of GSCI statements over (V, C). Then $\Sigma \vdash_{\mathfrak{S}} I(Y_1, \ldots, Y_k \mid X)$ if and only if for all $i = 1, \ldots, k$, $Y_i = \bigcup \mathcal{Y}$ for some $\mathcal{Y} \subseteq IDepB_{\Sigma,C}(X)$.*

Proof. Let $\Sigma \vdash_\mathfrak{G} I(Y_1, \ldots, Y_k \mid X)$. Then for all $i = 1, \ldots, k$, $\Sigma \vdash_\mathfrak{G} I(Y_i, S - XY_i \mid X)$ by the merging rule $\mathcal{M}$. Hence, for all $i = 1, \ldots, k$, $Y_i \in IDep_\Sigma(X)$. Since every element b of a Boolean algebra is the union over those atoms a with $a \subseteq b$ it follows that for all $i = 1, \ldots, k$, $Y_i = \bigcup \mathcal{Y}$ for $\mathcal{Y} = \{B \in IDepB_\Sigma(X) \mid B \subseteq Y_i\}$.

Vice versa, let for $i = 1, \ldots, k$, $Y_i = \bigcup \mathcal{Y}$ for some $\mathcal{Y} \subseteq IDepB_{\Sigma,C}(X)$. Let $IDepB_{\Sigma,C}(X) = \{B_1, \ldots, B_n\}$. Then $I(B_1, \ldots, B_n \mid X) \in \Sigma_\mathfrak{G}^+$ holds. Successive applications of the merging rule $(\mathcal{M})$ give $I(Y_1, \ldots, Y_k \mid X) \in \Sigma_\mathfrak{G}^+$.   □

For $\Sigma = \{\sigma_1, \sigma_2\}$ and $C = \{Cancer, Cavity, Gender\}$, we have $IDepB_{\Sigma,C}(Smoking) = \{\{Cancer, Cavity\}, \{Gender\}\}$. Hence, $\varphi \notin \Sigma_C^*$. If $Gender \in C$, then $\varphi \in \Sigma_C^*$.

Theorem 2. *The set $\mathfrak{G}$ is sound and complete for the implication of GSCI statements under controlled uncertainty.*

Proof. It remains to show completeness. Let $\Sigma \cup \{I(Y_1, \ldots, Y_m \mid X)\}$ be a set of GSCI statements over (V, C), and suppose that $I(Y_1, \ldots, Y_m \mid X) \notin \Sigma_\mathfrak{G}^+$. We will show that $I(Y_1, \ldots, Y_k \mid X) \notin \Sigma_C^*$.

Let $IDepB_{\Sigma,C}(X) = \{B_1, \ldots, B_n\}$, in particular $V = XB_1 \cdots B_n$. Since $I(Y_1, \ldots, Y_m \mid X) \notin \Sigma_\mathfrak{G}^+$ we conclude by Theorem 1 that there is some $j \in \{1, \ldots, m\}$ such that Y_j is not the union of some elements of $IDepB_{\Sigma,C}(X)$. Consequently, there is some $i \in \{1, \ldots, n\}$ such that $Y_j \cap B_i \neq \emptyset$ and $B_i - Y_j \neq \emptyset$ hold. Let $T := \bigcup_{j \neq i} B_j \cap C$, and $T' := \bigcup_{j \neq i} B_j - C$. In particular, V is the disjoint union of X, T, T', and B_i. For every $v \in V - C$ we define $dom(v) = \{0, 1, \mu\}$; and for every $v \in C$ we define $dom(v) = \{0, 1\}$. We define the following two assignments $\mathbf{a}_1$ and $\mathbf{a}_2$ of (V, C). We define $\mathbf{a}_1(v) = \mathbf{0}$ for all $v \in XB_iT$, $\mathbf{a}_1(v) = \mu$ for all $v \in T'$. We further define $\mathbf{a}_2(v) = \mathbf{a}_1(v)$ for all $v \in XTT'$, and $\mathbf{a}_2(v) = \mathbf{1}$ for all $v \in B_i$. As probability measure we define $P(\mathbf{a}_1) = P(\mathbf{a}_2) = 0.5$. The probability model is illustrated in Table 3. It follows from the construction that (dom, P) does not satisfy $I(Y_1, \ldots, Y_m \mid X)$.

Table 3. Probability model in the completeness proof of $\mathfrak{G}$

XT	T'	B_i	P
$0 \cdots 0$	$\mu \cdots \mu$	$0 \cdots 0$	0.5
$0 \cdots 0$	$\mu \cdots \mu$	$1 \cdots 1$	0.5

We show that (dom, P) satisfies every $I(V_1, \ldots, V_k \mid U) \in \Sigma$. Suppose for some complete assignment $\mathbf{u}$ of U, $P(\mathbf{u}) = 0$. Then equation (1) is satisfied.

If $P(\mathbf{u}, \mathbf{v}_o) = 0$ for some assignment $\mathbf{u}$ of U, and for some assignment $\mathbf{v}_o$ of V_o, then $P(\mathbf{u}, \mathbf{v}_1, \ldots, \mathbf{v}_k) = 0$. Then equation (1) is also satisfied. Suppose that for some complete assignment $\mathbf{u}$ of U, $P(\mathbf{u}) = 0.5$. If for some assignments $\mathbf{v}_l$ of V_l for $l = 1, \ldots, k$, $P(\mathbf{u}, \mathbf{v}_1) = \cdots = P(\mathbf{u}, \mathbf{v}_k) = 0.5$, then $P(\mathbf{u}, \mathbf{v}_1, \ldots, \mathbf{v}_k) = 0.5$, too. Again, equation (1) is satisfied.

It remains to consider the case where **u** is a complete assignment of U such that $P(\mathbf{u}) = 1$. In this case, the construction of the probability model tells us that $U \subseteq XT$. Consequently, we can apply the weak union ($\mathcal{W}$) and permutation rules ($\mathcal{P}$) to $I(V_1, \ldots, V_k \mid U) \in \Sigma$ to infer $I(V_1 - XT, \ldots, V_k - XT \mid XT) \in \Sigma_{\mathfrak{G}}^+$. Theorem 1 also shows that $I(B_i, TT' \mid X) \in \Sigma_{\mathfrak{G}}^+$. For $i = 1, \ldots, k$, let $Z_i = V_i - XTT'$ and $W_i = (V_i - XT) \cap T'$. Then, $B_i = Z_1 \cdots Z_k$, and $T' = W_1 \cdots W_k$. Thus, applying the restricted contraction rule ($\mathcal{R}$) to $I(Z_1 \cdots Z_k, TW_1 \cdots W_k \mid X)$ and $I(Z_1 W_1, \ldots, Z_k W_k \mid XT)$ we infer $I(Z_1, \ldots, Z_k, YW_1 \cdots W_k \mid X) \in \Sigma_{\mathfrak{G}}^+$, since $T \subseteq C$. It follows from Theorem 1 that $V_l - XT$, for every $l = 1, \ldots, k$, is the union of elements from $IDepB_{\Sigma,C}(X)$. Consequently, $V_o - XT = B_i$ for some $o \in \{1, \ldots, k\}$, and $V_l - XT = \emptyset$ for all $l \in \{1, \ldots, k\} - \{o\}$. Then, we are either in the previous case where $P(\mathbf{u}, \mathbf{v}_o) = 0$; or, $P(\mathbf{u}, \mathbf{v}_o) = 0.5$, and $P(\mathbf{u}, \mathbf{v}_l) = 1$ for every $l \in \{1, \ldots, k\} - \{o\}$, and $P(\mathbf{u}, \mathbf{v}_1, \ldots, \mathbf{v}_k) = 0.5$. Again, equation (1) is satisfied. This concludes the proof. $\qquad\square$

The specific probability model from the introduction is a counterexample for the implication of φ from $\Sigma = \{\sigma_1, \sigma_2\}$ with $C = \{\,Cancer,\ Cavity,\ Smoking\,\}$. It follows the construction in the proof of Theorem 2.

Specific instances of the implication problem for GSCI statements under controlled uncertainty are equivalent to sliced versions of the same instance under certainty. For a set Σ of GSCI statements over (V, C), let $\Sigma[XC]$ denote the set of those $I(V_1, \ldots, V_m \mid U) \in \Sigma$ where $U \subseteq XC$.

Theorem 3. *Let* $\Sigma \cup \{I(Y_1, \ldots, Y_k \mid X)\}$ *be a set of GSCI statements over* (V, C). *Then* $\Sigma \models_C I(Y_1, \ldots, Y_k \mid X)$ *iff* $\Sigma[XC] \models_V I(Y_1, \ldots, Y_k \mid X)$.

Proof. Assume that $\Sigma[XC] \models_V I(Y_1, \ldots, Y_k \mid X)$ does not hold. Then there is a special probability model $\pi' = (dom, \{\mathbf{a}_1', \mathbf{a}_2'\})$ over (V, V) that satisfies every $\sigma \in \Sigma[XC]$ and violates $I(Y_1, \ldots, Y_k \mid X)$. With the notation from the proof of Theorem 2, for all $v \in V$, $\mathbf{a}_1'(v) = \mathbf{a}_2'(v)$ iff $v \notin B_i$, in particular $T' = \emptyset$. Define now the special probability model $\pi = (dom, \{\mathbf{a}_1, \mathbf{a}_2\})$ over (V, C), where $\mathbf{a}_l(v) = \mu$, if $v \in T'$, and $\mathbf{a}_l(v) = \mathbf{a}_l'(v)$, otherwise, for $l = 1, 2$. It follows that π satisfies every $\sigma \in \Sigma$ and violates $I(Y_1, \ldots, Y_k \mid X)$. In particular, for any $I(V_1, \ldots, V_m \mid U) \in \Sigma$ with $U \nsubseteq XC$ it follows that $U \cap (B_i T') \neq \emptyset$ and thus $\mathbf{a}_1$ and $\mathbf{a}_2$ are not U-complete. Hence, $\Sigma \models_C I(Y_1, \ldots, Y_k \mid X)$ does also not hold. The same arguments also work in the other direction. If π over (V, C) satisfies Σ and violates $I(Y_1, \ldots, Y_k \mid X)$, then π' satisfies $\Sigma[XC]$ and violates $I(Y_1, \ldots, Y_k \mid X)$. $\qquad\square$

For example, $\Sigma[Smoking, Cavity, Cancer] = \{\sigma_1\}$ and $\{\sigma_1\} \not\models_V \varphi$ as the following probability model shows:

Smoking	Gender	Cancer	Cavity	P
true	true	true	true	0.5
true	true	false	false	0.5

Note that this probability model violates σ_2.

4   Equivalence to Database Decompositions

Let $\mathfrak{A} = \{v_1, v_2, \ldots\}$ be a countably infinite set of *attributes*. A *relation schema* is a finite non-empty subset R of $\mathfrak{A}$. Each attribute v of R is associated with a set $dom(v)$, called the *domain* of v. In order to encompass incomplete information the domain of each attribute contains the null marker, denoted by $\mu \in dom(v)$. A *tuple* over R is a function $t : R \rightarrow \bigcup_{v \in R} dom(v)$ with $t(v) \in dom(v)$ for all $v \in R$. The null marker occurrence $t(v) = \mu$ associated with an attribute v in a tuple t means that "no information" is available about the value $t(v)$ of t on v. For $X \subseteq R$ let $t(X)$ denote the restriction of the tuple t over R to X. A (partial) *relation* r over R is a finite set of tuples over R. Let t_1 and t_2 be two tuples over R. It is said that t_1 *subsumes* t_2 if for every attribute $v \in R$, $t_1(v) = t_2(v)$ or $t_2(v) = \mu$ holds. In consistency with previous work [1,10,18,22,37]: No relation shall contain two tuples t_1 and t_2 such that t_1 subsumes t_2. For a tuple t over R and a set $X \subseteq R$, t is said to be X-complete, if for all $v \in X$, $t(v) \neq \mu$. Similar, a relation r over R is said to be X-complete, if every tuple t of r is X-complete. A relation r over R is said to be a *complete relation*, if it is R-complete. We recall the definition of projection and join operations on partial relations [1,22]. Let r be some relation over R. Let X be some subset of R. The *projection* $r(X)$ of r on X is the set of tuples t for which (i) there is some $t_1 \in r$ such that $t = t_1(X)$ and (ii) there is no $t_2 \in r$ such that $t_2(X)$ subsumes t and $t_2(X) \neq t$. For $Y \subseteq X$, the Y-complete projection $r_Y(X)$ of r on X is $r_Y(X) = \{t \in r(X) \mid t \text{ is } Y\text{-complete}\}$. Given X-complete relations r over R and s over S such that $X = R \cap S$ the *natural join* $r \bowtie s$ of r and s is the relation over $R \cup S$ which contains those tuples t such that there are some $t_1 \in r$ and $t_2 \in s$ with $t_1 = t(R)$ and $t_2 = t(S)$ [22].

A *full first-order hierarchical decomposition* (FOHD) over the relation schema R is an expression $X : [Y_1 \mid \ldots \mid Y_k]$ with a non-negative integer k, $X, Y_1, \ldots, Y_k \subseteq R$ such that $Y_1, \ldots, Y_k$ form a partition of $R - X$. A relation r over R is said to *satisfy* the FOHD $X : [Y_1 \mid \cdots \mid Y_k]$ over R, denoted by $\models_r X : [Y_1 \mid \cdots \mid Y_k]$, iff $r_X = (\cdots (r_X(XY_k) \bowtie r_X(XY_{k-1})) \bowtie \cdots) \bowtie r_X(XY_1)$.

The FOHD $\emptyset : [Y_1 \mid \cdots \mid Y_k]$ expresses the fact that any relation over R is the Cartesian product over its projections to attribute sets in $\{Y_i\}_{i=1}^{k}$. For $k = 0$, the FOHD $X : [\,]$ is satisfied trivially, where $[\,]$ denotes the empty list.

Following Atzeni and Morfuni [1], a *null-free subschema* (NFS) over the relation schema R is an expression $nfs(C)$ where $C \subseteq R$. The NFS $nfs(C)$ over R is satisfied by a relation r over R, denoted by $\models_r nfs(C)$, if and only if r is C-complete. SQL allows the specification of attributes as NOT NULL. NFSs occur in everyday database practice: the set of attributes declared NOT NULL forms the NFS over the underlying relation schema.

For a set $\hat{\Sigma} \cup \{\hat{\varphi}\}$ of FOHDs, and an NFS $nfs(C)$ over relation schema R, we say that $\hat{\Sigma}$ C-implies $\hat{\varphi}$, denoted by $\hat{\Sigma} \models_C \hat{\varphi}$, if and only if every C-complete relation r that satisfies $\hat{\Sigma}$ also satisfies $\hat{\varphi}$. The *implication problem* for FOHDs in the presence of a null-free subschema is defined as follows.

PROBLEM:	Implication problem for FOHDs
INPUT:	Relation schema R with NFS $nfs(C)$, Set $\hat{\Sigma} \cup \{\hat{\varphi}\}$ of FOHDs over R
OUTPUT:	Yes, if $\hat{\Sigma} \models_C \hat{\varphi}$; No, otherwise

For a GSCI $\varphi = I(Y_1, \ldots, Y_k \mid X)$ over (V, C) we define the corresponding FOHD $\hat{\varphi} = X : [Y_1 \mid \cdots \mid Y_k]$, and for a set Σ of GSCI over (V, C) let $\hat{\Sigma} = \{\hat{\sigma} \mid \sigma \in \Sigma\}$ denote the corresponding set of FOHDs over $R = V$ with NFS $nfs(C)$. Let $(\hat{\mathcal{T}}), (\hat{\mathcal{P}}), (\hat{\mathcal{R}}), (\hat{\mathcal{W}}), (\hat{\mathcal{M}})$ denote the inference rules for FOHDs under NFS $nfs(C)$ over R that result from $(\mathcal{T}), (\mathcal{P}), (\mathcal{R}), (\mathcal{W}), (\mathcal{M})$ by the translation above. Following the lines of reasoning for GSCIs, one obtains an axiomatization for FOHD implication in the presence of an NFS. The special probability model π reduces to a two-tuple relation r with the two assignments in π. Completeness follows from the key observation that r satisfies an FOHD $U : [V_1, \ldots, V_m]$ if and only if $U \cap (T'B_i) \neq \emptyset$ or $B_i \subseteq V_o$ for some $o \in \{1, \ldots, m\}$.

Theorem 4. *The set $\hat{\mathfrak{S}} = \{(\hat{\mathcal{T}}), (\hat{\mathcal{P}}), (\hat{\mathcal{R}}), (\hat{\mathcal{W}}), (\hat{\mathcal{M}})\}$ is a finite axiomatization for the implication of FOHDs in the presence of a null-free subschema.* □

A C-complete two-tuple relation over R that satisfies $\hat{\Sigma}$ and violates $\hat{\varphi}$ becomes a special probability model over (V, C) that satisfies Σ and violates φ, simply by assigning probability one half to each tuple. Vice versa, removing the probabilities from the assignments of the special probability model results in a two-tuple relation with the desired characteristics.

Theorem 5. *Let $\Sigma \cup \{I(Y_1, \ldots, Y_k)\}$ denote a set of GSCI statements over (V, C). Then $\Sigma \models_C I(Y_1, \ldots, Y_k \mid X)$ if and only if $\hat{\Sigma} \models_C X : [Y_1 \mid \cdots \mid Y_k]$.* □

For $\hat{\Sigma} = \{\hat{\sigma}_1, \hat{\sigma}_2\}$ and $C = \{Cancer, Cavity, Smoking\}$, we have $\hat{\Sigma} \not\models_C \hat{\varphi}$. The relation with the two tuples from the introduction is not the natural join of the projections:

Smoking	Cancer
true	true
true	false

Smoking	Gender	Cavity
true	μ	true
true	μ	false

.

The results show that the implication problem of GSCI statements can be efficiently decided without considering probabilities at all. This extends findings for SCI statements [23,25,24].

5   Algorithm and Complexity

In practice it often suffices to know if an GSCI statement φ is C-implied by Σ. Then it is inefficient to determine Σ_C^*. Let $\Sigma \cup \{I(Y_1, \ldots, Y_k \mid X)\}$ be a set of GSCIs over (V, C), and let $\overline{\Sigma}$ result from Σ by replacing each $I(V_1, \ldots, V_m \mid U) \in \Sigma$ by the m multivalued dependencies $U \twoheadrightarrow V_i$. Indeed, the MVD $U \twoheadrightarrow V_i$ is equivalent to the binary FOHD $U : [V_i \mid V_1 \cdots V_m - V_i]$. Applying Theorems 3 and 5, $\Sigma \models_C \{I(Y_1, \ldots, Y_k \mid X)\}$ iff for all $i = 1, \ldots, k$, $\overline{\Sigma}[XC] \models X \twoheadrightarrow Y_i$.

Corollary 1. *Using the algorithm in [12], the implication problem* $\Sigma \models_C$ $\{I(Y_1, \ldots, Y_k \mid X)\}$ *can be decided in time* $\mathcal{O}(|\overline{\Sigma}| + \min\{n_{\overline{\Sigma[XC]}}, \log(k \cdot p_{\overline{\Sigma[XC]}})\} \times$ $|\overline{\Sigma[XC]}|)$. *Herein,* $|\overline{\Sigma}|$ *denotes the total number of attributes in* $\overline{\Sigma}$, $n_{\overline{\Sigma}}$ *denotes the cardinality of* $\overline{\Sigma}$, *and* $p_{\overline{\Sigma}}$ *denotes the maximum among the numbers of sets in* $IDepB_{\Sigma,C}(X)$ *that have non-empty intersection with* Y_i, $i = 1, \ldots, k$.   □

6   Non-extensibility of Findings

We would like to make a few remarks on the non-extensibility of the findings we have presented here. These are well documented in the research literature. While we have established an axiomatization of GSCI implication under controlled uncertainty by a finite set of Horn rules, this is impossible to achieve for GCI statements under controlled uncertainty. Indeed, already in the special case of CI statements under certainty, i.e. where we consider conditional independence between two sets of random variables only and where all variables are complete, it is known that an axiomatization by a finite set of Horn rules does not exist [33]. Recently, an axiomatization by a finite set of Horn rules was established for stable conditional independence statements [29], and the associated implication problem shown to be coNP-hard [28]. Therefore, GSCI statements form an expressive and efficient fragment of stable CI statements.

Similarly, one may consider first-order hierarchical decompositions $X : [Y_1 \mid \cdots \mid Y_k]$ where $XY_1 \cdots Y_k$ may not contain all underlying attributes. Again, no axiomatization by a finite set of Horn rules can exist since it is known that such an axiomatization does not exist for embedded multivalued dependencies [32], i.e. in the special case where $k = 2$ and where complete relations are considered only. It is further known that the implication problem of embedded multivalued dependencies is undecidable [19].

Our equivalence between the implication of GSCIs under controlled uncertainty and the implication of FOHDs in the presence of a null-free subschema cannot be extended to cover GCIs and arbitrary first-order hierarchical decompositions. Indeed, the implication problems of conditional independence statements and embedded multivalued dependencies do not coincide, already for the special where all variables (attributes) are complete [33].

7   Conclusion and Future Work

Real-world data samples contain missing values that cannot all be revealed. For this purpose we have introduced an appropriate semantics for statements about the pairwise conditional independence between finitely many sets of random variables. Random variables can be declared complete when missing data are known not occur for them, or when assignments with missing data on those variables are to be excluded from data samples. We have established characterizations of the associated implication problem for saturated independence statements in terms of axioms, an almost linear-time algorithm, and database logic. Our findings do not extend to more expressive statements which are not axiomatizable, not equivalent to database logic, or likely to be intractable. Our results therefore contribute to a solid foundation for probabilistic reasoning about uncertainty under the realistic assumption of missing values.

For future work the impact of our approach on Bayesian networks should be investigated in depth. Immediate open questions concern axiomatizations for (generalized) marginal and stable conditional independence statements under controlled uncertainty. It would be interesting to investigate the implication problem of GSCI statements under controlled uncertainty when the set of random variables is undetermined, similar to the implication problem of SCI statements under certainty [3].

References

1. Atzeni, P., Morfuni, N.: Functional dependencies and constraints on null values in database relations. Information and Control 70(1), 1–31 (1986)
2. Batista, G., Monard, M.: An analysis of four missing data treatment methods for supervised learning. Applied Artificial Intelligence 17(5-6), 519–533 (2003)
3. Biskup, J., Hartmann, S., Link, S.: Probabilistic conditional independence under schema certainty and uncertainty. In: Hüllermeier, E., Link, S., Fober, T., Seeger, B. (eds.) SUM 2012. LNCS, vol. 7520, pp. 365–378. Springer, Heidelberg (2012)
4. Chickering, D., Heckerman, D.: Efficient approximations for the marginal likelihood of Bayesian networks with hidden variables. Machine Learning 29(2-3), 181–212 (1997)
5. Darwiche, A.: Modeling and Reasoning with Bayesian Networks. Cambridge University Press (2009)
6. Dawid, A.: Conditional independence in statistical theory. Journal of the Royal Statistical Society 41(1), 1–31 (1979)
7. Delobel, C.: Normalization and hierarchical dependencies in the relational data model. ACM Trans. Database Syst. 3(3), 201–222 (1978)
8. Dempster, A., Laird, N., Rubin, D.: Maximum likelihood from incomplete data via the EM algorithm. Journal of the Royal Statistical Society B 39, 139 (1977)
9. Fayyad, U., Piatetsky-Shapiro, G., Smyth, P.: From data mining to knowledge discovery in databases. AI Magazine 17(3), 37–54 (1996)
10. Ferrarotti, F., Hartmann, S., Link, S.: Reasoning about functional and full hierarchical dependencies over partial relations. Inf. Sci. 235, 150–173 (2013)
11. Friedman, N.: Learning belief networks in the presence of missing values and hidden variables. In: ICML, pp. 125–133 (1997)
12. Galil, Z.: An almost linear-time algorithm for computing a dependency basis in a relational database. J. ACM 29(1), 96–102 (1982)
13. Geiger, D., Pearl, J.: Logical and algorithmic properties of conditional independence and graphical models. The Annals of Statistics 21(4), 2001–2021 (1993)
14. Geiger, D., Pearl, J.: Logical and algorithmic properties of independence and their application to Bayesian networks. Ann. Math. Artif. Intell. 2, 165–178 (1990)
15. Grädel, E., Väänänen, J.: Dependence and independence. Studia Logica 101(2), 399–410 (2012)
16. Graetzer, G.: General lattice theory. Birkhäuser, Boston (1998)
17. Halpern, J.: Reasoning about uncertainty. MIT Press (2005)

18. Hartmann, S., Link, S.: The implication problem of data dependencies over SQL table definitions: axiomatic, algorithmic and logical characterizations. ACM Trans. Database Syst. 37(2), Article 13 (2012)

19. Herrmann, C.: Corrigendum to "On the undecidability of implications between embedded multivalued database dependencies". Inf. Comput. 204(12), 1847–1851 (2006)

20. Kontinen, J., Link, S., Väänänen, J.: Independence in database relations. In: Libkin, L. (ed.) WoLLIC 2013. LNCS, vol. 8071, pp. 179–193. Springer, Heidelberg (2013)

21. Lauritzen, S.: The EM algorithm for graphical association models with missing data. Computational Statistics and Data Analysis 19, 191–201 (1995)

22. Lien, E.: On the equivalence of database models. J. ACM 29(2), 333–362 (1982)

23. Link, S.: Propositional reasoning about saturated conditional probabilistic independence. In: Ong, L., de Queiroz, R. (eds.) WoLLIC 2012. LNCS, vol. 7456, pp. 257–267. Springer, Heidelberg (2012)

24. Link, S.: Reasoning about saturated conditional independence under uncertainty: Axioms, algorithms, and Levesque's situations to the rescue. AAAI (2013)

25. Link, S.: Sound approximate reasoning about saturated probabilistic conditional independence under controlled uncertainty. J. Applied Logic (2013), http://dx.doi.org/10.1016/j.jal.2013.05.004

26. Marlin, B., Zemel, R., Roweis, S., Slaney, M.: Recommender systems, missing data and statistical model estimation. In: IJCAI, pp. 2686–2691 (2011)

27. Matúš, F.: Ascending and descending conditional independence relations. In: Transactions of the 11th Prague Conference on Information Theory, Statistical Decision Functions and Random Processes, pp. 189–200 (1992)

28. Niepert, M., Van Gucht, D., Gyssens, M.: Logical and algorithmic properties of stable conditional independence. Int. J. Approx. Reasoning 51(5), 531–543 (2010)

29. Niepert, M., Van Gucht, D., Gyssens, M.: On the conditional independence implication problem: A lattice-theoretic approach. In: UAI, pp. 435–443 (2008)

30. Pearl, J.: Probabilistic Reasoning in Intelligent Systems: Networks of Plausible Inference. Morgan Kaufmann, San Francisco (1988)

31. Singh, M.: Learning Bayesian networks from incomplete data. In: AAAI, pp. 534–539 (1997)

32. Stott Parker Jr., D., Parsaye-Ghomi, K.: Inferences involving embedded multivalued dependencies and transitive dependencies. In: SIGMOD, pp. 52–57 (1980)

33. Studený, M.: Conditional independence relations have no finite complete characterization. In: Transactions of the 11th Prague Conference on Information Theory, Statistical Decision Functions and Random Processes, pp. 377–396 (1992)

34. Thiesson, B.: Accelerated quantification of bayesian networks with incomplete data. In: KDD, pp. 306–311 (1995)

35. de Waal, P., van der Gaag, L.: Stable independence in perfect maps. In: UAI, pp. 161–168 (2005)

36. Wong, S., Butz, C., Wu, D.: On the implication problem for probabilistic conditional independency. IEEE Trans. Systems, Man, and Cybernetics, Part A: Systems and Humans 30(6), 785–805 (2000)

37. Zaniolo, C.: Database relations with null values. J. Comput. Syst. Sci. 28(1), 142–166 (1984)

Combinatorial Prediction Markets:
An Experimental Study

Walter A. Powell, Robin Hanson, Kathryn B. Laskey, and Charles Twardy

Volgenau School of Engineering,
George Mason University
4400 University Drive
Fairfax, VA 22030-4444 USA
{wpowell,klaskey,ctwardy}@gmu.edu

Abstract. Prediction markets produce crowdsourced probabilistic forecasts through a market mechanism in which forecasters buy and sell securities that pay off when events occur. Prices in a prediction market can be interpreted as consensus probabilities for the corresponding events. There is strong empirical evidence that aggregate forecasts tend to be more accurate than individual forecasts, and that prediction markets are among the most accurate aggregation methods. Combinatorial prediction markets allow forecasts not only on base events, but also on conditional events (e.g., "A if B") and/or Boolean combinations of events. Economic theory suggests that the greater expressivity of combinatorial prediction markets should improve accuracy by capturing dependencies among related questions. This paper describes the DAGGRE combinatorial prediction market and reports on an experimental study to compare combinatorial and traditional prediction markets. The experiment challenged participants to solve a "whodunit" murder mystery by using a prediction market to arrive at group consensus probabilities for characteristics of the murderer, and to update these consensus probabilities as clues were revealed. A Bayesian network was used to generate the "ground truth" scenario and to provide "gold standard" probabilistic predictions. The experiment compared predictions using an ordinary flat prediction market with predictions using a combinatorial market. Evaluation metrics include accuracy of participants' predictions and the magnitude of market updates. The murder mystery scenario provided a more concrete, realistic, intuitive, believable, and dynamic environment than previous empirical work on combinatorial prediction markets.

Keywords: Combinatorial prediction markets, crowdsourcing, Bayesian networks, combining expert judgment.

1   Crowdsourcing, Predictions, and Combinatorial Markets

Forecasting is important to business, national security, and society in general. Despite decades of research and immense resources dedicated to developing forecasting methods, getting better forecasts has proven elusive. Traditionally, forecasting has relied on judgments of a few experts. Measured on predictive accuracy, experts repeatedly

W. Liu, V.S. Subrahmanian, and J. Wijsen (Eds.): SUM 2013, LNAI 8078, pp. 283–296, 2013.
© Springer-Verlag Berlin Heidelberg 2013

disappoint, even when compared to simple statistical models like "no change" or linear models with equal or even random weights [1-6]. Furthermore, the data required for statistical models may be unavailable or inadequate. Recently, crowdsourcing has been shown to improve on the judgments of individual experts, or small groups of experts.

Common practice in crowdsourcing is to average judgments of a large group of individuals who have some knowledge of the problem [7]. Theory suggests that giving more weight to better forecasters should outperform a simple average, but in practice simple averaging has been surprisingly hard to beat. Recently, however, prediction markets have been shown to improve accuracy not only over individual or small groups of experts, but also over simple averaging [8-9]. A prediction market allows forecasters to aggregate information into a consensus probability distribution by purchasing assets that pay off contingent on an event of interest. Since resources for making predictions are limited, forecasters self-select to make forecasts for which they have the most information. Over time, the market gives greater weight to more successful forecasters. More accurate forecasters acquire greater resources with which to make further predictions; less accurate forecasters will lack the resources to have much influence on the consensus probabilities.

We are especially interested in forecasting many interrelated variables. For such problems, graphical models such as Bayesian networks provide a principled approach to modeling dependencies among variables. Pennock and Wellman [10] suggested the use of graphical models for belief aggregation. A combinatorial prediction market [8-11] increases the expressivity of an ordinary prediction market by allowing conditional forecasts (e.g., the probability of B given A is p) and/or Boolean combinations of events (e.g., the probability of B and A is q). Theory suggests that this greater expressivity, if appropriately captured in market prices, should give rise to more accurate forecasts. This is almost trivially true on joint forecasts, but should also hold for marginal forecasts when knowledge is distributed among participants and communication is primarily through the market. It should be particularly apparent if knowledge of correlations and knowledge of facts is held by different participants who communicate primarily via the market.

Specifying dependencies among forecasts using a graphical probability model allows tractable computation of a joint probability distribution among a large number of interdependent questions. If asset prices are set using a logarithmic market scoring rule (LMSR), then the assets can be factorized in a similar manner to probabilities, giving rise to similarly efficient algorithms for managing forecasters' assets [12].

For nearly two years the DAGGRE project [13-14] ran a public LMSR prediction market for geopolitical forecasting. The focus was on forecasting world events: usually questions with extended time horizons and significant irreducible uncertainty. These are the types of questions that have historically been the most vexing to intelligence analysts, economists and others. The DAGGRE market opened in October of 2011 as part of IARPA's Aggregate Contingent Estimation (ACE) program. The initial DAGGRE market was an ordinary ("flat") prediction market. In October of

2012, we launched a combinatorial prediction market. The market allowed users to forecast a question conditional on assumed values of another question. At any given time, there were on the order of 100 questions active on the market. Over time, some questions were removed as their outcomes became known, and new questions were added. Participants in the market were recruited from email solicitations, articles on blogs and newspapers, and personal recruiting at professional events. Participants received a small financial incentive for participating. Because of program restrictions, compensation did not depend on forecast accuracy, but the most accurate forecasters were recognized publicly on the DAGGRE site and listed on the leaderboard. Over the 20 months the market was open, more than 3000 participants contributed at least one forecast, with an average of about 150 forecasters per week. The market ran just over 400 total questions, about 200 of which were shared with four other teams in the IARPA-funded tournament.

Probability forecasts for the shared evaluation questions were reported daily to IARPA. When the outcome of an evaluation question became known, the question was scored by averaging the daily Brier score [15] over the period of time the target question was active. This approach has the benefit of rewarding forecasts that trend toward the correct outcome early during the period of time the question is being forecast. Forecasts were evaluated against a baseline system employing a uniformly weighted linear average of forecasts. Although early DAGGRE results were unreliable due to software issues, from February 2012 through May 2013, the DAGGRE market accuracy was about 38% greater than the baseline system. Accuracy was about the same before and after the launch of the combinatorial feature; however, usage of the combinatorial capability was low. About 10% of the users ever used the combinatorial feature, and only about 5% of the forecasts conditioned on another question. The DAGGRE prediction market closed in June of 2013 and will reopen in Fall 2013 with a change in focus to science and technology forecasting.

As a large-scale field study, the DAGGRE geopolitical market was not well suited to controlled experimentation. In this paper, we report a smaller-scale study that compares groups making the same predictions with the same information, one group using an ordinary flat market and the other using a combinatorial market.

2   Scope of the Experiment

The goal of the experiment was to investigate the effects on prediction market forecasts of allowing users to specify conditional probability links between questions. An experimental study [8] showed improved predictions with a combinatorial market, but on stylized forecasting problems with little face validity. The present experiment was designed to evaluate evidence and generate forecasts in a more concrete, realistic, intuitive, believable, and dynamic environment than previous work. Additionally, the experiment simulated the sequential nature of the flow of information in a prediction market.

3   Experimental Design

Using the actual DAGGRE market and real-world questions would have provided the most concrete, realistic, and believable environment, but using the actual market would cause experimental design challenges that could not easily be overcome. We therefore chose a "murder mystery" scenario to:

- Provide a concrete, realistic, intuitive, and believable, environment, in which relationships are based on statistical evidence familiar to the participants, e.g. men tend to be taller than women and people who wear bifocals tend to be older;
- Provide a common understanding of the basic relationships between the questions and clues upon which the belief structure (Bayes nets) could be built ;
- Use the same questions in a counterbalanced design;
- Control the delivery of information, providing clues sequentially (as in real prediction markets) and control the level of "expert" knowledge of the participants;
- Provide correct "gold standard" beliefs; and
- Control of the timing and order of the clues and outcomes of the questions, ensuring similar problems for different experimental runs.

The independent variables were:

- *Market* - the type of market, combinatorial or flat;
- *Market Order* - the order in which each type of market was used by the participants, combinatorial first or flat first.

The primary focus was on the effects of the *Market* variable. The *Market Order* variable allows analysis of interactions among the data due to learning effects. Each participant made predictions using both the combinatorial and flat markets.

In order to simulate the variation in levels of knowledge typical in prediction markets, different information concerning the relationships among the market questions and clues was distributed to the participants. In each session, each participant received conditional probability tables relating the market questions to each other and five of the ten conditional probability tables relating the clues (evidence) to the market questions. The conditional probability tables represented the expertise of each participant since the participants with a given table had the most accurate knowledge of the relationships between a specific clue and the market questions. Participants who didn't have access to specific tables had to rely on their general knowledge and the response of the market (including comments) to estimate the relationships between clues and questions.

The primary dependent variable was the accuracy of the predictions. For experiments on information aggregation, a common criterion is the ability to calculate ideal rational predictions given individual information and given the sum of all individual information. This ability allows us to define a mechanism's accuracy as the distance between an ideal distribution and the actual probability distribution produced by the mechanism. The measure of the accuracy of the predictions in each market was the Brier score. The Brier score is a proper scoring rule – that is, a forecaster minimizes his or her expected Brier score by accurately stating his/her true probability. The Brier score is often used as a measure to grade forecasts (Stevenson, et al. 2008). The Brier score is defined as

$$BrierScore = \frac{1}{N}\sum_{q=1}^{N}\sum_{s=1}^{R}(f_{qs} - o_{qs})^2 \ ,$$

where N is the number of questions, R is the number of possible outcomes for each question, f_{qs} is the probability forecast for outcome s of question q, and o_{qs} is an indicator (1 if yes; 0 if no) for whether the actual outcome for question q was s. Clearly, forecasts that predict the correct outcome with higher probabilities will result in lower Brier scores. The Brier score is a proper scoring rule, meaning that if outcomes are randomly generated according to a "gold standard" probability distribution, the Brier score is optimized by a forecaster who reports this "gold standard" distribution.

Also important in analyzing the participants' predictions in the expected Brier score, also known as the Brier prediction error. The expected Brier score is defined as

$$ExpectedBrierScore = \frac{1}{N}\sum_{q=1}^{N}\sum_{s=1}^{R}f_{qs}^{G}(f_{qs} - o_{qs})^2 \ ,$$

where f_{qs}^{BN} is the "gold standard" probability of a possible outcome. For our experiment, f_{qs}^{G} is the probability obtained from the Bayesian network used to generate the evidence, conditioned on the evidence that the forecaster has seen so far. The expected Brier score is measure of the inherent uncertainty in the problem – the best forecast that could be made given available evidence.

The experiment was conducted in sessions consisting of two paired trials designed so that each participant made predictions using both the combinatorial and flat markets. In each trial, the participants made predictions for five identical questions relating to characteristics of suspects in a "murder mystery." In the first trial of each session, half of the participants used an instance of the DAGGRE combinatorial market (Group A) and the remaining half used an instance of the DAGGRE flat market (Group B). In the second trial, each participant used the type of market he or she had not used on the first trial. In each trial, the instructions, training, market questions, and clue types were identical. Two scenarios – a series of clues and murderer characteristics, representing a specific murder scenario – were selected, one for each of the two trials. Each scenario used the same clues and murderer characteristics, but differed in the assigned values (e.g., female wearing heels vs. male wearing flats) and the order in which the clues were presented. The scenarios were constructed to be similar enough that effects due to scenario would be minimal.

Two sessions of the experiment have been conducted to date. The first session was conducted as part of the DAGGRE spring workshop in California for DAGGRE geopolitical market participants. Workshop attendees were all interested in geopolitical prediction markets, and ranged from novice to highly experienced. The second session was conducted at George Mason University using primarily third-year systems engineering students as participants. All GMU participants had passed a course in probability, and were considered to have the requisite critical thinking skills to understand quickly the functioning of the DAGGRE prediction markets. We wanted all participants to be familiar with the market, to reduce extraneous variation. All participants were given training such that they felt comfortable using both the flat and the combinatorial markets prior to beginning the experimental trials.

In order to provide the "gold standard" against which the participants' predictions could be compared, Bayesian networks representing the relationships between the market questions and clues were developed. The full Bayes net (Fig.1) was used as the basis for the relationships between market questions (dark) and clues (light). The flat Bayes net (Fig. 2) was obtained by removing links between market variables and setting their distributions to the marginal distributions obtained from the full Bayes net. The flat Bayes net was adopted as the "gold standard" for the non-combinatorial condition. Essentially, the combinatorial market and the flat market have identical market questions, clue types, and relationships between questions and clue types, but in the flat market the participants are unable to specify relationships between the market questions. The relationships among the market questions and the clue types were defined for the participants in conditional probability tables.

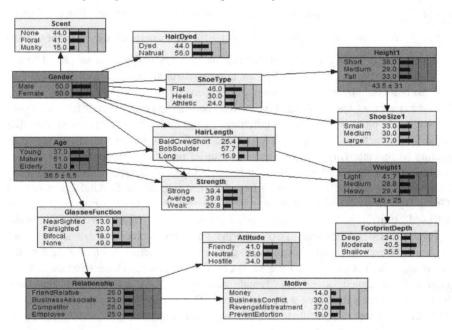

Fig. 1. The combinatorial (Full) Bayesian network captures relationships among market variables and clues

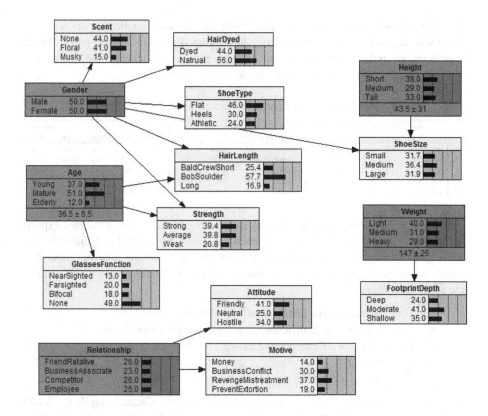

Fig. 2. The Flat Bayesian network is obtained by removing links between market variables

Market questions and clues were chosen to be intuitive, plausibly related to a murder mystery, and related to each other (to provide a valid test of a combinatorial prediction market). The relationships between physical clues were based on statistical evidence in the population of the US. For example, men are on average taller and heavier than women; shoe size is correlated with height. The strength of the relationships was exaggerated over the actual correlations in the U.S. population. This reflects the stereotypical nature of typical murder mysteries, and helped to ensure strong correlations among clues and market questions.

Once the relationships among the questions and clue types were established in the full Bayes net, the Bayes net was sampled to generate 100 simulated individuals who, according to the scenarios, were attendees at the New Year's eve party where the murder occurred. Participants were told that the murderer was one of these 100 suspects. Table 1 contains examples of the characteristics for each clue type and market question associated with each case (attendee).

Because "gold standard" predictions were required for a baseline against which the participants' predictions could be evaluated, after the 100 individuals were simulated, the marginal and conditional probabilities both the full and flat Bayes nets were replaced with actual frequencies taken from the simulated guest list for the party. These frequencies are reflected in Figs. 1 and 2, and were used to generate the "gold standard" predictions for the series of clues in each case.

Table 1. Partial Attendee Characteristic Table

Guest	Scent	Hair Dyed	Hair Length	Shoe Type	Strength	Glasses Function	Attitude	Motive	Footprint Depth	Shoe Size	Gender	Weight	Height	Age	Relation-ship
1	MSK	NAT	BCS	FLT	STR	NON	NUT	EXT	MOD	MED	M	MED	TAL	ELD	ASO
2	FLR	DYD	SHD	ATH	STR	BIF	NUT	EXT	DEP	SML	F	HVY	SHT	MAT	FOR
3	NON	NAT	BCS	FLT	WEK	FRS	FRN	EXT	MOD	LRG	M	MED	TAL	MAT	FOR
4	MSK	NAT	SHD	ATH	AVG	NON	HST	MNY	SHL	MED	M	HVY	TAL	MAT	CMP
5	NON	NAT	SHD	FLT	STR	NON	HST	EXT	MOD	LRG	M	MED	TAL	YNG	FOR

Of the 100 cases, two were selected as murderers, one for each of the two experimental trials. The differences between the Brier scores for the flat and full BNs, the differences expected Brier scores for the flat and full BNs, and the relative probability of each occurrence of each cases were used to select ten candidate cases. The candidate cases were those that:

- Had relatively large differences in the Brier scores;
- Had relatively large differences in the expected Brier scores;
- Had a variety of characteristics for clue types and outcomes for market questions; and
- Were above average in their probability of occurrence.

Larger differences in the Brier and Expected Brier scores indicated that there should be differences between the "gold standard" predictions and would provide opportunities for the participants to generate differences predictions in the combinatorial and flat markets. Cases with higher than average probability of occurrence were selected as representative of typical cases.

Once the candidate cases were selected, simulations using permutations of the ordering of clues were generated to evaluate the difference between the full and flat Bayes net "gold standard" predictions over time. The two cases used in the experiment and the ordering of the clues were chosen from those with larger average differences in Brier scores over time. Based on data gathered during a pilot test, the timing of the clues was established such that individuals would have sufficient time to analyze the impact of the clues and enter any updates to the market predictions.

At the beginning of the experiment, each group (combinatorial and flat) was subdivided into two subgroups, each of which was seated at a separate table. Each subgroup received marginal and conditional probability tables for the market variables, as well as conditional probability tables for a subset of the clues. Participants were given time to enter information from the probability tables into the market. Fig. 3 shows a screenshot of the interface used by participants to enter probabilities. The screen shows an assessment of the probability that the murder was a friend or relative, business associate, competitor, or employee of the victim. Participants see the current probability and a chart showing the history of probability values since the start of the experiment. Participants can use the "+" and "-" buttons to raise or lower the probability values. The interface also shows their expected score if each of the outcomes occurs. On the left-hand side of the screen, we see the current question highlighted. The top part of the screen shows assumptions. In the combinatorial condition, participants can drag other questions up into the assumption area and select an assumed value. In this case, the participant is assuming that the murderer is young; thus the probabilities shown to the right are conditional probabilities of relationship of the murder to the victim, given that the murderer was young.

Only those in the combinatorial condition could enter information about relationships among market questions. Dragging questions into the assumptions area was disabled in the flat condition. All participants could enter marginal probabilities. After about ten minutes, clues were handed out at intervals of a few minutes. Near the end of the experiment, as a way to keep up interest in the game, the guest list was handed out and subjects were challenged to identify the murderer. At this point, participants had enough to identify the murderer with certainty.

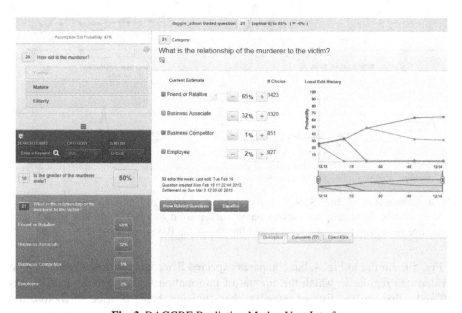

Fig. 3. DAGGRE Prediction Market User Interface

4   Results and Observations

The first clue was introduced about ten minutes after the probability tables were distributed. During those ten minutes, participants could use the market to establish the initial marginal and conditional probabilities for the market questions. Fig. 4 compares the time series of Brier scores for combinatorial and flat prediction markets starting from the time the first clue was distributed. The figure also shows the Brier scores for the full and flat Bayesian networks. Those in the combo condition made more edits, reflecting extra effort to correlate the market questions. Those in the flat market could not express those correlations in the market. Since the full Bayesian network represents all the available information, theoretically, on average it should have the lowest Brier score for a representative case drawn from the Bayes net. The flat Bayes net, since it does not capture the relationships among the market questions, should not as accurately predict the outcomes of the market questions. Indeed, in Fig. 4, the Brier scores for predictions from the flat Bayes net are always greater than those from the full Bayes net.

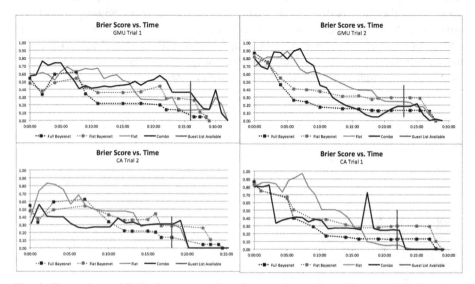

Fig. 4. Comparison of Brier Scores over time for experimental conditions and the Bayesian network models. Solid dark line shows Combo market; solid gray line shows Flat market; dark dotted line shows full Bayes net; light dotted line shows flat Bayes net.

Fig. 5 is similar to Fig. 4, but compares expected Brier scores. These figures can be divided into regions in which the amount of information available to the participants differed. Prior to the lists of attendee characteristics being distributed (before the vertical lines in the figures) the participants had available only the information contained in the clues and in the conditional probability tables. This information was reflected in the Bayes nets used to construct the scenarios and clues. Near the end of the experiment, to keep up interest, a list of party guests and their characteristics was distributed. At this point, the participants had access to information not captured in the Bayes nets, and this information was sufficient to identify the murderer with certainty before the clues resolving the questions were distributed. Therefore, our analysis stops when the guest list was distributed.

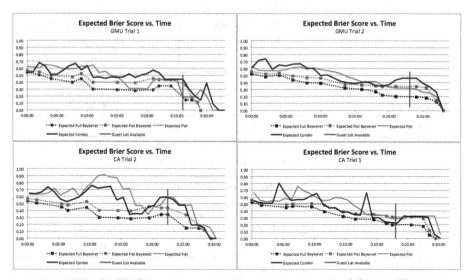

Fig. 5. Comparison of Expected Brier Scores over time for experimental conditions and the Bayesian network models. Solid dark line shows Combo market; solid gray line shows Flat market; dark dotted line shows full Bayes net; light dotted line shows flat Bayes net.

As can be seen in the figures, there was a lot of noise in the markets -- compared to the Bayes nets, and no clear tendency. Indeed, it is hard to tell from inspection which curve had the lower time-averaged Brier score. Table 2 summarizes the average Brier scores for the flat and combinatorial markets in each trial and compares them to the corresponding difference in the "gold standard" Brier scores obtained from the flat and full Bayes nets. In three of the trials, the average Brier scores from the combinatorial markets were lower than those from the flat markets indicating that, on average, the predictions made in the combinatorial market were more accurate than those made in the flat markets. The exception to this was GMU Trial 1 in which the average flat market predictions were more accurate. As expected, in three of the four trials, the average difference between the flat and combo participants' scores were less than those from the corresponding Bayes nets, indicating that on average the participants had not integrated all the available knowledge into their predictions. There was an exception to this trend also; in CA trial 2 the difference between the participants' Brier scores was greater than that between the Bayes net scores. Inspecting Fig. 4, it appears that the CA trial 2 combo participants were overconfident: their combinatorial Brier scores were below those of the full Bayes net, indicating that their predictions were stronger than they "should" have been with the information available; however, this overconfidence may have been warranted by the knowledge that they were participating in an experiment.

Table 2. Average Brier Scores

	Bayes Net Average Difference	Participant Average Difference
CA Trial 1	0.1085	0.0993
CA Trial 2	0.0281	0.1537
GMU Trial 1	0.0622	-0.0542
GMU Trial 2	0.1252	0.0951

Table 3. Average Expected Brier Scores

	Bayes Net Average Expected Difference	Participant Average Expected Difference
CA Trial 1	0.0569	0.0381
CA Trial 2	0.0764	0.0814
GMU Trial 1	0.0786	-0.0133
GMU Trial 2	0.0718	-0.0232

Although the results are suggestive, the effects seen in this experiment do not conclusively demonstrate an effect of combinatorial markets on accuracy.

Like the Brier scores, the expected Brier scores (Fig. 5 and Table 3) for the combinatorial market are not consistently lower than those for the flat market indicating that that the certainty in the participant's predictions was not consistently less for the combinatorial market than for the flat market. As can be seen in Table 3, at various time in the trials, the expected Brier scores in the combinatorial and flat markets approached the theoretical minimum generated by the full Bayes net expected Brier scores, though neither the combinatorial nor the flat market expected Brier score did so consistently. This lack of consistency is also evident in the average difference between the participants' expected Brier scores. Though the average expected Brier scores for the combinatorial market were less than those for the flat market in the California trials, they were slightly greater in the GMU trials.

5   Conclusion

The four trials in this experiment do not show a clear advantage for combo markets over flat markets on this "murder mystery" scenario. These results set limits on the conditions and range where a clear advantage may be seen. First, given the noise in the market estimates, the scenarios used in these trials provided insufficient theoretical difference (~10%) between the flat and combo Brier scores (as generated by the full and flat Bayes nets). Although each of our trials involved ~10 people working for several hours, the effective sample size is simply the number of trials, four (4). A scenario with a larger theoretical difference might show a consistent difference between the groups.

Second, to level the playing field, we provided direct evidence for all the market questions. But the most likely benefit to using a combinatorial market lies in the ability to propagate the effect of evidence through the market and influence predictions for market questions that are not directly related to the evidence. Examination of the data shows that changes in the predictions due to direct evidence seemed to overwhelm the changes in predictions due to evidence that was only indirectly related to each question. A clearer advantage for the combinatorial market might be seen if some questions could only be predicted from evidence relating to other correlated questions.

In designing future trials based on the experimental trials reported here, several modifications may increase the effects on the dependent variables (Brier scores and expected Brier scores). Possible modifications to the experimental design include making the correlations among the questions and between evidence and the question stronger; simulating more specialize knowledge i.e. lower percentage of the participants receive each conditional probability table; making it more difficult for participates to retain knowledge of specific relationships among the markets question and types of evidence; and adding questions for which no direct evidence is provided, but which are correlated with other questions for which there is evidence. Additionally, designing trials that take less time could result in more trials being run with the same number of participants and thus provide an overall increase in the statistical power of the experiment. Also, the experiment could be instrumented to provide data that would support the analysis of other metrics, e.g. joint probability distributions and conditionals. The basic design of these experimental trials seems sound, and improvements to the experimental design have been identified that should increase the ability of the Combinatorial Market experiment to determine the effects of using probabilistically linked questions on prediction markets.

Acknowledgements. This research was supported by the Intelligence Advanced Research Projects Activity (IARPA) via Department of Interior National Business Center contract number D11PC20062. The U.S. Government is authorized to reproduce and distribute reprints for Governmental purposes notwithstanding any copyright annotation thereon. Disclaimer: The views and conclusions contained herein are those of the authors and should not be interpreted as necessarily representing the official policies or endorsements, either expressed or implied, of IARPA, DoI/NBC, or the U.S. Government. The authors are grateful to Shou Matsumoto, Brandon Goldfedder and Jamie Ostheimer for software support.

References

1. Meehl, P.E.: Clinical Versus Statistical Prediction: A Theoretical Analysis and a Review of the Evidence. University of Minnesota Press (1954)
2. Dawes, R., Faust, D., Meehl, P.E.: Clinical Versus Actuarial Judgment. Science 243(4899), 1668–1674 (1989)

3. Marchese, M.C.: Clinical Versus Actuarial Prediction: a Review of the Literature. Perceptual and Motor Skills 75(2), 583–594 (1992)
4. Grove, W.M., Zald, D.H., Lebow, B.S., Snitz, B.E., Nelson, C.: Clinical Versus Mechanical Prediction: a Meta-Analysis. Psychological Assessment 12(1), 19–30 (2000)
5. Tetlock, P.: Expert Political Judgment: How Good Is It? How Can We Know? Princeton University Press (2005)
6. Silver, N.: The Signal and the Noise: Why So Many Predictions Fail — but Some Don't, 1st edn. Penguin Press HC (2012)
7. Surowiecki, J.: The wisdom of crowds. Anchor (2005)
8. Hanson, R.: Combinatorial information market design. Information Systems Frontiers 5(1), 107–119 (2003)
9. Hanson, R.: Logarithmic market scoring rules for modular combinatorial information aggregation. The Journal of Prediction Markets 1(1), 3–15 (2007)
10. Pennock, D.M., Wellman, M.P.: Graphical models for groups: Belief aggregation and risk sharing. Decision Analysis 3, 148–164 (2005)
11. Chen, Y., Pennock, D.M.: Designing markets for prediction. AI Magazine 31(4), 42–52 (2010)
12. Sun, W., Hanson, R., Laskey, K.B., Twardy, C.: Probability and Asset Updating using Bayesian Networks for Combinatorial Prediction Markets. In: Proceedings of the Twenty-Eighth Conference on Uncertainty in Artificial Intelligence, Catalina Island, USA (2012)
13. Berea, A., Maxwell, D., Twardy, C.: Improving Forecasting Accuracy Using Bayesian Network Decomposition in Prediction Markets. In: Proceedings of the AAAI Fall Symposium Series (2012)
14. Berea, A., Twardy, C.: Automated Trading in Prediction Markets. In: Greenberg, A.M., Kennedy, W.G., Bos, N.D. (eds.) SBP 2013. LNCS, vol. 7812, pp. 111–122. Springer, Heidelberg (2013)
15. Brier, G.W.: Verification of forecasts expressed in terms of probability. Monthly Weather Review 75, 1–3 (1950)

A Scalable Learning Algorithm for Kernel Probabilistic Classifier

Mathieu Serrurier and Henri Prade

IRIT - 118 route de Narbonne 31062, Toulouse Cedex 9, France
{serrurie,prade}@irit.fr

Abstract. In this paper we propose a probabilistic classification algorithm that learns a set of kernel functions that associate a probability distribution over classes to an input vector. This model is obtained by maximizing a measure over the probability distributions through a local optimization process. This measure focuses on the faithfulness of the whole probability distribution induced rather than only considering the probabilities of the classes separately. We show that, thanks to a pre-processing computation, the complexity of the evaluation of this measure with respect to a model is no longer dependent on the size of the training set. This makes the local optimization of the whole set of kernel functions tractable, even for large databases. We experiment our method on five benchmark datasets and the KDD Cup 2012 dataset.

1 Introduction

It is well known that machine learning algorithms are constrained by some learning bias (language bias, hypothesis bias, algorithm bias, etc.). In that respect, learning a precise model may be illusionary. Moreover, in case where security issues are critical for instance, predicting one class only, without describing the uncertainty about this prediction, may be unsatisfactory. Probabilistic classification aims at learning models that associate to an input vector a probability distribution over classes rather than a single class. K-nearest-neighbor methods [1] compute this distribution by considering the neighborhood of the input vector. Probabilities are then computed from the frequency of the classes. The quality of the distribution highly depends on the density of the data. Some other types of algorithms such as naive Bayes classifiers [7] and Gaussian processes [9,10,15] are based on Bayesian inference. Gaussian processes assume that the attribute values follow a Gaussian distribution, it uses kernels for describing the co-variance between such variables. Thus, these approaches suppose strong assumptions (high density data, independent attributes, priors about the type of the probability distribution that underlies the data , ...). Logistic regression has been also proposed as a probabilistic classifier [2,5] since it can be used for a direct estimation of the probability of the classes. This approach has been extended for the non linear case by the use of kernels logistic functions (KLR [4]) or kernel functions [16]. These last methods are based on minimization of the squared distance between the value of the class (0 or 1) and the predicted

W. Liu, V.S. Subrahmanian, and J. Wijsen (Eds.): SUM 2013, LNAI 8078, pp. 297–310, 2013.

value (between 0 and 1). Functions are learn independently for each class and a normalization post processing is needed. Moreover, these approaches are based on a costly optimization processes and are not tractable for large databases. Sugiyama [13] proposes an alternative reformulation of the calculus (still based on the minimization of squared distance) that partially overcomes this cost issue and skips the normalization step. Even if these methods are consistent with the maximum likelihood principle, there are based on the evaluation of the ability of the predicted distribution to identify the most probable class, but not on an evaluation of the faithfulness of the complete probability distribution with respect to the data.

In this paper we propose to learn a set of kernel functions as in this other kernel approaches. However, our method differs from them by many points. First, we constrain the parameters of the kernels in order to have the sum of the probabilities equal to 1. Second, the model is obtained through a local optimization process (we propose an implementation for two algorithms: the Nelder-Mead algorithm [8] and the particle swarm meta-heuristics [6]) by fixing the support vectors and maximizing a quality measure that estimates the faithfulness of a probability distribution with respect to a set of data. The kernel function are learned all together (rather than independently in classical kernel approaches). In this scope we have extended the loss function used in KLR to the whole distribution. Then, this measure relies on the squared distance between the optimal distribution (1 for the class, 0 for the other classes) and the proposed one. Moreover, we reformulate the computation of this measure for our set of kernel functions and make that the complexity of this computation no longer depend on the size of the dataset thanks to a pre-computation step. This allows us to handle very large databases, even when using costly meta-heuristics such as particle swarm.

The paper is structured as follows. Section 2 provides some definitions about probabilistic loss functions and a precise description of the proposed measure. In Section 3, we describe the model that we learn and we show how the measure can be reformulated in order to maximize performances. Section 4 is devoted to the description of the optimization process and the tuning of the parameters. Last, we validate our approach by experimentation on 5 benchmark datasets and on the KDD Cup 2012 dataset.

2   Probabilistic Loss Functions

Probabilistic loss functions are used for evaluating the adequateness of a probability distribution with respect to data. In this paper, we only consider the case of classification. A classification database is a set of n pairs $(\overrightarrow{x}_i, c_i)$, $1 \leq i \leq n$, where $\overrightarrow{x}_i$ is a vector of input variables in the feature space $\mathcal{X}$ and $c_i \in \{C_1, \ldots, C_q\}$ is the class variable. We note

$$\mathbb{1}_j(\overrightarrow{x}_i) = \begin{cases} 1 \text{ if } c_i = C_j \\ 0 \text{ otherwise.} \end{cases}$$

Given a probability distribution p on the discrete space $\Omega = \{C_1, \ldots, C_q\}$, we denote $p_1, \ldots, p_q$ the probability of being an element of Ω, i.e. $p(c_i = C_j) = p_i$. The values $p_1, \ldots, p_q$ entirely define p. The log-likelihood is a natural loss function for probability distributions. Formally the likelihood coincides with a probability value. The logarithmic-based likelihood is defined as follows (under the strict constraint $\sum_j^q p_j = 1$):

$$\mathcal{L}oss_{log}(p|\overrightarrow{x}_i) = -\sum_{j=1}^{q} \mathbb{1}_j(\overrightarrow{x}_i)log(p_j).$$

However, as a probabilistic loss function, the likelihood has some limitations. First, $\mathcal{L}oss_{log}$ is not defined when $p_j = 0$ and $\mathbb{1}_j = 1$. Second, it gives a very high weight to the error when probability is very low. These two issues are a strong limitation when the parameters are obtained through a local optimization process of a classifier. Indeed, avoiding the possibility to have $p_j = 0$ can be difficult when considering complex models. Moreover, exponential costs of error on low probability classes may have a too high effect on the whole model.

One common approach to overcome this problem is to turn the classification problem into a regression problem. The goal of kernel logistic regression and kernel regression is to minimize the least square between the value of the class (0 and 1) and the predicted probability. To this end, both methods independently learn a function (resp. kernel logistic function of kernel f_j for each class C_j). Then, for each x_i, f_j minimizes

$$LeastSquare(\overrightarrow{x}_i, f_j) = (\mathbb{1}_j(\overrightarrow{x}_i) - f_j(\overrightarrow{x}_i))^2.$$

It has been shown that minimizing this distance leads to an faithful estimation of the probability of being in the class C_j. However, since the function are obtained independently, the distribution encoded by the f_n' s does not necessarily satisfy $\sum_{j=1}^{q} f_j(\overrightarrow{x}_i) = 1$ for all $\overrightarrow{x}_i$. Thus, the predicted values have to be normalized in order to have a probability distribution and then, the faithfulness of the probability of being in the class C_j may be altered.

In this paper, we propose a method that learns the distribution directly. In order to achieve this goal we extend the previous expression in order to take into account the whole probability distribution p. Thus, we obtain:

$$LeastSquare(\overrightarrow{x}_i, p) = \sum_{j=1}^{q}(\mathbb{1}_j(\overrightarrow{x}_i) - p_j)^2$$

$$= 1 + \sum_{j=1}^{q} p_j^2 - 2 * \sum_{j=1}^{q} \mathbb{1}_j(\overrightarrow{x}_i) * p_j.$$

We build our loss function by removing the constant and normalizing the previous calculus. Then we have :

$$\mathcal{L}oss_{surf}(p|\overrightarrow{x}_i) = \frac{\sum_{j=1}^{q} \mathbb{1}_j(\overrightarrow{x}_i) * p_j - \frac{1}{2} * \sum_{i=1}^{q} p_i^2}{n}. \tag{1}$$

This loss function has been used in [11,12] for describing similar loss functions for possibility distributions and use it into regression process. We name it $\mathcal{Loss}_{surf}$ since maximizing this function is equivalent to minimize the square of the distance between p and the optimal probability distribution p^* (here we have for a given $\overrightarrow{x}_i$, $p_j^* = \mathbb{1}_j(\overrightarrow{x}_i)$. Then, contrarily to $(LeastSquare(\overrightarrow{x}_i, f_j)$, $\mathcal{Loss}_{surf}$ takes into account the whole distribution directly, and not the probability values independently.

3   Surface Probabilistic Kernel Classifier Learning

3.1   Definitions

We recall that a classification database is a set of n pairs, or examples, $(\overrightarrow{x}_i, c_i)$, $1 \leq i \leq n$, where $\overrightarrow{x}_i$ is a vector of input variables in the feature space $\mathcal{X}$ and $c_i \in \{C_1, \ldots, C_q\}$ is the class variable. A Surface Kernel probabilistic Classifier (Skc) associates a probability distribution over the classes to a vector of $\mathcal{X}$. A Skc is a set of q kernel functions $(Skc = \{f_1, \ldots, f_q\})$. A function f_j is a kernel function over r support vectors $\overrightarrow{s}_1, \ldots, \overrightarrow{s}_r$ (a support vector is a point in the feature space $\mathcal{X}$)), the same for all the functions, that encodes the probability of the example $\overrightarrow{x}_i$ pertaining to class C_j. Then we have:

$$f_j(\overrightarrow{x}) = \sum_{l=1}^{r}(\alpha_l^j * K(\overrightarrow{x}, \overrightarrow{s}_l)) + \alpha_{r+1}^j \tag{2}$$

where the α_l^j's and α_{r+1}^j are the parameters of the function and $K(.,.)$ is a kernel function. The probability of the example of pertaining to class C_i is then:

$$p_j(\overrightarrow{x}_i) = p(c_i = C_j) = f_j(\overrightarrow{x}_i). \tag{3}$$

We also have the following constraints:

1. $\forall j \in 1, \ldots, q, \forall l \in 1, \ldots, r, \sum_{j=1}^{q} \alpha_k^i = 0$
2. $\sum_{j=1}^{q} \alpha_{r+1}^j = 1$.
3. $\forall j \in 1, \ldots, q, \forall l \in 1, \ldots, r+1, -1 \leq \alpha_l^j \leq 1$

Constraints 1 and 2 guarantee that the probability distribution predicted for a vector $\overrightarrow{x}$ is normalized, i.e.:

$$\forall \overrightarrow{x} \in \mathcal{X}, \sum_{j=1}^{q} f_j(\overrightarrow{x}) = 1.$$

However, these constraints do not ensure that the distribution obtained is a genuine probability distribution. Indeed, we may have $f_j(\overrightarrow{x}) < 0$ or $f_j(\overrightarrow{x}) > 1$. This can be partially overcome with constraint 3, but it is not sufficient in general. This issue will be solved in the optimization process, as it will be explained in the following.

Once the probabilistic kernel functions defined, the goal is to find the Skc function that associate a probability distribution over classes that is as faithful as possible to each input vector of the training set. In the most favorable case, we will obtain a distribution that gives the probability value 1 to the right class. According to the previous definitions, the goal for learning Skc is to find the Skc that maximizes the surface loss function with respect to each example in the training set. This is formulated as follows: Find the α_l^j parameters of a Skc that maximizes the following expression:

$$\mathcal{L}oss_{surf}(Skc) = \sum_{i=1}^{n} \mathcal{L}oss_{surf}(Skc(\overrightarrow{x}_i)|\overrightarrow{x}_i). \tag{4}$$

The maximization of $\mathcal{L}oss_{surf}$ has several advantages:

- $\mathcal{L}oss_{surf}$ is defined even when the probability is equal to 0 and if the probability distribution is not normalized. $\mathcal{L}oss_{surf}$ is also defined for negative values or values greater than 1. Even if these values are not acceptable for probability prediction, it allows local optimizer to explore the whole feature space. Contrarily to the log-likelihood, it makes also the evaluation of the model possible when it performs well for a majority of examples and have aberrant values only for few examples. Moreover, $\mathcal{L}oss_{surf}$ always favors genuine probability functions. Thus, even if the definition of Skc permits such kind of abnormal distribution, this case never appears after the learning of a model in the experiments.
- $\mathcal{L}oss_{surf}$ evaluates the faithfulness of the probability distribution predicted and not only its ability to identify the most probable class. Moreover, even when only one example is considered, all the values of the probability distribution are taken into account (contrarily to log-likelihood) as it can be seen in Equation 1.

Then, this approach has some advantages with respect to the other Kernel approaches. First, the kernel functions have to be learned simultaneously and not one by one as it is done in Kernels approach (even for the binary case). Second, the combination of constraints and properties of the $\mathcal{L}oss_{surf}$ makes that the probability distribution predicted has not to be normalized. Finally, $\mathcal{L}oss_{surf}$ evaluates the faithfulness of the probability distribution without normalization when kernel approaches focus on maximizing the probability associated to the considered class (regardless of the values of the other classes). Even if these two approaches are acceptable in terms of accuracy maximization, the proposed approach seems best suited in terms of quality of the probability distributions learned. But on the contrary to other kernel approaches, finding the Skc function that maximizes $\mathcal{L}oss_{surf}(Skc)$ given a training set is a hard problem which has no simple analytical solution. Section 4 shows how this issue can be handled by local optimization algorithms. However, the computation time performance of these algorithms depends highly on the complexity cost of the evaluation of $\mathcal{L}oss_{surf}(Skc)$.

3.2 Complexity Evaluation and Reformulation of $\mathcal{L}oss_{surf}$

Given n examples, q classes, r support vectors and a Skc function, we have:

$$\mathcal{L}oss_{surf}(Skc) = \sum_{i=1}^{n}(\sum_{j=1}^{q}(\mathbb{1}_j(\vec{x}_i) * f_j(\vec{x}_i) - \frac{1}{2}f_j(\vec{x}_i)^2))$$

$$= \sum_{i=1}^{n}\sum_{j=1}^{q}(\mathbb{1}_j(\vec{x}_i) * \sum_{l=1}^{r}(\alpha_l^j * K(\vec{x}_i, \vec{s}_l)) + \alpha_{r+1}^j)$$

$$- \frac{1}{2}\sum_{i=1}^{n}\sum_{j=1}^{q}(\sum_{l=1}^{r}(\alpha_l^j * K(\vec{x}_i, \vec{s}_l)) + \alpha_{r+1}^j)^2.$$

Under this form, the complexity of the calculus is $\mathcal{O}(m * q * r)$. This can be problematic since local optimization algorithms require to evaluate the target function frequently. In this case, the optimization process will rapidly become too costly when the size of the training set increases. Fortunately, it can be reformulated in order to lead to a more tractable computation. For the sake of readability, we split $\mathcal{L}oss_{surf}(Skc)$ into two parts, namely

$$\mathcal{L}oss_{surf}(Skc) = Part1 - \frac{1}{2}Part2$$

such that:

$$Part1 = \sum_{i=1}^{n}(\sum_{j=1}^{q}(\mathbb{1}_j(\vec{x}_i) * f_j(\vec{x}_i)))$$

and

$$Part2 = \sum_{i=1}^{n}\sum_{j=1}^{q}f_j(\vec{x}_i)^2.$$

$Part1$ can be reformulated as follows:

$$Part1 = \sum_{i=1}^{n}\sum_{j=1}^{q}(\mathbb{1}_j(\vec{x}_i) * \sum_{l=1}^{r}(\alpha_l^j * K(\vec{x}_i, \vec{s}_l)) + \alpha_{r+1}^j)$$

$$= \sum_{i=1}^{n}\sum_{j=1}^{q}(\mathbb{1}_j(\vec{x}_i) * \alpha_{r+1}^j + \sum_{l=1}^{r}(\mathbb{1}_j(\vec{x}_i) * \alpha_l^j * K(\vec{x}_i, \vec{s}_l)))$$

$$= \sum_{j=1}^{q}\alpha_{r+1}^j * \sum_{i=1}^{n}\mathbb{1}_j(\vec{x}_i) + \sum_{j=1}^{q}\sum_{l=1}^{r}(\alpha_l^j * \sum_{i=1}^{n}(\mathbb{1}_j(\vec{x}_i) * K(\vec{x}_i, \vec{s}_l)))$$

$$= \sum_{j=1}^{q}(\alpha_{r+1}^j * NB_j) + \sum_{j=1}^{q}\sum_{l=1}^{r}(\alpha_l^j * K_j^l)$$

with $K_j^l = \sum_{i=1}^{n}(\mathbb{1}_j(\vec{x}_i) * K(\vec{x}_i, \vec{s}_l))$ and $NB_j = \sum_{i=1}^{n}\mathbb{1}_j(\vec{x}_i)$. It is interesting to remark that K_j^l and NB_j do not depend on the α_l^j's. Then, these values can be computed before the optimization process. During the optimization process, the complexity of the computation of $Part1$ goes down to $\mathcal{O}(q*r)$ which is independent from the size of the training set. In the same way, $Part_2$ can be reformulated as follows:

$$Part2 = \sum_{i=1}^{n}\sum_{j=1}^{q}(\sum_{l=1}^{r}(\alpha_l^j * K(\vec{x}_i, \vec{s}_l)) + \alpha_{r+1}^j)^2$$

$$= \sum_{i=1}^{n}\sum_{j=1}^{q}(\sum_{l=1}^{r}(\alpha_l^j * K(\vec{x}_i \vec{s}_l)))^2$$

$$+ \sum_{i=1}^{n}\sum_{j=1}^{q}((\alpha_{r+1}^j)^2 + 2 * \alpha_{r+1}^j * (\sum_{l=1}^{r}(\alpha_l^j * K(\vec{x}_i, \vec{s}_l))))$$

$$= \sum_{i=1}^{n}\sum_{j=1}^{q}\sum_{l=1}^{r}\sum_{t=1}^{r}(\alpha_l^j * \alpha_t^j * K(\vec{x}_i \vec{s}_l) * K(\vec{x}_i, \vec{s}_t))$$

$$+ n * \sum_{j=1}^{q}(\alpha_{r+1}^j)^2 + 2 * \sum_{j=1}^{q}(\alpha_{r+1}^j * \sum_{l=1}^{r}(\alpha_l^j * \sum_{i=1}^{n}K(\vec{x}_i, \vec{s}_l)))$$

$$= \sum_{j=1}^{q}\sum_{l=1}^{r}\sum_{t=1}^{r}(\alpha_l^j * \alpha_t^j * K_{s,l}) + n * \sum_{j=1}^{q}(\alpha_{r+1}^j)^2 + 2 * \sum_{j=1}^{q}(\alpha_{r+1}^j \sum_{l=1}^{r}(\alpha_l^j * K_l))$$

where $K_{s,l} = \sum_{i=1}^{n}(K(\vec{x}_i, \vec{s}_l) * K(\vec{x}_i, vs_t))$ and $K_l = \sum_{i=1}^{n}K(\vec{x}_i, \vec{s}_l)$. As previously, $K_{s,l}$ are independent from the α_l^j's. Then if we pre-compute the values $K_{s,l}$, the complexity of the computation of $Part2$ goes down to $\mathcal{O}(q*r^2)$. We obtain a complexity of $\mathcal{O}(q*r^2)$ for the calculus of $\mathcal{L}oss_{surf}(Skc)$ if we compute the values K_j^l, NB_j, $K_{s,l}$ and K_l before the optimization process. Then, we can perform an optimization process that is independent from the size of the database.

4   Optimization Process

As pointed out in the previous section, the fact learning f_j functions has to be done simultaneously makes that there is no simple analytical solution. Thanks to the offline computation of the values that depends on the size of the database, the evaluation of the target function of a model is not costly. In this context, the use of local optimization algorithm is possible. However, it requires to previously choose the number of support vectors and their values. The number of support vectors is a parameter of the algorithm. The vectors are then obtained with the k-means clustering algorithm. We use two different optimization algorithms. The first one is the Nelder-Mead algorithm [8] which is very fast but converges to local optimum. The second one is the particle swarm meta-heuristics [6] which is more costly but has better optimization performances.

4.1   Nelder-Mead Implementation

The Nelder-Mead algorithm is a heuristics for maximizing of a function F in a N dimensions space. It is based on the deformation of a simplex until it converges to a local optima (Algorithm 1). The algorithm stops after a fixed number of loops without increasing $F(e_1)$. In addition to its efficiency, the Nelder-Mead algorithm is very simple and does not require to derive the function F. However, it can be easily trapped into local optima and it depends on the starting configuration. Results can be improved by restarting the algorithm with different starting configurations. In our case a state e corresponds to the vectors that describe the parameters α_l^j of a Skc function given a kernel and a set of r support vectors. Then the dimension of the state space is $N = q * (r + 1)$. $\mathcal{L}oss_{surf}(Skc)$ corresponds to the function F. The starting configurations are chosen randomly and have to respect the constraint described in section 3.1. The operations on the space states ensure that the constraints are not violated during the algorithm.

Algorithm 1. Nelder-Mead

Choose $N + 1$ points $e_1, \ldots, e_{N+1}$
Order the points with respect to F
Compute e_0 the center of gravity of $e_1, \ldots, e_N$
$e_r = 2 * e_0 - e_{N+1}$
if $F(e_r) > F(e_N)$ **then**
  $e_t = e_0 + 2 * (e_0 - e_{N+1})$
  if $F(e_t) > F(e_r)$ **then**
    $e_{N+1} = e_t$
  else
    $e_{N+1} = e_r$
  end if
else
  $e_c = e_{N+1} + \frac{1}{2} * (e_0 - e_{N+1})$
  if $F(e_c) \geq F(e_n)$ **then**
    $e_{N+1} = e_c$
  else
    forall $i \geq 2$ $e_i = e_1 + \frac{1}{2} * (e_i - e_1)$
  end if
end if
return to step 2

4.2   Particle Swarm Implementation

In order to overcome the problem of local optima, we propose to use the particle swarm optimization algorithm. One of the advantages of the particle swarm optimization with respect to the other meta-heuristics is that it is particularly suitable for continuous problems. Particle swarm works in the same settings than Nelder-Mead algorithm. Particle swarm with N particles (N is no longer the dimension of the state space) is described in Algorithm 2.

Algorithm 2. Particle Swarm Optimization

Choose randomly N particles $e_1, \ldots, e_N$
for all i $be_i = e_i$
$eg = argmax_{e_i}(F(e_i))$ (best know position)
choose randomly N velocity vectors $v_1, \ldots, v_N$
repeat
  for $i = 1, \ldots, N$ **do**
    choose randomly r_p and r_g in $[0, 1]$
    $v_i = \omega * v_i + \phi_p * r_p * (be_i - ei) + \phi_g * r_g * (eg - ei)$
    $e_i = e_i + v_i$
    if $F(e_i) > F(be_i)$ **then**
      $be_i = e_i$
      if $F(be_i) > F(eg)$ **then**
        $eg = be_i$
      end if
    end if
  end for
until a chosen number of times

Here, one particle represents the parameters of Skc function (α_l^j). At each step of the algorithm, each particle is moved along its velocity vector (randomly fixed at the beginning). The velocity vectors are updated at each step by considering the current vectors, the vector from the current particle position to particle best known position and the vector from the current particle position to global swarm's best known position. In order to maintain the constraint 1 and 2 over the parameters α_l^j the values v_l^j of the velocity vector have to satisfy the following constraints:

- $\forall j \in 1, \ldots, q, \forall l \in 1, \ldots, r+1, \sum_{j=1}^{q} v_k^i = 0$
- $\forall j \in 1, \ldots, q, \forall l \in 1, \ldots, r+1, -1 \leq v_l^j \leq 1$

If we have $-1 > \alpha_l^j$ (resp. $\alpha_l^j > 1$) after the application of the velocity vector, we fix $\alpha_l^j = -1$ (resp. $\alpha_l^j = 1$).

The particle swarm algorithm is easy to tune. The three parameters for the updating of the velocity ω, ϕ_p and ϕ_g correspond respectively to the coefficient for the current velocity, the coefficient for the velocity to the particle best known position and the coefficient for the velocity to the global swarm's best known position. Based on [14], we use generic values that perform well in most of the cases ($\omega = 0.72$, $\phi_p = 1.494$, $\phi_g = 1.494$ and 16 particles).

5   Experimentation

In this section, we compare our algorithms with naive Bayes classifier (NBC) and the kernel approaches (SVM) based on least square minimization described in [16] and implement in the java version of LibSVM. We note Skc_{NM} for the maximization of $\mathcal{L}oss_{surf}$ with the Nelder-Mead algorithm and Skc_{PSO} for the maximization of $Loss_{surf}$ with the particle swarm optimization algorithm. We compare the results with respect to the accuracy, $Loss_{log}$ (mind that here the lower the value, the better) and $Loss_{surf}$. We also report time performance of SVM, Skc_{NM} and Skc_{PSO}. In the first experiments, we make 100000 steps of particle swarm movement and 5 restarts of Nelder-Mead algorithm. We empirically choose the number of support vectors with the formulas $r = 1 + 10log(n/3)$ where n is the number of examples. We use Gaussian kernels. All the experiments are done on a $3Ghz$ computer and all the algorithms are implemented with the JAVA language.

Table 1. Comparison of algorithms on 6 UCI dataset (10-cross validation)

db.	Alg.	Acc.	$Loss_{log}$	$Loss_{surf}$
Diab.	NBC	75.7[5.1]	0.57[0.15]	0.32[0.04]
	SVM	74.9[7.9]	0.52[0.09]	0.32[0.04]
	Skc_{NM}	76.2[8.1]	0.51[0.13]	**0.33[0.04]**
	Skc_{PSO}	**76.3[7.6]**	**0.49[0.1]**	**0.33[0.03]**
Breast.	NBC	95.7[2.1]	0.26[0.13]	0.45[0.02]
	SVM	95.1[2.4]	0.13[0.06]	0.46[0.01]
	Skc_{NM}	95.9[1.9]	0.12[0.04]	**0.46[0.01]**
	Skc_{PSO}	**95.9[1.8]**	**0.11[0.05]**	**0.46[0.01]**
Iono.	NBC	84.8[6.1]	0.7[0.34]	0.36[0.05]
	SVM	88.3[5.7]	0.29[0.06]	0.41[0.02]
	Skc_{NM}	90.8[2.7]	0.22[0.06]	**0.42[0.01]**
	Skc_{PSO}	**93.4[2.7]**	**0.2[0.06]**	**0.42[0.01]**
Mag. Tel.	NBC	72.7[0.8]	0.98[0.03]	0.26[0.0]
	SVM	**87.6[0.8]**	**0.3[0.01]**	**0.4[0.0]**
	Skc_{NM}	84.4[0.7]	0.38[0.01]	0.38[0.0]
	Skc_{PSO}	85.5[0.8]	0.38[0.01]	0.38[0.0]
Glass	NBC	50.3[15.4]	1.24[0.4]	0.22[0.05]
	SVM	**72.8[9.5]**	**0.81[0.18]**	**0.29[0.05]**
	Skc_{NM}	66.7[9.2]	0.86[0.19]	0.23[0.04]
	Skc_{PSO}	71.5[9.4]	0.79[0.13]	0.28[0.04]

5.1 Benchmark Dataset

In order to check the effectiveness of the algorithms, we used 5 benchmarks from UCI[1]. All the datasets have numerical attributes only. The Diabetes database describes 2 classes with 768 examples. The Breast cancer database contains 699 examples that describes 2 classes. The Ionosphere database describes 2 classes with 351 examples. The Magic telescope database contains 19020 examples that describes 2 classes. The Glass database describes 7 classes with 224 examples. The results presented in Table 1 are for 10-cross validation. Bold results correspond to the highest values. It shows that the Sks approaches outperform clearly NBC

---

[1] http://www.ics.uci.edu/~mlearn/MLRepository.html

Table 2. Computation time for the Skc algorithms on 6 UCI datasets

database	time		
	SVM	Skc_{NM}	Skc_{PSO}
Diabetes	0.2s	0.4s	7.9s
Breast c	0.1s.	0.2s	8s
Iono.	0.1s	1s	8s
Mag. Tel.	57s	1.9s	9.3s
Glass	0.1s	8s	23s

on all the databases (with a statistically significant difference for 3 databases) both for the accuracy and the $Loss_{surf}$. Skc_{PSO} outperforms SVM on 3 of the 5 databases (with a statistically significant difference for 1 database) and is outperformed on the two remaining ones (with a statistically significant difference for 1 database). This shows that our approaches compete with SVM probabilistic approach in terms of classification and have good performances for describing faithful probability distributions (even if we consider the log-likelihood). $Skc_{N}M$ and $Skc_{P}SO$ have close performances except when the number of classes increases. We can suppose that in this case $Skc_{N}M$ is more easily trapped in local optima.

Table 2 gives the time in seconds for performing the optimization of Skc. As expected $Skc_{N}M$ is around 10 times more efficient than $Skc_{P}SO$. Even if these times are larger than the SVM ones, they remain very low, and are not much sensitive to the size of the dataset (times for Ionosphere and Magic Telescope are closed for instance). The size of the database only matters for computing the support vectors and the pre-computed values (the number of support vector also increases slightly when the size increases). When the size of the database increases, as for magic telescope, our approaches become much faster than SVM approach.

5.2   KDD Cup 2012 Dataset

In order to check the scalability of our algorithms, we use our approaches on the KDD Cup 2012 database. This database describes a social network of microblogging. "Users" are people in the social network and "items" are famous people or objects that the users may follow. Users may be friend with other users. The task of this challenge is to predict if a user will accept or not to follow an item proposed by the system.

There are 10 millions of users described by their age, their genre, some keywords and their friends. There are 50000 items described by keywords and tags. The database contains 70 millions of propositions to follow an item with a label that indicates if the user has accepted the proposition or not. The problem is then a binary classification problem. We define 8 attributes based respectively on i)the percentage of users of the same genre as the target one, which follow the item, ii) the percentage of users of the same age category as the target one,

Table 3. Comparison of algorithms on KDD Cup 2012 dataset

database	Alg.	Acc.	$Loss_{log}$	$Loss_{surf}$	time
Size=1K	NBC	66.4	0	0.278	-
	SVM	70.8	0.580	0.302	0.4
	Skc_{NM} r=10	71.0	0.557	0.312	0.2s
	Skc_{NM} r=100	71.0	0.57	0.311	12s
	Skc_{PSO} r=10	71.4	0.557	0.313	8s
	Skc_{PSO} r=100	71.6	0.572	0.311	56s
Size=10K	NBC	68.0	0.663	0.285	-
	NBC	71.9	0.57	0.314	37s
	Skc_{NM} r=10	71.5	0.559	0.314	0.3s
	Skc_{NM} r=100	72.2	0.556	0.313	14s
	Skc_{PSO} r=10	71.6	0.56	0.313	8s
	Skc_{PSO} r=100	72.7	0.55	0.320	55s
Size=100K	NBC	69.0	0.662	0.292	-
	SVM	72	0.55	0.315	3.5 hours
	Skc_{NM} r=10	72.2	0.547	0.316	2s
	Skc_{NM} r=100	72.2	0.543	0.318	30s
	Skc_{PSO} r=10	72.2	0.547	0.316	10s
	Skc_{PSO} r=100	72.7	0.537	0.319	75s
Size=1M	NBC	68.3	0.665	0.29	-
	SVM	-	-	-	-
	Skc_{NM} r=10	72.2	0.551	0.315	19s
	Skc_{NM} r=100	72.5	0.543	0.318	97s
	Skc_{PSO} r=10	72.0	0.551	0.315	27s
	Skc_{PSO} r=100	73	0.534	0.321	119s

which follow the item, iii) the session time, iv) the number of friends of the target user that follow the item, v) the distance between the items followed by the user and the target item, vi) the number of users that follow the target items, vii) the number of items that are followed by the user, and viii) the number of times the item has been proposed to the user. We build a test dataset of around 1.9 millions of propositions.

Table 3 reports the result with different size of training sets (without any common tuple with the test dataset) and different number of support vectors. The results are computed on the test dataset. We can observe that performances increase when the size of the database increases, even if the dataset is summarized by the pre-computed values in the optimization process. $Skc_{P}SO$ performs slightly better than $Skc_{N}M$ and SVM approach. Last, time values confirm the efficiency of the approach and its low sensitivity with respect to the size of the dataset (less than 2 minutes for the $Skc_{P}SO$ with r = 100 and 1 million examples in the training set). It shows that our approaches are usable on very large database while SVM would have difficulties for managing databases with more than 10000 examples (and is intractable for more than 100000 examples).

6 Conclusion and Future Works

In this paper we have proposed a probabilistic classification method based on the maximization of a loss probabilistic function that takes into account the whole probability distribution and not only the probability of the class. We propose two algorithms that simultaneously learn a set of kernel functions that encodes a probability distribution over classes without any post-normalization process. Last, we show that the computation time of the approach is very little sensitive to the size of the dataset. Our method competes with the other kernel approaches on the used benchmark datasets. Experiments on the KDD Cup 2012 dataset confirm that the approach is efficient on very large datasets when kernel methods are not tractable. Moreover, the parameters of the algorithm can be tuned automatically as it has been done the whole experimentation.

In the future, the way of choosing the number of kernels and computing the support vectors has to be more deeply investigated and alternatives to clustering approach have to be explored. We will also study how the approach can be embedded into a gradient boosting process [3] in order to increase the performance when the number of attributes and classes is large. Lastly, we have to compare our algorithm more deeply with the other probabilistic approaches.

References

1. Cover, T.M., Hart, P.E.: Nearest neighbour pattern classification. IEEE Transactions on Information Theory 13, 21–27 (1967)
2. Fan, R.-E., Chang, K.-W., Hsieh, C.-J., Wang, X.-R., Lin, C.-J.: Liblinear: A library for large linear classification. Journal of Machine Learning Research 9, 1871–1874 (2008)
3. Friedman, J.H.: Greedy function approximation: A gradient boosting machine. Annals of Statistics 29, 1189–1232 (2000)
4. Jaakkola, T.S., Haussler, D.: Probabilistic kernel regression models. In: Proceedings of the 1999 Conference on AI and Statistics. Morgan Kaufmann (1999)
5. Jaakkola, T.S., Jordan, M.I.: A variational approach to bayesian logistic regression models and their extensions (1996)
6. Kennedy, J., Eberhart, R.: Particle swarm optimization. In: Proceedings of the IEEE International Conference on Neural Networks 1995, pp. 1942–1948 (1995)
7. Langley, P., Iba, W., Thompson, K.: An analysis of bayesian classifiers. In: Proceedings of AAAI 1992, vol. 7, pp. 223–228 (1992)
8. Nelder, J.A., Mead, R.: A simplex method for function minimization. The Computer Journal 7(4), 308–313 (1965)
9. Nickisch, H., Rasmussen, C.E.: Approximations for binary gaussian process classification. Journal of Machine Learning Research 9, 2035–2078 (2008)
10. Opper, M., Winther, O.: Gaussian processes for classification: Mean field algorithms. Neural Computation 12, 2000 (1999)

11. Serrurier, M., Prade, H.: Imprecise regression based on possibilistic likelihood. In: Benferhat, S., Grant, J. (eds.) SUM 2011. LNCS, vol. 6929, pp. 447–459. Springer, Heidelberg (2011)
12. Serrurier, M., Prade, H.: Maximum-likelihood principle for possibility distributions viewed as families of probabilities (regular paper). In: IEEE International Conference on Fuzzy Systems (FUZZ-IEEE), Taipei, Taiwan, pp. 2987–2993 (2011)
13. Sugiyama, M.: Superfast-trainable multi-class probabilistic classifier by least-squares posterior fitting. IEICE Transactions on Information and Systems 93-D(10), 2690–2701 (2010)
14. Trelea, I.C.: The particle swarm optimization algorithm: convergence analysis and parameter selection. Information Processing Letters 85(6), 317–325 (2003)
15. Williams, C.K.I., Barbe, D.: Bayesian classification with gaussian processes. IEEE Transactions on Pattern Analysis and Machine Intelligence 20(12), 1342–1351 (1998)
16. Wu, T.-F., Chih-Jen, C.-J., Weng, R.C.: Probability estimates for multi-class classification by pairwise coupling. Journal of Machine Learning Research 5, 975–1005 (2004)

Privacy-Preserving Social Network Publication Based on Positional Indiscernibility[*]

Tsan-sheng Hsu, Churn-Jung Liau, and Da-Wei Wang

Institute of Information Science
Academia Sinica, Taipei 115, Taiwan
{tshsu,liaucj,wdw}@iis.sinica.edu.tw

Abstract. In this paper, we address the issue of privacy preservation in the context of publishing social network data. The individuals in published social networks are typically anonymous; however, an adversary may be able to combine the released anonymous social network data with publicly available non-sensitive information to re-identify the individuals in a social network. In this paper, we consider the case that an adversary can query such publicly available databases with description logic(DL) concepts. To address the privacy issue, we utilize social position analysis techniques to determine the indiscernibility of individuals in a social network. Social position analysis attempts to find individuals that occupy the same position in a social network based on the pattern of their relationships to other actors. Recently, it was shown that social positions can be characterized by modal logics; thus, individuals occupying the same social position will satisfy the same set of modal formulas. Since DL has a close correspondences with modal logic, individuals occupying the same social position can not be distinguished by the knowledge expressed in DL formalisms. By partitioning a set of individuals into indiscernible classes in this way, we can easily test the safety of publishing the social network data.

Keywords: Data privacy, social network, description logic, indiscernibility, information granule.

1 Introduction

Indiscernibility is an important notion in uncertainty management and has been extensively studied in rough set theory[15]. The notion is also the basis of many approaches to privacy-preserving data publication. In the publication of data tables, simply maintaining the individuals' anonymity may not be sufficient to protect their privacy. The major threat to privacy is the re-identification of the individuals by linking the anonymous data to some external databases[16,19]. Although identifiers, such as names and social security numbers, are typically removed from released data sets, it has long been recognized that several quasi-identifiers (e.g., ZIP codes, age, and sex) can be used to re-identify individual records. The main reason is that the quasi-identifiers may appear with an individual's identifiers in

---

[*] This work was partially supported by NSC (Taiwan) Grant 98-2221-E-001-013-MY3.

W. Liu, V.S. Subrahmanian, and J. Wijsen (Eds.): SUM 2013, LNAI 8078, pp. 311–324, 2013.

another public database. Therefore, the problem is how to prevent adversaries inferring sensitive information about an individual by linking the released data set to some public databases[1]. To test if the publication of a data table cause privacy breach, individuals are grouped in a bin or an information granule based on the indiscernibility of their quasi-identifier values.

An analogous situation may arise with social network data as an increasing amount of non-sensitive information about personal social networks becomes publicly available. An adversary may combine the released anonymous social network with the publicly available non-sensitive information to re-identify the individuals in the social network. Unlike in the case of data tables, in a social network, two individuals with same quasi-identifier values may still be distinguishable by their relationships with other individuals. Thus, to formulate information granules for social networks, we have to consider the attributes of the individuals as well as the relationships between the individuals.

In this paper, we address the issue of privacy preservation in publishing social networks in the context that an adversary may be able to query publicly available databases by description logic(DL) formalisms[13]. To formulate information granules for social networks, we utilize social position analysis techniques to determine the indiscernibility of actors in a social network. Social position analysis attempts to find individuals that occupy the same position in a social network based on the patterns of their relationships with other actors. Different notions of positional equivalence have been proposed for different relationship patterns[9]. In this paper, we use the notions of *regular equivalence* and *exact equivalence* to define information granules. Recently, it was shown that social positions based on such equivalences can be characterized by modal logics[10,12]; thus, individuals occupying the same social position will satisfy the same set of modal formulas. Since DL has a close correspondences with modal logic, individuals occupying the same social position can not be distinguished by the knowledge expressed in DL formalisms. Hence, we can define information granules as social positions in a social network. Then, by generalizing the definition of information granules, we can extend the analytical techniques used for tabulated data to social network data.

The remainder of the paper is organized as follows. We review the basic notions of social position analysis and description logic in Sections 2 and 3 respectively. In Section 4, we use positional analysis techniques to address the privacy preservation issue in publishing social network data. Section 5 contains some concluding remarks.

2   Social Networks and Positional Analysis

Social networks are defined by actors (or individuals) and relations; or by nodes and edges in terms of graph theory[7]. A social network is generally defined

---

[1] In this paper, "adversary" or "adversaries" refers to anyone receiving data and having the potential to breach the privacy of individuals; while an individual or an actor refers to a person whose privacy must be protected.

as a relational structure $\mathfrak{N} = (U, (\alpha_i)_{i \in I})$, where U is the set of actors in the network, I is an index set and, for each $i \in I$, $\alpha_i \subseteq U^{k_i}$ is a k_i-ary relation on the domain U. If $k_i = 1$, then α_i is also called an attribute. Most SNA studies consider a simplified version of social networks with only binary relations. Hence, for ease of presentation, we focus on social networks with unary and/or binary relations. Thus, the social network considered in this article is a structure $\mathfrak{N} = (U, (p_i)_{i \in I}, (\alpha_j)_{j \in J})$, where U is a *finite set* of actors, $p_i \subseteq U$ for all $i \in I$, and $\alpha_j \subseteq U \times U$ for all $j \in J$. In terms of graph theory, $\mathfrak{N}$ is a labeled graph, where U is a set of nodes labeled with subsets of I, and each α_j denotes a set of (labeled) edges. For each $a \in U$, the out-neighborhood and in-neighborhood of a with respect to a binary relation α, denoted by $N_\alpha^+(a)$ and $N_\alpha^-(a)$ respectively, are defined as follows:

$$N_\alpha^+(a) = \{b \in U \mid (a, b) \in \alpha\},$$
$$N_\alpha^-(a) = \{b \in U \mid (b, a) \in \alpha\}.$$

If ρ is an equivalence relation on U and a is an actor, the ρ-equivalence class of a is equal to its neighborhood, i.e., $[a]_\rho = N_\rho^+(a) = N_\rho^-(a)$. Note that the latter equality holds because of the symmetry of ρ.

Several equivalence relations have been proposed for exploring the similarity between actors' roles. The simplest definition of positional equivalence is the concept of structural equivalence presented in [11], which states that two actors are positionally equivalent if they are related to the same individuals. Although structural equivalence is conceptually simple, it is sometimes restrictive. Regular equivalence relaxes the requirement that equivalent actors must be connected to identical actors, and suggests that actors occupy the same position if they are connected to positionally equivalent actors. According to Boyd and Everett's characterization [3], an equivalence relation ρ is a *regular equivalence* with respect to a binary relation α if it commutes with α, i.e.

$$\alpha \rho = \rho \alpha.$$

Let $\mathfrak{N} = (U, (p_i)_{i \in I}, (\alpha_j)_{j \in J})$ be a social network, and let ρ be an equivalence relation on U. Then, ρ is a regular equivalence with respect to $\mathfrak{N}$ if

1. $(a, b) \in \rho$ implies that $a \in p_i$ iff $b \in p_i$ for all $i \in I$; and
2. ρ is a regular equivalence with respect to α_j for all $j \in J$.

It is known that the coarsest regular equivalence of a network always exists. Thus, we can define that actors x and y are regularly equivalent, denoted by $x \cong_r y$, if (x, y) is in the coarsest regular equivalence of the network.

For regular equivalence, only the occurrence or non-occurrence of a position in the neighborhood of an actor is of interest. However, the number of occurrences is sometimes an important factor in positional analysis. In such cases, a restriction on the number can be added to the definition of regular equivalences. An equivalence relation ρ is an *exact equivalence* with respect to a binary relation α if for $a, b \in U$,

$$(a, b) \in \rho \Rightarrow N_\alpha^+(a) /\!/ \rho = N_\alpha^+(b) /\!/ \rho \text{ and } N_\alpha^-(a) /\!/ \rho = |N_\alpha^-(b) /\!/ \rho,$$

where for any $X \subseteq U$, $X/\!/\rho$ is the "quotient multiset" of X with respect to ρ; that is, $X/\!/\rho = \{\![x]_\rho \mid x \in X\!\}$, where $\{\!\!\{\cdot\}\!\!\}$ denotes a multiset. Thus, the number of equivalent neighbors must be the same for two actors to be considered exactly equivalent. Let $\mathfrak{N} = (U, (p_i)_{i \in I}, (\alpha_j)_{j \in J})$ be a social network, and let ρ be an equivalence relation on U. Then, ρ is an exact equivalence with respect to $\mathfrak{N}$ if

1. $(a, b) \in \rho$ implies that $a \in p_i$ iff $b \in p_i$ for all $i \in I$; and
2. ρ is an exact equivalence with respect to α_j for all $j \in J$.

The coarsest exact equivalence for a network exists and we can define two actors x and y as exactly equivalent, denoted by $x \cong_e y$, if they belong to the coarsest exact equivalence. The partition produced by an exact equivalence is also called an equitable partition or a divisor of a graph[9].

3   Description Logics

Description logic(DL)[2] is a fragment of first-order logic designed especially for knowledge representation. The connection between DL and modal logics has been explicated in [6,17]. Traditionally, modal logic has been considered the logic for reasoning about modalities, such as necessity, possibility, time, action, belief, knowledge, and obligation. However, semantically, it is essentially a language for describing relational structures[2]. A relational structure is simply a set accompanied by a collection of relations on that set. Thus, social networks are mathematically equivalent to relational structures, and DL is an appropriate formalism for representing social network knowledge.

The alphabet of a DL language consists of individual names, atomic concepts, atomic roles, concept constructors. and role constructors. Complex terms can be built inductively from atomic concepts and roles with constructors. Following the notations in [1], we use a, b, and c for individual names, A and B for atomic concepts, C and D for concept terms, and R and S for role terms. DL languages are differentiated by the constructors they provide.

Tarskian semantics for a DL is given by assigning the elements of the domain to individuals names, the sets to atomic concepts, and the binary relations to atomic roles. Formally, an interpretation is a pair $\mathfrak{I} = (\Delta^{\mathfrak{I}}, [\![\cdot]\!]_{\mathfrak{I}})$, where $\Delta^{\mathfrak{I}}$ is a non-empty set called the domain of the interpretation, and $[\![\cdot]\!]_{\mathfrak{I}}$ is an interpretation function that assigns an element in $\Delta^{\mathfrak{I}}$ to each individual name, a subset of $\Delta^{\mathfrak{I}}$ to each atomic concept, and a subset of $\Delta^{\mathfrak{I}} \times \Delta^{\mathfrak{I}}$ to each atomic role. For brevity, we usually omit the subscript and superscript $\mathfrak{I}$ if no confusion arises. The domain of $[\![\cdot]\!]$ can be extended to all concept and role terms by induction. We assume that distinct individual names denote distinct objects. Therefore, the unique name assumption (UNA) must be satisfied; that is, if a, b are distinct names, then $[\![a]\!] \neq [\![b]\!]$.

---

[2] Most of the notations and definitions in this section follow those introduced in [1].

In a knowledge base, one can distinguish between intensional knowledge, or general knowledge about the problem domain, and extensional knowledge, which is specific to a particular problem. Thus, a DL knowledge base is comprised of two components: a "TBox" and an "ABox". The TBox contains *terminological axioms* , which have the form

$$C \sqsubseteq D(R \sqsubseteq S) \quad \text{or} \quad C \equiv D(R \equiv S),$$

where C, D are concepts (and R, S are roles). Axioms of the first kind are called *inclusions,* while the second kind are called *equalities.* We only consider axioms involving concepts.

An equality whose left-hand side is an atomic concept is called a *definition.* Definitions are used to introduce symbolic names for complex descriptions. An atomic concept that does not occur on the left-hand side of any axiom is called *primitive.* A finite set of definitions Σ is called a terminology or TBox if no symbolic name is defined more than once; that is, if, for every atomic concept A, there is at most one axiom in Σ whose left-hand side is A. Let A, B be atomic concepts that occur in Σ. It is said that A *directly uses* B in Σ if B appears on the right-hand side of the definition of A, and "uses" is the transitive closure of the relation "directly uses". Then, Σ contains a *cycle* iff there exists an atomic concept in Σ that uses itself. Otherwise, Σ is *acyclic.*

The satisfaction of an axiom in an interpretation $\mathfrak{I} = \langle \Delta, \| \cdot \| \rangle$ is defined as follows:

$$C \equiv D \Leftrightarrow \|C\| = \|D\|,$$

$$C \sqsubseteq D \Leftrightarrow \|C\| \subseteq \|D\|.$$

If $\mathfrak{I}$ satisfies an axiom φ, it is written as $\mathfrak{I} \models \varphi$. A set of axioms Σ is satisfied by $\mathfrak{I}$, written as $\mathfrak{I} \models \Sigma$, if $\mathfrak{I}$ satisfies each axiom of Σ. Furthermore, Σ is satisfiable if it is satisfied by some $\mathfrak{I}$. An axiom φ is the logical consequence of a TBox Σ, denoted by $\Sigma \models \varphi$, if each interpretation that satisfies Σ also satisfies φ.

An ABox contains the description of the world in the form of *assertional axioms*

$$C(a) \quad \text{or} \quad R(a, b),$$

where C is a concept term , R is a role term, and a, b are individual names. The assertion $C(a)$ is called a *concept assertion*, whereas $R(a, b)$ is a *role assertion.* An interpretation $\mathfrak{I} = \langle \Delta, \| \cdot \| \rangle$ satisfies an assertion

$$C(a) \Leftrightarrow \|a\| \in \|C\|,$$

$$R(a, b) \Leftrightarrow (\|a\|, \|b\|) \in \|R\|.$$

$\mathfrak{I}$ satisfies the ABox Φ if it satisfies each assertion in Φ. It is also said that $\mathfrak{I}$ is a model of the assertion or the ABox. Finally, $\mathfrak{I}$ satisfies an assertion or an ABox Φ with respect to a TBox Σ if it satisfies both Σ and the assertion or the ABox. An assertion φ is a logical consequence of an ABox Φ, written as $\Phi \models \varphi$, if for every interpretation $\mathfrak{I}$, $\mathfrak{I} \models \Phi$ implies that $\mathfrak{I} \models \varphi$.

Let $\mathcal{L}$ be a DL language, and let $\mathfrak{I} = \langle \Delta, \| \cdot \| \rangle$ be an interpretation of $\mathcal{L}$. Then, two elements $u, v \in \Delta$ are *indiscernible* with respect to the $\mathcal{L}$-concepts (or simply indiscernible), written as $u \equiv_{\mathcal{L}} v$, if for any concept term C of $\mathcal{L}$, $u \in \|C\|$ iff $v \in \|C\|$. Two individual names a and b are *indiscernible* with respect to the interpretation $\mathfrak{I}$ (or simply indiscernible), written as $a \equiv_{\mathcal{L}}^{\mathfrak{I}} b$, if $\|a\|$ and $\|b\|$ are indiscernible. Let $\rho \subseteq \Delta \times \Delta$ be an equivalence relation on the domain of $\mathfrak{I}$. Then, we say that the $\mathcal{L}$-concepts are *preserved under* ρ if u, v are indiscernible for any $(u, v) \in \rho$. Obviously, a social network $(U, (p_i)_{i \in I}, (\alpha_j)_{j \in J})$ can be regarded as an interpretation of a DL language if each p_i corresponds to an atomic concept and each α_j corresponds to an atomic role. Thus, all definitions of interpretations can be applied to social networks. Next, we introduce two DL languages, called $\mathcal{ALCI}$ and $\mathcal{ALCQI}$, whose concepts are preserved under regular equivalence and exact equivalence respectively.

The $\mathcal{ALCI}$ and $\mathcal{ALCQI}$ languages belong to the well-known $\mathcal{AL}$ (attributive language) family, which was first presented in[18]. The constructors for the two languages and their semantics are shown in Table 1, where $\sharp(\cdot)$ denotes the cardinality of a set. Thus, the concept terms of $\mathcal{ALCI}$ are formed according to the following syntax rules:

$$A \mid \top \mid \bot \mid \neg C \mid C \sqcap D \mid \forall R : C \mid \forall R : C.$$

The syntax rules for $\mathcal{ALCQI}$ consist of those for $\mathcal{ALCI}$ and $\geq nR : C$ and $\leq nR : C$, where A is an atomic concept, C and D are concept terms, R is an atomic role or an inverse role, and n is a natural number.

4   Privacy-Preserving Social Network Publishing

4.1   Problem Formulation

We consider a scenario where a data owner wants to release a social network to the public. The basic requirement is that any identifying information must be removed from the data to be published; therefore, we can assume that the actors in the social network are anonymous. Formally, we consider an anonymous social network $\mathfrak{N} = (U, (p_i)_{i \in I}, (\alpha_j)_{j \in J})$, where U is a set of anonymous actors. The name of each actor in U can be regarded as a pseudonym. We use a DL language $\mathcal{L}_0$ characterized by $(X, (A_i)_{i \in I}, (R_j)_{j \in J})$ to describe the social network, where X is a set of individual names, each A_i is an atomic concept, and each R_j is an atomic role. The triple $(X, (A_i)_{i \in I}, (R_j)_{j \in J})$ is also called the *signature* of $\mathcal{L}_0$. The set X can be regarded as the set of real identifiers of the anonymous actors. Given the unique name assumption, we can assume that the cardinality of X and the cardinality of U are the same. Thus, the social network can be seen as an interpretation $\mathfrak{I} = (\Delta, \| \cdot \|)$ such that $\Delta = U$, $\|A_i\| = p_i (i \in I)$, and $\|R_j\| = \alpha_j (j \in J)$. Furthermore, since the data owner knows each anonymous actor's identifier, he knows the interpretation of all individual names; however, the adversary does not have such information. In other words, the restriction of the interpretation function $\| \cdot \|$ to X, denoted by $\| \cdot \| \downharpoonright X$, is known to the data owner, but not to the adversary.

Table 1. Constructors and semantics of description logics

Name	Syntax	Semantics
atomic concept	A	$\llbracket A \rrbracket \subseteq \Delta$
universal concept	$\top$	Δ
empty concept	$\bot$	$\emptyset$
complement (negation)	$\neg C$	$\Delta \backslash \llbracket C \rrbracket$
intersection	$C \sqcap D$	$\llbracket C \rrbracket \cap \llbracket D \rrbracket$
universal quantification	$\forall R : C$	$\{x \mid \forall y((x,y) \in \llbracket R \rrbracket \Rightarrow y \in \llbracket C \rrbracket)\}$
existential quantification	$\exists R : C$	$\{x \mid \exists y((x,y) \in \llbracket R \rrbracket \wedge y \in \llbracket C \rrbracket)\}$
qualified at-least restriction	$\geq nR : C$	$\{x \mid \sharp(\{y \in \llbracket C \rrbracket \mid (x,y) \in \llbracket R \rrbracket\}) \geq n\}$
qualified at-most restriction	$\leq nR : C$	$\{x \mid \sharp(\{y \in \llbracket C \rrbracket \mid (x,y) \in \llbracket R \rrbracket\}) \leq n\}$
atomic role	R	$\llbracket R \rrbracket \subseteq \Delta \times \Delta$
inverse role	R^-	$\{(x,y) \mid (y,x) \in \llbracket R \rrbracket\}$
counter domain	$\sim R$	$\{(x,x) \mid \forall y((x,y) \notin \llbracket R \rrbracket)\}$
role union	$R \sqcup S$	$\llbracket R \rrbracket \cup \llbracket S \rrbracket$
role composition	$R \circ S$	$\llbracket R \rrbracket \cdot \llbracket S \rrbracket = \{(x,y) \mid \exists z((x,z) \in \llbracket R \rrbracket \wedge (z,y) \in \llbracket S \rrbracket)\}$
transitive closure	R^+	$\bigcup_{k>0} \llbracket R \rrbracket^k$ where $\llbracket R \rrbracket^k = \llbracket R \rrbracket^{k-1} \cdot \llbracket R \rrbracket$
class identity	$C?$	$\{(x,x) \mid x \in \llbracket C \rrbracket\}$

Note that, according to our assumption, $\llbracket \cdot \rrbracket \downharpoonleft X$ is a bijection between X and U. In other words, we implicitly assumed that the adversary knows the set of real identifiers of all members of the network. This seems unrealistic at first glance, since it is much more often the case that the set of individuals who are members of the social network will not be available to the adversary. However, with a moderately expressive query language, the information can be easily revealed to the adversary. For example, if the query language contains the universal query $\top$ whose answer set is X or two mutually complementary queries φ and ψ whose answer sets are the complementary set of each other, then the adversary can easily know what X is, and consequently, it is impossible to hide the participation of any individual in the published social network.

Normally, a social network contains sensitive as well as non-sensitive information about the actors. Non-sensitive information can be obtained from external databases, and usually appears with the identifiers of individuals. To formalize the situation, we assume that atomic concepts and atomic roles can be partitioned into two subsets. That is, $I = I_s \cup I_n$ and $J = J_s \cup J_n$, where $I_s \cap I_n = \emptyset$ and $J_s \cap J_n = \emptyset$. An atomic concept (resp. role) A_i (resp. R_j) whose index $i \in I_n$ (resp. $j \in J_n$) is a non-sensitive concept (resp. role). Let $\mathcal{L}$ denote the sublanguage of $\mathcal{L}_0$ with the signature $(X, (A_i)_{i \in I_n}, (R_j)_{j \in J_n})$. Then, all $\mathcal{L}$-concepts and roles are non-sensitive. We assume that only non-sensitive concept terms about individuals can appear in an external database. Thus, an external database is an ABox comprised of assertions made about the individuals in X by using only $\mathcal{L}$-terms.

Domain knowledge may be available to the public in the form of a TBox. Since the axioms in a TBox do not involve any individuals, the TBox may contain sensitive and non-sensitive terms . However, a TBox should not allow an adversary to derive sensitive atomic concepts or roles from non-sensitive terms.

Thus, a sensitive atomic concept or role should not be defined by using only non-sensitive terms. The non-derivability of sensitive atomic concepts or roles from non-sensitive terms is a basic assumption about the publicly available TBox. If the assumption is not satisfied, then it may be possible to breach the privacy of individuals by using the external database alone. This aspect is not related to the publishing of the social network, and it is obviously beyond the scope of the current problem. In fact, the assumption can be further relaxed if the full contents of the Tbox are hidden from the public so that an adversary can only retrieve information from a public database through a limited query language. For simplicity, we assume that access to public databases is limited in this way.

Let Σ and Φ denote the publicly available TBox and ABox respectively. We consider the case that the data owner and the adversary can retrieve non-sensitive information about individuals by using $\mathcal{L}$-concept terms to query the public databases. The answer to a query C is defined as $Ans(C|\Sigma, \Phi) = \{a \in X \mid \Sigma \cup \Phi \models C(a)\}$. We also assume that the public databases are *truthful* with respect to the given social network (the verity assumption) and *complete* (the completeness assumption). The former means that $\mathfrak{I} \models \Sigma$ and $\mathfrak{I} \models \Phi$, and the latter means that $Ans(C|\Sigma, \Phi) \cup Ans(\neg C|\Sigma, \Phi) = X$ for any $\mathcal{L}$-concept term C. These two assumptions seem quite unrealistic because most databases are incomplete and/or contain incorrect information. However, we make the assumptions so that we can conduct the worst-case analysis of the privacy preservation issue. In other words, we consider the case where an adversary can obtain as much non-sensitive information about individuals as possible. If the individuals' privacy is not breached, even though the adversary can retrieve such truthful and complete information, then the privacy can still be preserved when only less reliable databases are available to the adversary.

In summary, the data owner possesses the following information:

1. the anonymous social network: $\mathfrak{N}$;
2. the vocabulary of the DL language $\mathcal{L}_0$: $(X, (A_i)_{i \in I}, (R_j)_{j \in J})$;
3. the partition of the vocabulary into sensitive and non-sensitive parts: $I = I_s \cup I_n$ and $J = J_s \cup J_n$;
4. the information retrieved from the public databases: $Ans(C|\Sigma, \Phi)$ for any $\mathcal{L}$-concept term C; and
5. the social network as the interpretation of $\mathcal{L}_0$: $\mathfrak{I}$.

If the anonymous social network is published, the adversary can obtain almost the same information as that of the data owner, but the information about $\mathfrak{I}$ is only partially known by the adversary. Because the released social network is anonymous, the adversary does not have $\| \cdot \| \downarrow X$, although he knows $\|A_i\|$ for $i \in I$ and $\|R_j\|$ for $j \in J$. Indeed, since $\|A_i\|$ is a subset of pseudonyms and $\|R_j\|$ is a binary relation between pseudonyms, the adversary can learn such information from the released anonymous network. However, $\| \cdot \| \downarrow X$ is the identification of real individuals with their corresponding pseudonyms which is exactly the target to be protected. Thus, we can formulate the issue of privacy-preserving social network publication as the following problem:

– *The data owner must decide if the privacy requirement would be violated if an adversary could access the above-mentioned information.*

Because we do not give a precise specification of the privacy requirement in the formulation, it actually represents a family of decision problems for the data owner. In the family, the *identity disclosure problem* and the *information disclosure problem* are particularly interesting. The privacy requirement for the former is that the identities of the individuals must be hidden from the adversary, whereas the requirement for the latter is that a predefined set of sensitive facts about the individuals must be kept confidential. The two problems are formulated more precisely as follows:

– Identity disclosure problem: *Could the adversary infer* $\| \cdot \| \downharpoonleft X$ *or* $\|a\|$ *for some* $a \in X$?
– Information disclosure problem: *Could the adversary infer* $C(a)$ *or* $R(a, b)$ *for some* $a, b \in X$, *where* C *is a sensitive concept and* R *is a sensitive role?*

4.2   Information Granules

As mentioned in Section 1, privacy preservation depends to a large extent on the indiscernibility of individuals from the information available to the adversary. Thus, to address the privacy preservation problem, the first step is to find the classes of indiscernible individuals. We call such classes *information granules*. They can be formed on the domain of pseudonyms U or on the domain of real individuals X. We choose the equivalence classes of $\equiv_{\mathcal{L}}$ and $\equiv_{\mathcal{L}}^?$ as the information granules of U and X respectively.

The following scenario explains the reason for this choice. Suppose the adversary wants to know which pseudonym in the social network corresponds to a particular individual $a \in X$. He will need to query the public databases with different $\mathcal{L}$-concept terms and compare a with the pseudonyms according to their memberships in the answer set, since this is the only way to link the actual individuals and the pseudonyms. Initially, the adversary considers all elements in U as possible pseudonyms for a because he does not have any information at that point. Next, he eliminates the pseudonyms that do not match a by considering the answers to the queries sequentially. When he reaches the point where no further eliminations are possible, the remaining elements are possible pseudonyms for a according to all the available information. Formally, let $C_1, C_2, C_3, \ldots$ be the enumeration of all $\mathcal{L}$-concepts and let $U_0 = U$. Then, we can define

$$U_k = \begin{cases} U_{k-1} \backslash \| C_k \|, & \text{if } a \notin Ans(C_k | \Sigma, \Phi), \\ U_{k-1} \backslash \| \neg C_k \|, & \text{if } a \in Ans(C_k | \Sigma, \Phi), \end{cases}$$

for $k \geq 1$ and $U_\infty^a = \bigcap_{k \geq 1} U_k$. In the same way, we can also define U_∞^b for any $b \in X$. Then, it is straightforward to derive the following result.

Proposition 1

1. *For any $a \in X$, $u \in U_\infty^a$ iff $u \equiv_{\mathcal{L}} \|a\|$ (or equivalently, $U_\infty^a = [\![\|a\|]\!]_{\equiv_{\mathcal{L}}}$).*
2. *For any $a, b \in X$, if $a \equiv_{\mathcal{L}}^{\exists} b$, then $U_\infty^a = U_\infty^b$.*

The proposition shows that, based on the information available to the adversary, he cannot differentiate $\|a\|$ from the other pseudonyms in the equivalence class $[\![\|a\|]\!]_{\equiv_{\mathcal{L}}}$. Thus, unless $[\![\|a\|]\!]_{\equiv_{\mathcal{L}}}$ is a singleton, the adversary cannot infer the value of $\|a\|$. The larger the equivalence class $[\![\|a\|]\!]_{\equiv_{\mathcal{L}}}$, the more difficult it will be for the adversary to identify the pseudonym of a. Therefore, we can use the well-known k-*anonymity* criterion[19,20] to assess the risk of identity disclosure. In the current context, the criterion is formulated as follows.

Definition 1. *For any $k > 0$, the released social network $\mathfrak{N} = (U, (p_i)_{i \in I}, (\alpha_j)_{j \in J})$ satisfies the k-anonymity criterion if for all $u \in U$, $\sharp([u]_{\equiv_{\mathcal{L}}}) \geq k$ (or equivalently, for all $a \in X$, $\sharp([a]_{\equiv_{\mathcal{L}}^{\exists}}) \geq k$).*

Note that 1-anonymity is always satisfied trivially. Although 2-anonymity is usually sufficient to prevent the risk of identity disclosure, k-anonymity is sometimes required for larger k when the released data contains highly sensitive information.

It has been shown that k-anonymity can not fully prevent the compromising of individual privacy because of homogeneity attacks[8,5,21,22]. In the current context, this means that all individuals in $[\![\|a\|]\!]_{\equiv_{\mathcal{L}}}$ satisfy the same sensitive property that can be derived from the released social network. Let $\mathcal{SC}$ and $\mathcal{SR}$ denote, respectively, a set of sensitive concepts and a set of sensitive roles in the language $\mathcal{L}_0$. Then, the *logical safety* criterion[8] can be used to prevent the disclosure of sensitive information.

Definition 2. *The released social network $\mathfrak{N} = (U, (p_i)_{i \in I}, (\alpha_j)_{j \in J})$ is logically safe with respect to a $\mathcal{L}_0$-concept C (resp. role R) if, for all $u \in U$, there exists $v \in [u]_{\equiv_{\mathcal{L}}}$ such that $v \notin \|C\|$ (resp. $v \notin \|\exists R : \top\|$ and $v \notin \|\exists R^- : \top\|$). The social network is simply logically safe if it is logically safe with respect to each concept in $\mathcal{SC}$ and each role in $\mathcal{SR}$.*

Once the logical safety requirement has been violated for some $\|a\| \in U$, the adversary can infer some sensitive information about a without identifying the node in the network that corresponds to a. The criterion guarantees the heterogeneity of the pseudonyms that the adversary can not distinguish from $\|a\|$ with respect to the sensitive concepts and roles; hence, the adversary cannot infer any sensitive information about a with certainty.

4.3 Computation with Positional Analysis

We defined information granules as equivalence classes of the indiscernibility relations $\equiv_{\mathcal{L}}$ and $\equiv_{\mathcal{L}}^{\exists}$, but we did not specify particular DL languages in the definition. For a given DL language $\mathcal{L}$, checking if $u \equiv_{\mathcal{L}} v$ amounts to evaluating all

$\mathcal{L}$-concepts in u and v. Since the set of all $\mathcal{L}$-concepts is usually infinite, a straight-forward calculation is impossible. Thus, we need more practical procedures for computing information granules. The procedures may be quite diverse for different DL languages. However, because $\equiv_{\mathcal{L}}$ is closely related to the bisimulation of interpretations of modal logic according to the well-known Hennessy-Milner Theorem[2], we can find good algorithms for the computation of indiscernibility relations in the literature on bisimulation. In particular, it has been shown that regular equivalence and exact equivalence correspond to the bisimulation of interpretations of multi-modal logic and graded modal logic respectively[12,10]. Since $\mathcal{ALCI}$ and $\mathcal{ALCQI}$ can be exactly translated into multi-modal logic and graded modal logic respectively, we can use existing algorithms for social position analysis to compute the information granules. More precisely, the computation is based on the following proposition.

Proposition 2. *Let* $\mathfrak{N}_n = (U, (p_i)_{i \in I_n}, (\alpha_j)_{j \in J_n})$ *be the non-sensitive sub-network of* $\mathfrak{N}$*. Then, for* $x, y \in U$*,*

1. $x \cong_r y$ *in* $\mathfrak{N}_n$ *iff* $x \equiv_{\mathcal{L}} y$ *when* $\mathcal{L}$ *is formed by using only* $\mathcal{ALCI}$ *constructors and the role constructors introduced in Table 1;*
2. $x \cong_e y$ *in* $\mathfrak{N}_n$ *iff* $x \equiv_{\mathcal{L}} y$ *when* $\mathcal{L}$ *is formed by using only* $\mathcal{ALCQI}$ *constructors.*

According to [9], there exist $O(m \log_2 n)$-time algorithms for computing the coarsest regular equivalence $\cong_r$ and the coarsest equitable $\cong_e$, where m and n are, respectively, the number of links and the number of nodes in a network[4,14]. Thus, an implication of the above theorem is that the computation of information granules based on the $\mathcal{ALCI}$ and $\mathcal{ALCQI}$ languages can be achieved with the same time complexity.

Example 1. *Let us consider a social network* $\mathfrak{N} = (U, (p_1, p_2), (\alpha_1, \alpha_2, \alpha_3))$*, where*

- $U = \{1, 2, \cdots, 11\}$*,*
- $p_1 = \{1, 2, 3, 4, 5\}$*,*
- $p_2 = \{4, 6, 7\}$*,*
- $\alpha_1 = \{(1,4), (2,4), (2,5), (3,5), (6,10), (6,11), (7,8), (7,9), (7,10)\}$*,*
- $\alpha_2 = \{(1,6), (2,7), (3,6), (3,7), (4,9), (4,10), (5,8), (5,11)\}$*, and*
- $\alpha_3 = \{(2,9), (4,8), (9,5)\}$*.*

Furthermore, p_1, α_1*, and* α_2 *are non-sensitive, while* p_2 *and* α_3 *are sensitive. The non-sensitive part of the network is shown in Figure 1. We assume the query language for the external database is an* $\mathcal{ALCI}$ *language extended with the role constructors introduced in Table 1. The signatures of* $\mathcal{L}_0$ *and* $\mathcal{L}$ *are therefore* $(X, A_1, A_2, R_1, R_2, R_3)$ *and* (X, A_1, R_1, R_2) *respectively, where* $X = \{a_1, a_2, \cdots, a_{11}\}$*. By viewing the social network as an interpretation of the language* $\mathfrak{I} = (U, \|\cdot\|)$*, we can define* $\|a_i\| = i (1 \le i \le 11)$*,* $\|A_i\| = p_i (i = 1, 2)$*, and* $\|R_i\| = \alpha_i (i = 1, 2, 3)$*. According to our problem formulation, the adversary does not know the evaluation* $\|a_i\|$*, but he can retrieve the information about the individuals from the database. For ease of presentation, we assume that all concepts* and roles are primitive. Thus, the TBox is empty and the ABox contains the full description of the graph in Figure 1.

Suppose the adversary retrieves the information by using two queries A_1 and $\exists R_1 : \top$ (and perhaps their negations). Then, he can find that a_1, a_2, and a_3 satisfy both queries; a_4 and a_5 satisfy A_1, but not $\exists R_1 : \top$; a_6 and a_7 satisfy $\exists R_1 : \top$, but not A_1; and a_8, a_9, a_{10}, and a_{11} do not satisfy either of the queries. Thus, the adversary can construct the following mapping:

$$\begin{aligned}
\{a_1, a_2, a_3\} &\longmapsto \{1, 2, 3\} \\
\{a_4, a_5\} &\longmapsto \{4, 5\} \\
\{a_6, a_7\} &\longmapsto \{6, 7\} \\
\{a_8, a_9, a_{10}, a_{11}\} &\longmapsto \{8, 9, 10, 11\},
\end{aligned}$$

which corresponds to the partition of the nodes into four blocks in Figure 1. Furthermore, since the partition is based on the regular equivalence of the network, the adversary can not obtain a finer partition by any further $\mathcal{L}$-concept queries. Therefore, the four blocks also correspond to the equivalence classes of the $\equiv_{\mathcal{L}}$ relation and are the information granules of the network. Since the minimum size of the blocks is 2, the release of the network satisfies the 2-anonymity requirement, but not the 3-anonymity requirement. Moreover, suppose $\mathcal{SC} = \{A_2\}$ and $\mathcal{SR} = \{R_3\}$. Then, the logical safety requirement is also violated because $\{6, 7\} \subseteq \|A_2\| = \{4, 6, 7\}$ and $\{4, 5\} \subseteq \|\exists R_3 : \top\| \cup \|\exists R_3^- : \top\| = \{2, 4, 5, 8, 9\}$. Once the privacy breach is detected, several sanitization strategies can be applied to the network to improve its privacy level, which we can not present here due to the space limit. ■

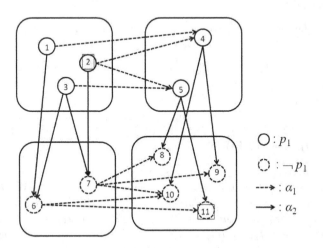

Fig. 1. The non-sensitive part of a social network

5   Concluding Remarks

In this paper, we address the issue of privacy-preserving social network publication by using social position analysis techniques. We assume the adversary's background knowledge is represented by concept assertions of DL. Based on the modal logic characterization of positional analysis, actors occupying the same social position are considered indiscernible in terms of the adversary's background knowledge. Thus, by partitioning a social network into equivalence classes based on the indiscernibility relation, we can generalize the privacy criteria developed for tabulated data, such as k-anonymity, logical safety, l-diversity, and t-closeness, to social network data. An important feature of the proposed approach is that it does not make any assumption about the content of the adversary's background knowledge as long as it can be represented as concept assertions in the given DL language. As a result, the approach is robust against a wide range of knowledge-based attacks, and it guarantees the privacy of individuals even though the adversary has access to all public concept assertions about them.

There are several restrictions on the knowledge representation language. This situation impacts the form of the adversary's background knowledge that can be represented. For example, we only allow the use of concept assertions to represent the adversary's background knowledge; however, sometimes relational assertions about some public relations between individuals are also available to the adversary. Moreover, in general social networks, we may have to represent binary relations as well as k-ary relations for any $k > 1$. These restrictions are not essential to our framework because we can employ full-fledged quantificational logic as the knowledge representation formalism, and define the indiscernibility relation as a congruence relation with respect to all non-private predicates available to the public. However, from a practical viewpoint, we would like to develop efficient algorithms to identify such congruence relations. In Section 4.3, we showed that efficient algorithms can be found for some classes of DL languages. In our future work, we will try to extend the algorithms to a more general setting.

References

1. Baader, F., Nutt, W.: Basic description logics. In: Baader, F., Calvanese, D., McGuinness, D.L., Nardi, D., Patel-Schneider, P.F. (eds.) Description Logic Handbook, pp. 47–100. Cambridge University Press (2002)
2. Blackburn, P., de Rijke, M., Venema, Y.: Modal Logic. Cambridge University Press (2001)
3. Boyd, J.P., Everett, M.G.: Relations, residuals, regular interiors, and relative regular equivalence. Social Networks 21(2), 147–165 (1999)
4. Cardon, A., Crochemore, M.: Partitioning a graph in $o(|a|\log_2|v|)$. Theoretical Computer Science 19, 85–98 (1982)
5. Chiang, Y.C., Hsu, T.-S., Kuo, S., Liau, C.J., Wang, D.W.: Preserving confidentiality when sharing medical database with the Cellsecu system. International Journal of Medical Informatics 71, 17–23 (2003)

6. de Rijke, M.: Description logics and modal logics. In: Proceedings of the 1998 International Workshop on Description Logics, DL 1998 (1998)
7. Hanneman, R.A., Riddle, M.: Introduction to Social Network Methods. University of California, Riverside (2005)
8. Hsu, T.-S., Liau, C.-J., Wang, D.-W.: A logical model for privacy protection. In: Davida, G.I., Frankel, Y. (eds.) ISC 2001. LNCS, vol. 2200, pp. 110–124. Springer, Heidelberg (2001)
9. Lerner, J.: Role assignments. In: Brandes, U., Erlebach, T. (eds.) Network Analysis. LNCS, vol. 3418, pp. 216–252. Springer, Heidelberg (2005)
10. Liau, C.J.: Social networks and granular computing. In: Meyers, R.A. (ed.) Encyclopedia of Complexity and Systems Science, pp. 8333–8345. Springer (2009)
11. Lorrain, F., White, H.C.: Structural equivalence of individuals in social networks. Journal of Mathematical Sociology 1, 49–80 (1971)
12. Marx, M., Masuch, M.: Regular equivalence and dynamic logic. Social Networks 25(1), 51–65 (2003)
13. Nardi, D., Brachman, R.J.: An introduction to description logics. In: Baader, F., Calvanese, D., McGuinness, D.L., Nardi, D., Patel-Schneider, P.F. (eds.) Description Logic Handbook, pp. 5–44. Cambridge University Press (2002)
14. Paige, R., Tarjan, R.E.: Three partition refinement algorithms. SIAM Journal on Computing 16(6), 973–989 (1987)
15. Pawlak, Z.: Rough Sets–Theoretical Aspects of Reasoning about Data. Kluwer Academic Publishers (1991)
16. Samarati, P.: Protecting respondents' identities in microdata release. IEEE Transactions on Knowledge and Data Engineering 13(6), 1010–1027 (2001)
17. Schild, K.: A correspondence theory for terminological logics: Preliminary report. In: Proceedings of the 12th International Joint Conference on Artificial Intelligence, pp. 466–471 (1991)
18. Schmidt-Schauß, M., Smolka, G.: Attributive concept descriptions with complements. Artificial Intelligence 48(1), 1–26 (1991)
19. Sweeney, L.: Achieving k-anonymity privacy protection using generalization and suppression. International Journal on Uncertainty, Fuzziness and Knowledge-based Systems 10(5), 571–588 (2002)
20. Sweeney, L.: k-anonymity: a model for protecting privacy. International Journal on Uncertainty, Fuzziness and Knowledge-based Systems 10(5), 557–570 (2002)
21. Wang, D.W., Liau, C.J., Hsu, T.-S.: Medical privacy protection based on granular computing. Artificial Intelligence in Medicine 32(2), 137–149 (2004)
22. Wang, D.W., Liau, C.J., Hsu, T.-S.: An epistemic framework for privacy protection in database linking. Data and Knowledge Engineering 61(1), 176–205 (2007)

On the Implementation of a Fuzzy DL Solver over Infinite-Valued Product Logic with SMT Solvers[*]

Teresa Alsinet[1], David Barroso[1], Ramón Béjar[1], Félix Bou[2,3],
Marco Cerami[4], and Francesc Esteva[3]

[1] Department of Computer Science – University of Lleida
C/Jaume II, 69 – 25001 Lleida, Spain
{tracy,david,ramon}@diei.udl.cat
[2] Faculty of Mathematics, University of Barcelona
Gran Vía 585 – 08007 Barcelona, Spain
bou@ub.edu
[3] Artificial Intelligence Research Institute (IIIA-CSIC)
Campus UAB - 08193 Bellaterra, Barcelona, Spain
{fbou,esteva}@iiia.csic.es
[4] Department of Computer Science – Palacký University in Olomouc
17. listopadu 12 – CZ-77146 Olomouc, Czech Republic
marco.cerami@upol.cz

Abstract. In this paper we explain the design and preliminary implementation of a solver for the positive satisfiability problem of concepts in a fuzzy description logic over the infinite-valued product logic. This very solver also answers 1-satisfiability in quasi-witnessed models. The solver works by first performing a direct reduction of the problem to a satisfiability problem of a quantifier free boolean formula with non-linear real arithmetic properties, and secondly solves the resulting formula with an SMT solver. We show that the satisfiability problem for such formulas is still a very challenging problem for even the most advanced SMT solvers, and so it represents an interesting problem for the community working on the theory and practice of SMT solvers.

Keywords: description logics, fuzzy product logic, SMT solvers.

1 Introduction

In the recent years, the development of solvers for reasoning problems over description logics (DLs) has experienced an important growth, with very succesful approaches. We have two main approaches, the most traditional one, able to handle very expressive DLs, is the one based on Tableaux-like algorithms [1]. For certain DLs, the approach based on translations of the problem to more basic logical reasoning problems, like the ones

[*] Research partially funded by the Spanish MICINN projects ARINF (TIN2009-14704-C03-01/03) and TASSAT (TIN2010-20967-C04-01/03), MINECO project EdeTRI (TIN2012-39348-C02-01), Agreement Techologies (CONSOLIDER CSD 2007- 0022), Catalan Government (2009SGR-1433/34) and ESF project POST - UP II No. CZ.1.07/2.3.00/30.0041 that is co-financed by the European Social Fund and the state budget of the Czech Republic.

W. Liu, V.S. Subrahmanian, and J. Wijsen (Eds.): SUM 2013, LNAI 8078, pp. 325–330, 2013.
© Springer-Verlag Berlin Heidelberg 2013

based on a translation to propositional clausal forms has shown to be very sucessful [8]. Very recently, the approach based on doing translations to less simple knowledge representation formalisms and then using Sat Modulo Theory (SMT) solvers, has started to receive high interest [6].

In the case of DLs over fuzzy logics (Fuzzy DLs), the state-of-the-art on solvers can be summarized mainly on the work of Straccia and Bobillo with *fuzzyDL*, their solver for Fuzzy DL over Łuckasiewicz Logic [9] (available in Straccia's home page), that is based on a mixture of tableau rules and Mixed Integer Linear Programming. Note that the problem faced in [9], called *concept satisfiability w.r.t. knowledge bases with GCI*, is more general than the one faced in the present paper. Unfortunately, the general problem of concept satisfiability w.r.t. knowledge bases with GCI over infinite-valued Łukasewicz semantics has been proved to be undecidable in [3]. Nevertheless, as proved in [3], the solver proposed in [9] can solve the concept satisfiability problem without knowledge bases.

In this work, we present a solver for the concept satisfiability problem without knowledge bases, in the Fuzzy DL $\mathcal{ALE}$ over the infinite-valued product logic. This problem has been studied in [2] and the FDL under exam has been denoted $\Pi\text{-}\mathcal{ALE}$. Our approach is based on the last work, where the authors show that the positive and 1-satisfiability problems in $\Pi\text{-}\mathcal{ALE}$ limited to quasi-witnessed models are decidable.

To prove the above result the authors give a reduction (inspired by the one given by Hájek for witnessed models [7]) of the concept satisfiability problem in $\Pi\text{-}\mathcal{ALE}$ with respect to quasi-witnessed models to an entailment problem between two sets of propositional formulas. The algorithm presented in [2] takes a description concept C_0 as input and recursively produces a pair of propositional theories as output. The propositional theories produced as output jointly represent a description of an FDL interpretation (a kind of Kripke model) that is supposed to satisfy concept C_0 (in the case, obviously that C_0 is satisfiable) in the sense that C_0 is satisfiable if and only if it can be proved that one of the propositional theories is not entailed by the other. The novelty of the algorithm presented in [2] is that it can describe possibly infinite models by means of a finite set of propositional formulas.

For this reason, the algorithm is more complex than the one of Hájek. The algorithm proposed by Hájek, indeed, just produces one propositional theory with the property of being satisfiable if and only if the concept C_0 is satisfiable with respect to witnessed models. The advantage of dealing with only witnessed models is that then the finite model property holds. But in the case of quasi-witnessed models this property fails and there can be the case of dealing with infinite models of a certain shape. In this sense the two propositional theories in the output of the algorithm introduced in [2] represent positive and negative constraints that this kind of structures must respect in order to be models for the concepts considered. Hence, the problem of finding a propositional evaluation that satisfies the set of propositions $QWT(C_0)$ but not the set Y_{C_0}, is exactly the problem of deciding whether Y_{C_0} is not entailed by $QWT(C_0)$.

Moreover they prove that positive satisfiability in first order product logic (and, as a consequence, in $\Pi\text{-}\mathcal{ALE}$) coincide with positive satisfiability with respect to the quasi-witnessed models of first order product logic. If the same completeness result holds for the notion of 1-satisfiability is still an open problem.

In this paper we present a solver that works by first performing a direct reduction of the problem to a satisfiability problem of a boolean formula with real valued variables and non-linear terms, more concretely boolean formulas valid in $(\mathbb{R}, +, -, \cdot, /, \{q : q \in \{\mathbb{Q}\})$, and secondly solves the resulting formula with a SMT solver. Solving such formulas is still a very challenging problem for even the most advanced SMT solvers, and in this work we show results suggesting that this satisfiability problem for Π-$\mathcal{ALE}$ is a real challenging problem for SMT solvers, and so it represents an interesting problem for the community working on the theory and practice of SMT solvers.

2   An SMT-Based Solver for the Π-$\mathcal{ALE}$ Description Logic

2.1   System Architecture Design

For solving the r-satisfiability problem (where $r \in \mathbb{Q} \cap [0, 1]$), with witnessed or quasi-witnessed models, of an input concept C_0 in Π-$\mathcal{ALE}$, our system follows the next steps:

1. The user introduces the expression of the concept C_0 to be solved, and selects a class of models to search: witnessed or quasi-witnessed.
2. From the parsing tree of C_0, we either generate the set WT_{C_0} (when selecting witnessed models) or the set QWT_{C_0} (when selecting quasi-witnessed models).
3. We obtain a corresponding formula F_{C_0}, from WT_{C_0} or QWT_{C_0}, such that it will have a solution in $(\mathbb{R}, +, -, \cdot, /, \{q : q \in \{\mathbb{Q}\})$ if C_0 is satisfiable with the class of models we have selected. This is explained in more detail in the next subsection.
4. The formula F_{C_0} is solved with a suitable SMT solver.

In our current implementation we use the SMT solver Z3 [5] although the formula F_{C_0} to be solved is generated in SMT 2.0 format, so we can use any SMT solver able to solve formulas in $(\mathbb{R}, +, -, \cdot, /, \{q : q \in \{\mathbb{Q}\})$. There is an on-line version of our solver available at the URL: http://arinf.udl.cat/fuzzydlsolver.

2.2   Translation of Fuzzy Propositional Axioms to Non-linear Real Arithmetic Formulas

In [2] the authors showed a translation of the r-satisfiability problem with respect to quasi-winessed models of a concept C_0 over the logic Π-$\mathcal{ALE}$ to an entailment problem of a propositional theory (called QWT_{C_0}) in Product Logic. Instead on trying to solve directly QWT_{C_0}, our approach is based on a reduction to the problem of solving the satisfiability of a corresponding formula F_{C_0} built over quantifier-free real non-linear arithmetic logic such that F_{C_0} is satisfiable if and only if the concept C_0 is r-satisfiable in a quasi-witnessed model over Π-$\mathcal{ALE}$.

We explain the reduction for the particular case of witnessed models, presented in the work of Hájek, that is based on a simpler fuzzy propositional theory (called WT_{C_0}). For every proposition p, we generate a corresponding formula $f(p)$ over quantifier-free non-linear real arithmetic logic. See Definition 3 in [7] for a detailed explanation of all the axioms in WT_{C_0} obtained from an input concept C_0 or Definition 10 in [2] for the corresponding explanation of the axioms in QWT_{C_0} for the more general case of quasi-witnessed models. The formulas to generate depend on the form of the proposition p, and are indicated in Table 1. In the table, $ite(C, A, B)$ is a shorthand for: if condition C is true, then A must be true, else B must be true and $it(C, A)$ is a shorthand for: if condition C is true, then A must be true. For example, the formula of the first row indicates that real value assigned to the propositional variable of an universal concept, $pr(\forall R.C(d_\sigma))$, must be equal to 1 if $pr(R(d_\sigma, d_{\sigma,n})) \leq pr(C(d_{\sigma,n}))$ and $pr(C(d_{\sigma,n}))/pr(R(d_\sigma, d_{\sigma,n}))$ otherwise.

Table 1. Reduction of formulas from the propositional theory WT_{C_0} to formulas in the corresponding set of non-linear arithmetic boolean formulas F_{C_0}

p	translation $f(p)$
$(\forall R.C(d_\sigma) \equiv (R(d_\sigma, d_{\sigma,n}) \rightarrow C(d_{\sigma,n})))$	$ite(pr(R(d_\sigma, d_{\sigma,n})) \leq pr(C(d_{\sigma,n})), pr(\forall R.C(d_\sigma)) = 1,$ $pr(\forall R.C(d_\sigma)) \cdot pr(R(d_\sigma, d_{\sigma,n})) = pr(C(d_{\sigma,n})))$
$(\exists R.C(d_\sigma) \equiv (R(d_\sigma, d_{\sigma,n}) \boxdot C(d_{\sigma,n})))$	$pr(\exists R.C(d_\sigma)) = pr(R(d_\sigma, d_{\sigma,n})) * pr(C(d_{\sigma,n}))$
$\forall R.C(d_\sigma) \rightarrow (R(d_\sigma, d_{\sigma,m}) \rightarrow C(d_{\sigma,m}))$	$it(pr(R(d_\sigma, d_{\sigma,m}) > pr(C(d_{\sigma,m})),$ $pr(\forall R.C(d_\sigma)) \leq \frac{pr(C(d_{\sigma,m}))}{pr(R(d_\sigma, d_{\sigma,m}))})$
$(R(d_\sigma, d_{\sigma,m}) \boxdot C(d_{\sigma,m}))) \rightarrow \exists R.C(d_\sigma)$	$pr(R(d_\sigma, d_{\sigma,m})) \cdot pr(C(d_{\sigma,m})) \leq pr(\exists R.C(d_\sigma))$

Then, to solve the r-satisfiability problem of concept C_0 we must determine whether:

$$F_{C_0} \cup \{0 \leq pr(E) \leq 1 \mid pr(E) \in Vars(F_{C_0})\} \cup \{pr(C_0) = r\}$$

is satisfiable in $(\mathbb{R}, +, -, \cdot, /, \{q : q \in \{\mathbb{Q}\}\})$, where $Vars(F_{C_0})$ denotes the set of all the propositional variables used in formulas of F_{C_0}.

When we ask instead to solve the problem over quasi-witnessed models, we consider then the theory $QWT(C_0)$. In this case, we change the formula produced in the first row of Table 1 for:

$$(pr(\forall R.C(d_\sigma)) = 0) \vee (ite(pr(R(d_\sigma, d_{\sigma,n})) \leq pr(C(d_{\sigma,n})), \; pr(\forall R.C(d_\sigma)) = 1,$$
$$pr(\forall R.C(d_\sigma)) \cdot pr(R(d_\sigma, d_{\sigma,n})) = pr(C(d_{\sigma,n}))))$$

And we have also to consider the additional set of propositions in Y_{C_0} of Definition 10 in [2], that are of the form:

$$\neg \forall R.C(d_\sigma) \boxdot (R(d_\sigma, d_{\sigma,n}) \rightarrow C(d_{\sigma,n}))$$

that must not be equal to 1 in any solution of the satisfiability problem in order to encode valid quasi-witnessed models. The idea used in [2] to add the constraint "$pr(R(d_\sigma, d_{\sigma,n}))$ $> pr(C(d_{\sigma,n}))$" to Y_{C_0} is that when $pr(\forall R.C(d_\sigma)) = 0$ it is possible to finitely encode an infinite model with (infinite) individuals $d^1_{\sigma,n}, d^2_{\sigma,n}, \ldots$ which satisfies

$$lim_{i \rightarrow \infty} \frac{pr(C(d_{\sigma,n}))^i}{pr(R(d_\sigma, d_{\sigma,n}))^i} = 0.$$

So, for each such proposition we introduce this additional formula in F_{C_0}:

$$it(pr(\forall R.C(d_\sigma)) = 0, \; pr(R(d_\sigma, d_{\sigma,n})) > pr(C(d_{\sigma,n})))$$

which translates the fact that propositions in Y_{C_0} should not be satisfied in terms of satisfiability of non-linear arithmetic boolean formulas.

3   Preliminary Evaluation

Consider the following family of 1-satisfiable concepts, indeed satisfiable with witnessed models, in our logic $\Pi\text{-}\mathcal{ALE}$, that use the relation symbol $friend$ and the atomic concept symbol $popular$, determined by the following regular expression:

$$\overbrace{\forall friend.....\forall friend.\,popular}^{n+1} \sqcap \overbrace{\exists friend.....\exists friend.\,\neg popular}^{n} \qquad (1)$$

where n is an integer parameter with $n \geq 1$.

Consider also the following family of $1-$satisfiable concepts, but only with quasi-witnessed models, determined by the regular expression:

$$\overbrace{\forall friend.....\forall friend.\,popular}^{n} \sqcap \neg\overbrace{\forall friend.....\forall friend.(popular \,\boxdot\, popular)}^{n} \qquad (2)$$

where n is as before an integer parameter with $n \geq 1$.

Table 2. Formula size (in Kbytes) and solving times (in seconds) for F_{C_0} obtained with our two benchmarks of concepts with Z3 SMT solver. The generation time of the formula F_{C_0} was less than 0.08 seconds up to $n = 8$ and less than 0.2 seconds for the other sizes.

| | Benchmark for concepts (1) | | | | | Benchmark for concepts (2) | | | |
| | WT_{C_0} | | QWT_{C_0} | | | WT_{C_0} | | QWT_{C_0} | |
n	size	solving time	size	solving time	n	size	solving time	size	solving time
3	20	0.033	24	0.029	3	16	0.023	20	0.036
4	40	0.041	44	0.063	4	32	0.055	36	0.105
5	80	0.118	92	0.216	5	64	0.143	76	0.450
6	164	0.379	184	0.806	6	132	0.465	152	2.101
7	332	1.327	372	3.094	7	264	1.639	304	> 1200
8	672	5.080	756	> 1200	8	528	11.301	616	> 1200
9	1400	> 1200	1500	> 1200	9	1100	> 1200	1300	> 1200
10	2700	> 1200	3100	> 1200	10	2200	> 1200	2500	> 1200

Table 2 shows the computation times[1], obtained with the SMT solver Z3 (version 4.3.2), when solving the instances from our benchmarks in the range $n \in \{3, 4, \ldots, 10\}$. We have solved the instances with both encodings, the one for only witnessed models and the one for quasi-witnessed models. The table also shows the size of the resulting formulas F_{C_0} obtained from each encoding. We observe that on the first benchmark, with both encodings we solve the instances within the time limit of 20 minutes up to $n = 7$, but with the quasi-witnessed encoding is always harder to solve it. For the second benchmark, the situation is even more different between both encodings. The witnessed

---

[1] The results were obtained with a Linux PC with four Intel i7 2.67 GhZ processors, and with a memory limit of 7GB per execution. A time equal to > 1200 means that the execution was aborted after 20 minutes without being able to solve the instance.

encoding solves the instances (find that they are not satisfiable with witnessed models) up to $n = 8$. By contrast, the quasi-witnessed encoding solves the instances only up to $n = 6$ and always in more time.

4   Conclusions and Future Work

Our results show that the performance of our SMT-based approach, that works by solving a non-linear real arithmetic boolean formula is really problematic. So, we are now developing a version of our tool that will consider a translation of the problem to a satisfiability problem over a linear real arithmetic problem. This new tool is based on some results shown in [4,2] and follows a similar approach to the one proposed in [10] to develop a satisfiability solver for different many-valued propositional logics.

References

1. Baader, F.: Tableau algorithms for description logics. In: Dyckhoff, R. (ed.) TABLEAUX 2000. LNCS, vol. 1847, pp. 1–18. Springer, Heidelberg (2000)
2. Cerami, M., Esteva, F., Bou, F.: Decidability of a description logic over infinite-valued product logic. In: Proceedings of KR 2010 (2010)
3. Cerami, M., Straccia, U.: On the (un)decidability of fuzzy description logics under łukasiewicz t-norm. Information Sciences 227, 1–21 (2013)
4. Cignoli, R., Torrens, A.: An algebraic analysis of product logic. Multiple-Valued Logic 5, 45–65 (2000)
5. de Moura, L., Bjørner, N.: Z3: An Efficient SMT Solver. In: Ramakrishnan, C.R., Rehof, J. (eds.) TACAS 2008. LNCS, vol. 4963, pp. 337–340. Springer, Heidelberg (2008)
6. Haarslev, V., Sebastiani, R., Vescovi, M.: Automated Reasoning in $\mathcal{ALCQ}$ via SMT. In: Bjørner, N., Sofronie-Stokkermans, V. (eds.) CADE 2011. LNCS, vol. 6803, pp. 283–298. Springer, Heidelberg (2011)
7. Hájek, P.: Making fuzzy description logic more general. Fuzzy Sets and Systems 154(1), 1–15 (2005)
8. Hustadt, U., Motik, B., Sattler, U.: Reducing SHIQ-description logic to disjunctive Datalog programs. In: Proceedings of KR 2004, pp. 152–162 (2004)
9. Straccia, U., Bobillo, F.: Mixed integer programming, general concept inclusions and fuzzy description logics. In: Proceedings of EUSFLAT 2007, pp. 213–220 (2007)
10. Vidal, A., Bou, F., Godo, L.: An SMT-based solver for continuous t-norm based logics. In: Hüllermeier, E., Link, S., Fober, T., Seeger, B. (eds.) SUM 2012. LNCS, vol. 7520, pp. 633–640. Springer, Heidelberg (2012)

On the Merit of Selecting Different Belief Merging Operators

Pilar Pozos-Parra*, Kevin McAreavey, and Weiru Liu

Queen's University Belfast, Northern Ireland
{p.pozos-parra,kmcareavey01,w.liu}@qub.ac.uk

Abstract. Belief merging operators combine multiple belief bases (a profile) into a collective one. When the conjunction of belief bases is consistent, all the operators agree on the result. However, if the conjunction of belief bases is inconsistent, the results vary between operators. There is no formal manner to measure the results and decide on which operator to select. So, in this paper we propose to evaluate the result of merging operators by using three ordering relations (fairness, satisfaction and strength) over operators for a given profile. Moreover, a relation of conformity over operators is introduced in order to classify how well the operator conforms to the definition of a merging operator. By using the four proposed relations we provide a comparison of some classical merging operators and evaluate the results for some specific profiles.

1 Introduction

Belief merging looks at strategies for combining belief bases from different sources, which in conjunction may be inconsistent, in order to obtain a consistent belief base representing the group. Logic-based belief merging has been studied extensively [1,14,10,11,8,9]. A well known strategy is the use of an operator Δ which takes as input the belief bases (profile) E and outputs a new consistent merged belief base $\Delta(E)$. Often operators require additional information such as a priority relationship between the bases or numbers representing base weights. However, in many applications this information does not exist and we must accord equal importance to each of the beliefs and bases. Among existing operators which are independent of additional information, we can mention: Δ_{MCS}, Δ_{Σ}, Δ_{Gmax} and DA^2 operators. In each case the belief bases are described using a finite number of propositional symbols; there is no hierarchy, nor priority, nor any difference in reliability of the sources. Prioritized belief bases or weighted bases, such as in possibilistic logic [12], will be consider in future work.

Considering flat profiles, there are two main families of belief merging operators: the formula-based operators (also called syntax-based operators) and the model-based operators (also called distance-based operators). The operators belonging to the former family select subsets of consistent formulae from the profile E. While the variety depends on the selection criterion, there is no formal way

---

* The first author was supported by CONACyT and UJAT.

W. Liu, V.S. Subrahmanian, and J. Wijsen (Eds.): SUM 2013, LNAI 8078, pp. 331–345, 2013.

to compare different criteria. The operators belonging to the latter family define a distance between worlds, a distance from worlds to bases, and a distance from worlds to profiles with the help of an aggregation function. Then, the operators take as models of the merged result, those worlds which are closest to the belief profile. While the distances allow us to define a notion of closeness in any framework, we miss a general measure that indicates how close a profile E is to the merged base $\Delta(E)$. A general measure will allow us to compare, in a formal manner, the results from different operators.

In [2] the notion of a base satisfaction index (individual index) was introduced, where such an index measures the closeness from a base K to a merged base $\Delta(E)$. The index is a function that takes as input two belief bases: a belief base $K \in E$ and the merged base $\Delta(E)$. The output is a numeral value $i(K, \Delta(E))$ which represents the degree of satisfaction of the base given the merged base. While this notion allows us to measure the satisfaction of every member of a profile, there is no measure for the satisfaction of the whole profile. So, in this paper we propose to evaluate the result of merging operators by using three ordering relations (fairness, satisfaction and strength) over the operators for a given profile. Moreover, a relation of conformity over the operators is introduced in order to classify the degree to which an operator conforms to the definition of a merging operator. By using the four proposed relations we provide a comparison for some classical merging operators and we evaluate the results of these operators for some specific profiles.

The objective of this paper is to draw a comparative landscape through criterion based on a degree of satisfaction and notions of conformity and strength for many merging operators from the literature. We focus on operators for merging bases represented as sets of propositional formulae, where no priorities or weights are given. The rest of the paper is organized as follows. After some preliminaries, in Section 3 we recall some of the main merging operators from the literature. Then, we introduce the degree of satisfaction and three relations as well as introducing a method for measuring the result of the operators. In Section 5 we compare some results for operators through our proposal for specific profiles. Finally, we conclude mentioning some future work.

2   Preliminaries

We consider a language $\mathcal{L}$ of propositional logic using a finite set of propositional variables $P := \{p_1, p_2, ..., p_m\}$, the standard connectives, and the boolean constants $\top$ and $\bot$ representing always $true$ and $false$, respectively. $|A|$ denotes the cardinality of a set A or the absolute value of a number A.

An interpretation or world w is a function from P to $\{0, 1\}$, the set of worlds of the language is denoted by $\mathcal{W}$, its elements will be denoted by boolean vectors of the form $(w(p_1), ..., w(p_m))$, where $w(p_i) = 1$ (representing $true$) or $w(p_i) = 0$ (representing $false$) for $i = 1, ..., m$. A world w is a model of $\phi \in \mathcal{L}$ if and only if ϕ is true under w in the classical truth-functional manner. The set of models of a formula ϕ is denoted by $mod(\phi)$. The formula ϕ is consistent if and only if there exists a model of ϕ. The formula ϕ is a logical consequence of a formula ψ, denoted $\psi \models \phi$ if and only if $mod(\psi) \subseteq mod(\phi)$. For any set of models $M \subseteq \mathcal{W}$, let $form(M)$ denote a formula whose set of models are precisely M (up to logical equivalence), i.e., $mod(form(M)) = M$.

A *belief base* K is a finite set of propositional formulae of $\mathcal{L}$ representing the beliefs from a source. Some approaches identify K by the conjunction of its elements so each knowledge base can be treated as a single formula. For this reason we use L rather than $2^{\mathcal{L}}$ or $\mathcal{L}$ to denote the set of all belief bases. A belief profile E is a multiset (bag) of n belief bases $E = \{K_1, ..., K_n\}$ $(n \geq 1)$. The profile represents the set of information sources to be processed. We denote the conjunction of bases in E by $\bigwedge E$ and the disjunction of bases in E by $\bigvee E$. A profile E is consistent if and only if $\bigwedge E$ is consistent. The multi-set union between E_1 and E_2 is denoted by $E_1 \sqcup E_2$.

In [6] eight postulates have been proposed to characterize the process of belief merging with integrity constraints in a propositional setting. This characterization is rephrased, without reference to integrity constraints, producing the following M1–M6 postulates.

Definition 1. *Let E, E_1, E_2 be belief profiles, K_1 and K_2 be consistent belief bases. Let Δ be an operator which assigns to each belief profile E a belief base $\Delta(E)$. Δ is a merging operator if and only if it satisfies the following postulates:*

(M1) $\Delta(E)$ is consistent
(M2) if $\bigwedge E$ is consistent then $\Delta(E) \equiv \bigwedge E$
(M3) if $E_1 \equiv E_2$, then $\Delta(E_1) \equiv \Delta(E_2)$
(M4) $\Delta(\{K_1, K_2\}) \wedge K_1$ is consistent if and only if $\Delta(\{K_1, K_2\}) \wedge K_2$ is consistent
(M5) $\Delta(E_1) \wedge \Delta(E_2) \models \Delta(E_1 \sqcup E_2)$
(M6) if $\Delta(E_1) \wedge \Delta(E_2)$ is consistent, then $\Delta(E_1 \sqcup E_2) \models \Delta(E_1) \wedge \Delta(E_2)$

The postulates describe the principles that a belief merging operator should satisfy. Among them, syntax irrelevance (M3), and fairness (M4) are key postulates. In the literature [6] we can find some operators which are considered merging operators even though they do not satisfy the six postulates. Therefore a relation, based on the number of postulates for which operators conform, may be a first attempt at comparing operators. Formally:

Conformity Relation. An operator Δ_1 is *more conforming than* an operator Δ_2, denoted $\Delta_1 \geq \Delta_2$, if Δ_1 satisfies more postulates than Δ_2.

This relation is strictly numeric in that we do not consider the satisfaction of any one postulate to be more desirable than the satisfaction of another.

3   Belief Merging Operators

As we stated before, there are two main families of merging operators: formula-based and model-based operators. The former selects some formulae from the union of the bases with the help of a selection criterion. The latter selects some interpretation with the help of some distances and aggregation functions.

3.1   Formula-Based Operators

Formula-based operators are based on the selection of consistent subsets of formulae in the union of the members of a profile E. In [1], the operators aim to find all maximally consistent subsets (MCS) of the inconsistent union of belief bases. When an integrity constraint is imposed the operator only selects the MCS which are consistent w.r.t. the integrity constraint. The operators are defined w.r.t. a function MCS, whose input is a belief base K and an integrity constraint μ, and the output is the set of maximal (w.r.t. inclusion) consistent subsets of $K \cup \{\mu\}$ that contains μ, formally, $MCS(K, \mu)$ is the set of all F s.t.:

1. F is consistent,   2. $F \subseteq K \cup \{\mu\}$,
3. $\mu \in F$ and   4. if $F \subset F' \subseteq K \cup \{\mu\}$, then F' is inconsistent.

MCS is extended for a profile E as follows: $MCS(E, \mu) = MCS(\bigcup_{K \in E} K, \mu)$.

Another function that helps to define some operators is $|MCS|$, which can be defined by replacing inclusion with cardinality in 4: if $|F| < |F'|$, s.t. $F' \subseteq K \cup \{\mu\}$, then F' is not consistent. $|MCS|$ is extended for a profile E in a similar manner. The following operators have been defined in [1,5]:

1. $\Delta_{MCS_1}(E, \mu) = \bigvee MCS(E, \mu)$.
2. $\Delta_{MCS_3}(E, \mu) = \bigvee\{F : F \in MCS(E, \top) \text{ and } F \cup \{\mu\} \text{ consistent}\}$.
3. $\Delta_{MCS_4}(E, \mu) = \bigvee |MCS|(E, \mu)$.
4. $\Delta_{MCS_5}(E, \mu) = \begin{cases} \bigvee \{F \cup \{\mu\} : F \in MCS(E, \top) & \text{if } \exists F \in MCS(E, \top) \\ \text{and } F \cup \{\mu\} \text{ consistent}\} & \text{s.t. } F \cup \{\mu\} \neq \emptyset, \\ \mu & \text{otherwise.} \end{cases}$

The first three operators correspond respectively to operators $Comb1(E, \mu)$, $Comb3(E, \mu)$ and $Comb4(E, \mu)$ proposed in [1]. In order to assure consistency, Δ_{MCS_3} was modified in [5] as Δ_{MCS_5}. These operators are syntax sensitive.

Example 1. From [11]. Let $E = \{K_1, K_2, K_3\}$ where $K_1 = \{a\}$, $K_2 = \{a \to b\}$ and $K_3 = \{a, \neg b\}$. Then, $\Delta_{MCS}(E) = \{a, a \to b\} \vee \{a, \neg b\} \vee \{\neg b, a \to b\}$.

3.2   Model-Based Operators

In most model-based frameworks an operator Δ is defined by a function $m : L^n \to 2^{\mathcal{W}}$ from the set of profiles to the power set of $\mathcal{W}$ s.t. $\Delta(E) = form(m(E))$. For simplicity we use the standard notation $mod(\Delta(E))$ rather than $m(E)$. The process is defined using three distances: a distance from one world to another $d(w, w')$, a distance from a world to a belief base $d(w, K)$ based on $d(w, w')$ and a distance from a world to a profile $d(w, E)$ based on $d(w, K)$. The latter distance is usually defined by aggregation functions and allows us to define a pre-order $\leq_E$. The closest worlds to the profile are the models of the merging process.

We summarize the definitions as follows. A distance[1] between worlds is a function $d : \mathcal{W} \times \mathcal{W} \to \mathbb{R}^+$ from the Cartesian square of $\mathcal{W}$ to the set of positive real numbers s.t. for all $w, w' \in \mathcal{W}$:

1. $d(w, w') = d(w', w)$ and   2. $d(w, w') = 0$ iff $w = w'$.

The distance between a world and a belief base is a function $d : \mathcal{W} \times L \to \mathbb{R}^+$ from the Cartesian product of $\mathcal{W}$ and the set of belief bases to the set of positive real numbers. Some methods define this distance as the minimal distance between world w and any model of base K, i.e., $d(w, K) = min_{w' \in mod(K)} d(w, w')$. Finally, the distance between a world and a profile is a function $d_a : \mathcal{W} \times L^n \to \mathbb{R}^+$ from the Cartesian product of $\mathcal{W}$ and the set of profiles to the set of positive real numbers, defined as the result of applying the aggregation function $a : \mathbb{R}^{+^n} \to \mathbb{R}^+$ to the distances between w and every profile member, i.e. $d_a(w, E) = a(d(w, K_1), ..., d(w, K_n))$ s.t. $E = \{K_1, ..., K_n\}$.

Definition 2. *An aggregation function a is a total function associating a positive real number to every finite n-tuple of positive real numbers s.t. for all $x_1, ..., x_n, x, y \in \mathbb{R}^+$:*

1. if $x \leq y$, then $a(x_1, ..., x, ..., x_n) \leq a(x_1, ..., y, ..., x_n)$,
2. $a(x_1, ..., x_n) = 0$ iff $x_1 = ... = x_n = 0$ and
3. $a(x) = x$.

Any aggregation function induces a total pre-order $\leq_E$ on the set $\mathcal{W}$ w.r.t. the distances to a given profile E. Thus, the merging operator $\Delta_{d,a}$ for a profile E is defined as a belief base (up to logical equivalence) whose models are the set of all worlds with the minimal distance d_a to the profile E, i.e.,

$$mod(\Delta_{d,a}(E)) = min(\mathcal{W}, \leq_E).$$

Every framework consists of a distance and an aggregation function. The distance between worlds most widely used in the literature is Hamming distance[2], which is the number of propositional variables on which two worlds differ, i.e.,

$$d(w, w') = \sum_{p \in P} |w(p) - w'(p)|.$$

Two outstanding aggregation functions are maximum and sum, their corresponding distance are defined, respectively, as follows:

$$d_{max}(w, E) = \max_{K \in E} d(w, K) \quad \text{and} \quad d_{\Sigma}(w, E) = \sum_{K \in E} d(w, K).$$

In both cases, the induced pre-order is defined with the help of $\leq$ over real numbers as follows:

$$w \leq_E w' \quad iff \quad d_a(w, E) \leq d_a(w', E).$$

---

[1] As in [9], the triangle inequality is not required.
[2] From now, if a belief merging operator $\Delta_{d,a}$ uses Hamming distance, in order to avoid heavy notations, we identify it by Δ_a.

Another well known operator is Δ_{Gmax}, introduced in [7], where the aggregation function does not output a number but instead outputs a vector of numbers, which is the result of sorting the input distances in descending order, i.e.,

$$d_{Gmax}(w, E) = sort(d(w, K_1), ..., d(w, K_n)).$$

The operator Δ_{Gmax} uses the lexicographic ordering $\leq_{lex}$ for comparing vectors, the pre-order induced is defined as follows.

$$w \leq_E w' \quad iff \quad d_{Gmax}(w, E) \leq_{lex} d_{Gmax}(w', E).$$

Example 2. Given profile E from Example 1, and variables a and b in that order, then: $mod(\Delta_{max}(E)) = \{(0,0), (1,0), (1,1)\}$; $mod(\Delta_{\Sigma}(E)) = \{(1,0), (1,1)\}$; and $mod(\Delta_{Gmax}(E)) = \{(1, 0), (1,1)\}$.

4   On the Measure of Merging Operators

Comparing the number of postulates (from Definition 1) for which merging operators conform, provides a means to generally evaluate and compare operators. Only Δ_{Σ} and Δ_{Gmax} satisfy all six postulates. Δ_{max} satisfies the first five postulates. However, given a naive operator $\Delta_{\top}$ (typical of Yager's rule for merging belief functions [15]) s.t. $\Delta_{\top} = \bigwedge E$ if $\bigwedge E$ is consistent and $\Delta_{\top} = \top$ otherwise. This operator satisfies the first five postulates and satisfies the last postulate when both profiles E_1 and E_2 are either consistent or inconsistent. Under this characterization we can consider $\Delta_{\top}$ to be more conforming than operators such as Δ_{max} which satisfy fewer postulates. However $\Delta_{\top}$ does not help to make decisions since the result is a tautology when the sources of information are inconsistent and the information of a tautology is neither useful nor informative. For this reason, we also need to classify operators based on their merging result in order to select the best operator for a given profile.

We propose to classify operators based on: (1) conformity; (2) the degree of satisfaction of their merging result w.r.t the given profile and two relations over operators; and (3) a relation of strength over operators. The degree of satisfaction of a belief base is formally defined as follows:

Definition 3 (Degree of satisfaction of belief bases). *Function SAT :* $L \times L \to [0, 1]$ *is called a the degree of satisfaction of belief bases iff for any belief base K and K', it satisfies the following postulates:*

Reflexivity: $SAT(K, K') = 1$ *iff* $mod(K') \cap mod(K) \neq \emptyset$.
Monotonicity: $SAT(K, K') \geq SAT(K, K^*)$ *iff* $mod(K') \subseteq mod(K^*)$.

Semantically, a degree of satisfaction for a belief base K in a given profile E, is a measure of how satisfied the belief base K is by the merged base $K' = \Delta(E)$ resulting from the application of a merging operator Δ on the profile E. Notice that the definition considers a general case where K' may be a belief base which is not necessarily the result of a merging operator. Two stronger variants are:

Definition 4. *A rational degree of satisfaction is a degree of satisfaction which satisfies the Rationality postulate: $SAT(K, K') = 0$ if $mod(K') \cap mod(K) = \emptyset$.*

Definition 5. *A symmetric degree of satisfaction is a degree of satisfaction which satisfies the Symmetry postulate: $SAT(K, K') = SAT(K', K)$.*

Based on this degree of satisfaction we can define the degree of satisfaction of a profile as follows:

Definition 6 (Degree of satisfaction of belief profiles). *Let E be a profile, SAT be a degree of satisfaction of belief bases and a be an aggregation function. The degree of satisfaction of E by K' based on SAT and a, denoted $SAT_a(E, K')$, is defined as follows: $SAT_a(E, K') = a_{K \in E} SAT(K, K')$.*

Then we can define a maximum and minimum degree of satisfaction for a profile E as follows:

Definition 7. *Let E be a profile and K' be a belief base. Then $SAT_{max}(E, K')$ is the maximum degree of satisfaction of E by K' iff $SAT_{max}(E, K') = max_{K \in E} SAT(K, K')$. Also, $SAT_{min}(E, K')$ is the minimum degree of satisfaction of E by K' iff $SAT_{min}(E, K') = min_{K \in E} SAT(K, K')$.*

4.1   Instantiation of the Degree of Satisfaction (Base Satisfaction Index)

Notice that Definition 3 is about properties of a measure, no specific measures are actually given. This section and the next provide these measures. In the literature we can find a way to define the satisfaction of a base given the merged base: in [2] the notion of a base satisfaction index is the degree of satisfaction of $K \in E$, given $\Delta(E)$, as a total function i from $L \times L$ to $[0, 1]$. Then $i(K, \Delta(E))$ indicates how close a base K is to the merged base $\Delta(E)$. In [2] four indexes are proposed when no additional information about the sources is available: i_w, i_s, i_p and i_d. These base satisfaction indexes satisfy Definition 3, so they can be considered as degrees of satisfaction of a belief base[3]. Formally:

Definition 8 (weak drastic index). *This boolean index takes value 1 if the merging result is consistent with the base and 0 otherwise, formally:*

$$i_w(K, \Delta(E)) = \begin{cases} 1 & \text{if } K \wedge \Delta(E) \text{ is consistent,} \\ 0 & \text{otherwise.} \end{cases}$$

Definition 9 (strong drastic index). *This boolean index takes value 1 if the belief base is a logical consequence of the merging result and 0 otherwise, formally:*

$$i_s(K, \Delta(E)) = \begin{cases} 1 & \text{if } \Delta(E) \models K, \\ 0 & \text{otherwise.} \end{cases}$$

---

[3] For the sake of readability, we use 'base satisfaction index' and 'degree of satisfaction of a belief base' as synonyms, however, notice that a belief satisfaction index was defined without imposing properties.

Definition 10 (probabilistic index). *This index takes the value of the probability of getting a model of K among the models of $\Delta(E)$, formally:*

$$i_p(K, \Delta(E)) = \begin{cases} 0 & \text{if } |mod(\Delta(E))| = 0 \\ \frac{|mod(K) \cap mod(\Delta(E))|}{|mod(\Delta(E))|} & \text{otherwise.} \end{cases}$$

So, i_p takes its minimal value 0 when no model of K is in the models of the merged base $\Delta(E)$ and its maximal value when each model of the merged based is a model of K. The fact that i_p is based on model counting allows some granularity in the notion of satisfaction. Notice, i_s can be obtained by truncating or dropping the decimal numbers of the i_p result. In fact, i_p can be seen as the probability of getting the belief base as a logical consequence of the merged result.

Definition 11 (Dalal index). *This index grows antimonotonically with the Hamming distance between the two bases under consideration, i.e., the minimal distance between a model of the base K and a model of base $\Delta(E)$, formally:*

$$i_d(K, \Delta(E)) = 1 - \frac{\min_{w \in mod(K), w' \in mod(\Delta(E))} d(w, w')}{|P|}.$$

This index takes its minimal value when every variable must be flipped to obtain a model of $\Delta(E)$ from a model of K, while takes its maximal value whenever K is consistent with $\Delta(E)$ and no flip is required.

Examples for these indexes are shown in Tables 2 and 3. We can propose other indexes such as $i'_p = \frac{|mod(K) \cap mod(\Delta(E))|}{|mod(\Delta(K))|}$: the probability of getting the merged result as a logical consequence of the belief base. However, there is no background theory to support this proposal. Next we introduce a new base satisfaction index based on inconsistency measures for propositional belief bases [3,4]. Considering the level in which the inconsistency is measured, there are two classes of measures: Base-level measures and Formula-level measures. Those in the former class measure the inconsistency of the belief base as a whole. While those in the latter class measure the degree to which each formula in the belief base is responsible for the inconsistency of the base. The output of the former is a number while the output of the latter is a numerical vector with elements representing each formula in the belief base. This work considers solely the former class. Another classification found in the literature considers *how* inconsistency is measured. In this case there are two main types of measures: Formula-centric measures that count the number of formulae required for creating the inconsistency: the more formulae required to produce an inconsistency, the less inconsistent the base; and atom-centric measures, that take into account the proportion of the language affected by inconsistency: the more propositional variables affected, the more inconsistent the base.

Definition 12 (Base-level measure of inconsistency). *An inconsistency measure on a belief base is a function $I : L \to \mathbb{R}$.*

Diverse measures are defined in [3,4], however, we will choose the measures which satisfy two properties: syntax-insensitivity, i.e. the measures of two equivalent belief bases are equal; and normalization, i.e. the measure is a real number between 0 and 1. The former is required in order to assure fairness of evaluation w.r.t. the way of writing formulae. The latter is required to assure uniformity in the evaluation. Moreover, we consider degrees of satisfaction between 0 and 1, representing 0 and 100% satisfaction, respectively. As far as we know the only measure that satisfies both properties is I_{LP_m} [4].

The inconsistency measure I_{LP_m} is defined as the normalized minimum number of inconsistent truth values in the LP_m models of the belief base. Formally:

$$I_{LP_m}(K) = \frac{min_{w \in mod_{LP}(K)}(|w!|)}{|P|}$$

where K is a belief base and LP_m extends the notion of worlds considering three truth values $\{0, 1, \frac{1}{2}\}$, representing $true$, $false$ and the additional truth value $both$ meaning both "true and false". Then a world is a function from P to $\{0, 1, \frac{1}{2}\}$. 3^P is the set of all worlds for LP_m. Truth values are ordered as $0 <_t \frac{1}{2} <_t 1$ and $w(\top) = 1$, $w(\bot) = 0$, $w(\neg\phi) = \frac{1}{2}$ iff $w(\phi) = \frac{1}{2}$, $w(\neg\phi) = 1$ iff $w(\phi) = 0$, $w(\phi \wedge \psi) = min_{\leq_t}(w(\phi), w(\psi))$ and $w(\phi \vee \psi) = max_{\leq_t}(w(\phi), w(\psi))$. The LP_m models of the belief base are defined as: $mod_{LP}(K) = \{w \in 3^P \mid w(K) \in \{1, \frac{1}{2}\}\}$ and $w! = \{x \in P \mid w(x) = \frac{1}{2}\}$. The minimum models of a formula are: $min(mod_{LP}(\phi)) = \{w \in mod_{LP}(\phi) \mid \nexists w' \in mod_{LP}(K) \text{ s.t. } w'! \subset w!\}$.

Definition 13 (Base-level inconsistency index). *The base-level inconsistency index is defined as:* $i_i(K, \Delta(E)) = 1 - I(K \cup \Delta(E))$.

This index grows antimonotonically with the base-level measure of inconsistency I between the union of the two bases under consideration. This index takes its minimal value when the degree of inconsistency of the union of the bases is the maximum, while it takes its maximal value whenever the union of the bases is consistent. We consider only the instance: $i_L(K, \Delta(E)) = 1 - I_{LP_m}(K \cup \Delta(E))$.

Proposition 1. *The five satisfaction indexes are degrees of satisfaction. More specifically i_s, i_w and i_p are rational degrees of satisfaction. Also i_d and i_L are symmetric degrees of satisfaction.*

4.2   Instantiation of the General Degree of Satisfaction (Profile Satisfaction Index)

Using the base satisfaction indexes, one can define a satisfaction index for the whole profile. The profile satisfaction indexes are instantiations of degrees of satisfaction of belief profiles, in this work we will use both notions indistinctly. The notion of a profile satisfaction index is the degree of satisfaction of E, given $\Delta(E)$. The index is defined as a total function i from $L^n \times L$ to $\mathbb{R}$. Thus, $i(E, \Delta(E))$ indicates how close a profile is to the merged base $\Delta(E)$, formally:

Definition 14 (Profile satisfaction index). *Let E be a profile, i be a base satisfaction index and a be an aggregation function, the profile satisfaction index based on i and a is defined as follows:* $i_a(E, \Delta(E)) = a_{K \in E} i(K, \Delta(E))$.

There are many ways to measure the satisfaction of the profile given the merged base. The following measure says that a profile is as satisfied as the satisfaction of its least satisfied element, it is an instantiation of the minimum degree of satisfaction of a profile.

$$i_{min}(E, \Delta(E)) = min_{K \in E} i(K, \Delta(E))$$

Alternatively, another measure says that a profile is satisfied holistically, as the sum of the satisfaction of its elements.

$$i_{\Sigma}(E, \Delta(E)) = \Sigma_{K \in E} i(K, \Delta(E)).$$

4.3   Evaluating Merging Operators

Postulate M4 only says that no preference should be given to either belief base if they are inconsistent, however this is questioned in the literature. It is possible for us to define a more refined postulate (relation) of fairness using the degree of satisfaction, s.t. we can assign a relative degree of fairness to an operator.

Fairness relation. An operator Δ_1 is *fairer than* an operator Δ_2, denoted $\Delta_1 \succeq \Delta_2$ iff for all E, $SAT_{max}(E, \Delta_1(E)) - SAT_{min}(E, \Delta_1(E)) \leq SAT_{max}(E, \Delta_2(E)) - SAT_{min}(E, \Delta_2(E))$.

This means that a fairer operator minimizes the difference between degrees of satisfaction among bases. We can also define a satisfaction relation between operators based on the degree of satisfaction SAT as follows:

Satisfaction relation. An operator Δ_1 is *more satisfactory than* an operator Δ_2, denoted $\Delta_1 \sqsupseteq \Delta_2$ if for all E, $SAT_a(E, \Delta_2(E)) \leq SAT_a(E, \Delta_1(E))$.

This means that a more satisfactory operator maximizes the degree of satisfaction of belief profiles. Both types of ordering relations (fairness and satisfaction) can be used to select the best operators in terms of these criterion. See Table 1.

However, operators such as $\Delta_\top$ will have the highest degree of fairness and satisfaction, in comparison to the remaining operators. Moreover, in relation to M1–M6 postulates, the operator $\Delta_\top$ is considered more (or at least equally) conforming than Δ_{MCS} and Δ_{max}. We can conclude that $\Delta_\top$ is a good choice. However, $\Delta_\top$ does not produce useful and informative results since they will be a tautology when the profile is inconsistent, so, the conformity, fairness and satisfaction relations are insufficient. For this reason we need some way to classify the degree of "useful and informative" merging results and so we propose to use the notion of *strength* introduced in [13]. With this notion we can say that $\Delta_\top$ is weaker than the other operators since its merging results are weaker.

Strength relation. An operator Δ_1 is *stronger than* an operator Δ_2, denoted $\Delta_1 \sqsupseteq \Delta_2$, if for all E, $mod(\Delta_1(E)) \subseteq mod(\Delta_2(E))$.

Using this notion, we can conclude that the merging operator Δ_{Gmax} is stronger that Δ_{max} and the operator Δ_T is the weakest (see Table 1).

In short, the postulates M1–M6 allow us to define a conformity relation between operators s.t. an operator which satisfies more postulates is considered more conforming. Additionally, a degree of satisfaction allows us to define another relation between operators s.t. an operator with a higher degree of satisfaction is 'better' than an operator with a lower degree of satisfaction, i.e., the operator is closer to the original information in comparison to other possible merging results (assuming different merging operators are available). Based on this degree of satisfaction we define another relation of fairness over operators. Finally, we define a strength relation over operators. Unfortunately, these 4 relations over operators cannot identified the best operator in a general case, i.e. for every profile (see Table 1). However, we can combine the strength relation with the fairness and satisfaction relations to define a method to classify the operators results for a given profile. Notice the relations of fairness, satisfaction and strength can be used for particular cases of E, where we can say, for example, that the operator Δ_1 is stronger than Δ_2 for a given E if $mod(\Delta_1(E)) \subseteq mod(\Delta_2(E))$.

Table 1. Comparison of operators in terms of operators being more conforming ($\Delta_1 \geq \Delta_2$), fairer ($\Delta_1 \succeq \Delta_2$), more satisfactory ($\Delta_1 \sqsupseteq \Delta_2$) or stronger ($\Delta_1 \supseteq \Delta_2$), where n/a means not comparable or not found

Δ_1 \ Δ_2	Δ_{MCS}	Δ_{max}	Δ_{Σ}	Δ_{Gmax}	Δ_T
Δ_{MCS}	$\geq, \succeq, \sqsupseteq, \supseteq$	n/a	$\not\geq$	$\not\geq$	$\not\geq$
Δ_{max}	$\geq$	$\geq, \succeq, \sqsupseteq, \supseteq$	$\not\geq$	$\not\geq, \sqsupseteq$	$\geq$
Δ_{Σ}	$\geq$	$\geq$	$\geq, \succeq, \sqsupseteq, \supseteq$	$\geq$	$\geq$
Δ_{Gmax}	$\geq$	$\geq, \supseteq$	$\geq$	$\geq, \succeq, \sqsupseteq, \supseteq$	$\geq$
Δ_T	$\geq, \succeq, \sqsupseteq, \not\supseteq$	$\geq, \succeq, \sqsupseteq, \not\supseteq$	$\not\geq, \succeq, \sqsupseteq, \not\supseteq$	$\not\geq, \succeq, \sqsupseteq, \not\supseteq$	$\geq, \succeq, \sqsupseteq, \supseteq$

Example 3. From [14,7]. Let $E = \{K_1, K_2, K_3\}$ where $K_1 = \{(S \vee O) \wedge \neg D\}$, $K_2 = \{(\neg S \wedge D \wedge \neg O) \vee (\neg S \wedge \neg D \wedge O)\}$ and $K_3 = \{S \wedge D \wedge O\}$.

Using i_p for Example 3, we have $i_{p,max}(E, \Delta_{max}(E)) = 0.33$, $i_{p,max}(E, \Delta_{Gmax}(E)) = 1$, and $i_{p,min}(E, \Delta_{max}(E)) = i_{p,min}(E, \Delta_{Gmax}(E)) = 0$ (see Table 3). So, for this E, Δ_{max} is fairer than Δ_{Gmax} but using $i_{p,\Sigma}$, Δ_{Gmax} is more satisfactory than Δ_{max}. Moreover, as stated previously, Δ_{Gmax} is stronger than Δ_{max} and Δ_{Gmax} is more conforming than Δ_{max} since Δ_{Gmax} conforms to all six postulates while Δ_{max} only conforms to five.

Even for a particular E the selection of a "best result" is not always evident. In order to classify operators for any profile we must generalize two relations. For this reason we extend the fairness and satisfaction relations for belief bases rather than for the result of operators , as follows:

Fairness relation over belief bases. A belief base K_1 is *fairer than* a base K_2, denoted $K_1 \succeq K_2$, for every profile E if $SAT_{max}(E, K_1) - SAT_{min}(E, K_2) \leq SAT_{max}(E, K_1) - SAT_{min}(E, K_2)$.

Satisfaction relation over belief bases. A belief base K_1 is *more satisfactory than* a base K_2, denoted $K_1 \sqsupseteq K_2$, for every profile E if $SAT_a(E, K_2) \leq SAT_a(E, K_1)$.

Now, notice that if Δ_1 is stronger than Δ_2 for a given E then there exists a set of worlds Ω s.t. $mod(\Delta_1(E)) \cup \Omega = mod(\Delta_2(E))$, i.e. some worlds appearing in $\Delta_2(E)$ may be 'erased' in the process of merging with Δ_1. If $\Delta_1(E)$ is fairer than $form(\Omega)$ and $\Delta_1(E)$ is more satisfactory than $form(\Omega)$, we can conclude that the worlds which have been 'eliminated' by Δ_1 do not affect the properties of fairness and satisfaction of Δ_1 w.r.t. the extra worlds in $mod(\Delta_2(E))$; and given that Δ_1 is stronger than Δ_2 we can conclude that the result of Δ_1 is better that the result of Δ_2. In selecting a result, we can say that Δ_1 offers less choice than Δ_2 and so it is more useful for making decisions.

5   Comparing Operators Results

In this section we demonstrate instantiations of the degrees of satisfaction (i_w, i_s, i_p, i_d and i_L) and their corresponding satisfaction profile indexes as applied to two profiles selected from the literature. Satisfaction indexes for Example 1 (resp. Example 3) are shown in Table 2 (resp. Table 3).

Table 2. Satisfaction indexes for Example 1

	$\Delta_{MCS}(E)$	$\Delta_{max}(E)$	$\Delta_\Sigma(E)$	$\Delta_{Gmax}(E)$
$i_w(K_1, \Delta_a(E))$	1	1	1	1
$i_w(K_2, \Delta_a(E))$	1	1	1	1
$i_w(K_3, \Delta_a(E))$	1	1	1	1
$i_{w,min}(E, \Delta_a(E))$	1	1	1	1
$i_{w,\Sigma}(E, \Delta_a(E))$	3	3	3	3
$i_s(K_1, \Delta_a(E))$	0	0	1	1
$i_s(K_2, \Delta_a(E))$	0	0	0	0
$i_s(K_3, \Delta_a(E))$	0	0	0	0
$i_{s,min}(E, \Delta_a(E))$	0	0	0	0
$i_{s,\Sigma}(E, \Delta_a(E))$	0	0	1	1
$i_p(K_1, \Delta_a(E))$	0.66	0.66	1	1
$i_p(K_2, \Delta_a(E))$	0.66	0.66	0.5	0.5
$i_p(K_3, \Delta_a(E))$	0.33	0.33	0.5	0.5
$i_{p,min}(E, \Delta_a(E))$	0.33	0.33	0.5	0.5
$i_{p,\Sigma}(E, \Delta_a(E))$	1.66	1.66	2	2
$i_d(\ldots, \Delta_a(E))$	same as i_w			
$i_L(\ldots, \Delta_a(E))$	same as i_w			

Using the strong drastic index i_s for Example 1, the results are (in almost all cases) 0, meaning the bases are inconsistent with the merged base. i_p produces a greater degree of granularity in the results which means it is a more discriminative index. In both examples, the new base satisfaction index i_L shows that

Table 3. Satisfaction indexes for Example 3

	$\Delta_{MCS}(E)$	$\Delta_{max}(E)$	$\Delta_\Sigma(E)$	$\Delta_{Gmax}(E)$
$i_w(K_1, \Delta_a(E))$	1	1	1	1
$i_w(K_2, \Delta_a(E))$	1	0	1	0
$i_w(K_3, \Delta_a(E))$	1	0	0	0
$i_{w,min}(E, \Delta_a(E))$	1	0	0	0
$i_{w,\Sigma}(E, \Delta_a(E))$	3	1	2	1
$i_s(K_1, \Delta_a(E))$	0	0	1	0
$i_s(K_2, \Delta_a(E))$	0	0	1	0
$i_s(K_3, \Delta_a(E))$	0	0	0	0
$i_{s,min}(E, \Delta_a(E))$	0	0	0	0
$i_{s,\Sigma}(E, \Delta_a(E))$	0	0	2	0
$i_p(K_1, \Delta_a(E))$	0.5	0.33	1	1
$i_p(K_2, \Delta_a(E))$	0.5	0	0.5	0
$i_p(K_3, \Delta_a(E))$	0.5	0	0	0
$i_{p,min}(E, \Delta_a(E))$	0.5	0	0	0
$i_{p,\Sigma}(E, \Delta_a(E))$	1.5	0.33	1.5	1
$i_d(K_1, \Delta_a(E))$	1	1	1	1
$i_d(K_2, \Delta_a(E))$	1	0.66	1	0.66
$i_d(K_3, \Delta_a(E))$	1	0.66	0.66	0.66
$i_{d,min}(E, \Delta_a(E))$	1	0.66	0.66	0.66
$i_{d,\Sigma}(E, \Delta_a(E))$	3	2.33	2.66	2.33
$i_L(\dots, \Delta_a(E))$	same as i_d			

the Δ_{MCS} merging operator will be maximally satisfied for each belief base K_i as long as K_i is consistent. Likewise, the profile satisfaction indexes $i_{L,min}$ and $i_{L,\Sigma}$ will be maximally satisfied, as long as $\forall K_i \in E$, K_i is consistent. In both examples, the i_L and i_d indexes produce the same results. The reason is: firstly, they are both normalized with the number of variables in the merged base; and secondly, in these examples, the number of inconsistent variables is equal to the minimum distance between models in K_i and the merged base.

In [7] the authors claim for Example 3 that Δ_{Gmax} selects the interpretations chosen by both Δ_{max} and Δ_Σ, showing its good behavior, however they do not provide a formal definition of 'good behavior'. Our proposal, on the other hand, allows us to provide this definition: using i_d and $i_{d,\Sigma}$, we can conclude that Δ_{Gmax} is stronger than Δ_{max} and Δ_Σ, moreover, $\Delta_{Gmax}(E)$ is fairer and more satisfactory than $form(\Omega_{max})$ and $form(\Omega_\Sigma)$ (the 'extra' worlds of $\Delta_{max}(E)$ and $\Delta_\Sigma(E)$, respectively). So, we can conclude that the result given by Δ_{Gmax} is better than the results given by Δ_{max} and Δ_Σ.

6   Conclusion

We proposed a method for measuring the result of different merging operators. Firstly, we defined a relation of conformity over operators in order to classify the degree to which an operator conforms to six postulates describing the principles that a belief merging operator should satisfy. Next, we introduced the notion of a degree of satisfaction of belief bases. We discovered that some base satisfaction indexes found in the literature satisfy the definition of a degree of satisfaction of belief bases, so we use them to define a profile satisfaction index. Based on the notion of a degree of satisfaction and a profile satisfaction index, we defined two more ordering relations over merging operators: fairness and satisfaction. However, by using these relations the measure of operators does not give intuitive classifications, for example operators such as $\Delta_\top$ are well placed, even though the result is neither informative nor useful. So, a fourth relation over operators was introduced, called strength, in order to address this issue. Even while using the four proposed relations, some operators are not fully comparable. This means that we cannot find a best operator for every profile. However the relations do allow us to find the best operators for a given profile.

The proposed method is as follows: first, determine the conformity of an operator, next, if an operator Δ_1 is stronger than an operator Δ_2 for a profile E we can continue, otherwise stop since comparison is not possible. Choose degree of satisfactions SAT and SAT_a in order to compare the operators. Find Ω: the worlds that are included in $\Delta_2(E)$ but not in $\Delta_1(E)$. If $\Delta_1(E)$ is fairer and more satisfactory than Ω in terms of SAT and SAT_a then Δ_1 provides a better result than Δ_2 for the fixed profile E given SAT and SAT_a. While the method is in a preliminary phase, the application on some examples from the literature allows us to formally demonstrate claims such as the 'good behavior' of Δ_{Gmax}.

Our proposed method does not work with integrity constraints however these will be considered in future work. Also, currently we only consider flat belief bases, but we intend to extend this for prioritized bases. In terms of aggregation functions, we analyzed min and Σ for generating the satisfaction index of a profile, however there are other functions available, such as $Gmin$, which could be analyzed. We also intend to propose a profile satisfaction index based on formula-level inconsistency measures.

References

1. Baral, C., Kraus, S., Minker, J., Subrahmanian, V.S.: Combining knowledge bases consisting of first-order analysis. Com. Int. 8, 45–71 (1992)
2. Everaere, P., Konieczny, S., Marquis, P.: The strategy-proofness landscape of merging. J. of Art. Int. Research 28, 49–105 (2007)
3. Hunter, A., Konieczny, S.: Approaches to measuring inconsistent information. In: Bertossi, L., Hunter, A., Schaub, T. (eds.) Inconsistency Tolerance. LNCS, vol. 3300, pp. 191–236. Springer, Heidelberg (2005)
4. Hunter, A., Konieczny, S.: On the measure of conflicts: Shapley inconsistency values. Artificial Intelligence 174(14), 1007–1026 (2010)
5. Konieczny, S.: On the difference between merging knowledge bases and combining them. In: KR 2000, pp. 135–144 (2000)
6. Konieczny, S., Lang, J., Marquis, P.: DA^2 merging operators. Artif. Intell. 157(1-2), 49–79 (2004)
7. Konieczny, S., Pino-Pérez, R.: On the logic of merging. In: KR 1998, pp. 488–498 (1998)

8. Konieczny, S., Pino-Pérez, R.: Merging information under constraints: a logical framework. J. of Logic. and Computation 12(5), 773–808 (2002)
9. Konieczny, S., Pino-Pérez, R.: Logic based merging. Journal of Philosophical Logic 40(2), 239–270 (2011)
10. Liberatore, P., Schaerf, M.: Arbitration (or how to merge knowledge bases). IEEE Transactions on Knowledge and Data Engineering 10(1), 76–90 (1998)
11. Lin, J., Mendelzon, A.: Knowledge base merging by majority. In: Pareschi, R., Fronhoefer, B. (eds.) Dynamic Worlds: From the Frame Problem to Knowledge Management. Kluwer Academic (1999)
12. Liu, W., Qi, G., Bell, D.A.: Adaptive merging of prioritized knowledge bases. Fundam. Inform. 73(3), 389–407 (2006)
13. Marchi, J., Bittencourt, G., Perussel, L.: Prime forms and minimal change in propositional belief bases. Ann. Math. Artif. Intell. 59(1), 1–45 (2010)
14. Revesz, P.Z.: On the Semantics of Arbitration. Journal of Algebra and Computation 7(2), 133–160 (1997)
15. Yager, R.R.: On the dempster-shafer framework and new combination rules. Inf. Sci. 41(2), 93–137 (1987)

Possibilistic DL-Lite

Salem Benferhat and Zied Bouraoui

Université Lille - Nord de France
CRIL - CNRS UMR 8188
Artois, F-62307 Lens
{benferhat,bouraoui}@cril.fr

Abstract. *DL-Lite* is one of the most important fragment of description logics that allows a flexible representation of knowledge with a low computational complexity of the reasoning process. This paper investigates an extension of *DL-Lite* to deal with uncertainty associated with objects, concepts or relations using a possibility theory framework. Possibility theory offers a natural framework for representing uncertain and incomplete information. It is particularly useful for handling inconsistent knowledge. We first provide foundations of possibilistic *DL-Lite*, denoted by π-*DL-Lite*, where we present its syntax and its semantics. We then study the reasoning tasks and show how to measure the inconsistency degree of a knowledge base using query evaluations. An important result of the paper is that the extension of the expressive power of *DL-Lite* is done without additional extra computational costs.

1 Introduction

Description Logics (DLs, for short) [2] are well-known logics based on first order logic, introduced for representing knowledge. Nowadays, DLs have regained an important place in various domain areas and especially in the Semantic Web. DLs provide the foundations of the Web Ontology Language (*OWL*). According to *W3C* [1] three profiles of *OWL2* are proposed as sub-languages of the full *OWL2* language, to offer important advantages in particular application scenarios. One of these profiles is *OWL2-QL* dedicated to applications that use huge volumes of data where query answering is the most important reasoning task. *OWL2-QL* is based on *DL-Lite* which is a family of tractable DLs investigated by [4]. Indeed, Knowledge Bases (KB) consistency and all DLs standard reasoning services are polynomial for combined complexity (*i.e.* the overall size of the KB) [1]. In these logics, the most important task of reasoning is answering complex queries (especially conjunctive queries) where the reasoning complexity is in *LogSpace* for data complexity (*i.e.* the size of the data) [1].

Now, in real world applications, knowledge is usually affected with uncertainty and imprecision. Recently, several works have been proposed to deal with probabilistic and non-probabilistic uncertainty [8] on one hand and to deal with fuzzy information [11] on the other hand. A particular attention was given to fuzzy

---

[1] http://www.w3.org/TR/owl2-overview/

W. Liu, V.S. Subrahmanian, and J. Wijsen (Eds.): SUM 2013, LNAI 8078, pp. 346–359, 2013.

extensions of DLs (*e.g.* [18,3]) and *DL-Lite* (*e.g.* [19,12]). Besides, some works are devoted to possibilistic extensions of DLs (*e.g.* [10,8,14]) which are basically based on standard DLs reasoning services. However, there is no work on possibilistic extension of *DL-Lite* and there is no work that has been proposed to extend query answering within a possibility theory setting. This paper concerns the development of uncertainty-based *DL-Lite* using possibility theory. Possibility theory [9] is a very natural framework to deal with ordinal and qualitative uncertainty. It deals with non-probabilistic information as it is particularly appropriate when the uncertainty scale only reflects a priority relation between different pieces of information.

Possibilistic Description Logics (*Possibilistic-DLs* for short) are frameworks introduced to deal with uncertainty and to ensure reasoning under inconsistent KB. Originally, the use of possibility theory to extend DLs has been proposed by [10] then has been discussed by [8]. In these works the syntax and the semantics of DLs has been extended in the possibility theory framework by attaching to every axiom a confidence degree to encode its certainty. This confidence degree first reflects to what extent an axiom can be considered as certain (priority, important, etc) in the available knowledge. And then it is used to determine the inconsistency of a KB and to ensure inference services. However, there are no algorithms to compute inconsistency of a *Possibilistic-DLs* KB. In addition, only some inference services have been defined. Such limitation has constituted the main topics of the works proposed by [17,16] where the authors first redefine the syntax and semantics of *Possibilistic-DLs*, and then investigate several inference services that can be done on a *Possibilistic-DLs* KB. Furthermore, they provided an algorithm to compute inconsistency degree and possibilistic inference services. It has been shown that checking the consistency degree and several inference services can be done with classical DLs reasoning services through consistent sub-sets of the *Possibilistic-DLs* KB. An implementation of a reasoner called *DL-Poss*, has been provided in [13]. A deeper discussion on *Possibilistic-DLs* has been provided in [14]. Finally, it is important to point out that another method has been introduced in [6,15] for checking inconsistency of *possibilistic-DLs* as a direct extension of the tableau algorithm.

An important question addressed in this paper is : "How one can extend the expressive power of *DL-Lite,* to deal with possibilistic uncertain information, without increasing the computational cost?". This paper provides a positive answer to this question. Such a good result is possible when we restrict ourself to *DL-Lite.* Note first that most extensions of possibilistic DLs [17,16,6] all need some extra computation costs. In these existing approaches, computing inconsistency degree comes down to achieve a number of calls (at least $log_2 N$ calls, where N is the size of the uncertainty scale) to the inconsistency checking in standard (without uncertainty) DLs.

This paper departs from existing approaches and follows another direction to achieve reasoning tasks in possibilistic *DL-Lite*. The idea is to modify the inconsistency computation algorithm used in standard *DL-Lite* by simply propagating the uncertainty degrees associated with axioms. In fact, we will see that the uncertainty propagation does not generate any extra computational cost, and hence the computational complexity of possibilistic *DL-Lite* is the same as the one of *DL-Lite*.

The rest of this paper is organized as follows: Section 2 briefly recalls preliminaries on *DL-Lite*. Section 3 rephrases possibility theory framework over *DL-Lite* interpretations. Section 4 discusses the possibilistic extension of *DL-Lite*, denoted π-*DL-Lite*, where we present its syntax and its semantics. Section 5 introduces the so-called π-*negated* closure of a π-*DL-Lite* knowledge bases. Section 6 gives a method to compute inconsistency of the π-*DL-Lite* KB using query evaluations. Section 7 studies deferent possibilistic inferences. Section 8 concludes the paper.

2   DL-Lite Logic

The vocabulary of DLs is based on concepts which correspond to unary predicates to denote sets of individuals, and roles, which correspond to binary predicates, to denote binary relations among individuals. A description language is characterized by a set of constructs used to build complex concepts and roles form atomic ones and it is employed to structure a domain of interest. Each description language allows different sets of constructs. A DLs knowledge base is specified through several inclusions between concepts and roles.

In this paper, we focus on *DL-Lite* one of the most important fragment of DLs. For sake of simplicity, we only consider $DL\text{-}Lite_{core}^{H}$ (originally $DL\text{-}Lite_R$) that underlies *OWL2-QL* language as *DL-Lite* logic. For more details about the different logics in *DL-Lite* family see [1]. However, results of this paper are valid for other logics of the *DL-Lite* family.

The language of $DL\text{-}Lite_{core}$ is the core language for $DL\text{-}Lite_R$ and it is ensured by a description language defined as follow [5]:

$$
\begin{array}{llll}
B \longrightarrow & A \mid \exists R & C \longrightarrow & B \mid \neg B \\
R \longrightarrow & P \mid P^- & E \longrightarrow & R \mid \neg R
\end{array}
$$

where A is an atomic concept, P is an atomic role, Concepts B (*resp.* C) are called basic (*resp.* complex) concepts and roles R (*resp.* E) are called basic (*resp.* complex) roles. Note that *DL-Lite* language does not allows the use of the conjunctive and the disjunctive operators. However, one can easily add conjunctions (*resp.* disjunction) in the right-hand side (*resp.* left-hand side) of inclusion axioms. Indeed, the conjunction of the form $B \sqsubseteq C_1 \sqcap C_2$ is equivalent to the pair of inclusion axioms $B \sqsubseteq C_1$ and $B \sqsubseteq C_2$, while the disjunction of the form $B_1 \sqcup B_2 \sqsubseteq C$ is equivalent to the pair of inclusion axioms $B_1 \sqsubseteq C$ and $B_2 \sqsubseteq C$.

A *DL-Lite* knowledge base is a pair $\mathcal{K} = \langle \mathcal{T}, \mathcal{A} \rangle$ where $\mathcal{T}$ is a TBox and $\mathcal{A}$ is an ABox. The $DL\text{-}Lite_{core}$ TBox is constituted by a finite set of *inclusion axioms* of the form $B \sqsubseteq C$. Let use a_i and a_j to denote two individuals (constants), the

DL-Lite$_{core}$ ABox is constituted by a finite set of *membership assertion* on atomic concepts and on atomic roles of the form $A(a_i)$ and $P(a_i, a_j)$. The *DL-Lite$_R$* extends *DL-Lite$_{core}$* with the ability of specifying inclusion assertions between roles of the form $R \sqsubseteq E$. For more detailed description on *DL-Lite* family, see [5].

As usual in DLs, the *DL-Lite* semantics is given by an interpretation $I = (\Delta, .^I)$ which consists of a non-empty domain Δ and an interpretation function $.^I$. The function $.^I$ assigns to each individual a an element $a^I \in \Delta^I$, to each atomic concept A a subset $A^I \subseteq \Delta^I$ and to each atomic role P a subset $P^I \subseteq \Delta^I \times \Delta^I$ over the domain. Furthermore, the interpretation function $.^I$ is extended to complex concepts and roles (*e.g.* $(P^-)^I = \{(y, x) \in \Delta^I \times \Delta^I \,|\, (x, y) \in P^I\}$ and $(\exists R)^I = \{x \in \Delta^I \,|\, \exists y \in \Delta^I \text{ such that } (x, y) \in R^I\}$).

For the TBox, we say that an interpretation I is a model of an inclusion axiom, denoted by $I \models B \sqsubseteq C$ (*resp.* $I \models R \sqsubseteq E$) iff $B^I \subseteq C^I$ (*resp.* $R^I \subseteq E^I$). For the ABox, we say that an interpretation I is a model of membership assertion, denoted by $I \models A(a_i)$ (*resp.* $I \models P(a_i, a_j)$) iff $a_i^I \in A^I$ (*resp.* $(a_i^I, a_j^I) \in P^I$). Note that we only consider *DL-Lite* with unique name assumption (*i.e.* $a_i \neq a_j$ where $i \neq j$). Thus, I is a model of knowledge base $\mathcal{K} = \langle \mathcal{T}, \mathcal{A} \rangle$, denoted by $I \models \mathcal{K}$, iff $I \models \mathcal{T}$ and $I \models \mathcal{A}$. A KB $\mathcal{K}$ is said to be consistent (or satisfiable) if it admits at least one model.

3   Possibility Distribution over DL-Lite Interpretations

Possibility theory (e.g. [9]) offers an important framework for representing and reasoning with uncertain, partial and inconsistent pieces of information. In what follows, we rephrase possibility theory framework over *DL-Lite* interpretations. Let $\mathcal{L}$ be a finite *DL-Lite* description language, Ω be a universe of discourse and $I = (\Delta, .^I) \in \Omega$ be a *DL-Lite* interpretation.

3.1   Possibility Distribution

A possibility distribution is considered as one of the main block of possibility theory. It is a mapping, denoted by π, from the universe of discourse Ω to the unit interval $[0, 1]$. It assigns to each interpretation $I \in \Omega$ a possibility degree $\pi(I) \in [0, 1]$ that represents its compatibility or consistency relative to the available knowledge. When $\pi(I) = 1$, we say that I is totally possible and it is fully consistent with the available knowledge. When $\pi(I) = 0$, we say that I is impossible and it is fully inconsistent with the available knowledge. Then, two special cases exist: a total ignorance when $\forall I \in \Omega, \pi(I) = 1$ and a complete knowledge when $\exists I' \in \Omega, \pi(I') = 1$ and $\forall I \in \Omega, I' \neq I, \pi(I) = 1$. By convention, a possibility distribution π is said to be normalized if there exists at least one totally possible interpretation, namely $\exists I \in \Omega, \pi(I) = 1$, otherwise, we say that π is sub-normalized. For two events I and I', we say that I is more consistent or compatible than I' if $\pi(I) > \pi(I')$.

3.2 Possibility and Necessity Measures

Let us consider φ be a subset of Ω. Let $\neg\varphi$ be the complementary of φ, namely $\neg\varphi = \Omega \setminus \varphi$. In standard possibility theory, given a possibility distribution π, one can define two measures from 2^Ω to the interval $[0,1]$ which discriminate between the plausibility and the certainty of a subset φ. These two measure are:

Possibility Measure. A possibility measure, denoted by Π, is a function of the form $\Pi(\varphi) = max\{\pi(I) : I \in \varphi\}$

$\Pi(\varphi)$ evaluates to what extent the subset φ is compatible with the available knowledge encoded by π. When $\Pi(\varphi) = 1$, we say that φ is certainty true if $\Pi(\neg\varphi) = 0$ and we say that the φ is somewhat certain if $\Pi(\neg\varphi) \in]0,1[$. When $\Pi(\varphi) = 1$ and $\Pi(\neg\varphi) = 1$, we say that there is a total ignorance about φ. The possibility measure satisfies the following properties for normalized possibility distributions:

$$\forall\varphi \in \Omega, \forall\psi \in \Omega, \Pi(\varphi \cup \psi) = max(\Pi(\varphi), \Pi(\psi))$$

$$\forall\varphi \in \Omega, \forall\psi \in \Omega, \Pi(\varphi \cap \psi) \leq min(\Pi(\varphi), \Pi(\psi))$$

Necessity Measure. A necessity measure, denoted by N, is a function of the form $N(\varphi) = 1 - \Pi(\neg\varphi)$

$N(\varphi)$ evaluates to what extent φ is certainty entailed from available knowledge encoded by π. When $N(\varphi) = 1$, we say that φ is certain. When $N(\varphi) \in]0,1[$, we say that φ is somewhat certain. When $N(\varphi) = 0$ and $N(\neg\varphi) = 0$, we say that there is a total ignorance. The necessity measure satisfies the following properties for normalized possibility distributions:

$$\forall\varphi \in \Omega, \forall\psi \in \Omega, N(\varphi \cap \psi) = min(N(\varphi), N(\psi))$$

$$\forall\varphi \in \Omega, \forall\psi \in \Omega, N(\varphi \cup \psi) \geq max(N(\varphi), N(\psi))$$

Now, clearly not all subsets of Ω represent axioms of *DL-Lite* language. For instance, assume that our vocabulary is composed of one concept A and two individuals a_1 and a_2. Assume that we have two interpretations $I_1 = (\Delta = \{a_1, a_2\}, .^{I_1})$ and $I_2 = (\Delta = \{a_1, a_2\}, .^{I_2})$ such that $A^{I_1} = \{a_1\}$ and $A^{I_2} = \{a_2\}$. Clearly, $\{I_1, I_2\}$ does not correspond to any axiom of our *DL-Lite* language, since $\{I_1, I_2\}$ intuitively encodes the axiom $A(a_1) \vee A(a_2)$, while the disjunction operator is not allowed in *DL-Lite* language.

In the following, possibility and necessity measures are assumed to only be defined over a *DL-Lite* language. Namely, if ϕ is an axiom, we define its associated possibility measures as: $\Pi(\phi) = \max_{I \in \Omega}\{\pi(I) : I \models \phi\}$ and its associated necessity measures as: $N(\phi) = 1 - \max_{I \in \Omega}\{\pi(I) : I \not\models \phi\}$ where $I \not\models \phi$ means that I is not a model of ϕ.

4   Possibilistic DL-Lite

In this section we go one step further in the definition of possibilistic extension of *DL-Lite*, denoted by π-*DL-Lite* by presenting its syntax and how to generate a possibility distribution associated with a π-*DL-Lite* KB.

4.1   Syntax

We consider $\mathcal{L}$ the description language *DL-Lite* recalled in Section 2.

Definition 1. *A π-DL-Lite KB $\mathcal{K} = \{\langle \phi_i, \alpha_i \rangle : 1, ..., n\}$ is a set of possibilistic axioms of the form $\langle \phi, \alpha \rangle$ where ϕ is an axiom expressed in $\mathcal{L}$ and $\alpha \in \ ]0, 1]$ is the degree of certainty of ϕ.*

Only somewhat certain information (namely $\alpha > 0$) are explicitly represented in π-*DL-Lite* KB. $\langle \phi, \alpha \rangle$ means that the uncertainty degree of ϕ is at least equal to α. The higher is the degree α the more important is the axiom or the fact. The degree α can be associated either with an inclusion axiom between concepts or roles (TBox), or with facts (ABox). A π-*DL-Lite* $\mathcal{K}$ will also be represented by a couple $\mathcal{K} = \langle \mathcal{T}, \mathcal{A} \rangle$ where both elements in $\mathcal{T}$ and $\mathcal{A}$ may be uncertain. Note that, if we consider $\alpha = 1$ then we represent a classical *DL-Lite* KB: $\mathcal{K}^* = \{\phi_i : \langle \phi_i, \alpha_i \rangle \in \mathcal{K}\}$.

Example 1. Let *Teacher*, *PhdStudent* and *Student* be three atomic concepts and *TeachesTo* be an atomic role. The following possibilistic TBox $\mathcal{T}$ and the possibilistic ABox $\mathcal{A}$ will be used in the rest of the paper:

$\mathcal{T} = \{\ \langle Teacher \sqsubseteq \neg Student, .8 \rangle,$
$\langle PhdStudent \sqsubseteq Student, .7 \rangle,$
$\langle PhdStudent \sqsubseteq Teacher, .9 \rangle,$
$\langle \exists teachesTo \sqsubseteq Teacher, .6 \rangle,$
$\langle \exists teachesTo^- \sqsubseteq Student, .5 \rangle\}.$
$\mathcal{A} = \{\langle Student(b), .95 \rangle, \langle teachesTo(b,c), 1 \rangle\}.$

In π-*DL-Lite* KB, the necessity degree attached to an axiom reflects its confidence and evaluates to what extent this axiom is considered as certain. For instance the axiom $\langle teachesTo(b,c), 1 \rangle$ states that we are absolutely certain that the "*the Teacher b teachesTo the Student c*". However the axiom $\langle PhdStudent \sqsubseteq Student, .7 \rangle$ simply states that a *PhdStudent* may be a *Student* with a certainty degree equal or greater than .7.

4.2   From π-*DL-Lite* Knowledge Base to π-*DL-Lite* Possibility Distribution

The semantics of π-*DL-Lite* is given by a possibility distribution, denoted $\pi_{\mathcal{K}}$, defined over the set of all interpretations $I = (\Delta, \cdot^I)$ of a *DL-Lite* language (see Section 3). As in standard possibilistic logic [7], given a π-*DL-Lite* knowledge base $\mathcal{K}$ the possibility distribution induced by $\mathcal{K}$ is defined as follow:

Definition 2. *For every $I \in \Omega$*

$$\pi_\mathcal{K}(I) = \begin{cases} 1 & if \; \forall \langle \phi_i, \alpha_i \rangle \in \mathcal{K}, I \models \phi_i \\ 1 - max\{\alpha_i : (\phi_i, \alpha_i) \in \mathcal{K} | I \nvDash \phi_i\} & otherwise \end{cases}$$

where $\models$ is the satisfaction relation between *DL-Lite* formulas recalled in Section 2. $\langle \phi_i, \alpha_i \rangle \in \mathcal{K}$ means that $\langle \phi_i, \alpha_i \rangle \in \mathcal{K}$ either belongs to the TBox $\mathcal{T}$ or the ABox $\mathcal{A}$ of $\mathcal{K}$.

Example 2. (Example 1 continued) Using Definition 2, we compute the following possibility degree of three interpretations where $\Delta = \{b, c\}$:

I	$\cdot^I$	$\pi_\mathcal{K}$
I_1	$(Student)^I = \{b, c\}, (PhdStudent)^I = \{b\}, (Teacher)^I = \{b\}$ $(teachesTo)^I = \{(b, c)\}$	.2
I_2	$(Student)^I = \{b, c\}, (PhdStudent)^I = \{\}, (Teacher)^I = \{\}$ $(teachesTo)^I = \{(b, c)\}$	.4
I_3	$(Student)^I = \{b\}, (PhdStudent)^I = \{\}, (Teacher)^I = \{c\}$ $(teachesTo)^I = \{(c, b)\}$	0

In this example, we can see that the interpretation I_1 does not satisfy $\langle Teacher \sqsubseteq \neg Student, .8 \rangle$, the interpretation I_2 does not satisfy $\langle \exists teachesTo \sqsubseteq Teacher, .6 \rangle$ and the interpretation I_3 does not satisfy $\langle teachesTo(b, c), 1 \rangle$. Hence, no one of these interpretations is a model of $\mathcal{K}$.

A π-*DL-Lite* KB is said to be consistent if the possibility distribution $\pi_\mathcal{K}$ is normalized, namely there exists an interpretation I such that $\pi_\mathcal{K}(I) = 1$. If not, $\mathcal{K}$ is said to be inconsistent and its inconsistency degree is defined semantically as follow:

Definition 3. *The inconsistency degree of a π-DL-Lite KB, denoted by $Inc(\mathcal{K})$, is semantically defined as follow:* $Inc(\mathcal{K}) = 1 - \max_{I \in \Omega}\{\pi_\mathcal{K}(I)\}$

If $Inc(\mathcal{K}) = 1$ then $\mathcal{K}$ is fully inconsistent and if $Inc(\mathcal{K}) = 0$ then it is consistent.

Example 3. (Example 2 continued), in fact, one can check that the inconsistency degree of $\mathcal{K}$ according to $\pi_\mathcal{K}$ is : $Inc(\mathcal{K}) = 1 - \max_{I \in \Omega}\{\pi_\mathcal{K}(I)\} = .6$, and hence $\mathcal{K}$ is inconsistent (in fact, there is no way to find an interpretation that satisfy $\mathcal{K}$ with a degree greater than .6).

Remark 1. In propositional possibilistic logic, each possibilistic KB induces a joint possibility distribution and conversely. Although each π-*DL-Lite* KB induces a unique joint possibility distribution, the converse does not hold. Consider again the example where we only have one concept A and two individuals a_1 and a_2. Consider four interpretations I_1, I_2, I_3 and I_4 having the same domain $\Delta = \{a_1, a_2\}$ and where $A^{I_1} = \{a_1\}, A^{I_2} = \{a_1\}, A^{I_2} = \{a_1, a_2\}$ and $A^{I_4} = \emptyset$.

Assume that $\pi(I_1) = \pi(I_2) = 1$ and $\pi(I_3) = \pi(I_4) = .5$. One can check that there is no π-*DL-Lite* KB such that $\pi_{\mathcal{K}} = \pi$.

5   Possibilistic Closure in π-*DL-Lite*

Let us first point out that one can easily add conjunctions in the right side of inclusion axioms. Proposition 1 shows that a complex inclusion axiom of the form $\langle B_1 \sqsubseteq B_2 \sqcap B_3, \alpha \rangle$ can be splitted into elementary inclusion axioms that can be added to $\mathcal{K}$ without modifying its possibility distribution. Proposition 1 can be derived from Proposition 5 in [14] for general DLs.

Proposition 1. *Let* $\mathcal{K} = \{IP \cup \{\langle B_1 \sqsubseteq B_2 \sqcap B_3, \alpha \rangle\}, \mathcal{A}\}$ *and* $\mathcal{K}' = \{IP \cup \{ \langle B_1 \sqsubseteq B_2, \alpha \rangle, \langle B_1 \sqsubseteq B_3, \alpha \rangle\}, \mathcal{A}\}$ *then* $\mathcal{K}$ *and* $\mathcal{K}'$ *induces the same possibility distribution.*

Hence, the language given in Section 2 is a simplification of the one based on conjunctions used in the right side (*resp.* disjunction in left side) of inclusion axioms.

The aim of this section is to define the so-called π-*negated* closure of a π-*DL-Lite* KB. This notion is crucial for defining the concepts of consistency and inference from a π-*DL-Lite* KB. A possibilistic TBox $\mathcal{T} = \{IP, IN\}$ can be viewed as composed of positive inclusions (*PI*) of the form $\langle B_1 \sqsubseteq B_2, \alpha \rangle$ or $\langle R_1 \sqsubseteq R_2, \alpha \rangle$ and negative inclusions (*NI*) of the form $\langle B_1 \sqsubseteq \neg B_2, \alpha \rangle$ or $\langle R_1 \sqsubseteq \neg R_2, \alpha \rangle$. Conceptually, the *PI* axioms (*resp. NI* axioms) represent subsumption (*resp.* disjunction) between concepts or roles. Roughly speaking, this closure denoted π-$neg(\mathcal{T})$, will contain possibilistic negated axioms of the form $\langle B_1 \sqsubseteq \neg B_2, \alpha \rangle$ or $\langle R_1 \sqsubseteq \neg R_2, \alpha \rangle$ that can be derived from $\mathcal{T}$. The set π-$neg(\mathcal{T})$ is obtained by applying a set of rules that extends the ones defined in standard *DL-Lite* when axioms are weighted with uncertainty degrees.

At the beginning π-$neg(\mathcal{T})$ is set to an empty set.

Rule 1. *Let* $\mathcal{T} = \{IP, IN\}$ *then* $IN \subseteq \pi - neg(\mathcal{T})$.

This rule simply means that negated axioms explicitly stated in $\mathcal{T}$ can be trivially derived from $\mathcal{T}$.

Example 4. (Example 1 continued): Using Rule 1, we add $\langle Teacher \sqsubseteq \neg Student , .8 \rangle$ as *NI* to $\pi - neg(\mathcal{T})$.

Rule 2. *If* $\langle B_1 \sqsubseteq B_2, \alpha_1 \rangle \in \mathcal{T}$ *and* $\langle B_2 \sqsubseteq \neg B_3, \alpha_2 \rangle \in \pi - neg(\mathcal{T})$ *or* $\langle B_3 \sqsubseteq \neg B_2 , \alpha_2 \rangle \in \pi - neg(\mathcal{T})$ *then add* $\langle B_1 \sqsubseteq \neg B_3, min(\alpha_1, \alpha_2) \rangle$ *to* $\pi - neg(\mathcal{T})$.

Rule 3. *If* $\langle R_1 \sqsubseteq R_2, \alpha_1 \rangle \in \mathcal{T}$ *and* $\langle R_2 \sqsubseteq \neg R_3, \alpha_2 \rangle \in \pi - neg(\mathcal{T})$ *or* $\langle R_3 \sqsubseteq \neg R_2 , \alpha_2 \rangle \in \pi - neg(\mathcal{T})$ *then add* $\langle R_1 \sqsubseteq \neg R_3, min(\alpha_1, \alpha_2) \rangle$ *to* $\pi - neg(\mathcal{T})$.

Rules 2 and 3 simply state that transitivity holds with a weight equal to the least weight of premises axioms.

Rule 4. *if* $\langle R_1 \sqsubseteq R_2, \alpha_1 \rangle \in \mathcal{T}$ *and* $\langle \exists R_2 \sqsubseteq \neg B, \alpha_2 \rangle \in \pi - neg(\mathcal{T})$ *or* $\langle B \sqsubseteq \neg \exists R_2, \alpha_2 \rangle \in \pi - neg(\mathcal{T})$ *then add* $\langle \exists R_1 \sqsubseteq \neg B, min(\alpha_1, \alpha_2) \rangle$ *to* $\pi - neg(\mathcal{T})$.

Rule 5. *If* $\langle R_1 \sqsubseteq R_2, \alpha_1 \rangle \in \mathcal{T}$ *and* $\langle \exists R_2^- \sqsubseteq \neg B, \alpha_2 \rangle \in \pi - neg(\mathcal{T})$ *or* $\langle B \sqsubseteq \neg \exists R_2^-, \alpha_2 \rangle \in \pi - neg(\mathcal{T})$ *then add* $\langle \exists R_1^- \sqsubseteq \neg B, min(\alpha_1, \alpha_2) \rangle$ *to* $\pi - neg(\mathcal{T})$.

Rule 6. *If* $\langle R \sqsubseteq \neg R, \alpha \rangle \in \pi - neg(\mathcal{T})$ *or* $\langle \exists R \sqsubseteq \neg \exists R, \alpha \rangle \in \pi - neg(\mathcal{T})$ *or* $\langle \exists R^- \sqsubseteq \neg \exists R^-, \alpha \rangle \in \pi - neg(\mathcal{T})$ *then add* $\langle R \sqsubseteq \neg R, \alpha \rangle$ *and* $\langle \exists R \sqsubseteq \neg \exists R, \alpha \rangle$ *and* $\langle \exists R^- \sqsubseteq \neg \exists R^-, \alpha \rangle$ *to* $\pi - neg(\mathcal{T})$.

Proposition 2. *Let* $\mathcal{T} = \{IP, IN\}$ *and* $\pi - neg(\mathcal{T})$ *be the closure of* $\mathcal{T}$ *obtained using Rules (1-6). Then* $\mathcal{K} = \{\mathcal{T}, \mathcal{A}\}$ *and* $\mathcal{K}' = \{\mathcal{T} \cup \pi - neg(\mathcal{T}), \mathcal{A}\}$ *induce the same possibility distribution, namely* $\forall I$, $\pi_{\mathcal{K}}(I) = \pi_{\mathcal{K}'}(I)$.

Example 5. From the $\mathcal{K} = \{\mathcal{T}, \mathcal{A}\}$ of Example 1, one can check that applying Rule 1 - Rule 6 gives the following $\pi - neg(\mathcal{T})$ where :

$$\pi - neg(\mathcal{T}) = \{\langle Teacher \sqsubseteq \neg Student, .8 \rangle,$$
$$\langle PhdStudent \sqsubseteq \neg Teacher, .7 \rangle,$$
$$\langle PhdStudent \sqsubseteq \neg Student, .8 \rangle,$$
$$\langle \exists teachesTo \sqsubseteq \neg Student, .6 \rangle,$$
$$\langle \exists teachesTo^- \sqsubseteq \neg Teacher, .5 \rangle\}$$
$$\mathcal{A} = \{\langle Student(b), .95 \rangle , \langle teachesTo(b, c), 1 \rangle\}$$

6   Checking Inconsistency

From now on, $\pi - neg(\mathcal{T})$ denotes the result of applying Rules 1-6 until reaching the closure (namely, no negated axioms can be added using Rules 1-6). An important result is that computing inconsistency of $\mathcal{K} = \{\mathcal{T}, \mathcal{A}\}$ comes down to compute inconsistency degree of $\mathcal{K}' = \{\pi - neg(\mathcal{T}), \mathcal{A}\}$.

Proposition 3. *Let* $\mathcal{K} = \{\mathcal{T}, \mathcal{A}\}$ *and let* $\mathcal{K}' = \{\pi - neg(\mathcal{T}), \mathcal{A}\}$ *then* $Inc(\mathcal{K}) = Inc(\mathcal{K}')$.

Proposition 3 is important since it provides a way to compute the inconsistency degree of a π-*DL-Lite* KB. Indeed, verifying inconsistency of $\mathcal{K} = \{\mathcal{T}, \mathcal{A}\}$ is reduced to verifying the inconsistency of $\mathcal{K}' = \{\pi - neg(\mathcal{T}), \mathcal{A}\}$. A contradiction is presented when a same individual (*resp.* two individuals) belongs to two negated concepts (*resp.* negated roles) (*i.e.* NI in $\pi - neg(\mathcal{T})$). Then, checking inconsistency is done by means a set of weighted queries issued from $\pi - neg(\mathcal{T})$. Subsection 6.1 formalizes this concept of weighted queries while subsection 6.2 provides the algorithm to compute inconsistency degrees using a set of weighted queries.

6.1   Weighted Queries

The idea is to evaluate over $\mathcal{A}$ suitable weighted queries expressed from $\pi - neg(\mathcal{T})$ to exhibit whether the ABox $\mathcal{A}$ contains or not contradictions and to compute the inconsistency degree. To obtain the set of weighted queries q_c from $\pi - neg(\mathcal{T})$, we propose a translation function ψ. ψ has an argument a possibilistic NI $\langle B_1 \sqsubseteq \neg B_2, \alpha \rangle$ or $\langle R_1 \sqsubseteq \neg R_2, \alpha \rangle$ and produces a weighted first order formula.

Definition 4. ψ *is a function that transforms all axioms in* $\pi - neg(\mathcal{T})$ *to weighted query* q_c:

- $\psi(\langle B_1 \sqsubseteq \neg B_2, \alpha \rangle) = \langle (x, \gamma_1, \gamma_2).\lambda_1(x, \gamma_1) \wedge \lambda_2(x, \gamma_2), \alpha \rangle$ *with*
  - $\lambda_i(x, \gamma_i) = A_i(x, \gamma_i)$ *if* $B_i = A_i$
  - $\lambda_i(x, \gamma_i) = \exists y_i.P_i(x, y_i, \gamma_i)$ *if* $B_i = \exists P_i$
  - $\lambda_i(x, \gamma_i) = \exists y_i.P_i(y_i, x, \gamma_i)$ *if* $B_i = \exists P_i^-$
- $\psi(\langle R_1 \sqsubseteq \neg R_2, \alpha \rangle) = \langle (x, y, \gamma_1, \gamma_2).\nu_1(x, y, \gamma_1) \wedge \nu_2(x, y, \gamma_2), \alpha \rangle$ *with*
  - $\nu_i(x, y, \gamma_i) = P_i(x, y, \gamma_i)$ *if* $R_i = P_i$
  - $\nu_i(x, y, \gamma_i) = P_i(y, x, \gamma_i)$ *if* $R_i = P_i^-$

Intuitively, if $\langle B_1 \sqsubseteq \neg B_2, \alpha \rangle$ belongs in $\pi - neg(\mathcal{T})$, then a query associated to $B_1 \sqsubseteq \neg B_2$ simply means return all $\{B_1(x, \gamma_1), B_2(x, \gamma_2)\}$ that are present in the ABox.

Example 6. From Example 5, we obtain the following weighted queries using Definition 4:

$$q_c = \langle (x, \gamma_1, \gamma_2).Teacher(x, \gamma_1) \wedge Student(x, \gamma_2), .8 \rangle$$
$$q_c = \langle (x, \gamma_1, \gamma_2).PhdStudent(x, \gamma_1) \wedge Teacher(x, \gamma_2), .7 \rangle$$
$$q_c = \langle (x, \gamma_1, \gamma_2).PhdStudent(x, \gamma_1) \wedge Student(x, \gamma_2), .8 \rangle$$
$$q_c = \langle (x, \gamma_1, \gamma_2). (\exists y.teachesTo(x, y, \gamma_1)) \wedge Student(x, \gamma_2), .6 \rangle$$
$$q_c = \langle (x, \gamma_1, \gamma_2). (\exists y.teachesTo(y, x, \gamma_1)) \wedge Teacher(x, \gamma_2), .5 \rangle$$

6.2 An Algorithm for Computing Inconsistency Degrees

Now, we provide the algorithm *Inconsistency*, which takes as input a $\mathcal{K}' = \{\pi - neg(\mathcal{T}), \mathcal{A}\}$ and computes $Inc(\mathcal{K})$, the inconsistency degree of $\mathcal{K}$.

Algorithmus 1. *Inconsistency* $(\mathcal{K})$

Input: $\mathcal{K}' = \{\pi - neg(\mathcal{T}), \mathcal{A}\}$
Output: $Inc(\mathcal{K})$
1: $cont := \{0\}$
2: **for all** $(\phi_i, \alpha_i) \in \pi - neg(\mathcal{T}); i = 1..|\pi - neg(\mathcal{T})|$ **do**
3:     $(q_c, \alpha_q) := (\psi(\phi_i, \alpha_i))$
4:     **if** $Eval(q_c, \mathcal{A}) \neq \emptyset$ **then**
5:         $\beta := max(Eval(q_c, \mathcal{A}))$
6:         **if** $\beta > \alpha_q$ **then**
7:             $cont := cont \cup \{\alpha_q\}$
8:         **else**
9:             $cont := cont \cup \{\beta\}$
10: **return** $max(cont)$

In this algorithm, the set *cont* stores the inconsistency degrees founded during the algorithm. $Eval(q_c, \mathcal{A})$ denotes the evaluation of a weighted query q_c over $\mathcal{A}$ obtained by transforming $\pi - neg(\mathcal{T})$ with the function given in Definition 4. For (a, α_i) and (a, α_j) presented in a query result, we only consider one individual $(a, min(\alpha_i, \alpha_j))$. $\beta = max(Eval(q_c, \mathcal{A}))$ represents the maximum weight of all tuples in $Eval(q_c, \mathcal{A})$. At this point, if the weight of the query is less than β (*i.e.* $\alpha_q < \beta$) then the contradiction is issued from the query and implicitly form the TBox corresponding axioms. Otherwise (*i.e.* $\alpha_q \geq \beta$) then the contradiction is issued from the result of the query evaluation and implicitly from ABox assertions. Finally, the inconsistency degree of $\mathcal{K}$ ($Inc(\mathcal{K})$) is the maximum of all contradiction degrees of the *cont*. In case of consistency, the "if part" of the algorithm (lines 4-9) is never used, and the algorithm returns the value 0 (namely, $Inc(\mathcal{K}) = 0$). This explain why *cont* is initialized to $\{0\}$(line 1).

Example 7. From Example 6, only the query $\langle(x, \gamma_1, \gamma_2). (\exists y.teachesTo(x, y, \gamma_1)) \wedge Student(x, \gamma_2), .6\rangle$ presents a contradiction: $\langle q_c, .6\rangle = \{(b, .95, 1)\}$. Thus, the inconsistency degree of the KB is $Inc(\mathcal{K}) = .6$.

We now provide two propositions that show on one hand that our π-*DL-Lite* extends standard *DL-Lite* and on other hand that the computational complexity of Algorithm 1 is the same as the one in standard *DL-Lite*.

Proposition 4. *Let* $\mathcal{K}_s = \{\mathcal{T}_s, \mathcal{A}_s\}$ *be a standard DL-Lite. Let* $\mathcal{K}_\pi = \{\mathcal{T}_\pi, \mathcal{A}_\pi\}$ *where* $\mathcal{T}_\pi$ *(resp.* $\mathcal{A}_\pi$) *is defined from* $\mathcal{T}_s$ *(resp.* $\mathcal{A}_s$) *by assigning a degree 1 to each axiom of* $\mathcal{T}_s$ *(resp.* $\mathcal{A}_s$), *namely :* $\mathcal{T}_\pi = \{\langle \phi_i, 1 \rangle : \phi_i \in \mathcal{T}_s\}$ *and* $\mathcal{A}_\pi = \{\langle \phi_i, 1 \rangle : \phi_i \in \mathcal{A}_s\}$. *Then* $\mathcal{K}_s$ *is consistent (in the sense of standard DL-Lite) iff* $Inc(\mathcal{K}_\pi) = 0$ *and* $\mathcal{K}_s$ *is inconsistent iff* $Inc(\mathcal{K}_\pi) = 1$.

Proposition 5. *The complexity of Algorithm 1 is the same as the one used in standard DL-Lite ([5], section 3.3 , Theorem 26)*

To see why proposition 5 holds it is enough to see the differences between Algorithm 1 and the one used in ([5], section 3.1.3) for standard *DL-Lite*. The first remarks, concerns the returned result. On our algorithm, results of queries are weighted while in standard DL-Lite, they are not. This does not change the complexity. The difference concerns lines 4-9, where in standard *DL-Lite* algorithm they are replaced by:

1: **if** $Eval(q_c, \mathcal{A}) \neq \emptyset$ **then**
2:      **return** $True$
3: **else**
4:      **return** $False$

It is easy first to see that in case of consistency both algorithms perform same steps, because the " if part of the algorithm" is never considered. Now in case of inconsistency, the worst case appears when the whole "loop" is used, namely inconsistency appears with the last element of $\pi - neg(\mathcal{T})$. In both cases let $\mathcal{A}$ be the result of the evaluation of $Eval(q_c)$. This needs at least $O(|\mathcal{A}|)$ steps. Algorithm 1 (contrary to the algorithm in standard *DL-Lite* [5]) computes also

$max\left\{\alpha_i : \langle\phi_i, \alpha_i\rangle \in \mathcal{A}\right\}$ which needs again $O\left(|\mathcal{A}|\right)$. Since trivially, $O\left(2\left|\mathcal{A}\right|\right) = O\left(|\mathcal{A}|\right)$, our algorithm has the same complexity as in standard *DL-Lite*. Hence we increase the expressive power of *DL-Lite* while keeping the complexity as low as the one of standard *DL-Lite*.

7 Inference in Possibilistic DL-Lite

In this section, we first present classical inference problem (*i.e.* subsumption and instance checking). First, we define an $\alpha - cut$ of $\mathcal{T}$ (*resp.* $\mathcal{A}$ and $\mathcal{K}$), denoted $\mathcal{T}_{>\alpha}$ (resp. $\mathcal{A}_{>\alpha}$, $\mathcal{K}_{>\alpha}$), a sub base of $\mathcal{T}$ (*resp.* $\mathcal{A}$ and $\mathcal{K}$) composed of formulas having a weight greater than alpha (α). In possibilistic *DL* inference problems such as subsumption and instance checking can be reduced to the task of computing the inconsistency degree of the KB [14]. We present in the following inference services in π-*DL-Lite*:

- Flat subsumption: Let $\mathcal{T}$ be a possibilistic TBox, B_1 and B_2 be two general concepts, A be an atomic concept not appearing in $\mathcal{T}$, and a be a constant. Then, $\mathcal{K} \models_{\pi} B_1 \sqsubseteq B_2$ iff the KB $\mathcal{K}_1 = \{\mathcal{T}_1, \mathcal{A}_1\}$ where $\mathcal{T}_1 = \mathcal{T}_{>Inc(\mathcal{K})} \cup \{\langle A \sqsubseteq B_1, 1\rangle, \langle A \sqsubseteq \neg B_2, 1\rangle\}$ and $\mathcal{A}_1 = \{\langle A(a), 1\rangle\}$ is inconsistent whatever is the degree ($\exists \alpha > 0$ such that $Inc(\mathcal{K}_1) = \alpha$).
- Subsumption with a necessity degree: Let $\mathcal{T}$ be a possibilistic TBox, B_1 and B_2 be two general concepts, A be an atomic concept not appearing in $\mathcal{T}$, and be a a constant. Then, $\mathcal{K} \models_{\pi} \langle B_1 \sqsubseteq B_2, \alpha\rangle$ iff the KB $\mathcal{K}_1 = \{\mathcal{T}_1, \mathcal{A}_1\}$ where $\mathcal{T}_1 = \mathcal{T}_{\geq\alpha} \cup \{\langle A \sqsubseteq B_1, 1\rangle, \langle A \sqsubseteq \neg B_2, 1\rangle\}$ and $\mathcal{A}_1 = \{\langle A(a), 1\rangle\}$ is inconsistent where $Inc(\mathcal{K}_1) = \alpha$ and $\alpha > Inc(\mathcal{K})$.
- Flat instance checking: Let $\mathcal{K}$ be a π-*DL-Lite* KB, B be a concept, A be an atomic concept not appearing in $\mathcal{T}$, and a be a constant. Then, $\mathcal{K} \models_{\pi} B(a)$ iff the KB $\mathcal{K}_1 = \{\mathcal{T}_1, \mathcal{A}_1\}$ where $\mathcal{T}_1 = \mathcal{T}_{>Inc(\mathcal{K})} \cup \{\langle A \sqsubseteq \neg B, 1\rangle\}$ and $\mathcal{A}_1 = \{\langle A(a), 1\rangle\}$ is inconsistent (whatever is the degree).
- Instance checking with a necessity degree: Let $\mathcal{K}$ be a π-*DL-Lite* KB, B be a concept, A be an atomic concept not appearing in $\mathcal{T}$, and a be a constant. Then, $\mathcal{K} \models_{\pi} \langle B(a), \alpha\rangle$ iff the KB $\mathcal{K}_1 = \{\mathcal{T}_1, \mathcal{A}_1\}$ where $\mathcal{T}_1 = \mathcal{T}_{>\alpha} \cup \{\langle A \sqsubseteq \neg B, 1\rangle\}$ and $\mathcal{A}_1 = \{\langle A(a), 1\rangle\}$ is inconsistent where $Inc(\mathcal{K}_1) = \alpha$ and $\alpha > Inc(\mathcal{K})$.

KB consistency is verified by Algorithm *Inconsistency*, presented above, where $Inc(\mathcal{K}) = 0$. Hence, all these basic inferences can be obtained using Algorithm 1. Note the difference between flat subsumption (*resp.* instance checking) and subsumption with a necessity degree (*resp.* instance checking with a necessity degree) is that in the first case we only check whether the subsumption holds whatever is the degree, while is the second case, subsumption should be satisfied to some degree.

8 Conclusions and Future Works

In this paper, we investigated a possibilistic extension of *DL-Lite*. We first introduced the syntax and the semantics of such extensions. We provided properties of π-DL-$Lite$ and show how to compute the inconsistency degree of π-DL-$Lite$ KB having a complexity identical to the one used in standard DL-$Lite$. This is done by defining π-DL-$Lite$ negative closure that extends the one of standard DL-$Lite$. Then, we gave a method to check consistency for π-DL-$Lite$. Finally, we discussed inference problems. In particular we distinguish different inference tasks depending whether we use flat inferences or weighted inferences. Results of this paper are important since they extended DL-$Lite$ languages to deal with priority (between TBox axioms or ABox axioms) or uncertainty without changing the computational complexity. Future works concern the revision of π-DL-$Lite$ KB in presence of new pieces of information.

Acknowledgment. This work has been supported by the french Agence Nationale de la Recherche for the ASPIQ project ANR-12-BS02-0003.

References

1. Artale, A., Calvanese, D., Kontchakov, R., Zakharyaschev, M.: The dl-lite family and relations. J. Artif. Intell. Res., JAIR (2009)
2. Baader, F., McGuinness, L., Nardi, D., Patel-Schneider, P.F.: The description logic handbook: Theory, implementation, and applications. Cambridge University Press (2003)
3. Bobillo, F., Straccia, U.: fuzzydl: An expressive fuzzy description logic reasoner. In: FUZZ-IEEE, pp. 923–930. IEEE (2008)
4. Calvanese, D., Giacomo, G.D., Lembo, D., Lenzerini, M., Rosati, R.: Dl-lite: Tractable description logics for ontologies. In: Proceedings, The Twentieth National Conference on Artificial Intelligence and the Seventeenth Innovative Applications of Artificial Intelligence Conference, AAAI 2005, pp. 602–607. AAAI Press / The MIT Press (2005)
5. Calvanese, D., De Giacomo, G., Lembo, D., Lenzerini, M., Rosati, R.: Tractable reasoning and efficient query answering in description logics: The dl-lite family. J. Autom. Reasoning 39(3), 385–429 (2007)
6. Couchariere, O., Lesot, M.-J., Bouchon-Meunier, B.: Consistency checking for extended description logics. In: Proceedings of the 21st International Workshop on Description Logics, DL 2008. Description Logics, vol. 9, pp. 602–607. CEUR-WS.org / CEUR Workshop Proceedings (2008)
7. Dubois, D., Lang, J., Prade, H.: Possibilistic logic. In: The Handbook of Logic in Artificial Intelligence and Logic Programming, vol. 3, pp. 439–513. Clarendon Press, Oxford (1994)
8. Dubois, D., Mengin, J., Prade, H.: Possibilistic uncertainty and fuzzy features in description logic. a preliminary discussion. In: Sanchez, E. (ed.) Fuzzy Logic and the Semantic Web. Capturing Intelligence, vol. 1, pp. 101–113. Elsevier (2006)
9. Dubois, D., Prade, H.: Possibility theory. Plenum Press, New-York (1988)
10. Hollunder, B.: An alternative proof method for possibilistic logic and its application to terminological logics. International Journal of Approximate Reasoning 12(2), 85–109 (1995)
11. Lukasiewicz, T., Straccia, U.: Description logic programs under probabilistic uncertainty and fuzzy vagueness. Int. J. Approx. Reasoning 50(6), 837–853 (2009)

12. Pan, J.Z., Stamou, G.B., Stoilos, G., Thomas, E.: Expressive querying over fuzzy dl-lite ontologies. In: Proceedings of the 2007 International Workshop on Description Logics, DL 2007. Description Logics, vol. 250, pp. 602–607. CEUR-WS.org / CEUR Workshop Proceedings (2007)

13. Qi, G., Ji, Q., Pan, J.Z., Du, J.: PossDL — A possibilistic DL reasoner for uncertainty reasoning and inconsistency handling. In: Aroyo, L., Antoniou, G., Hyvönen, E., ten Teije, A., Stuckenschmidt, H., Cabral, L., Tudorache, T. (eds.) ESWC 2010, Part II. LNCS, vol. 6089, pp. 416–420. Springer, Heidelberg (2010)

14. Qi, G., Ji, Q., Pan, J.Z., Du, J.: Extending description logics with uncertainty reasoning in possibilistic logic. Int. J. Intell. Syst. 26(4), 353–381 (2011)

15. Qi, G., Pan, J.Z.: A tableau algorithm for possibilistic description logic $\mathcal{ALC}$. In: Domingue, J., Anutariya, C. (eds.) ASWC 2008. LNCS, vol. 5367, pp. 61–75. Springer, Heidelberg (2008)

16. Qi, G., Pan, J.Z., Ji, Q.: Extending description logics with uncertainty reasoning in possibilistic logic. In: Mellouli, K. (ed.) ECSQARU 2007. LNCS (LNAI), vol. 4724, pp. 828–839. Springer, Heidelberg (2007)

17. Qi, G., Pan, J.Z., Ji, Q.: A possibilistic extension of description logics. In: Proceedings of the 2007 International Workshop on Description Logics, DL 2007, vol. 4724, pp. 602–607. CEUR-WS.org / CEUR Workshop Proceedings (2007)

18. Straccia, U.: A fuzzy description logic. In: Mostow, J., Rich, C. (eds.) AAAI/IAAI, pp. 594–599. AAAI Press / The MIT Press (1998)

19. Straccia, U.: Towards top-k query answering in description logics: The case of dl-lite. In: Fisher, M., van der Hoek, W., Konev, B., Lisitsa, A. (eds.) JELIA 2006. LNCS (LNAI), vol. 4160, pp. 439–451. Springer, Heidelberg (2006)

Group Preferences for Query Answering in Datalog+/− Ontologies

Thomas Lukasiewicz, Maria Vanina Martinez,
Gerardo I. Simari, and Oana Tifrea-Marciuska

Department of Computer Science, University of Oxford, UK
firstname.lastname@cs.ox.ac.uk

Abstract. In the recent years, the Web has been changing more and more towards the so-called Social Semantic Web. Rather than being based on the link structure between Web pages, the ranking of search results in the Social Semantic Web needs to be based on something new. We believe that it can be based on ontological background knowledge and on user preferences. In this paper, we thus propose an extension of the Datalog+/− ontology language that allows for dealing with partially ordered preferences of groups of users. We focus on answering k-rank queries in this context. In detail, we present different strategies to compute group preferences as an aggregation of the preferences of a collection of single users. We then provide algorithms to answer k-rank queries for DAQs (disjunctions of atomic queries) under these group preferences. We show that such DAQ answering in Datalog+/− can be done in polynomial time in the data complexity, as long as query answering can also be done in polynomial time (in the data complexity) in the underlying classical ontology.

1 Introduction

In the recent years, several important changes are taking place on the classical Web. First, the so-called Web of Data is increasingly being realized as a special case of the Semantic Web. Second, as a part of the Social Web, users are acting more and more as first-class citizens in the creation and delivery of contents on the Web. The combination of these two technological waves is called the *Social Semantic Web* (or also *Web 3.0*), where the classical Web of interlinked documents is more and more turning into (i) semantic data and tags constrained by ontologies, and (ii) social data, such as connections, interactions, reviews, and tags. The Web is thus shifting away from data on linked Web pages towards less such interlinked data in social networks on the Web relative to underlying ontologies. This requires new technologies for search and query answering, where the ranking of search results is not based on the link structure between Web pages anymore, but on the information available in the Social Semantic Web, in particular, the underlying ontological knowledge and the preferences of the users.

Modeling the preferences of a group of users is also an important research topic in its own right. With the growth of social media, people post their preferences and expect to get personalized information. Moreover, people use social networks as a tool to organize events, where it is required to combine the individual preferences and suggest items obtained from aggregated user preferences. For example, if there is a movie night of

W. Liu, V.S. Subrahmanian, and J. Wijsen (Eds.): SUM 2013, LNAI 8078, pp. 360–373, 2013.

friends, family trip, or dinner with working colleagues, one has to decide which is the ideal movie or location for the group, given the preferences of each member.

To address this problem, individual user preferences can be adopted and then aggregated to group preferences. However, this comes along with two additional challenges. The first challenge is to define a group preference semantics that solves the possible *disagreement* among users (a system should return results in such a way that each individual benefits from the result). For example, people (even friends) often have different tastes on movies. The second challenge is to allow for efficient algorithms, e.g., to compute efficiently the answers to queries under aggregated group preferences [2].

There are many studies that are addressing the area of group modeling. Indirectly, it is related to the area of social choice (group decision making, i.e., aiming at the decision that is best for a user given the opinion of individuals), which was studied in mathematics, economics, politics, and sociology [14,17]. Other areas related to social choice are meta-search [11], collaborative filtering [9], and multi-agent systems [18].

Current approaches that deal with group preferences have been studied in the area of recommender systems [2,13], which focus on quantitative preferences. However, in many real-world scenarios, the ordering of preferences is incomplete. This appears due to privacy issues or an incomplete elicitation process (users may not want to be asked too many questions). Furthermore, it is often difficult to determine the appropriate numerical preferences and weights that maximize the utility of a decision [4]. For example, it is difficult for a user to determine a numerical value (i.e., 0.7 or 0.9) to rate a movie. Therefore, there is a growing interest in formalisms for representing and reasoning with qualitative incomplete preferences [15,8,1,10]. In [10], we have introduced an extension of Datalog+/– with preferences for a single user, while we now focus on extending Datalog+/– with preferences of a group of users that comes with the two additional aforementioned challenges. This paper solves them by providing aggregated answers for DAQs (disjunctions of atomic queries) in polynomial time. The main contributions of this paper can be summarized as follows.

- We introduce the G-PrefDatalog+/– framework, which is a combination of the Datalog+/– ontology language with group preferences. To our knowledge, this is the first combination of ontology languages with group preferences.
- We present several strategies to compute group preferences as an aggregation of sets of single-user preferences, based on social choice theory [12]. Using these aggregated group preferences, we give algorithms for answering k-rank queries for DAQs , based on a novel augmented chase procedure.
- We analyze the complexity of these algorithms, showing that answering k-rank queries for DAQs in G-PrefDatalog+/– can be done in polynomial time in the data complexity, as long as query answering in the underlying classical ontology can also be done in polynomial time in the data complexity.

The rest of this paper is organized as follows. In Section 2, we briefly recall some basic concepts from classical Datalog+/– and its extension by single-user preferences. Section 3 introduces the syntax and semantics of G-PrefDatalog+/–. In Section 4, we present query answering algorithms, using different strategies to obtain group preferences from individual user preferences, and focusing especially on k-rank queries to DAQs. Section 5 summarizes our main results and gives an outlook on future research.

2   Preliminaries

In this section, we briefly recall some necessary background concepts.

2.1   Datalog+/–

We first recall some basics on Datalog+/– [6], namely, on relational databases, (Boolean) conjunctive queries ((B)CQs), tuple- and equality-generating dependencies (TGDs and EGDs, respectively), negative constraints, the chase, and ontologies in Datalog+/–.

Databases and Queries. We assume (i) an infinite universe of *(data) constants* Δ (which constitute the "normal" domain of a database), (ii) an infinite set of *(labeled) nulls* Δ_N (used as "fresh" Skolem terms, which are placeholders for unknown values, and can thus be seen as variables), and (iii) an infinite set of variables $\mathcal{V}$ (used in queries, dependencies, and constraints). Different constants represent different values (*unique name assumption*), while different nulls may represent the same value. We assume a lexicographic order on $\Delta \cup \Delta_N$, with every symbol in Δ_N following all symbols in Δ. We denote by $\mathbf{X}$ sequences of variables $X_1, \dots, X_k$ with $k \geqslant 0$. We assume a *relational schema* $\mathcal{R}$, which is a finite set of *predicate symbols* (or simply *predicates*). A *term* t is a constant, null, or variable. An *atomic formula* (or *atom*) a has the form $P(t_1, ..., t_n)$, where P is an n-ary predicate, and $t_1, ..., t_n$ are terms.

A *database (instance)* D for a relational schema $\mathcal{R}$ is a (possibly infinite) set of atoms with predicates from $\mathcal{R}$ and arguments from Δ. A *conjunctive query (CQ)* over $\mathcal{R}$ has the form $Q(\mathbf{X}) = \exists \mathbf{Y}\, \Phi(\mathbf{X}, \mathbf{Y})$, where $\Phi(\mathbf{X}, \mathbf{Y})$ is a conjunction of atoms (possibly equalities, but not inequalities) with the variables $\mathbf{X}$ and $\mathbf{Y}$, and possibly constants, but without nulls. A *Boolean CQ (BCQ)* over $\mathcal{R}$ is a CQ of the form $Q()$, often written as the set of all its atoms, without quantifiers. Answers to CQs and BCQs are defined via *homomorphisms*, which are mappings $\mu \colon \Delta \cup \Delta_N \cup \mathcal{V} \to \Delta \cup \Delta_N \cup \mathcal{V}$ such that (i) $c \in \Delta$ implies $\mu(c) = c$, (ii) $c \in \Delta_N$ implies $\mu(c) \in \Delta \cup \Delta_N$, and (iii) μ is naturally extended to atoms, sets of atoms, and conjunctions of atoms. The set of all *answers* to a CQ $Q(\mathbf{X}) = \exists \mathbf{Y}\, \Phi(\mathbf{X}, \mathbf{Y})$ over a database D, denoted $Q(D)$, is the set of all tuples $\mathbf{t}$ over Δ for which there exists a homomorphism $\mu \colon \mathbf{X} \cup \mathbf{Y} \to \Delta \cup \Delta_N$ such that $\mu(\Phi(\mathbf{X}, \mathbf{Y})) \subseteq D$ and $\mu(\mathbf{X}) = \mathbf{t}$. The *answer* to a BCQ $Q()$ over a database D is *Yes*, denoted $D \models Q$, iff $Q(D) \neq \emptyset$.

Given a relational schema $\mathcal{R}$, a *tuple-generating dependency (TGD)* σ is a first-order formula of the form $\forall \mathbf{X} \forall \mathbf{Y}\, \Phi(\mathbf{X}, \mathbf{Y}) \to \exists \mathbf{Z}\, \Psi(\mathbf{X}, \mathbf{Z})$, where $\Phi(\mathbf{X}, \mathbf{Y})$ and $\Psi(\mathbf{X}, \mathbf{Z})$ are conjunctions of atoms over $\mathcal{R}$ (without nulls), called the *body* and the *head* of σ, denoted $body(\sigma)$ and $head(\sigma)$, respectively. Such σ is satisfied in a database D for $\mathcal{R}$ iff, whenever there exists a homomorphism h that maps the atoms of $\Phi(\mathbf{X}, \mathbf{Y})$ to atoms of D, there exists an extension h' of h that maps the atoms of $\Psi(\mathbf{X}, \mathbf{Z})$ to atoms of D. All sets of TGDs are finite here. Since TGDs can be reduced to TGDs with only single atoms in their heads, in the sequel, every TGD has w.l.o.g. a single atom in its head. A TGD σ is *guarded* iff it contains an atom in its body that contains all universally quantified variables of σ. The leftmost such atom is the *guard atom* (or *guard*) of σ.

Query answering under TGDs, i.e., the evaluation of CQs and BCQs on databases under a set of TGDs is defined as follows. For a database D for $\mathcal{R}$, and a set of TGDs

Σ on $\mathcal{R}$, the set of *models* of D and Σ, denoted $mods(D, \Sigma)$, is the set of all (possibly infinite) databases B such that (i) $D \subseteq B$ and (ii) every $\sigma \in \Sigma$ is satisfied in B. The set of *answers* for a CQ Q to D and Σ, denoted $ans(Q, D, \Sigma)$, is the set of all tuples $\mathbf{a}$ such that $\mathbf{a} \in Q(B)$ for all $B \in mods(D, \Sigma)$. The *answer* for a BCQ Q to D and Σ is *Yes*, denoted $D \cup \Sigma \models Q$, iff $ans(Q, D, \Sigma) \neq \emptyset$. Note that query answering under general TGDs is undecidable [3], even when the schema and TGDs are fixed [5]. Decidability of query answering for the guarded case follows from a bounded tree-width property. The data complexity of query answering in this case is P-complete.

Negative constraints (or simply *constraints*) γ are first-order formulas of the form $\forall \mathbf{X} \Phi(\mathbf{X}) \rightarrow \bot$, where $\Phi(\mathbf{X})$ (called the *body* of γ) is a conjunction of atoms (without nulls). Under the standard semantics of query answering of BCQs in Datalog+/– with TGDs, adding negative constraints is computationally easy, as for each constraint $\forall \mathbf{X} \Phi(\mathbf{X}) \rightarrow \bot$, we only have to check that the BCQ $\Phi(\mathbf{X})$ evaluates to false in D under Σ; if one of these checks fails, then the answer to the original BCQ Q is true, otherwise the constraints can simply be ignored when answering the BCQ Q.

Equality-generating dependencies (or *EGDs*) σ, are first-order formulas of the form $\forall \mathbf{X} \Phi(\mathbf{X}) \rightarrow X_i = X_j$, where $\Phi(\mathbf{X})$, called the *body* of σ, denoted $body(\sigma)$, is a (without nulls) conjunction of atoms, and X_i and X_j are variables from $\mathbf{X}$. Such σ is satisfied in a database D for $\mathcal{R}$ iff, whenever there exists a homomorphism h such that $h(\Phi(\mathbf{X})) \subseteq D$, it holds that $h(X_i) = h(X_j)$. Adding EGDs over databases with TGDs along with negative constraints does not increase the complexity of BCQ query answering as long as they are *non-conflicting* [6]. Intuitively, this ensures that, if the chase (see below) fails (due to strong violations of EGDs), then it already fails on the database D, and if it does not fail, then the EGDs do not have any impact on the chase with respect to query answering.

We usually omit the universal quantifiers in TGDs, negative constraints, and EGDs, and we implicitly assume that all sets of dependencies and/or constraints are finite.

The Chase. The *chase* was first introduced to enable checking implication of dependencies, and later also for checking query containment. By "chase", we refer both to the chase procedure and to its output. The TGD chase works on a database via so-called TGD *chase rules* (see [6] for an extended chase with also EGD chase rules).

TGD Chase Rule. Let D be a database, and σ a TGD of the form $\Phi(\mathbf{X}, \mathbf{Y}) \rightarrow \exists \mathbf{Z} \Psi(\mathbf{X}, \mathbf{Z})$. Then, σ is *applicable* to D if there exists a homomorphism h that maps the atoms of $\Phi(\mathbf{X}, \mathbf{Y})$ to atoms of D. Let σ be applicable to D, and h_1 be a homomorphism that extends h as follows: for each $X_i \in \mathbf{X}$, $h_1(X_i) = h(X_i)$; for each $Z_j \in \mathbf{Z}$, $h_1(Z_j) = z_j$, where z_j is a "fresh" null, i.e., $z_j \in \Delta_N$, z_j does not occur in D, and z_j lexicographically follows all other nulls already introduced. The *application of σ on D* adds to D the atom $h_1(\Psi(\mathbf{X}, \mathbf{Z}))$ if not already in D.

The chase algorithm for a database D and a set of TGDs Σ consists of an exhaustive application of the TGD chase rule in a breadth-first (level-saturating) fashion, which outputs a (possibly infinite) chase for D and Σ. Formally, the *chase of level up to 0* of D relative to Σ, denoted $chase^0(D, \Sigma)$, is defined as D, assigning to every atom in D the *(derivation) level* 0. For every $k \geqslant 1$, the *chase of level up to k* of D relative to Σ, denoted $chase^k(D, \Sigma)$, is constructed as follows: let $I_1, \ldots, I_n$ be all possible images of bodies of TGDs in Σ relative to some homomorphism such that (i)

$I_1, \ldots, I_n \subseteq chase^{k-1}(D, \Sigma)$ and (ii) the highest level of an atom in every I_i is $k - 1$; then, perform every corresponding TGD application on $chase^{k-1}(D, \Sigma)$, choosing the applied TGDs and homomorphisms in a (fixed) linear and lexicographic order, respectively, and assigning to every new atom the *(derivation) level* k. The *chase* of D relative to Σ, denoted $chase(D, \Sigma)$, is defined as the limit of $chase^k(D, \Sigma)$ for $k \to \infty$.

The (possibly infinite) chase relative to TGDs is a *universal model*, i.e., there exists a homomorphism from $chase(D, \Sigma)$ onto every $B \in mods(D, \Sigma)$ [6]. This implies that BCQs Q over D and Σ can be evaluated on the chase for D and Σ, i.e., $D \cup \Sigma \models Q$ is equivalent to $chase(D, \Sigma) \models Q$. For guarded TGDs Σ, such BCQs Q can be evaluated on an initial fragment of $chase(D, \Sigma)$ of constant depth $k \cdot |Q|$, which is possible in polynomial time in the data complexity.

Datalog+/– Ontologies. A *Datalog+/– ontology* $O = (D, \Sigma)$, where $\Sigma = \Sigma_T \cup \Sigma_E \cup \Sigma_{NC}$, consists of a database D, a set of TGDs Σ_T, a set of non-conflicting EGDs Σ_E, and a set of negative constraints Σ_{NC}. We say O is *guarded* iff Σ_T is guarded.

Example 1. A simple Datalog+/– ontology $O = (D, \Sigma)$ for movies (which is inspired by http://www.movieontology.org) is given below. Intuitively, D encodes that m_1, m_2, and m_3 are science fiction, documentary, and horror movies, and have the actor sets $\{a_1\}$, $\{a_2\}$, and $\{a_1, a_3\}$, respectively, while Σ encodes that documentary and adventure movies are movies, that science fiction and horror movies are adventure movies, and that every movie has an actor.

$$D = \{scifi\_movie(m_1),\ doc\_movie(m_2),\ thrilling(m_3),\ has\_actor(m_1, a_1),$$
$$has\_actor(m_2, a_2),\ has\_actor(m_3, a_1),\ has\_actor(m_3, a_3)\};$$
$$\Sigma = \{doc\_movie(T) \to movie(T),\ imag\_movie(T) \to movie(T),$$
$$scifi\_movie(T) \to imag\_movie(T),\ thrilling(T) \to imag\_movie(T),$$
$$movie(T) \to \exists A\ has\_actor(T, A)\}. \qquad \blacksquare$$

2.2   Preference Datalog+/–

We now recall the PrefDatalog+/– language introduced in [10], which is a generalization of Datalog+/– by preferences. For a more general survey of preferences in the context of databases, we refer the reader to [16]. The approach to define preferences logically was pursued in [7].

In the following, we denote by Δ_{Ont}, $\mathcal{V}_{Ont}$, and $\mathcal{R}_{Ont}$ the infinite sets of constants, variables, and predicates, respectively, of standard Datalog+/– ontologies as described in the previous section. For the preference extension, we assume a finite set of constants Δ_{Pref}, an infinite set of variables $\mathcal{V}_{Pref}$, and a finite set of predicates $\mathcal{R}_{Pref}$ such that $\mathcal{R}_{Pref} \subseteq \mathcal{R}_{Ont}$, $\Delta_{Pref} \subseteq \Delta_{Ont}$, and $\mathcal{V}_{Pref} \subseteq \mathcal{V}_{Ont}$. These sets give rise to corresponding Herbrand bases $\mathcal{H}_{Ont}$ and $\mathcal{H}_{Pref}$, as well as Herbrand universes $\mathcal{U}_{Ont}$ and $\mathcal{U}_{Pref}$, respectively, consisting of all constructible ground atoms and terms, where $\mathcal{H}_{Pref} \subseteq \mathcal{H}_{Ont}$.

A *preference formula* pf: $C(a, b)$ is a first-order formula defining a preference relation $\prec_C$ on $\mathcal{H}_{Pref}$ as follows: $a \prec_C b$ if $C(a, b)$. We call $C(a, b)$ the *condition* of pf, denoted $cond(pf)$. A *preference-based Datalog+/– (PrefDatalog+/–) ontology* $KB = (O, P)$ consists of a Datalog+/– ontology O and a set of preferences formulas P.

Example 2. Consider again the ontology $O = (D, \Sigma)$ from Example 1. Then, the movie preferences of a user may be represented by the following preference formulas in P, encoding that science-fiction movies are preferred over documentaries, and that documentaries are preferred over horror movies:

$$P : \begin{cases} C_1 : doc\_movie(X) \prec scifi\_movie(Y) \text{ if } X \neq Y; \\ C_2 : movie(X) \prec movie(Y) \text{ if } doc\_movie(X) \land scifi\_movie(Y) \land X \neq Y; \\ C_3 : movie(X) \prec movie(Y) \text{ if } thrilling(X) \land doc\_movie(Y) \land X \neq Y. \end{cases} \blacksquare$$

3   Group Preference Datalog+/–

In this section, we introduce G-PrefDatalog+/–, which generalizes PrefDatalog+/– to group preferences; we first define its syntax and then its semantics.

3.1   Syntax

Intuitively, we now have a group of n users, and each of these users has a set of preferences. Thus, in *G-PrefDatalog+/–*, rather being combined with the set of preferences of a single user (like in PrefDatalog+/–), the Datalog+/– ontology is now combined with a collection of sets of preferences (denoted *group profile*), one for each user. We first define the notion of group profile for $n \geq 1$ users as follows.

Definition 1 (Group Profile). A *group profile* $\mathcal{P} = (P_1, \ldots, P_n)$ for $n \geq 1$ users is a collection of n sets of preference formulas, one such set P_i for every user $i \in \{1, \ldots, n\}$.

We next define the notion of a G-PrefDatalog+/– ontology, which consists of a Datalog+/– ontology and a group profile for $n \geq 1$ users.

Definition 2 (G-PrefDatalog+/–). A *group preference-based Datalog+/– (G-PrefDatalog+/–) ontology* $KB = (O, \mathcal{P})$ consists of a Datalog+/– ontology O and a group profile $\mathcal{P} = (P_1, \ldots, P_n)$ for $n \geq 1$ users. For every user $i \in \{1, \ldots, n\}$, we denote by $User(KB, i)$ the PrefDatalog+/– ontology $KB_i = (O, P_i)$.

The following notion aggregates single-user preferences via counting preference pairs. Let $a, b \in \mathcal{H}_{Pref}$ such that (i) $O \models a, b$ and (ii) $cond(pf_i)(a, b)$ for some $i \in \{1, \ldots, n\}$, $pf_i \in P_i$, and P_i in $\mathcal{P}$. Then, $\#(a, b)$ is the number of users that prefer a over b.

Example 3. Consider again the Datalog+/– ontology $O = (D, \Sigma)$ from Example 1. Then, a G-PrefDatalog+/– ontology $KB = (O, \mathcal{P})$ for $n = 3$ users is given by the following group profile $\mathcal{P} = (P_1, P_2, P_3)$:

$$P_1 : \begin{cases} C_{1,1} : doc\_movie(X) \prec scifi\_movie(Y) \text{ if } X \neq Y; \\ C_{1,2} : movie(X) \prec movie(Y) \text{ if } doc\_movie(X) \land scifi\_movie(Y) \land X \neq Y; \\ C_{1,3} : movie(X) \prec movie(Y) \text{ if } thrilling(X) \land doc\_movie(Y) \land X \neq Y; \end{cases}$$

$$P_2 : \begin{cases} C_{2,1} : movie(m_1) \prec movie(m_2) \text{ if } \top; \end{cases}$$

$$P_3 : \begin{cases} C_{3,1} : movie(X) \prec movie(Y) \text{ if } doc\_movie(X) \land scifi\_movie(Y) \land X \neq Y; \\ C_{3,2} : movie(X) \prec movie(Y) \text{ if } has\_actor(X, A_1) \land has\_actor(Y, A_2) \land \\ \qquad A_1 = a_3 \land A_2 = a_2 \land X \neq Y. \end{cases}$$

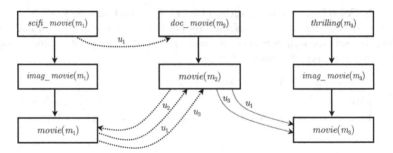

Fig. 1. G-PrefDatalog+/– representing preferences semantically, where the chase is augmented by preferences user preferences edges (dotted edges)

Observe also that $\#(movie(m_1), movie(m_2)) = 2$, since $movie(m_1)$ is preferred over $movie(m_2)$ by users 1 and 2, while $\#(scifi\_movie(m_1), doc\_movie(m_2)) = 1$, since scifi_movie$(m_1)$ is preferred over $doc\_movie(m_2)$ by user 2.   ■

3.2 Semantics

The semantics of G-PrefDatalog+/– is an extension of the semantics of PrefDatalog+/–. Given a G-PrefDatalog+/– ontology $KB = (O, \mathcal{P})$ for $n \geqslant 1$ users, we say that user $i \in \{1, \ldots, n\}$ prefers b over a in KB, denoted $KB \models a \prec_i b$, iff (i) $O \models a, b$ and (ii) $cond(pf_i)(a, b)$ for some $pf_i \in P_i$.

Intuitively, the consequences of a G-PrefDatalog+/– ontology $KB = (O, \mathcal{P})$ are computed in terms of the chase for the classical Datalog+/– ontology O, and the collection of sets of preference formulas $\mathcal{P}$ describes the preference relation over pairs of atoms in $\mathcal{H}_{Ont}$. The extended chase forest for PrefDatalog+/– can thus be extended to groups by labeling the preference edges with the ids of the users (see Fig. 1).

4 Query Answering Using Group Preferences

In this section, we introduce aggregation strategies that are suitable for sets of partially ordered preferences. We have analyzed strategies developed for quantitative preferences [12] (which yield a weak order) and adapted them to our model focusing on avoiding the addition of extra information (i.e., without introducing new preferences). The study of properties of this strategies (e.g., from social choice theory) is out of the scope of this paper. These aggregation strategies are grouped into two categories: the first is called *collapse to single user* (CSU), and it reduces the group modeling problem to a single user problem by creating a virtual user that is constructed by aggregating the preferences of the individuals from the group; the second class of strategies is called *score aggregation*, and it first computes rankings relative to a query for each user, and then aggregates these rankings using a voting-based strategy. Moreover, for each of the two classes of strategies, we provide algorithms for answering disjunctive k-rank queries and complexity results for them. The choice for k-rank comes as a natural extension of skyline where incomparable elements appear.

To define answers to queries, we adopt the following notion from classical logic. A *substitution* is a function from variables to variables and constants. Two sets S and T *unify* via a substitution θ iff $\theta S = \theta T$, where θA denotes the application of θ to all variables in all elements of A. The *most general unifier* (mgu) is a unifier θ such that for all other unifiers ω, there exists a substitution σ such that $\omega = \sigma\theta$.

4.1 Collapse to Single User

In the collapse to single user (CSU) approach, the preference relations for each user are taken into account in the generation of a new relation that encodes the dominant preferences. This single user preference relation is then used to answer queries.

To compute k-rank answers to queries Q over a G-PrefDatalog+/– ontology $KB = ((D, \Sigma), \mathcal{P})$, we use the chase forest for D, Σ, and Q [6]. This is defined as the directed graph consisting of: (i) for every $a \in D$, one node labeled with a, and (ii) for every node labeled with $a \in chase(D, \Sigma)$ and for every atom $b \in chase(D, \Sigma)$ that is obtained from a and possibly other atoms by a one-step application of a TGD $\sigma \in \Sigma$ with a as guard, one node labeled with b along with an edge from the node labeled with a. We propose a *group preference augmented chase* forest, comprised of the necessary finite part of the chase forest relative to a given query that is augmented with two additional kinds of labeled edges: *individual user preference edges* (*u-edges*) and *aggregated group preference edge* (*g-edges*). A u-edge (a, b), with $a, b \in chase(D, \Sigma)$, occurs iff $b \prec_P^u a$. The edge is labeled with u representing the identity of the user that makes the claim. The g-edge from a to b is an edge labeled with $\#(a, b)$. We apply the *CSU-PrefChase*(KB, Q) algorithm that constructs the g-edges as follows: for every u-edge $a \prec_P^u b$, we check if there is a g-edge between a and b. If there is no such edge, then we add the g-edge (b, a), make a marked, and set the label to 1. If there is a g-edge (b, a), then we increase the label by 1. If there is a g-edge (a, b) and *label* > 1, then we decrease the label by 1, otherwise we delete the g-edge, and make a unmarked whenever a has no other incoming g-edge (see Figure 2). Finally, any cycles that appeared (this can happen even when none of the $\prec_P^u$ relations have cycles) need to be removed—there are many different ways in which this can be done, such as preferring edges that have higher labels. The following result shows that this structure can be computed in polynomial time in the data complexity.

Theorem 1. *If $KB = (O, \mathcal{P})$ is a guarded G-PrefDatalog+/– ontology and Q a DAQ, CSU-PrefChase(KB, Q) can be computed in time $O(n^2)$ in the data complexity.*

Proof (sketch). The *CSU-PrefChase*(KB, Q) structure consists of the chase forest for (D, Σ) with additional u-edges. The construction of the classical chase forest can be done in time $O(n)$ in the data complexity for guarded TGDs [6]. Now, computing the u-edges can be done in time $O(n^2)$ in the data complexity, as this involves, for each user u, iterating through all pairs (a_i, a_j) of nodes and testing in turn all formulas $pf_u \in P_u$ to see if $\models Cond(pf_u)(a_i, a_j)$—note that $|\mathcal{P}|$ is fixed in the data complexity. While computing the u-edges, we construct the g-edges and mark the nodes; finally, removing cycles and computing the transitive closure is also accomplished in $O(n^2)$ time. □

We now define k-rank answers based on the CSU strategy; we restrict the query language to DAQs to ensure tractability.

Definition 3 (*k*-**Rank Answers**). Consider a G-PrefDatalog+/– ontology $KB = (O, P)$ and a DAQ $Q(\mathbf{X}) = q_1(\mathbf{X}_1) \vee \ldots \vee q_n(\mathbf{X}_n)$. The set of 1-rank answers to Q is defined as: $\{\theta q_i \mid O \models \theta q_i \text{ and } \nexists \theta' \text{ such that } O \models \theta' q_j \text{ and there is no edge in } CSU\text{-}PrefChase(KB, Q) \text{ of the form } (\theta' q_j, \theta q_i)\}$, where θ, θ' are mgu's for the variables in $\mathbf{X}$. A *k*-rank answer to Q is defined for transitive relations as a sequence of maximal length of mgu's for the variables in $\mathbf{X}$: $S = \langle \theta_1, \ldots \theta_{k'} \rangle$ built by subsequently appending the 1-rank answers to Q, removing these atoms from consideration, and repeating the process until either $S = k$ or no more 1-rank answers to Q remain.

An Algorithm for Answering *k*-Rank Queries. Algorithm *k*-rank (Algorithm 1) begins by computing the group preference-augmented chase forest relative to the input ontology and query. The main while-loop iterates through the process of computing the 1-rank answers to Q by using the markings in the forest. Once the loop is finished, the algorithm returns the first k results. In each iteration of the loop, the node markings have done almost all of the work towards answering the query; all the algorithm needs to do is to go through the structure and find the nodes whose labels satisfy Q and, if unmarked, add them to the output. Once nodes are in the output list, the algorithm eliminates these nodes and the associated edges from the augmented group chase forest. If there are marked nodes with no incoming g-edges (after eliminating the nodes and edges) these nodes are changed to unmarked. The variable *noEdgeTop* deals with the nodes that do not have any preference edges. Such nodes give rise to two alternatives: to consider them as top favorite, *noEdgeTop* = *true* (since no other item is preferred to them) or to consider them as least favorite. Note that this algorithm returns one possible answer—incomparable elements may produce different answers.

Example 4. Consider the running example and the group forest depicted in Figure 2 for $Q(X) = movie(X) \vee thrilling(X)$. The call *k-rank*$(KB, Q(X), 1, false)$ returns $\langle movie(m_1) \rangle$; note, however, that if *noEdgeTop* is true, then $\langle thrilling(m_3) \rangle$ is also a 1-rank answer to Q. Call *k-rank*$(KB, Q(X), 2, true)$ returns $\langle movie(m_1), thrilling(m_3) \rangle$ (the order can be reversed, as they are incomparable), and *k-rank*$(KB, Q(X), 3, false)$ returns $\langle movie(m_1), movie(m_2), movie(m_3) \rangle$.   ∎

The following result establishes correctness and complexity for Algorithm *k*-rank.

Theorem 2. *Let KB be a G-PrefDatalog+/– ontology, Q be a DAQ, and $k \geqslant 0$. Then, (i) Algorithm k-rank correctly computes a k-rank answer to Q, and (ii) if CSU-Pref-Chase(KB, Q) is computable in $O(n^2)$ time in the data complexity, then Algorithm k-rank runs in time $O(n^2)$ in the data complexity.*

As a consequence of Theorem 2, we have the following:

Corollary 1. *Let $KB = (O, P)$ be a G-PrefDatalog+/– ontology, Q be a disjunctive query, and $k \geqslant 0$. If O is a guarded Datalog+/– ontology, then a k-rank answer to Q can be computed in polynomial time in the data complexity.*

4.2   Score Aggregation

Instead of adopting the approach described in the previous section, specific strategies can be used to combine the answers to k-rank queries computed individually for each

---

Algorithm 1. k-rank$(KB = ((D, \Sigma), \mathcal{P}), Q, k, noEdgeTop)$

Input: G-PrefDatalog+/− ontology KB and DAQ $Q(X)$, $k \geqslant 0$, $noEdgeTop$.
Output: k-rank answers $\{a_1, ..., a_{k'}\} \in Result$ to Q.
```
 1: Result = ⟨⟩;   i:= k;
 2: Chase:= CSU-PrefChase(KB, Q);
 3: while i > 0 do
 4:     GA:= ∅;                          ▷ Initialize GA as an empty set of ground atoms
 5:     for each atom a labeling node v ∈ Chase do
 6:         if a ⊨ Q and v is unmarked then
 7:             if (v has has at least one outgoing g-edge) or noEdgeTop then
 8:                 GA:= GA ∪ {a};
 9:     Result:= Result ∪ GA;
10:     Remove GA from Chase along with all associated edges;
11:     if no g-edges remain then
12:         noEdgeTop:= true;
13:     for each node v ∈ Chase do
14:         if v is marked and has no incoming g-edges then
15:             make v unmarked;
16:     i:= i − |GA|;
17: return truncate(Result, k).
```

---

user based on voting mechanisms from the social choice literature. The main difference is that, in the CSU approach, a single k-ranking is computed from a preference relation obtained from all the users individual preferences. We consider the following strategies:

– *Plurality Voting.* This strategy calculates the top preferred items for each user independently. The items' frequency for all the users are summed up, and the items with the highest number of votes win.

– *Least Misery Strategy.* This strategy first removes from consideration the elements that are the least preferred by each of the users, and then applies plurality voting. The idea behind it is that a group is as happy as its least happy member. A disadvantage of this approach is that a minority opinion can dominate the result: if an element is highly preferred by all users except one, then this element will never be chosen.

– *Average Without Misery Strategy.* This strategy is a generalization of the *Least Misery Strategy* in which the first t least preferred elements for each member are removed.

– *Fairness Strategy.* This strategy is often applied when people want to fairly divide a set of items: each person chooses in turn, until everyone has made one choice. Next, everybody chooses a second item, often starting with the person who had to choose last in the previous round. Using this strategy, top items from all individuals are selected.

– *Dictatorship.* With this strategy, groups are dominated by one person—the dictator— who casts the only vote that counts.

An Algorithm for Aggregated k-Rank Query Answering. We now present an algorithm that integrates the aggregation strategies presented above to obtain an aggregated k-rank, leveraging Algorithm k-rank (Algorithm 1). Algorithm Agg-k-rank

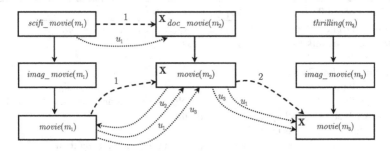

Fig. 2. Collapse to single user: *CSU-PrefChase*(KB, *movie*(X)). Dashed arrows are the g-edges, labeled with a natural number (how many users prefer node a over node b minus how many users prefer node b over a). The dotted edges are the u-edges, labeled with the id of the user that claims the preference order). The marks in the upper left corner represent marked nodes.

(Algorithm 2) works as follows: first, if *Misery* is *true*, it computes all the nodes that are undesired for each user; if *useFairness* is *true*, it initializes a vector of iterators that it will use to compute the next 1-rank answers to Q for each user. The *nextElement* procedure computes the next i-rank for the user as done in Algorithm 1 but only considering elements that have not already been added to *Result* and do not appear in *Misery*. Finally, if *useFairness* is *false*, the algorithm computes the k-rank answers to Q for each user and applies plurality voting. We define informally the following variables, used to determine the strategy of the algorithm:

useFairness: Boolean variable indicating whether the fairness strategy is used or not.

useMisery: if an item is t-least liked by a member then this item should not be in the top-k for the overall group. To calculate the least t-liked for each for the member, we create first $KB'(O', P')$ from $KB(O, P)$, that contains the same $O' = O$ but the P' contains the set P with inverse relationships (if we have $a \prec b$ if $cond(a, b) \in P$ for some user, then we have $b \prec a$ if $cond(a, b) \in P'$). We compute the k-rank for each member u of the group, $C_u = k\text{-}rank(User(KB', u), Q, t, false)$—the last parameter indicates that if a node does not have any incoming edges from user u, then this node is not part of the solution. At the end, C_u contains the least preferred t-nodes.

t: threshold for *useMisery* $= true$ (the t-least favorite items should not be part of top-k).

Algorithm *Agg-k-rank* can thus be used to compute result with respect to all of the aggregation strategies discussed above:

– For Plurality Voting, we call *Agg-k-rank*($KB, Q, k, false, false, 0$); Figure 3 provides an example for the query *movie*(X) $\vee$ *thrilling*(X); for $k = 1$, the algorithm returns *movie*(m_1) (*thrilling*(m_3) is also a possible answer).

– For Least Misery, we call *Agg-k-rank*($KB, Q, k, true, false, 1$); the last parameter is set to t for the Average Without Misery strategy; in the running example, for $k = 2$, one possible answer is $\langle thrilling(m_3), movie(m_1) \rangle$. The first atom is in the 2-rank for each user, and the second is in the 2-rank for users u_1 and u_3.

– For Fairness, we call *Agg-k-rank*($KB, Q, k, false, true, 0$); the top items for each individual are selected in turn in line 16 and added to the result until k answers are obtained.

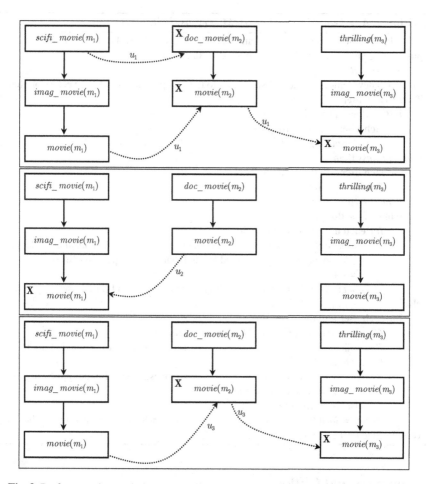

Fig. 3. Preference chase relative to Plurality voting and query $movie(X) \lor thrilling(X)$

In the running example, for $k = 3$, we get $\langle movie(m_1), movie(m_2), thrilling(m_3)\rangle$, assuming that users went in order $\langle u_1, u_2, u_3\rangle$.

– For Dictatorship, we must ignore every user's preferences except for that of the dictator (called d), so we call $Agg\text{-}k\text{-}rank(User(KB, d), Q, k, false, true, 0)$. In our case, if u_2 is the dictator, then $movie(m_2)$, $movie(m_3)$, or $thrilling(m_3)$ can all be answers.

Corollary 2. *Let $KB = (O, \mathcal{P})$ be a G-PrefDatalog+/– ontology, Q be a DAQ, and $k \geqslant 0$. If O is a guarded Datalog+/– ontology, then an Aggregated-k-rank answer to Q can be computed in polynomial time in the data complexity.*

Proof (sketch). Generating KB' can be done in polynomial time. The rest of the algorithm computes the k-rank answers for each user, which we showed to be possible in polynomial time. $\square$

Algorithm 2. *Agg-k-rank*(*KB*, *Q*, *k*, *useMisery*, *useFairness*, *t*)

Input: G-PrefDatalog+/− ontology *KB* and DAQ $Q(X)$, $k \geqslant 0$, *useMisery*, *useFairness*, *t*.
Output: Aggregated k-answers $\{a_1, ..., a_{k'}\} \in Result$ to Q.

1: *Misery*:= ∅;   *Result*:=⟨⟩;
2: *PrefIter*:= new (empty) vector, of size equal to the group, of iterators over vectors of atoms
3: **if** *useMisery* **then**
4:      $KB' = Inverse(KB)$;
5:      **for each** User u in the Group **do**
6:          C_u:= *k-rank*(*User*(KB', u), Q, t, *false*);   *Misery*:= *Misery* ∪ C;
7: **if** *useFairness* **then**
8:      **for each** user u in the group **do**
9:          *PrefIter*[u]:= Initialize k-rank iterator
10:     *done*:= *false*; i:= 0;
11:     **while** !*done* **do**
12:         **for each** user u in the group **do**
13:             *Result*:= *Result* ∪ *PrefIter*[u].*nextElement*;   ▷ *nextElement* considers only
                    elements not previously added and that do not appear in *Misery*
14:             i:= $i + 1$;
15:             *done*:= ($i = k$) or iterators in *PrefIter* reached the end for all users;
16: **else**
17:     **for each** user u in the group **do**
18:         C_u:= *k-rank*(*User*(*KB*, u), Q, k, *true*) − *Misery*;
19:         $Res = ∅$;   ▷ set of ⟨*node*, *vote*⟩ (ground atom, integer)
20:         **for each** $c \in C_u$ **do**
21:             **if** $c \in Res.nodes$ **then**   ▷ takes the set of nodes appearing in *Res*
22:                 $Res.node[c]$:= $Res.node[c] + 1$;
23:             **else** *Res*:= *Res* ∪ {⟨c, 1⟩};
24:         *Result*:= *Sort*(*Res*).*nodes*;   ▷ Sort by vote and return the nodes
25: **return** *truncate*(*Result*, k).

5   Summary and Outlook

In this paper, we have proposed an extension of the Datalog+/− ontology language that allows for dealing with partially ordered preferences of groups of users. To our knowledge, this is the first combination of ontology languages with group preferences. We have focused on answering k-rank queries in this context. In detail, we have presented different strategies to compute group preferences as an aggregation of the preferences of a collection of single users. We have then provided algorithms to answer k-rank queries for DAQs (disjunctions of atomic queries) under these group preferences. We have shown that such DAQ answering in Datalog+/− can be done in polynomial time in the data complexity, as long as query answering can also be done in polynomial time (in the data complexity) in the underlying classical ontology.

Current and future work involves implementing and testing the G-PrefDatalog+/− framework. Furthermore, we want to explore which of the aggregation strategies is similar to human judgment and thus well-suited as a general default aggregation strategy for search and query answering in the Social Semantic Web.

Acknowledgments. This work was supported by the UK EPSRC grant EP/J008346/1 "PrOQAW: Probabilistic Ontological Query Answering on the Web", the ERC (FP7/ 2007-2013)/ERC grant 246858 ("DIADEM"), and by a Yahoo! Research Fellowship, and by a Google European Doctoral Fellowship.

References

1. Ackerman, M., Choi, S.Y., Coughlin, P., Gottlieb, E., Wood, J.: Elections with partially ordered preferences. Public Choice (2012)
2. Amer-Yahia, S., Roy, S.B., Chawla, A., Das, G., Yu, C.: Group recommendation: Semantics and efficiency. Proc. VLDB Endow. 2(1), 754–765 (2009)
3. Beeri, C., Vardi, M.Y.: The implication problem for data dependencies. In: Even, S., Kariv, O. (eds.) ICALP 1981. LNCS, vol. 115, pp. 73–85. Springer, Heidelberg (1981)
4. Brafman, R.I., Domshlak, C.: Preference handling — An introductory tutorial. AI Mag. 30(1), 58–86 (2009)
5. Calì, A., Gottlob, G., Kifer, M.: Taming the infinite chase: Query answering under expressive relational constraints. In: Proc. KR 2008, pp. 70–80. AAAI Press (2008)
6. Calì, A., Gottlob, G., Lukasiewicz, T.: A general Datalog-based framework for tractable query answering over ontologies. J. Web Sem. 14, 57–83 (2012)
7. Chomicki, J.: Preference formulas in relational queries. ACM Trans. Database Syst. 28(4), 427–466 (2003)
8. Lang, J., Pini, M.S., Rossi, F., Salvagnin, D., Venable, K.B., Walsh, T.: Winner determination in voting trees with incomplete preferences and weighted votes. Auton. Agent. Multi-Ag. 25(1), 130–157 (2012)
9. Linden, G., Smith, B., York, J.: Industry report: Amazon.com recommendations: Item-to-item collaborative filtering. IEEE Distributed Systems Online 4(1) (2003)
10. Lukasiewicz, T., Martinez, M.V., Simari, G.I.: Preference-based query answering in Datalog+/– ontologies. In: Proc. IJCAI (in press, 2013)
11. Manoj, M., Jacob, E.: Information retrieval on internet using meta-search engines: A review. Journal of Scientific and Industrial Research 67(10), 739–746 (2008)
12. Masthoff, J.: Group modeling: Selecting a sequence of television items to suit a group of viewers. User Modeling and User-Adapted Interaction 14(1), 37–85 (2004)
13. Ntoutsi, I., Stefanidis, K., Norvag, K., Kriegel, H.-P.: gRecs: A group recommendation system based on user clustering. In: Lee, S.-g., Peng, Z., Zhou, X., Moon, Y.-S., Unland, R., Yoo, J. (eds.) DASFAA 2012, Part II. LNCS, vol. 7239, pp. 299–303. Springer, Heidelberg (2012)
14. Pattanaik, P.K.: Voting and Collective Choice: Some Aspects of the Theory of Group Decision-making. Cambridge University Press (1971)
15. Pini, M.S., Rossi, F., Venable, K.B., Walsh, T.: Aggregating partially ordered preferences. J. Log. Comput. 19(3), 475–502 (2009)
16. Stefanidis, K., Koutrika, G., Pitoura, E.: A survey on representation, composition and application of preferences in database systems. ACM TODS 36(3), 19:1–19:45 (2011)
17. Taylor, A.D.: Social Choice and the Mathematics of Manipulation. Cambridge University Press (2005)
18. Wooldridge, M.: An Introduction to Multiagent Systems. Wiley (2009)

Reasoning with Semantic-Enabled Qualitative Preferences

Tommaso Di Noia[1], Thomas Lukasiewicz[2], and Gerardo I. Simari[2]

[1] Dipartimento di Ingegneria Elettrica e dell'Informazione, Politecnico di Bari, Italy
t.dinoia@poliba.it
[2] Department of Computer Science, University of Oxford, UK
firstname.lastname@cs.ox.ac.uk

Abstract. Personalized access to information is an important task in all real-world applications where the user is interested in documents, items, objects or data that match her preferences. Among qualitative approaches to preference representation, CP-nets play a prominent role: with their clear graphical structure, they unify an easy representation of user desires with nice computational properties when computing the best outcome. In this paper, we explore how to reason with CP-nets in the context of the Semantic Web, where preferences are linked to formal ontologies. We show how to compute Pareto optimal outcomes for a semantic-enabled CP-net by solving a constraint satisfaction problem, and we present complexity results related to different ontological languages.

1 Introduction

During the recent years, several revolutionary changes are taking place on the classical Web. First, the so-called Web of Data is more and more being realized as a special case of the Semantic Web. Second, as a part of the Social Web, users are acting more and more as first-class citizens in the creation and delivery of contents on the Web. The combination of these two technological waves is called the *Social Semantic Web* (or also *Web 3.0*), where the classical Web of interlinked documents is more and more turning into (i) semantic data and tags constrained by ontologies, and (ii) social data, such as connections, interactions, reviews, and tags.

The Web is thus shifting away from data on linked Web pages towards less such interlinked data in social networks on the Web relative to underlying ontologies. This requires new technologies for search and query answering, where the ranking of search results is not based on the link structure between Web pages anymore, but on the information available in the Social Semantic Web, in particular, the underlying ontological knowledge and the preferences of the users. Given a query, these latter play a fundamental role when a crisp yes/no answer is not enough to satisfy a user's needs, since there is a certain degree of uncertainty in possible answers [8].

We have two main ways of modeling preferences: (a) *quantitative preferences* are associated with a number representing their worth or they are represented as an ordered set of objects (e.g., "my preference for WiFi connection is 0.8" and "my preference for cable connection is 0.4"), while (b) *qualitative preferences* are related to each other via pairwise comparisons (e.g., "I prefer WiFi over cable connection").

W. Liu, V.S. Subrahmanian, and J. Wijsen (Eds.): SUM 2013, LNAI 8078, pp. 374–386, 2013.

Actually, the qualitative approach is a more natural way of representing preferences, since humans are not very comfortable in expressing their "wishes" in terms of a numerical value. To have a quantitative representation of her preferences, the user needs to explicitly determine a value for a large number of alternatives usually described by more than one attribute. It is generally much easier to provide information about preferences as pairwise qualitative comparisons [8]. One of the most powerful qualitative frameworks for preference representation and reasoning are perhaps CP-nets [3]. They are a graphical language that unifies an easy representation of user desires with nice computational properties when computing the best outcome. Most of the work done on CP-nets and more generally on preference representation mainly deals with a propositional representations of preferences. In this paper, we propose an enhancement of CP-nets by adding ontological information associated to preferences. This is an initial step towards a new type of semantic search techniques able to go far beyond PageRank and similar algorithms. They will be able to exploit social information, e.g., information coming from social networks, and model it as semantic-enabled user preferences.

The rest of this paper is organized as follows. In Section 2, we briefly recall CP-nets and description logics (DLs), which are the two main technologies that we use in our approach. Section 3 introduces ontological CP-nets, i.e., CP-nets enriched with ontological descriptions, and it describes how to compute optimal outcomes. In Sections 4 and 5, we provide complexity results and discuss related work, respectively. Finally, we give a summary of the results in this paper and an outlook on future work.

2   Preliminaries

We start by introducing some notions and formalisms that are necessary to present our framework. Given a set of variables $\mathcal{V}$, an *outcome* is an assignment to all the variables in $\mathcal{V}$. A *preference relation* $\succeq$ is a total pre-order over the set of all outcomes. We write $o_1 \succ o_2$ (resp., $o_1 \succeq o_2$) to denote that o_1 is strictly preferred (resp., strictly or equally preferred) to o_2. If $o_1 \succ o_2$, then o_2 is *dominated* by o_1. If there is no outcome o such that $o \succ o_1$, then o_1 is *undominated*.

A *conditional preference* is an expression $(\alpha \succ \beta \mid \gamma)$, where α, β, and γ are formulas. It intuitively means that "*given* γ, *I prefer* α *over* β". In the following, we often write $(\alpha \mid \gamma)$ and $(\neg \alpha \mid \gamma)$ to denote $(\alpha \succ \neg \alpha \mid \gamma)$ and $(\neg \alpha \succ \alpha \mid \gamma)$, respectively, and we use $\tilde{\alpha}$ to represent one of the elements among α and $\neg \alpha$.

2.1   CP-Nets

Conditional preferences networks (CP-nets) [3] are a formalism to represent and reason about qualitative preferences. They are a compact but powerful language, which allows the specification of preferences based on the notion of *conditional preferential independence (CPI)* [13]. Let $A, B \in \mathcal{V}$ be two variables and $\mathcal{R} \subset \mathcal{V}$ be a set of variables such that A, B, and $\mathcal{R}$ partition $\mathcal{V}$, and $Dom(A)$, $Dom(B)$, and $Dom(\mathcal{R})$ represent all possible assignments for A, B, and all the variables in $\mathcal{R}$, respectively. We say that A is *conditionally preferentially independent (CPI)* of B given an assignment $\rho \in Dom(\mathcal{R})$ iff, for every $\alpha_1, \alpha_2 \in Dom(A)$ and $\beta_1, \beta_2 \in Dom(B)$, we have that $\alpha_1\beta_1\rho \succ \alpha_2\beta_1\rho$ iff $\alpha_1\beta_2\rho \succ \alpha_2\beta_2\rho$. Here, $\succ$ represents the preference order on assignments to sets of variables. CP-nets are a graphical language to model CPI statements. Formally, a *CP-net* N consists of a directed graph G over a set of variables $\mathcal{V} = \{A_i \mid i \in \{1,\dots,n\}\}$ as nodes, along with a conditional preference table $CPT(A_i)$ for every variable A_i, which contains a preference for each pair of values of A_i conditioned to all possible assignments to the parents of A_i in G. Given a CP-net N, we denote by $\mathcal{CPT}_i$ the set of all conditional preferences represented by $CPT(A_i)$, and we define $\mathcal{CPT}_N = \{\mathcal{CPT}_i \mid i \in \{1,\dots,n\}\}$. An example of a CP-net (over only binary variables) is shown in Fig. 1.

Example 1 (Hotel). A CP-net for representing preferences for hotel accommodations is shown in Fig. 1. Note that this is a toy example, whose purpose is to show the representational expressiveness of CP-nets in modeling user profiles. In this simple case, we use five binary variables of the following meaning:

  α_1: the hotel is located near the sea;
  α_2: the hotel is located in the city center;
  α_3: scooters for rent;
  α_4: parking available;
  α_5: bikes for rent.

Looking at $CPT(A_3)$, e.g., we see that the user prefers to have a scooter for rent in case the hotel is located near the sea or in the city center, and that she prefers to not have a scooter for rent if the hotel is neither near the sea nor in the city center. ∎

To establish an order among possible outcomes of a CP-net, we introduce the notion of *worsening flip*, which is a change in the value of a variable that worsens the satisfaction of user preferences. As an example, if we consider the CP-net in Fig. 1, we have a worsening flip moving from $\alpha_1\alpha_2\alpha_3\neg\alpha_4\neg\alpha_5$ to $\alpha_1\alpha_2\neg\alpha_3\neg\alpha_4\neg\alpha_5$. Indeed, given $\alpha_1\alpha_2\neg\alpha_4\neg\alpha_5$, we see that α_3 is preferred over $\neg\alpha_3$. Based on this notion, we can state that $\alpha_1\alpha_2\alpha_3\neg\alpha_4\neg\alpha_5 \succ \alpha_1\alpha_2\neg\alpha_3\neg\alpha_4\neg\alpha_5$.

Given a CP-net, the two main queries that one may ask are:

 – *dominance query*: given two outcomes o_1 and o_2, does $o_1 \succ o_2$ hold?
 – *outcome optimization*: compute an optimal (i.e., undominated) outcome for the preferences represented by a given CP-net.

Given an acyclic CP-net, one can compute the best outcome in linear time. The algorithm just follows the order among variables represented by the graph and assigns values to the variables A_i from top to bottom satisfying the preference order in the corresponding $CPT(A_i)$. For example, in the CP-net in Fig. 1, the optimal outcome is $\alpha_1\alpha_2\alpha_3\alpha_4\alpha_5$. Finding optimal outcomes in cyclic CP-nets is NP-hard.

2.2 Constrained CP-Nets

In constrained CP-nets [18,4], constraints among variables are added to the basic formalism of CP-nets. Adding constraints among variables may reduce the set of possible outcomes $\mathcal{O}$. The approach to finding the optimal outcomes proposed in [18] relies on a

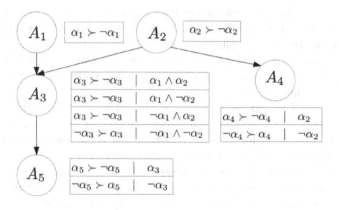

Fig. 1. An example of a CP-net over five binary variables

reduction of the preferences represented in the CP-net to a set of hard constraints (which can be represented in clause form for binary variables), taking into account the variables occurring in the preferences. Given a CP-net N and a set of constraints C, an outcome o is *feasible* iff it satisfies all the constraints in C. A feasible outcome is *Pareto optimal* [4] iff it is undominated. These optimal outcomes now correspond to the solutions of a constraint satisfaction problem. For binary variables, given a conditional preference $(\alpha_{n+1} \mid \bigwedge_{i=1...n} \alpha_i)$, the corresponding constraint is the clause

$$\bigwedge_{i=1...n} \alpha_i \to \alpha_{n+1}. \tag{1}$$

Given a CP-net N and a set of constraints C, a feasible Pareto optimal outcome is exactly an assignment satisfying the corresponding set of clauses and all constraints in C. We refer the reader to [18,4] for further details, including examples.

2.3 Description Logics

Description logics (DLs) are a family of formalisms that are well-established in knowledge representation and reasoning [1]. We assume that the reader is familiar with DLs, and we here only briefly recall the elements that we use in the presented approach. Basic elements of DLs are *concept names*, *role names*, and *individuals*. DLs are usually endowed with a model-theoretic formal semantics. A semantic *interpretation* is a pair $\mathcal{I} = (\Delta^{\mathcal{I}}, \cdot^{\mathcal{I}})$, where $\Delta^{\mathcal{I}}$ represents the *domain*, and $\cdot^{\mathcal{I}}$ is the *interpretation function*. This function maps every concept name to a subset of $\Delta^{\mathcal{I}}$, and every role name to a subset of $\Delta^{\mathcal{I}} \times \Delta^{\mathcal{I}}$. The symbols $\top$ and $\bot$ are used, respectively, to represent the most generic concept and the most specific concept. Hence, their formal semantics correspond to $\top^{\mathcal{I}} = \Delta^{\mathcal{I}}$ and $\bot^{\mathcal{I}} = \emptyset$. The interpretation function also applies to complex concepts: $(\alpha \sqcap \beta)^{\mathcal{I}} = \alpha^{\mathcal{I}} \cap \beta^{\mathcal{I}}$, $(\alpha \sqcup \beta)^{\mathcal{I}} = \alpha^{\mathcal{I}} \cup \beta^{\mathcal{I}}$, $(\neg \alpha)^{\mathcal{I}} = \Delta^{\mathcal{I}} \setminus \alpha^{\mathcal{I}}$. In the following, we use $\alpha \to \beta$ and $\alpha \leftrightarrow \beta$ to denote $\neg \alpha \sqcup \beta$ and $(\alpha \to \beta) \sqcap (\beta \to \alpha)$, respectively. Given a generic formula α, we use $\mathcal{I} \models \alpha$ to say that $\alpha^{\mathcal{I}} \neq \emptyset$.

Here, we use $\mathcal{T}$ (for "terminology") to denote a DL ontology, i.e., a set of axioms of the form $\alpha \sqsubseteq \beta$ (*inclusion*) and $\alpha \equiv \beta$ (*definition*), where α and β are concepts. We say that α is *subsumed* by β relative to $\mathcal{T}$ iff $\mathcal{T} \models \alpha \sqsubseteq \beta$, denoted $\alpha \sqsubseteq_\mathcal{T} \beta$; α is *not satisfiable* relative to the ontology $\mathcal{T}$ iff it is under $\mathcal{T}$ subsumed by the most specific concept, i.e., $\mathcal{T} \models \alpha \sqsubseteq \bot$, denoted $\alpha \sqsubseteq_\mathcal{T} \bot$; α is *not subsumed* by β relative to $\mathcal{T}$ iff $\mathcal{T} \not\models \alpha \sqsubseteq \beta$, denoted $\alpha \not\sqsubseteq_\mathcal{T} \beta$.

We write $\mathcal{I} \models \mathcal{T}$ to denote that for each axiom $\alpha \sqsubseteq \beta \in \mathcal{T}$, it holds $\alpha^\mathcal{I} \subseteq \beta^\mathcal{I}$. Similarly, $\mathcal{I} \models_\mathcal{T} \alpha \sqsubseteq \beta$ with $\alpha \sqsubseteq \beta \notin \mathcal{T}$ denotes that both $\mathcal{I} \models \mathcal{T}$ and $\mathcal{I} \models \alpha \sqsubseteq \beta$.

Definition axioms of the form $CN \equiv \alpha$ can be used to define a new concept name CN, then used as synonym for the formula α.

Without loss of generality, we write an *interpretation* $\mathcal{I}$ as a full conjunction of concept names and negated concept names. We say that $\mathcal{I}$ *satisfies* a concept α under $\mathcal{T}$, denoted $\mathcal{I} \models_\mathcal{T} \alpha$, iff $\mathcal{T} \models \mathcal{I} \sqsubseteq \alpha$. We say that α is *satisfiable* under $\mathcal{T}$ iff an interpretation $\mathcal{I}$ exists such that $\mathcal{I} \models_\mathcal{T} \alpha$.

There are many different DLs with different expressiveness [1]. The approach that we present here does not depend on a specific DL and can be applied to the very expressive $\mathcal{SROIQ}(D)$, which is the DL behind OWL 2 (Web Ontology Language) [11]. For further details on DLs, we refer the reader to [1].

3   Ontological CP-Nets

We now introduce a framework for preference representation that is harnessing the technologies described in the previous section. The idea is to combine CP-nets and DLs. In the framework that we propose here, variable values are satisfiable DL formulas.

We consider only binary variables here. Two conditional preferences $(\alpha \mid \gamma)$ and $(\alpha' \mid \gamma')$ are *equivalent* under an ontology $\mathcal{T}$ iff $\gamma \equiv_\mathcal{T} \gamma'$ and $\alpha \equiv_\mathcal{T} \alpha'$.

Definition 1 (ontological CP-net). An *ontological CP-net* $(N, \mathcal{T})$ consists of a CP-net N and an ontology $\mathcal{T}$ such that:

(i) for each variable $A \in \mathcal{V}$, it holds that $Dom(A) = \{\alpha, \neg\alpha\}$, where both α and $\neg\alpha$ are DL formulas that are satisfiable relative to $\mathcal{T}$;

(ii) all the conditional preferences in $\mathcal{CPT}_N$ are pairwise not equivalent.

Note that even if we do not have any explicit hard constraint expressed among the variables of the CP-net, due to the underlying ontology, we have a set of implicit constraints among the values of the variables $\mathcal{V}$ in the CP-net. We show in Section 3.2 how to explicitly encode such constraints to compute an optimal outcome.

Example 2 (Hotel cont'd). Consider a simple ontology, describing the services offered by a hotel, containing the following set of axioms:

$$functional(rent);$$
$$Scooter \sqsubseteq Motorcycle;$$
$$Motorcycle \sqsubseteq \neg Bike;$$
$$\exists rent.Scooter \sqsubseteq \exists facilities.(Parking \sqcap$$
$$\exists payment \sqcap \forall payment.Free).$$

Suppose that we have the variables A_3, A_4, and A_5 of the CP-net of Fig. 1 with the domains $Dom(A_3) = \{\alpha_3, \neg\alpha_3\}$, $Dom(A_4) = \{\alpha_4, \neg\alpha_4\}$, and $Dom(A_5) = \{\alpha_5, \neg\alpha_5\}$, respectively, where:

$$\alpha_3 = \exists rent.Scooter;$$
$$\alpha_4 = \exists facilities.Parking;$$
$$\alpha_5 = \exists rent.Bike.$$

It is then not difficult to verify that $\alpha_3 \sqcap \alpha_5 \sqsubseteq_\mathcal{T} \bot$ and $\alpha_3 \sqsubseteq_\mathcal{T} \alpha_4$. Hence, A_3 and A_5 constrain each other, as well as A_3 and A_4.   ∎

Following [18], to compute the outcomes of a CP-net N, we can transform N into a set of constraints represented in clausal form. For each preference $\Phi = (\tilde{\alpha} \mid \gamma) \in \mathcal{CPT}_N$, we write the following clause:

$$\gamma \to \tilde{\alpha}. \tag{2}$$

In a constrained CP-net, if we had propositional *true/false* variables, an outcome would be a model, i.e., a *true/false* assignment that satisfies all the constraints and some of the clauses built, starting from the preferences represented in $\mathcal{CPT}_N$. For ontological CP-nets, we reuse this notion and say that an outcome $\mathcal{I}$ that satisfies a preference Φ, denoted $\mathcal{I} \models \Phi$, is an interpretation such that:

$$\mathcal{I} \models_\mathcal{T} \gamma \to \tilde{\alpha}.$$

Using a notation similar to the one proposed in [18], we call *DL-opt(N)* the set of DL clauses corresponding to all the preferences in $\mathcal{CPT}_N$.

Definition 2 (feasible outcome and dominance). Given an ontological CP-net $(N, \mathcal{T})$, an outcome $\mathcal{I}$ is *feasible* iff $\mathcal{I} \models \mathcal{T}$. A feasible outcome $\mathcal{I}$ is *undominated* iff no feasible outcome $\mathcal{I}'$ exists such that $\mathcal{I}' \succ \mathcal{I}$.

3.1 Propositional Compilation of DL Formulas

Given a set of satisfiable DL formulas $\mathcal{F} = \{\phi_i \mid i \in \{1, \ldots, n\}\}$, some of them may *constrain* others, because of their logical relationships. For example, we may have $\phi_i \sqcap \neg\phi_j \sqsubseteq \phi_k$ or $\phi_i \sqcap \phi_j \sqsubseteq \bot$. By the equivalence $\alpha \sqsubseteq \beta \equiv \top \sqsubseteq \neg\alpha \sqcup \beta$, we can always represent each constraint in its logically equivalent clausal form. The previous constraints are then equivalent to $\top \sqsubseteq \neg\phi_i \sqcup \phi_j \sqcup \phi_k$ and $\top \sqsubseteq \neg\phi_i \sqcup \neg\phi_j$, respectively. In the following, we represent a clause ψ either as a logical disjunctive formula $\psi = \tilde{\phi}_1 \sqcup \cdots \sqcup \tilde{\phi}_n$ or as a set of formulas $\psi = \{\tilde{\phi}_1, \ldots, \tilde{\phi}_n\}$. Moreover, we often write $\tilde{\phi}_1 \sqcup \cdots \sqcup \tilde{\phi}_n$ to denote $\top \sqsubseteq \tilde{\phi}_1 \sqcup \cdots \sqcup \tilde{\phi}_n$.

A DL ontology can be seen as a set of logical constraints that reduces the set of models for a formula. Given a set of DL formulas $\mathcal{F}$, in the following, we show how to compute a compact representation of an ontology $\mathcal{T}$ as a set of clauses whose variables have a one-to-one mapping to the formulas in $\mathcal{F}$.

Definition 3 (ontological constraint). Given an ontology $\mathcal{T}$ and a set of formulas $\mathcal{F} = \{\phi_i \mid i \in \{1, \dots, n\}\}$ satisfiable w.r.t. $\mathcal{T}$, we say that $\mathcal{F}$ is *minimally constrained* w.r.t. $\mathcal{T}$ iff

1. there exists a formula $\tilde{\phi}_1 \sqcup \dots \sqcup \tilde{\phi}_n$ such that $\mathcal{T} \models \top \sqsubseteq \tilde{\phi}_1 \sqcup \dots \sqcup \tilde{\phi}_n$;
2. there is no proper subset $\mathcal{E} \subset \mathcal{F}$ such that the previous condition holds.

The formula $\top \sqsubseteq \tilde{\phi}_1 \sqcup \dots \sqcup \tilde{\phi}_n$ is called an *ontological constraint*.

An ontological constraint is an explicit representation of the constraints existing among a set of formulas, due to the information encoded in the ontology $\mathcal{T}$.

Definition 4 (ontological closure). Given an ontology $\mathcal{T}$ and a set of formulas $\mathcal{F} = \{\phi_i \mid i \in \{1, \dots, n\}\}$ satisfiable w.r.t. $\mathcal{T}$, we call *ontological closure* of $\mathcal{F}$, denoted $\mathcal{OCL}(\mathcal{F}, \mathcal{T})$, the set of ontological constraints built, if any, for each set in $2^{\mathcal{F}}$.

The ontological constraint is an explicit representation of all the logical constraints considering also an underlying ontology. If we are interested only in the relationships between predefined formulas (due to $\mathcal{T}$), then the corresponding ontological closure is a compact and complete representation.

Example 3 (Hotel cont'd). Given the set $\mathcal{F} = \{\alpha_3, \alpha_4, \alpha_5\}$, due to the axioms in the ontology, we have the two minimally constrained sets $\mathcal{F}' = \{\alpha_3, \alpha_5\}$ and $\mathcal{F}'' = \{\alpha_3, \alpha_4\}$ and the two corresponding ontological constraints $\neg\alpha_3 \sqcup \neg\alpha_5$ (indeed $\alpha_3 \sqcap \alpha_5 \sqsubseteq_{\mathcal{T}} \bot$) and $\neg\alpha_3 \sqcup \alpha_4$ (indeed $\alpha_3 \sqsubseteq_{\mathcal{T}} \alpha_4$). The corresponding ontological closure is then $\mathcal{OCL}(\mathcal{F}, \mathcal{T}) = \{\neg\alpha_3 \sqcup \neg\alpha_5, \neg\alpha_3 \sqcup \alpha_4\}$. ∎

Proposition 1. *If $\mathcal{T} \models \bigsqcap \tilde{\phi}_i \sqsubseteq \bot$, then $\mathcal{OCL}(\mathcal{F}, \mathcal{T}) \models \bigsqcap \tilde{\phi}_i \sqsubseteq \bot$.*

Proof. Since $\mathcal{T} \models \bigsqcap \tilde{\phi}_i \sqsubseteq \bot$, this means that we have the corresponding clause $\psi = \bigsqcup \neg\tilde{\phi}_i$ such that $\mathcal{T} \models \top \sqsubseteq \psi$. If $\mathcal{F} = \{\phi_i \mid i \in \{1, \dots, n\}\}$ is minimally constrained, then $\psi \in \mathcal{OCL}(\mathcal{F}, \mathcal{T})$, otherwise, by definition of $\mathcal{OCL}(\mathcal{F}, \mathcal{T})$, there will be a clause $\psi' \in \mathcal{OCL}(\mathcal{F}, \mathcal{T})$ such that $\psi' \subset \psi$. $\square$

Given a set $\mathcal{F} = \{\phi_i \mid i \in \{1, \dots, n\}\}$ of satisfiable formulas, we say that the set $\tilde{\mathcal{F}} = \{\tilde{\phi}_i \mid i \in \{1, \dots, n\}\}$ is a *feasible* assignment for $\mathcal{F}$ iff

$$\mathcal{OCL}(\mathcal{F}, \mathcal{T}) \not\models \bigsqcap_i \tilde{\phi}_i \sqsubseteq \bot.$$

Note that by Proposition 1, we have that if $\tilde{\mathcal{F}}$ is a feasible assignment for $\mathcal{F}$, then we have $\mathcal{T} \not\models \bigsqcap_i \tilde{\phi}_i \sqsubseteq \bot$, i.e., $\bigsqcap_i \tilde{\phi}_i$ is satisfiable w.r.t. $\mathcal{T}$.

Proposition 2. *For each set of satisfiable formulas $\mathcal{F}$, there always exists a feasible assignment.*

Proof. For each interpretation $\mathcal{I}$ such that $\mathcal{I} \models \mathcal{T}$, since $\mathcal{T} \models \mathcal{OCL}(\mathcal{F}, \mathcal{T})$, we have $\mathcal{I} \models \mathcal{OCL}(\mathcal{F}, \mathcal{T})$. Hence, we always have an interpretation for each clause $\psi^j \in \mathcal{OCL}(\mathcal{F}, \mathcal{T})$ such that $(\psi^j)^{\mathcal{I}} \neq \emptyset$. Given a clause ψ^j, by the semantics of $\sqcup$, we have that $\mathcal{I} \models \psi_j$ iff for at least one $\tilde{\phi}_i^j \in \psi^j$ the relation $\mathcal{I} \models \top \sqsubseteq \tilde{\phi}_i^j$ holds. $\square$

3.2 Computing Optimal Outcomes

The main task that we want to solve with our framework is finding an undominated feasible outcome. In this section, we show how to compute it, given an ontological CP-net. The approach mainly relies on the HARD-PARETO algorithm of [18] (see Algorithm 1).

If we have an ontological CP-net $(N, \mathcal{T})$, the variable values (formulas) in a set $\mathcal{F}$ may constrain each other, and the corresponding constraints are encoded in $\mathcal{OCL}(\mathcal{F}, \mathcal{T})$. The ontological closure of a set of formulas explicitly represents all the logical constraints among them with respect to an underlying ontology. The computation of all feasible Pareto optimal solutions for an ontological CP-net goes through the Boolean encoding of both the ontology $\mathcal{T}$ and of the clauses corresponding to the preferences represented in $\mathcal{CPT}_N$ for each variable $A_j \in V$. To use HARD-PARETO, we need a few pre-processing steps. Given the ontological CP-net $(N, \mathcal{T})$:

1. for each $A_j \in V$ with $Dom(A_j) = \{\alpha_j, \neg\alpha_j\}$, choose a fresh concept name V_j;
2. define the ontology $\mathcal{T}' = \mathcal{T} \cup \{V_j \equiv \alpha_j \mid j \in \{1, \ldots, |V|\}\}$;
3. define the ontological CP-net $(N', \mathcal{T}')$, where N' is the same CP-net as N but for the domain of its variables. In particular, in N', we have $Dom(A_j) = \{V_j, \neg V_j\}$;
4. define $\mathcal{F} = \{V_j \mid j \in \{1, \ldots, |V|\}\}$, where V_j are the concept names introduced in step 1;
5. compute $\mathcal{OCL}(\mathcal{F}, \mathcal{T}')$;
6. introduce a Boolean variable v_j for each $V_j \in \mathcal{F}$;
7. transform $\mathcal{OCL}(\mathcal{F}, \mathcal{T}')$ into the corresponding set of Boolean clauses $\mathcal{C}$ by replacing V_j with the corresponding binary variable v_j;
8. transform $DL\text{-}opt(N')$ into the set of Boolean clauses $opt(N')$ by replacing $V_j \in Dom(A_j)$ with the corresponding variable v_j.

Note that $\mathcal{T}$ is logically equivalent to $\mathcal{T}'$. Indeed, we just introduced equivalence axioms to define new concept names V_j used as synonyms of complex formulas α_j. The same holds for $(N, \mathcal{T})$ and $(N', \mathcal{T}')$, since we just rewrite formulas in $Dom(A_j)$ with an equivalent concept name.

Example 4 (Hotel cont'd). With respect to the CP-net in Fig. 1, if we consider

$$\alpha_1 = \exists location.OnTheSea,$$
$$\alpha_2 = \exists location.CityCenter,$$

then we obtain:

- $\mathcal{T}' = \mathcal{T} \cup \{V_1 \equiv \exists location.OnTheSea, \ldots, V_5 \equiv \exists rent.Bike\}$;
- $\mathcal{C} = \{\neg v_3 \vee \neg v_5, \neg v_3 \vee v_4\}$;
- $opt(N') = \{v_1, v_2, v_1 \wedge v_2 \rightarrow v_3, \ldots, v_2 \rightarrow v_4, \ldots, v_3 \rightarrow v_5, \ldots\}$. ∎

Once we have $\mathcal{C}$ and $opt(N')$, we can compute the optimal outcome of $(N, \mathcal{T})$ by using the slightly modified version of HARD-PARETO represented in Algorithm 1. The function $sol(\cdot)$ used in Algorithm 1 computes all the solutions for the Boolean constraint satisfaction problem represented by $\mathcal{C}$, *opt(N')* and $\mathcal{C} \cup$ *opt(N')*. Differently from the original HARD-PARETO, by Proposition 2, we know that $\mathcal{C}$ is always consistent, and so we do not need to check its consistency at the beginning of the algorithm. Moreover, note that the algorithm works with propositional variables although we are computing Pareto optimal solutions for an ontological CP-net. This means that the dominance test in line 11 can be computed using well-known techniques for Boolean problems.

Input: *opt(N')* and $\mathcal{C}$

1   $S_{opt} \leftarrow sol(\mathcal{C} \cup opt(N'))$;
2   **if** $S_{opt} = sol(\mathcal{C})$ **then**
3 | **return** S_{opt};
4   **end**
5   **if** $sol(opt(N')) \neq \emptyset$ *and* $S_{opt} = sol(opt(N'))$ **then**
6 | **return** S_{opt};
7   **end**
8   $S \leftarrow sol(\mathcal{C}) - S_{opt}$;
9   **repeat**
10 | choose $o \in S$;
11 | **if** $\forall o' \in sol(\mathcal{C}) - o, o' \not\succ o$ **then**
12 | | $S_{opt} \leftarrow S_{opt} \cup \{o\}$;
13 | **end**
14 | $S \leftarrow S - \{o\}$;
15   **until** $S = \emptyset$;
16   **return** S_{opt}.

Algorithm 1. Algorithm HARD-PARETO adapted to ontological CP-nets.

The outcomes returned by Algorithm 1 in S_{opt} are *true/false* assignments to the Boolean variables v_j. To compute undominated outcomes for the original ontological CP-net $(N, \mathcal{T})$, we need to revert to a DL setting. Hence, we build the set $DL\text{-}S_{opt}$, where for each outcome $o_i \in S_{opt}$, we add to $DL\text{-}S_{opt}$ the following formula:

$$\mathcal{I}_i = \bigsqcap \{V_j \mid v_j = true \text{ in } o_i\} \sqcap \bigsqcap \{\neg V_j \mid v_j = false \text{ in } o_i\}.$$

Theorem 1. *Given an ontological CP-net $(N, \mathcal{T})$, the formulas $\mathcal{I}_i \in DL\text{-}S_{opt}$ are undominated outcomes for $(N, \mathcal{T})$.*

4 Computational Complexity

We now explore the complexity of the main computational problems in ontological CP-nets for underlying ontological languages with typical complexity of deciding knowledge base satisfiability, namely, tractability and completeness for EXP and NEXP. We also provide some special tractable cases of dominance testing in ontological CP-nets.

4.1 General Results

For tractable ontology languages (i.e., those for which deciding knowledge base satsfiability is tractable), the complexity of ontological CP-nets is dominated by the complexity of CP-nets. That is, deciding (a) consistency, (b) whether a given outcome is undominated, and (c) dominance of two given outcomes are all PSPACE-complete. Here, the lower bounds follow from the fact that ontological CP-nets generalize CP-nets, for which these problems are all PSPACE-complete [10]. As for the upper bounds, compared to standard CP-nets, these problems additionally involve knowledge base satisfiability checks, which can all be done in polynomial time and thus also in polynomial space. Note that in (a) (resp., (b)), one has to go through all outcomes o' and check that it is not the case that $o \succ o'$ (resp., $o' \succ o$), which can each and thus overall be done in polynomial space. Note also that the same complexity results hold for ontology languages with PSPACE-complete knowledge base satisfiability checks and that even computing the set of all undominated outcomes (generalizing (b)) is PSPACE-complete under the condition that there are only polynomially many of them.

Theorem 2. *Given an ontological CP-net* $(N, \mathcal{T})$ *over a tractable ontology language,*

(a) deciding whether $(N, \mathcal{T})$ *is consistent,*
(b) deciding whether a given outcome o is undominated,
(c) deciding whether $o \prec o'$ *for two given outcomes o and o'*

are all PSPACE-complete.

In particular, if the ontological CP-net is defined over a DL of the *DL-Lite* family [7] (which all allow for deciding knowledge base satisfiability in polynomial time, such as *DL-Lite$_\mathcal{R}$*, which stands behind the important OWL 2 QL profile [16]), deciding (a) consistency, (b) whether a given outcome is undominated, and (c) dominance of two given outcomes are all PSPACE-complete.

Corollary 1. *Given an ontological CP-net* $(N, \mathcal{T})$ *over a DL from the DL-Lite family,*

(a) deciding whether $(N, \mathcal{T})$ *is consistent,*
(b) deciding whether a given outcome o is undominated,
(c) deciding whether $o \prec o'$ *for two given outcomes o and o'*

are all PSPACE-complete.

For EXP (resp., NEXP) complete ontology languages (i.e., those for which knowledge base satisfiability is complete for EXP (resp., NEXP)), the complexity of ontological CP-nets is dominated by the complexity of the ontology languages. That is, deciding (a) inconsistency, (b) whether a given outcome is dominated, and (c) dominance of two given outcomes are all complete for EXP (resp., NEXP). Here, the lower bounds follow from the fact that all three problems in ontological CP-nets can be used to decide knowledge base satisfiability in the underlying ontology language. As for the upper bounds, in (a) and (b), we have to go through all outcomes, which is in EXP (resp., NEXP). Then, we have to perform knowledge base satisfiability checks, which are also in EXP (resp., NEXP), and dominance checks in standard CP-nets, which are in PSPACE, and thus also in EXP (resp., NEXP). Overall, (a) to (c) are thus in EXP (resp., NEXP). Note that computing the set of all undominated outcomes (generalizing (b)) is also EXP-complete for EXP-complete ontology languages.

Theorem 3. *Given an ontological CP-net $(N, \mathcal{T})$ over an EXP (resp., NEXP) complete ontology language,*

(a) deciding whether $(N, \mathcal{T})$ is inconsistent,
(b) deciding whether a given outcome o is dominated,
(c) deciding whether $o \prec o'$ for two given outcomes o and o'

are all complete for EXP (resp., NEXP).

In particular, if the ontological CP-net is defined over the expressive DL $\mathcal{SHIF}(\mathbf{D})$ (resp., $\mathcal{SHOIN}(\mathbf{D})$) [12] (which stands behind OWL Lite (resp., OWL DL) [15,11], and allows for deciding knowledge base satisfiability in EXP [12,20] (resp., NEXP, for both unary and binary number encoding; see [17,20] and the NEXP-hardness proof for $\mathcal{ALCQIO}$ in [20], which implies the NEXP-hardness of $\mathcal{SHOIN}(\mathbf{D})$)), deciding (a) inconsistency, (b) whether a given outcome is dominated, and (c) dominance of two given outcomes are all complete for EXP (resp., NEXP).

Corollary 2. *Given an ontological CP-net $(N, \mathcal{T})$ over the DL $\mathcal{SHIF}(\mathbf{D})$ (resp., $\mathcal{SH}$-$\mathcal{OIN}(\mathbf{D})$),*

(a) deciding whether $(N, \mathcal{T})$ is inconsistent,
(b) deciding whether a given outcome o is dominated,
(c) deciding whether $o \prec o'$ for two given outcomes o and o'

are all complete for EXP (resp., NEXP).

4.2   Tractability Results

If the ontological CP-net is a polytree and defined over a tractable ontology language, deciding dominance of two outcomes can be done in polynomial time, which follows from the fact that for standard polytree CP-nets, dominance can be decided in polynomial time [3]. Note that polytree ontological CP-nets are always consistent.

Theorem 4. *Given an ontological CP-net $(N, \mathcal{T})$ over a tractable ontology language, where N is a polytree, deciding whether $o \prec o'$ for two given outcomes o and o' can be done in polynomial time.*

In particular, if the ontological CP-net is a polytree and defined over a DL of the *DL-Lite* family, deciding dominance of two outcomes can be done in polynomial time.

Corollary 3. *Given an ontological CP-net $(N, \mathcal{T})$ over a DL from the DL-Lite family, where N is a polytree, deciding whether $o \prec o'$ for two given outcomes o and o' can be done in polynomial time.*

5   Related Work

Constrained CP-nets were originally proposed in [4], along with the algorithm SEARCH-CP, which uses branch and bound to compute undominated outcomes. The algorithm has an anytime behavior: it can be stopped at any time, and the set of computed solutions are a subset of the set containing all the undominated outcomes. This means that in case we are interested in any undominated outcome, we can use the first one returned by SEARCH-CP. In [18], HARD-PARETO is presented. The most notable difference is that HARD-PARETO does not rely on topological information like SEARCH-CP, but it exploits only the CP-statements, thus allowing to work also with cyclic CP-nets. Differently from the previous two papers, in our work, we allow the variable domains to contain DL formulas constrained via ontological axioms.

There are a very few papers describing how to combine Semantic Web technologies with preference representation and reasoning using CP-nets. To our knowledge, the most notable work is [2]. Here, in an information retrieval context, Wordnet is used to add a semantics to CP-net variables. Another interesting approach to mixing qualitative preferences with a Semantic Web technology is presented in [19], where the authors describe an extension of SPARQL, which can encode user preferences in the query.

A combination of conditional preferences (very different from CP-nets) with DL reasoning for ranking objects is introduced in [14]. A ranking function is described that exploits conditional preferences to perform a semantic personalized search and ranking over a set of resources annotated via an ontological description.

6   Summary and Outlook

In classical decision theory and analysis, the preferences of decision makers are modeled by utility functions. Unfortunately, the effort needed to obtain a good utility function requires a significant involvement of the user [9]. This is one of the main reasons behind the success obtained by CP-nets since they were originally proposed [5]: they are compact, easily understandable and well-suited for combinatorial domains, such as multi-attribute ones. In this paper, we have described how to reason with CP-nets whose variable values are DL formulas that refer to a common ontology. The proposed framework is very useful in many semantic retrieval scenarios such as semantic search.

After the introduction of CP-nets, other related formalisms have been proposed such as TCP-nets (Trade-off CP-nets) [6] or CP-theories [21]. TCP-nets extend CP-nets by allowing to express also relative important statements between variables. With TCP-nets, the user is allowed to express her preferences over compromises that sometimes may be required. CP-theories generalize (T)CP-nets allowing conditional preference statements on the values of a variable, along with a set of variables that are allowed to vary when interpreting the preference statement. In future work, we plan to enrich these frameworks by introducing ontological descriptions and reasoning, thus allowing the development of more powerful semantic-enabled preference-based retrieval systems.

Acknowledgments. This work was supported by the UK EPSRC grant EP/J008346/1 "PrOQAW: Probabilistic Ontological Query Answering on the Web", the ERC (FP7/ 2007-2013) grant 246858 ("DIADEM"), and by a Yahoo! Research Fellowship.

References

1. Baader, F., Calvanese, D., Mc Guinness, D., Nardi, D., Patel-Schneider, P.F. (eds.): The Description Logic Handbook. Cambridge University Press (2002)
2. Boubekeur, F., Boughanem, M., Tamine-Lechani, L.: Semantic information retrieval based on CP-nets. In: Proc. FUZZ-IEEE 2007, pp. 1–7. IEEE Computer Society (2007)
3. Boutilier, C., Brafman, R.I., Domshlak, C., Hoos, H.H., Poole, D.: CP-nets: A tool for representing and reasoning with conditional ceteris paribus preference statements. J. Artif. Intell. Res. 21, 135–191 (2004)
4. Boutilier, C., Brafman, R.I., Domshlak, C., Hoos, H.H., Poole, D.: Preference-based constrained optimization with CP-nets. Computat. Intell. 20(2), 137–157 (2004)
5. Boutilier, C., Brafman, R.I., Hoos, H.H., Poole, D.: Reasoning with conditional ceteris paribus preference statements. In: Proc. UAI 1999, pp. 71–80. Morgan Kaufmann (1999)
6. Brafman, R.I., Domshlak, C., Shimony, S.E.: On graphical modeling of preference and importance. J. Artif. Intell. Res. 25(1), 389–424 (2006)
7. Calvanese, D., De Giacomo, G., Lembo, D., Lenzerini, M., Rosati, R.: Tractable reasoning and efficient query answering in description logics: The DL-Lite family. J. Autom. Reasoning 39(3), 385–429 (2007)
8. Domshlak, C., Hüllermeier, E., Kaci, S., Prade, H.: Preferences in AI: An overview. Artif. Intell. 175(7/8), 1037–1052 (2011)
9. French, S.: Decision Theory: An Introduction to the Mathematics of Rationality. In: Ellis Horwood Series in Mathematics and its Applications. Prentice Hall (1988)
10. Goldsmith, J., Lang, J., Truszczynski, M., Wilson, N.: The computational complexity of dominance and consistency in CP-nets. J. Artif. Intell. Res. 33, 403–432 (2008)
11. Hitzler, P., Krötzsch, M., Parsia, B., Patel-Schneider, P.F., Rudolph, S.: OWL 2 Web Ontology Language Primer, 2nd edn. W3C Recommendation (December 11, 2012), http://www.w3.org/TR/owl2-primer/
12. Horrocks, I., Patel-Schneider, P.F.: Reducing OWL entailment to description logic satisfiability. In: Fensel, D., Sycara, K., Mylopoulos, J. (eds.) ISWC 2003. LNCS, vol. 2870, pp. 17–29. Springer, Heidelberg (2003)
13. Keeney, R.L., Raiffa, H.: Decisions with Multiple Objectives: Preferences and Value Tradeoffs. Cambridge University Press (1993)
14. Lukasiewicz, T., Schellhase, J.: Variable-strength conditional preferences for ranking objects in ontologies. J. Web Sem. 5(3), 180–194 (2007)
15. McGuinness, D.L., van Harmelen, F.: OWL Web Ontology Language Overview. W3C Recommendation (February 10, 2004), http://www.w3.org/TR/2004/REC-owl-features-20040210/
16. Motik, B., Cuenca Grau, B., Horrocks, I., Wu, Z., Fokoue, A., Lutz, C.: OWL 2 Web Ontology Language Profiles, 2nd edn. W3C Recommendation (December 11, 2012), http://www.w3.org/TR/owl2-profiles/
17. Pratt-Hartmann, I.: Complexity of the two-variable fragment with counting quantifiers. Journal of Logic, Language and Information 14(3), 369–395 (2005)
18. Prestwich, S.D., Rossi, F., Brent Venable, K., Walsh, T.: Constraint-based preferential optimization. In: Proc. AAAI/IAAI 2005, pp. 461–466. AAAI Press / MIT Press (2005)
19. Siberski, W., Pan, J.Z., Thaden, U.: Querying the Semantic Web with preferences. In: Cruz, I., Decker, S., Allemang, D., Preist, C., Schwabe, D., Mika, P., Uschold, M., Aroyo, L.M. (eds.) ISWC 2006. LNCS, vol. 4273, pp. 612–624. Springer, Heidelberg (2006)
20. Tobies, S.: Complexity Results and Practical Algorithms for Logics in Knowledge Representation. Doctoral Dissertation, RWTH Aachen, Germany (2001)
21. Wilson, N.: Extending CP-nets with stronger conditional preference statements. In: Proc. AAAI 2004, pp. 735–741. AAAI Press (2004)

Author Index